CRITICAL PRAISE FOR *ARTICLES OF FAITH*

"An amply rewarding journey. . . . Presents both sides of a complex issue and brings it alive through character, detail, context, and a narrative so compellingly written that at times it reads like a novel."
— Susan Lydon, *San Jose Mercury News*

"An enthralling, skillfully balanced narrative about this important and complicated national debate."
— Paul Galloway, *Chicago Tribune*

"*Articles of Faith* illuminates the two distinct faces of the post-*Roe* drive against abortion, those of protestors and those of legislators. Gorney's portrait of Sam Lee is impressive."
— David J. Garrow, *The New York Times*

"In Cynthia Gorney, this subject has at last found its ideal chronicler. . . . Best nonfiction book of 1998. More than that, it is a book that deserves to last for its evenhanded and immensely readable quality. . . . Gorney's reporting, while balanced, is anything but blandly 'objective.' She brings equal passion, vigor, and empathy to each character, making their battle of principles rage in a reader's brain."
— Samuel G. Freedman, *Newsday*

"Gorney can write with great power."
— Beryl Lieff Benderly, *The Washington Post Book World*

"The book abounds in poignant detail."
— Michael J. Ybarra, *Los Angeles Times*

"*Articles of Faith* gives a balanced history of the post–*Roe* v. *Wade* battle over abortion."
— Elissa Schapple, *Vanity Fair*

"Compassionate and eloquent."
— *The Economist*

"A thorough and thoroughly excellent account of the modern abortion conflict. Cynthia Gorney accomplishes what precious little other reporting about the issue has. She imbues her story with the subtle and contrasting shades of individual emotion and intellect grappling with this difficult question."
— Louise Kiernan, *Chicago Tribune*

"Admirably balanced, this chronicle of abortion politics tracks two passionate partisans, one pro-choice and the other pro-life, from the 1960s to the present."
—Paul Hond, *Glamour* (Chosen as a 1998 "Great Read.")

"Eminently readable. . . . Gorney [is able] to capture the nuances."
—Sara Engram, *The Sun* (Baltimore)

"Prodigiously researched . . . gracefully written. . . . A welcome reminder that evenhandedness can carry the day even when the subject matter is as multifaceted, controversial, and inflammatory as abortion."
—David Andrusko, *First Things*

"Gorney's *Articles of Faith* looks in vivid detail at the abortion struggle in Missouri that culminated in the 1989 Supreme Court decision *Webster* v. *Reproductive Health Services*. . . . Well written. . . . Provides an engaging historical narrative by creating intimate, multidimensional portraits of the key players and issues on both sides of the struggle."
—Vivien Labaton, *Ms. Magazine*

"Utterly absorbing. . . . A wealth of facts."
—Karen Durbin, *Elle*

"Thoughtful and evenhanded."
—Joe Dirck, *The Plain Dealer* (Cleveland)

"An important and engaging work that should help further the conversation in constructive ways for both sides."
—*Publishers Weekly*

"In *Articles of Faith*, Cynthia Gorney accomplishes what might seem a mission impossible. She has won the ear and the trust of both sides in the abortion debate. With considerable skill, sympathy and respect, she weaves a tale of two movements: one to legalize abortion, the other to ban it. If you have time for only one book on abortion, this is it."
—Susan C. Hegger, *St. Louis Post-Dispatch*

"Gorney's achievement in the book is extracting the reality from 25 years of rhetoric and stereotype. She shows people grappling with emotions that serve as a backdrop to the ongoing political battle."
—Angela Hill, *The Oakland* (CA) *Tribune*

"Gorney is able to get inside the conflict."
—Sara Nelson, *Glamour*

"The beauty of Gorney's tale is that she chooses not the most extreme activists to carry the story, but the most active believers."
—Kimberly Conniff, *Brill's Content*

"Gorney's *Articles of Faith* offers an unprecedented fair-minded account of the passion and conviction that inspired people on both sides."
—Wendy Murray Zoba, *Christianity Today*

"Unbiased and riveting. . . . She doesn't tell us about abortion so much as she lets the men and women on the front lines show the impact of the issue as it evolved, the quiet heroes, the leaders and the followers, the self-righteous and ordinary folks."
—Lisa Petrillo, *The San Diego Union-Tribune*

"Dense and compelling detail. . . . A largely unknown history . . . *Articles of Faith* is essential reading."
—Duane Davis, *Rocky Mountain News*

"The finest and most evenhanded work on the subject. . . . Ms. Gorney not only peers into the nooks and crannies of circumstances that led up to *Webster*, but breathes life into the personalities so deeply engaged in the politics of persuasion."
—John Gamino, *The Dallas Morning News*

"*Articles of Faith* portrays decent people on both sides with respect, humanizing them and illuminating their convictions."
—*Booklist*

"A book for those who seek a deeper comprehension of one of the great political and moral conflicts of our time. With grace and expertise, Cynthia Gorney has provided us with a riveting excursion into the minds of activists on both sides of a debate that seems far from ended. *Articles of Faith* is living history at its most timely."
—Scott Ellsworth, *The Oregonian*

"Insightful. . . . [*Articles of Faith*] will move readers on either side of the issue."
—*Beatrice* Interview: 1998 (beatrice.com/interviews)

"Cynthia Gorney's *Articles of Faith* makes the reader privy to the soul searching of the protagonists on both sides of the abortion struggle."
—Lorraine Gengo, *Hartford* (CT) *Advocate*

"Magisterial . . . researched with both journalistic precision and scholarly care."
—Diane Roberts, *The Atlanta Journal-Constitution*

"Fascinating."
—Pamela Schaeffer, *National Catholic Reporter*

"Gorney casts an unflinching eye on the abortion process itself."
—Phyllis Orrick, *Berkeley (CA) Express*

"Inspiring."
—Denise Shannon, *Conscience*

"A must-read for anyone interested in a thoughtful, balanced, fascinating living history of a contemporary crisis, the Abortion Wars."
—Marilyn LaRocque, *The Napa Valley Register*

"With her remarkable skills as a reporter and as a storyteller, [Gorney] has written a book that is all the more profound for providing us with vivid, complicated human beings rather than familiar rhetoric. It's been a long time since *Roe* v. *Wade* and the libraries are filled with books on the subject. *Articles of Faith* is the most vivid of all, an essential book."
—David Remnick, author of *Lenin's Tomb*

"This is contemporary history at its very best. A gifted reporter has taken one of the most emotional and divisive of today's issues, abortion, and, with what seems to me scrupulous fairness and very considerable literary skill, has traced its path to the very top of today's conflicted social and political agenda."
—David Halberstam

"Remarkable. . . . [Gorney] does exactly what we expect good journalists should do—report the facts and tell the story as it happened. . . . Her focus on two advocates, Judith Widdicombe and Samuel Lee, displays an empathy so rare in this debate that we grasp not only their competing moral and political arguments, but the religious beliefs and social experiences that move each to give their lives for over two decades to the opposing camps in the abortion wars."
—Margaret O'Brien Steinfels, editor, *Commonweal*

"Cynthia Gorney's *Articles of Faith* is an extraordinary accomplishment. . . . Comparable to Normal Mailer's re-creation of Gary Gilmore and his world in *The Executioner's Song*, this is a book that makes our times come alive and issues come clear through the stories of people we have met previously only as blurred figures on the evening news or vague partisan fighters behind the headlines."
—Eugene Kennedy, author of *My Brother Joseph: The Spirit of a Cardinal and the Story of a Friendship*

"This is a breathtaking piece of journalism: Gorney's reporting is dazzling, her writing elegant, her perspective remarkably fair."
—Peggy Orenstein, author of *Schoolgirls: Young Women, Self-Esteem, and the Confidence Gap*

ARTICLES
OF FAITH

A Frontline History
of the Abortion Wars

Cynthia Gorney

A Touchstone Book
Published by Simon & Schuster

FOR BILL

 TOUCHSTONE
Rockefeller Center
1230 Avenue of the Americas
New York, NY 10020

Copyright © 1998 by Cynthia Gorney
Revised epilogue copyright © 2000 by
Cynthia Gorney
All rights reserved,
including the right of reproduction
in whole or in part in any form.
First Touchstone Edition 2000
TOUCHSTONE and colophon are
registered trademarks of Simon & Schuster Inc.
Designed by Edith Fowler
Manufactured in the United States of America

10 9 8 7 6 5 4 3 2 1

The Library of Congress has cataloged the
Simon & Schuster edition as follows:

Gorney, Cynthia.
    Articles of faith : a frontline history of the
abortion wars / Cynthia Gorney.
        p. cm.
    Includes bibliographical references (p.   ) and
index.
        1. Abortion—Missouri—History.   2. Pro-life
movement—Missouri—History.   3. Pro-choice
movement—Missouri—History.   I. Title.
HQ767.5.U5G68   1998
363.46'0973—DC21   97-35065   CIP
ISBN 0-684-80904-4
        0-684-86747-8 (pbk)

Author's Note
and Acknowledgments

The story that takes place in these pages begins and ends in the state of Missouri, where for a certain period at the close of the 1980s reporters and photographers converged from many parts of the United States—there were some from other countries, too, from Australia and Germany and Japan—with urgent, important questions about induced abortion and the protections guaranteed under law. At that time I was a staff writer for *The Washington Post,* asked by my editors to produce a Sunday magazine article about the Supreme Court case that was drawing so much attention to Missouri, and as I prepared for my trip to St. Louis I looked in libraries and bookstores for the background reading that would help me trace the modern American abortion conflict, which I assumed dated back sixteen years to the famous January 1973 ruling called *Roe* v. *Wade.* I was wrong about that; I was wrong about a great many things, as I was to discover, and one of them was the assumption that a rich assortment of written reference works would explain for me, in language not intended to persuade or demean, what had happened in the years of conflict over legalized abortion—what events had led Americans to react as they did, to group as they did, and to settle with such ferocity into warring camps.

On the abortion shelf I found dozens of books of argument, some of them shrill, some of them eloquent. I found religious theses, essay collections, sociological studies, and medical texts, but I could not find a single volume that told me in narrative fashion what I wanted to know, which I began to understand was at once simple and alarmingly complex: How did we come to this? Who were the partisans? What were the stories of women and men whose personal understanding of the events around them was so different that they had come to describe these events in entirely different vocabularies?

I have incurred many debts in the six years of work on this book, but none are as deep as my enduring obligation to the people inside and outside Missouri who listened to my explanation of what I wished to do and agreed to help, giving generously of their time in interviews, making available personal correspondence and private memoranda, rooting through attics and basements for file folders and scrapbooks that had lain undisturbed for ten or twenty years—all the while keeping to themselves what must have been real misgivings about the nature of the enterprise. I tried to describe my intentions the same way at the outset of each of the five hundred interviews that took place for this book: I hoped to write a history, set largely within a single state and told from the perspective of people on both sides, of what I came to regard as the crucial two decades in the American abortion conflict, the period of extraordinary change that began in the late 1960s with the first state efforts to modify old criminal abortion statutes. I supposed that in Missouri I would find guides into a drama that has unfolded in similar fashion in every state in the country; what I had not anticipated was the power, conscience, and personal sacrifice in the story these guides had to tell. Nor had I anticipated the candor with which so many of them would tell it, all the while knowing that their listener was collecting parallel stories from the enemy camp.

My greatest thanks go to Judith Widdicombe and Samuel Lee, whose patience and good humor withstood scores of lengthy interviews and telephone conversations, and to the unfailingly friendly and helpful members of their families. Special thanks also to Loretto Wagner, Frank Susman, Andrew Puzder, Sylvia Hampton, Matt Backer, Vivian Diener Meyer, Louis DeFeo, Mary Bryant, the late Kathy Edwards Wise, Michael Freiman, B. J. Isaacson, and the other Missouri men and women who accommodated multiple interviews and telephone inquiries; to the pre-*Roe* abortion volunteers, pro and con, who allowed a stranger to use their home file cabinets and bookshelves as lending libraries; and to the national leaders and activists who made time to talk to me at length even though my narrative focused on their lesser-known counterparts in Missouri.

I owe a different kind of thanks to new friends from St. Louis: to Etta and Ted Taylor, who welcomed me into their home as an itinerant houseguest, offering lodgings, wonderful company, and the best food I have ever eaten in my life; to Martha Shirk and William Woo, former reporter for and editor of the *St. Louis Post-Dispatch,* who were extraordinarily generous about introducing a newcomer to territory they knew so well; and to William Freivogel and the other *Post-Dispatch* reporters and librarians who humored my occasional requests for arcane information. I am also indebted to the legal writer and former Supreme Court clerk Edward Lazarus, who interrupted his own work again and again to explain fine points of appellate procedure and constitutional law; to the attorneys Bobbie Welling, Nancy Stearns, and Robert Post, who offered suggestions and corrections on portions of the manuscript; and to Frances Kissling and the many other people who volunteered their political, medical, religious, or regional expertise in reviewing certain passages.

Colleagues kept me going; a triple measure of gratitude is due to two. Jon Carroll hauled me through the difficult spots chapter by chapter, bullying, exhorting, improving one draft after another with his gifted editor's eye. Bob Thompson, the *Washington Post* editor who first sent me to Missouri, kept reading abortion narrative and providing wise editorial advice long after his magazine-article duties were over. Mark A. Gorney spent more than a year helping with research; for research assistance I am also grateful to Carol Command in Columbia, Pamela Williams in St. Louis, and Elizabeth DeVita in Washington, D.C.

Commiseration, prodding, and unflagging optimism came from Nora Gallagher, Lynn Darling, Anna Quindlen, and Carole Robin; thanks also to Douglas Foster, Joe Kane, John Jacobs, Lydia Chavez, Paul Hendrickson, Joseph Nocera, Laurie Leiber, Catherine Hunter, Henry Mayer, and the extended family from Soda Springs. My agent, Amanda Urban, was confident about this project long before I was, for which I remain in her debt. At Simon & Schuster, my editor, Alice Mayhew, and her associate Elizabeth Stein proved as energetic and supportive as their formidable reputations had predicted they would; my deepest appreciation to both of them, especially to Alice for plucking the title so nimbly from the air, and to the meticulous Steve Messina and his copyediting staff. I also feel fortunate to have had access for a time to the thoughtful and sympathetic counsel of the late J. Anthony Lukas, whose brilliant 1985 book *Common Ground* remains a model for any reporter trying to tell the stories of people caught up on opposing sides of great social conflicts.

To my family—my parents, my brothers, my children, Aaron and Joanna, and my heroic husband, Bill Sokol—my thanks and my love: you carried me all the way here.

CYNTHIA GORNEY
July 1997

Contents

PART ONE: THE PHONE ON THE PORCH   13

1   Codes   15
2   The Wedge   38
3   Heaven and Earth   62
4   The Writing in the Heart   95
5   Exhibit A   122
6   The Ruling   161
7   The Target   194

PART TWO: A HOUSE AFIRE   227

8   The Education of the Pharisee   229
9   A Lesser of Evils   265
10 A House Afire   308

PART THREE: THE SEVENTH QUESTION   377

11 The Litmus Test   379
12 The Count   414

13   The Seventh Question   450
14   Zealots   486
     Epilogue   518

     Sources   527
     Notes   539
     Index   555

PART ONE

The Phone on the Porch

ONE

Codes

1968

WHAT HAPPENED in Sedalia was not a thing that Robert Duemler talked about: not the place, or the smell, or the low sick feeling that came to his stomach when he thought about it afterward. Doctors see things they wish they had not seen, he understood that about being a doctor, and Robert Duemler was a quiet man who delivered babies for a living at a big Catholic hospital where they were cordial to him even though he belonged to the United Church of Christ. The hospital was outside St. Louis, which people liked to call a Catholic city, as though the great domed cathedral somehow spanned the length and breadth of the city limits, and when Duemler was at work there were nurses in nuns' habits who bustled briskly down the corridors and nodded to him in their deferential nurses' way: Good morning, doctor.

Among the obstetrical nurses in this hospital was a big smart opinionated woman named Judith Widdicombe, and sometimes after deliveries Duemler and Widdicombe would lean against a wall in their scrubs and watch the commotion around them and talk. Duemler was in his thirties then, but he was older than she was, and it amused him to see Judy Widdicombe irritate the senior physicians by pointing out when and where she thought they had bungled their work. She was not deferential. She had no concept of deference. She thought some of the older nurses had used themselves up and should be let go, and she thought the hospital

cheated its nurses by forcing them to take unpaid after-shift time to prepare the next crew, and she would say these things in a resonant contralto
that carried through the halls of the hospital. There was some undiplomatic clarity of vision to Judy Widdicombe that Duemler admired, as
though she didn't care what she clipped as she swung the machete before
her, and he thought she saw as he did the full and splendid sweep of
obstetrical practice: they might be attending the birth of a patient's sixth
or seventh child and know, without exactly saying it to each other, that
sometimes this felt like bringing forth the miracle of life and sometimes it
just felt like handing a small squalling package to a tired woman who was
not visibly interested in receiving it.

Judy Widdicombe would say casually to Duemler, not looking at him:
"I bet she doesn't really want that kid."

Maybe not, Duemler would say. And they would talk about something else. But he understood that Widdicombe, who was not a nun, and
who apparently was not a Roman Catholic either, was signaling him. There
was a code at work, and Duemler was not sure what it was, but he knew
that he was in on it somehow, and that it had to do with ideas that might
be dangerous even for the undiplomatic to discuss in a public place.

After a while, when no one else was listening, he told her what had
happened in Sedalia.

He was in the Air Force, a young Air Force doctor, and in 1962 he was
dispatched to central Missouri to serve on the base at Sedalia. He was the
first obstetrician the base had ever had, and every morning he would get
up in his military housing and go over to the little Air Force hospital to
examine patients. Sometimes they would call him in the middle of the
night and he would switch on the light and fumble into his trousers and
go out the door in the unsteady gait of a man working hard to mimic
alertness. That was when this call came, in the middle of the night, and
when Duemler walked into the emergency room what he saw, in more
places than he would have thought possible, was blood. There was blood
on the walls. There was blood on the floor. There was blood on the gurney
and on the towels and on the hands and arms of the emergency crew, who
were silent now, and no longer moving rapidly. Beneath them lay a woman
whose skin had gone pallid and slack, and when Duemler lifted her legs
into the stirrups and cleaned some of the blood away, he saw that someone
had pushed inside her vagina with a sharp instrument and aimed it toward
the cervix and thrust straight up. The blood vessels to either side of the
cervix had emptied all over the Air Force emergency room and the car in
which this woman's husband had driven her twenty miles, which was the
distance between the hospital and the abortionist.

The husband told Duemler they had five children already.

Five children already: Duemler remembered that for a long time afterward, when he was no longer able to summon up the husband's height or the color of his hair or anything except the flat bewildered look on his face as his wife was pronounced dead on the examining table. *Expired* was what Duemler wrote, in his precise physician's notation; nobody ever told him why they could not have six children or what kind of person she might have been when she was alive in the house with the five they had. He had learned in medical school that you pull down the shade when you have to, that you cannot climb into the life of every patient and still show up for work in the morning; but after it was over he would think about Sedalia and the words *stupid, stupid* would press up inside him. It was a stupid death. When he was a resident, two years before the Air Force, an older doctor had asked him to assist at a dilation and curettage, and Duemler had stood there and handed over the cervical dilators while the older doctor cleaned out a woman's uterus under general anesthesia in the sterile operating room of a fine St. Louis hospital. The older doctor told him that the patient was a nice young woman whose boyfriend had come from a family of high social standing and that they had all invented a plausible diagnosis requiring termination of pregnancy to save her life. Duemler understood that his job was to hold the dilators and shut up, but when he thought about Sedalia he always thought about this other woman also, the one with the fancy boyfriend.

So he told Judy Widdicombe that story, too.

Then she knew she had been right about Bob Duemler, and she came to him one day and said: Listen. There is a group of us, and we are doing something illegal. We need your help.

In 1968, the year Judith McWhorter Widdicombe began recruiting medical doctors into a conspiracy to commit repeated felonies in the state of Missouri, she was thirty years old and lived in a modestly proportioned pea green house with her two sons and her husband Arthur, who ran a *St. Louis Post-Dispatch* delivery route. At that time Judy had three different occupations, each of which interested her deeply and made her feel useful in the world: from morning until early afternoon she cared for children, her own two sons and various children of neighbors who needed part-time baby-sitting; from early afternoon until eleven o'clock at night she worked the hospital swing shift in labor and delivery, attending to patients and supervising less experienced nurses; and one evening each week she stayed home with her husband to answer telephone calls from people who were threatening to commit suicide.

The telephone-answering work was a volunteer job and was sometimes very difficult, but Judy liked it and worked hard at it, especially during the early months when she was first trying to accustom herself to the distraught voices on the other end of the line. She was a tall, broad-boned woman who thought of herself as forthright, perhaps tactless on occasion, but not exceptionally brave or imaginative; since childhood she had felt that she was overweight, which made her self-conscious about her appearance, although she had blue eyes and strong arms and light brown hair, which she wore back in a sprayed chignon, and her husband Art thought her smile was beautiful. Art was by nature shyer than Judy (their first date, by his recollection, had taken place after she marched up to him in church and invited him to a dance), and he loved listening to Judy's singing voice, which was so strong and clear that she was often invited to sing alone at weddings, or to solo for the Methodist church choir to which both Judy and Art belonged.

Before the time at which this story commences, in fact, Judy and Arthur Widdicombe had lived rather quietly and conventionally in suburban St. Louis, their attentions divided among the competing demands of work, neighbors, church, and family. Their home was decorated with Early American furniture and stood one mowed lawn from the house where Art Widdicombe had grown up, so that Judy could walk outside and see the white lace curtains of her mother-in-law's kitchen window; there was a barbecue, and a maple tree with branches thick enough to hold a scrapwood treehouse, and when the neighborhood children were underfoot Judy would go into the kitchen and make caramel popcorn, or chocolate chip cookies, or donuts she fried in a deep steel pan and then sprinkled with powdered sugar while they were still warm. The minister from church liked visiting the Widdicombe house because it seemed to him that Judy propped the front door open and invited in any local friends who happened by. "In the wintertime I remember the house had a wood-fire smell because she kept the fireplace going," recalls the minister, whose name was Ken Gottman. "She more or less held open house. She kept the coffee-pot on."

She was a caretaker. She liked being regarded as a caretaker; Judy sometimes thought she was such a capable nurse because she had found a job that obliged her to spend her working hours ministering to people who needed her. When she was a child, on her father's farm in Indiana, Judy had tended to her baby brother while her mother disappeared into farm chores; when the family moved to St. Louis, after Judy's father had sold the farm and become a traveling salesman, Judy still kept a motherly watch over her brother, and although she thought for a while that she might try to become a professional singer—she left St. Louis after high school to enroll at Texas Christian University, which had offered her a voice scholarship—it took Judy only one semester at Texas Christian to switch to the university's school of nursing, where she felt comfortable at once and excelled in her studies. She was good at the science. She liked the feeling of confidence that swelled up in her when she put on her uniform. She transferred back to a more accelerated nursing program at a hospital in St. Louis and graduated with honors; the hospital gave Judy an achievement award for pediatric nursing and asked her to stay on and teach.

When her second son was born, in 1964, Judy tried to retire so she could care for both her children full time, but without her nursing work she felt restless and incomplete, and after a while she reported back to the hospital, pretending the family finances were forcing her to work. But she understood that the money was a kind of excuse for something she wanted to do anyway. She liked balancing her family life with her nursing shifts, and she also began volunteering regularly at church, where the young Reverend Gottman was trying to update the youth fellowship program. By 1968, Judy was one of the Kirkwood United Methodist Church's principal youth counselors, leading Sunday discussion groups and occasional church retreats, and when Gottman decided to add a sex education course, Judy offered to help the minister look for suitable pamphlets and educational films. She placed an inquiring telephone call one day to what she had been told was the Social Health Association, but she reached instead the administrative offices of the new St. Louis telephone-counseling service called Suicide Prevention. "You're calling the wrong place," said the Suicide Prevention director, a psychologist named Gwyndolyn Harvey, after Judy explained what she was looking for. "But while I have you on the telephone, we're recruiting volunteers too."

And that was how Judy came to know Gwyn Harvey and to volunteer one evening a week for Suicide Prevention. She completed the obligatory training in crisis intervention, which Judy could see required a special delicacy and alertness when the personal contact was only by voice, and eventually during Judy's shift all St. Louis Suicide Prevention calls were routed to a special telephone on the Widdicombes' back porch. Judy could close the door to this porch, shutting out the noise from the main part of the house. The telephone sat on a plain wood desk. The calls would begin after dinner, as the boys were being tucked away into their rooms, and when Judy heard the porch telephone ring she was always frightened at first about what she might hear when she picked up the receiver, sobbing or slow thick threats or once in a while a man declaring that he had a gun in his hand as they spoke. Judy would press the receiver to her ear and try to work with the voice on the other end—What's been happening, That sounds awful, You must feel very lonely—try to draw the person out, give him someone to talk to. Art would look in on her from time to time so that Judy could signal him when she thought he needed to use their regular telephone to make an emergency call to a psychiatrist or have the telephone company start a trace so the police could intervene, but more often Judy was able to talk the callers through their panic and convince them to go for help themselves.

She was very good at it. She liked sitting out there on the porch at night, with their breakfast table and their squishy plaid couch and outside in the garden, beyond the screened porch windows, the shadowed bulk of the big maple tree. All of this surrounded her and made her feel embraced while Judy held the telephone receiver and led her caller carefully away from harm, just as she might do up in the hospital wards. A patient needed something, and Judy was able to give it to him directly: counsel, comfort, like medications in the IV line.

Then as the Suicide Prevention service was publicized and the telephone number passed from one social service agency to another, pregnant women began to call the number too. But the pregnant women were not suicidal—not the way the other callers were. They were direct. They would say to Judy: This is a telephone number for desperate people, right? Then fine. I am desperate. *I want an abortion.*

Sometimes they would add, as though in deference to the name before the telephone listing, as though this might produce otherwise unavailable information: If I don't get an abortion, I will kill myself.

And Judy didn't know what to say.

She was a registered nurse in the state of Missouri and she knew a very great deal about the human body and American medical procedure. She knew how to operate complicated machinery, and how to pound a dying person's chest to make a heart start beating again, and how to irrigate bowels and press a hypodermic needle into human flesh and put a tube down an unconscious person's throat. She knew how to monitor the heartbeat lines on a fetoscope during labor, how to feel for a prolapsed umbilical cord, and how to select without guidance the proper surgical clamps for controlling heavy bleeding during a cesarean section.

But she didn't know where a pregnant woman could get an abortion.

Call your doctor, Judy said to the first of the Suicide Prevention abortion callers, feeling foolish as she said it. If their doctors had been able to get these women abortions, the women would not have been calling Suicide Prevention. Judy had never had occasion to open the Missouri statute book and read the paragraphs under Section 559, where the crime of abortion was described as "felony manslaughter," but she knew that it was illegal and that in the hospital where she worked it was also unspeakable. This was a word that no one repeated aloud, unless it was part of the medical terminology for what lay people think of as miscarriage: *spontaneous abortion,* a chart in the obstetrics and gynecology wing of Judy's hospital might read, but that was all.

"Nobody talked about it," Judy Widdicombe recalls. "In retrospect, I know that through the years there were women who'd come in for 'diagnostic D&C's' who were really pregnant. But *nobody* talked about it."

Gwyn Harvey told Judy that the other Suicide Prevention volunteers were reporting abortion callers too, that apparently the telephone number was being passed along as a last-resort resource even though nobody at Suicide Prevention had any useful abortion information at all. At one time Gwyn had known a gynecological resident who told vivid stories about women being wheeled into the emergency room after illegal abortions, but all Gwyn remembered from these stories were the descriptions of lacerations and puncture wounds and infections and uncontrolled bleeding. She had no idea where the women had gone for their abortions, or what kind of instruments had been used on them, or whether it was possible, in a Midwestern state at the close of the 1960s, to obtain an abortion that was clean and safe. "Suicidology" had become a modern field of study, and Gwyn had bulging files of reprints about the care of persons who were threatening to kill themselves. But when she listened to the detailed accounts of the pregnant women's telephone calls, it seemed to her that most of these women were not really looking for crisis intervention or a referral to a psychologist for extended treatment. What they wanted was something more practical than that. What they wanted, as far as Gwyn Harvey could see, was an abortionist.

Lysol douche
Hexol douche
bleach douche
green soap and glycerine douche
powdered kitchen mustard douche
hydrogen peroxide douche
potassium permanganate corrosive tablets
intrauterine installation of kerosene and vinegar
gauze packing
artist's paintbrush
curtain rod
slippery elm stick
garden hose
rubber tube
polyethylene tube
glass cocktail stirrer
ear syringe
telephone wire
copper wire
coat hanger (wire)
nut pick
pencil
cotton swabs
clothespin
knitting needle
rubber catheter
woven silk catheter
catheter with stylette
chopsticks

bicycle pump and tube
football pump and plastic straw
plastic tube with soap solution
gramophone needle
bulb syringe
castor oil by mouth
quinine by mouth
ergot by mouth
Humphries No. 11 tablets by mouth
turpentine by mouth

A roster of terrible specificity accumulated, in the years leading up to the late 1960s, and was committed to print in the journals of American medicine. *This was what she did it with,* the examining physicians must have murmured to each other as the bedside curtains were drawn: this was found in the cervix, or identified in the bloodstream, or extracted with forceps from the uterine wall in which it had lodged. Sometimes a woman could be persuaded to describe the device she had pushed inside herself or the fluid she had drunk or the particular tools she believed the abortionist had used; sometimes the doctors could look inside the vaginal area, or open an abdomen at autopsy, and recognize the markers left behind.

Air pumped into the uterus left the large blood vessels distended. Turpentine, when ingested or introduced by douche, gave the urine an odor reminiscent of violets; lower abdominal tenderness was a sign that soap or detergent might have been forced up the cervix. Potassium permanganate tablets, pushed into the vagina to stimulate bleeding that emergency-room doctors might then misidentify as a miscarriage already underway, left craters of corroded tissue along the vaginal walls, so that suturing them was like trying to put fine surgical stitches into a softened stick of butter. At Kansas City General Hospital, where a doctor named Samuel Montello was called for gynecological emergencies while he was establishing his private practice in the 1950s, he learned to recognize the tiny indentations he could see through the speculum: cervical bruises, still fresh, where metal instruments had clamped.

A generation earlier the popular press had made elliptical reference, when scandal or police raid appeared to demand it, to "an illegal operation." But the medical journals were blunt in their clinical reports: Acute Renal Failure Following Hexol-Induced Abortion. Pulmonary Embolism by Powdered Household Mustard Incident to Self-Induced Abortion. Foreign Bodies Lost in the Pelvis During Attempted Abortion, with Special Reference to Urethral Catheters. Single-hospital surveys would appear, or tabulations compiled by state maternal health committees: 927 abortions treated at the Kansas University Medical Center, 1948 to 1957; 32 criminal abortion deaths in Ohio, 1955 to 1959; 233 abortion deaths in California, 1957 to 1965.

The precision of these numbers was to a certain extent illusory. As late as the mid-1960s, sporadic regional studies notwithstanding, there was no practical way to accumulate genuinely accurate documentation on the annual number of American women who deliberately ended their own pregnancies, or tried to. For some years the figure of 681,600 was repeated from time to time; this was a statistic first printed in 1936 by a St. Louis obstetrics and gynecology professor named Frederick J. Taussig, whose lengthy book about abortion was for two decades the only serious abortion reference work on the shelves. Taussig spent five full pages explaining his calculations, which were extrapolated from four regional surveys and adjusted for what he supposed was the difference between rural and urban women. Then, with his equations laid out like schoolbook arithmetic, he presented his numbers: 681,600 abortions induced annually in the United States, and—directly attributable to these abortions, Taussig asserted— 8,179 deaths.

As estimates these might have been reasonable or unreasonable; on a national scale there was little else to go by. Extrapolation is an uncertain business, and Taussig himself observed that even recorded abortion data were highly unreliable. "The desire to avoid any public record of an event that is so often tied up with the moral life of the individual is bound to result in failure to get accurate information," he wrote. "The same lack of honesty appears in our mortality records, where we find cases of abortion death registered as pneumonia, kidney disease, or heart failure."

Because it was illegal, and because it could be made to look like something that was not illegal, the practice of abortion defied standard accounting even for those aborted women who did turn up at reputable medical centers—which many, probably most, did not. If a woman was brought to a hospital emergency room with a bleeding vagina and none of the evident markers, even the most diligent record keeper might elect to register the admission under "spontaneous abortion" or "obstetrical hemorrhage" or some other category that did not contain the words "criminal abortion." The Kansas University Medical Center survey, for example, which according to its author encompassed every abortion (every interrupted pregnancy that resulted in the loss of the fetus, in other words) treated at the hospital between January 1948 and December 1957, listed fewer than one tenth of those as "probable" criminal abortions. Other categories of explanation included traumatic falls, blighted ova, and premature separation of the placenta; but the largest category by far, covering 654 of the survey's 927 cases, was "unknown"—women had ended their pregnancies at the hospital, in other words, but no one ever wrote a reason into the charts.

By 1955, when the Planned Parenthood Federation of America medical director, Mary Steichen Calderone, convened a national conference whose proceedings would be published as the first major American post-Taussig abortion reference book, the conference participants' estimates of illegal abortions in this country ranged from 200,000 to 1.2 million per year. The death-from-abortion tables were similarly blurred, although it was widely agreed that the introduction of strong antibiotics—first the sulfa drugs in the late 1930s, and then penicillin after World War II—had provided physicians with an effective weapon against the internal infections that had killed many women of Taussig's era. If one believed Taussig's 1936 death rates, which some doctors thought were so inflated as to be fanciful even for that pre-antibiotic era, illegal abortion looked on paper to be considerably safer twenty years later; National Center for Health Statistics figures for the late 1950s, limited though they surely still were by the inadequacies of abortion reporting, listed fewer than three hundred deaths per year in the category that included illegal abortions.

"Abortion is no longer a dangerous procedure," Calderone wrote in 1960, stumping the proceedings of her conference in the *American Journal of Public Health*. "One can say, only 260 deaths from all types of abortions —that is a low mortality rate." Then, rhetorically, preparing to answer her own question: "Why should illegal abortion be a public health problem?"

Dr. Mary Calderone, who at the time of that writing had spent seven years with Planned Parenthood, was probably more responsible than any other American medical figure of her era for urging the subject of induced abortion into the lexicon of modern-day public health. She was a strong-minded woman of what in later years might have been called intense feminist inclinations (Calderone used to say she had been offered the medical directorship of Planned Parenthood because no man would take the job), and she had declared from the onset of her tenure that she intended to address illegal abortion as directly and strategically as a public health official might address a broad-scale contagious illness. "I have begun to think of abortion as a preventable disease," Calderone wrote to a colleague in 1954, "with an epidemiology quite similar in many respects to that of tuberculosis."

The venereal disease experts had taught them all something about pulling a scandalous illness into the serious arena of public health study, Calderone wrote, and she mentioned a friend chiding her for the birth control community's reluctance to look squarely at illegal abortion and its effects on the population at large. "You people in Planned Parenthood are where the tuberculosis people were about 30 years ago—you're still trying to cure the disease by using clinical individual application instead of by using Public Health methods," Calderone wrote that her friend had said. "To which I say Amen!"

The very notion of a national conference was sufficiently alarming, in 1955, that the participants were anxious that the press not get wind of it beforehand. The only precedent for such a formal and quasi-public abortion discussion had come more than a decade earlier, when the National Committee on Maternal Health had sponsored a 1942 conference that included studies of "spontaneous abortion," or miscarriage—and still made the organizers so nervous that they refused at the time to tell inquiring newspaper reporters who would participate or what papers would be presented. It was the word that panicked people: "abortion," "abortionist"; for many years, Planned Parenthood and its organizational precursors had gone to considerable effort publicly to disassociate abortion from birth control, which during the early days of the contraception campaigns was considered inflammatory enough on its own merits. The Calderone archive includes an undated draft of an early Planned Parenthood brochure in which the caption, ABORTION IS THE WRONG WAY TO LIMIT FAMILY SIZE, is accompanied by a drawing that someone has nixed as not desperate-looking enough. "Gal depicted looks much too detached," reads the hand-written note. "Artist promises revise so she is anguished but still attractive. I want her walking on street glancing fearfully off right, as if being watched."

Thus Calderone's 1955 abortion conference was in itself a kind of declaration of intent, a gathering of men and women with honorifics and worthy institutions attached to their names, all of them soberly discussing into microphones the problem of induced abortion in the United States. "Problem" was a word that appeared frequently in the published transcript: tragic problem, basic problem, problem in epidemiology. It was a problem because nobody could obtain respectable data. It was a problem because illegal abortion made a lot of women sick. It was a problem because there were urban public hospitals, like Philadelphia General and Los Angeles County, in which entire wards had been ceded to patients trying to recover from illegal abortion; at Los Angeles County, on any given afternoon during the late 1950s and early 1960s, fifty to one hundred patients at a time were separated off into what the doctors referred to as Infected OB.

Every one of these wards, over the years leading into the mid-1960s, produced physicians whose personal encounters with criminal abortion complications were to haunt them for many years afterward. "Infected OB is what they called it, but it was mostly infected abortions," recalls Gail Anderson, the Los Angeles medical professor who was hired in 1958 as head physician for the obstetrical and gynecological service at Los Angeles County General Hospital. "It looked like a set of intensive-care units, all full of abortion patients. If you can imagine walking into a room where there's anywhere from five to ten patients all attached to tubes or whatever —many times they were jaundiced from infections. You've got foul-smelling stuff coming from their uteruses. You've got shock. And in some cases you'd have patients in congestive heart failure. They'd die, in congestive heart failure, foaming at the mouth."

Gail Anderson walked the Infected OB ward every working day of his tenure at County's obstetrical and gynecological service, and when he took over the hospital's emergency medicine department and people wondered how an ob/gyn man could switch so seamlessly to trauma, he would always say: Well, if you had been where I have for the last thirteen years, you wouldn't need to ask. And around the country there were other physicians like Gail Anderson, doctors who had seen enough abortion patients to accumulate detailed casualty accountings of their own: *bicycle spokes, umbrella spokes, Lysol, burned holes in the rectum, feces passing through the vagina, the sickest women I ever saw.* Ira Gerstley, an obstetrician-gynecologist who joined the staff of Philadelphia General Hospital in 1956, used to recite this list when people asked him about complications from illegal abortion; he tried to care for one patient who was delivered to the hospital with no pulse, Gerstley would say, but who revived for a while and was lucid long enough to plead with him aloud after he took out her abscessed uterus and tried to sew her back together.

"She said, 'Doctor, help me, I'm dying,' and I knew it," Gerstley recalls. "There was nothing more I could do. She never made it back to her room."

That was the abortion story that appeared to have settled most indelibly into Gerstley's memory: he treated scores of infected abortion patients, he would recall in later years, but this one young woman was so lovely, and he worked on her for six hours straight, and she died. When Gerstley told that story, it was as though the great sorry array of Infected OB disasters had compressed themselves into a single hospital tableau, as though the image of this one young woman's collapsing organs (Gerstley still possesses a photograph, taken after the emergency hysterectomy, showing the opened abscessed uterus and the dirty catheter coiled inside) was enough to make any sensible person wonder at the wisdom of criminalizing abortion. In this way, during the first decade after Mary Calderone's abortion conference, the Philadelphia doctor was like many physicians who never worked an abortion ward but were haunted nonetheless; by the mid-1960s, every state in the country harbored at least a scattering of doctors who had come to believe that something was wrong with American abortion law, and many of them traced their moment of conversion not to the battlefield of a hospital ward but instead to a single patient, a single detail of a medical history, a single moment of helplessness or rage.

So it was in Missouri, during the first half of the 1960s, and so it was more specifically in greater St. Louis, where the future president of the St. Louis County Medical Society was himself such a physician, an obstetrician-gynecologist with some bad memories from his internship days. The ob/gyn's name was Melvin Schwartz, and Schwartz had been a novice hospital intern when he watched his first infected abortion patient die, bleeding, her abdomen distended, her temperature 106°. The infected patient was very young, still a teenager, and this had taken place back in the 1940s, when penicillin was difficult to come by, and Schwartz had kept running IV's and giving the girl sulfa drugs and following the family doctor's instructions, trying so hard, and the girl lasted four days before she died. Over the years after that Schwartz had treated perhaps a dozen other extremely sick abortion patients, but his conversion commenced during those four days and was completed—this was how Schwartz came to think of it afterward—one morning during the early 1960s, when a woman brought her pregnant teenaged daughter to Schwartz's medical office and told him the girl had been raped.

The rape account was true, Schwartz believed; the family had already reported it to the local police. "And the mother said: 'Can't you do something?' " Schwartz recalled in an interview before his death in 1992. "And I had to say: 'Legally, I can't do anything. I cannot do *anything*. And legally I cannot refer you to anybody.' "

But Schwartz made a referral anyway—illegally, by pointing to a certain listing in the Yellow Pages of the telephone directory. St. Louis in the 1960s maintained its share of inept abortionists, like every urban area in the United States; popular culture and the motion pictures had for many years advanced the image of unsavory men preying on pregnant women in cheap motel rooms, and most urban police officers could confirm that this was true as far as it went. But in some parts of the country a quiet equilibrium had also asserted itself: the inept abortionists were arrested every so often and put out of business, and the skilled abortionists went about their work. The mining town of Ashland, Pennsylvania, had Robert Spencer, a general practitioner whose telephone number was carried on scraps of paper in wallets up and down East Coast college campuses and whose office mail was sometimes delivered addressed: "The Doctor, Ashland." Maryland had G. Lotrell Timanus, a Baltimore physician who maintained friendly relations with the hospital faculty at Johns Hopkins University and for twenty-five years ran a busy medical office abortion practice until he was finally arrested in 1950.* Oklahoma had W. J. Bryan Henrie, an osteopath who had included abortions in his general practice for more than fifteen years by the time he was convicted of manslaughter after one of his patients died; on the day he left his hometown of Grove City to serve his two years in the penitentiary, five hundred people came to his going-away party to wish Henrie well.

And St. Louis had Mrs. Vinyard, who kept a Yellow Pages advertisement in the city telephone book, under the listings for Midwives: WHEN IN NEED OF SERVICE. For some years in St. Louis there were apparently two Mrs. Vinyards, Mrs. Gertrude and Mrs. Martha; nobody seemed to know precisely how they might be related, whether they operated under their

---

* University of Illinois historian Leslie Reagan, whose 1997 book *When Abortion Was a Crime* includes close examination of early to mid-twentieth-century abortion documentation, such as inquest records and medical society transcripts, argues that Timanus' arrest and conviction in 1950 (he served a four-and-a-half-month prison sentence) were part of a 1940s- and 1950s-era upsurge in the prosecution of competent abortionists who had operated for decades without police interference. For additional detail on the medical practice of well-regarded abortionists who operated illegally, see Carol Joffee, *Doctors of Conscience: The Struggle to Provide Abortions Before and After Roe v. Wade,* and Rickie Solinger, *The Abortionist: A Woman Against the Law.*

legal names, or even whether they were in fact two different women. But as far back as 1928 the local directory included a listing for a Gertrude Vinyard, Midwife, along with instructions on how to use a dial telephone ("Place the receiver to your ear, as you would with any other telephone. You will hear a hum-m-ming sound"), and for at least thirty years St. Louis medical people had been discreetly passing the Vinyard name from one generation to the next. A St. Louis doctor named Marvin Camel, who began his obstetrics residency at the city's massive Barnes Hospital in 1951, was told about Mrs. Vinyard by women who had already found themselves In Need of Service; she treated nurses for free, Camel was told, which he assumed was a form of professional courtesy, and when he stood just outside Barnes, he could see her sign, in blue neon, on the second floor of a shabby office building nearby: MIDWIFE. And when Robert Duemler was first establishing his obstetrical practice in St. Louis, an older physician showed him how to accommodate insistent pregnant patients without actually pronouncing the name of an abortionist: "He said, 'Tell them to look in the Yellow Pages under Midwives, and there will be one advertisement that has a thicker, darker line than the other,' " Duemler recalls. "And he was absolutely right."

The Vinyards were efficient, expensive, and reassuringly mysterious; physicians conducting follow-up examinations on their patients agreed that the midwives appeared to be using sterile technique, and they never showed up at doctors' offices to alarm anybody with their felonious presence. From time to time Melvin Schwartz had steered women to the Vinyards during the early years of his practice in St. Louis, but the act of doing so—the illicit, hypocritical nature of the referral, shunting women off to a midwife for a procedure he was too law-abiding to do himself— had never upset him the way it did the day he displayed the address in the telephone book to the pregnant teenager and her mother. "This hit me between the eyes," Schwartz recalled. "This was a *teenaged rape victim.* And that was what disturbed me about her."

After the abortion Schwartz examined the teenager and found her medically healthy—no infection or scarring. The midwife had done her work well. But he was angry about it for many months afterward, and in 1966, when he was associate editor of the Medical Society's semimonthly *Bulletin,* Schwartz and a medical colleague wrote what was for its time an extraordinarily provocative editorial to which they gave the capitalized title "DECISION . . . IN THE WRONG HANDS."

"The decision in Missouri to perform an abortion is not in the hands of the members of the medical profession," the editorial began,

> but is lodged in the law written by a group of elected public officials (some farmers, some lawyers, some businessmen, some professional politicians). That is wrong.

The decision should be a medical one. The result of the decision can have a tremendous effect on the physical and psychological well-being of the pregnant woman involved.

In Missouri we live under many antiquated laws written by non-medical zealots who have confused medical necessity with their own interpretation of the moral values of their day. Today pregnancy resulting from rape or incest must go on to its physical conclusion: birth. Medical science does not have the right to solve the psychological problem of the pregnant woman. She, in the fulfillment of her own logic and reason, must resort to the quack, the untrained, or the village midwife. . . .

The decision is a medical one—not theological—not political. The time has come for the medical profession to make all the decisions in this field, decisions that may (and often do) affect the physical and mental welfare of the patient involved. Now the patient's future is decided not by logic, not by reason, but by an obsolete law written by some who (rather than make enlightened progress) would fall back on the old bromide, "It was good enough for mother, and it's good enough for me."

Well, it isn't. It should not be. Missouri could be the first State in the Union to move into the twentieth century. WHY DOESN'T IT?

In the meantime, while this inflammatory editorial was being composed and readied for print (it ran in October 1966, and for some weeks afterward the *Bulletin*'s editors received ardent letters of both praise and condemnation), another St. Louis physician had advanced somewhat farther than that and was regularly sending pregnant women to the skilled abortionists in town—sending them surreptitiously but almost formally, with private written records noting the date of each referral. A United Church of Christ minister worked with him. They operated on the campus of St. Louis' Washington University; college students knew how to find them.

"This issue had come up in conversation—whispered abortion, and so forth," recalls the United Church of Christ minister, James W. Ewing, who had been assigned in the late 1950s to help run a campus ministry at Washington University and had taken to walking the university grounds every day so that students might feel free to approach him with their problems. By the early 1960s Ewing enjoyed a measure of trust among many of the students; he urged them into community action programs, helped organize early local involvement in the Southern civil rights movement, and developed a reputation for listening without recoiling or moralizing when a young woman confided that she was pregnant and intended to find an abortionist. At first Ewing had nothing but a sympathetic ear to offer these women. He would tell them that abortion was a very serious matter, that it could be dangerous—"and that there were other options," Ewing recalls, "like having the baby, working with their parents, whatever

—but if they chose to have an abortion, let's be very, very careful about this."

At that point Ewing would give vague, sensible advice: take someone along with you, watch for dirty instruments, be sure to see a doctor afterward. But for a long time he had no idea where the young women were going when they thanked him and left. Then in the fall of 1965, when Ewing asked about a particular student he had liked—a quiet girl, pretty, part of a bus contingent Ewing had mobilized the previous summer to register black voters in Alabama—someone told him she had gotten pregnant and had an illegal abortion and died.

He never found out if it was true. Nobody seemed to know. The story was that she had gone to a Kansas City abortionist and begun bleeding on the train on the way back, and that by the time someone delivered her to a hospital she had lost too much blood to be revived. In the story the girl was alone as she made her trip home, bleeding to death by herself in a Missouri passenger car, and Ewing found himself imagining this part of it over and over. "I was just sick inside, and angry, absolutely angry," he recalls. "It felt to me like a person wasting herself. So I went to John Vavra. I said, 'John, these women are getting butchered. There has to be a better way.' "

Dr. John Vavra was a Washington University medical professor who wore bow ties and wire-rimmed glasses, taught classes in Presbyterian Sunday school, served on the board of campus ministries, and was known around the university as a reader and lecturer of dizzying range: Vavra could give a single talk in which he managed to work in phenomenology, existentialism, Communism, salvation, and the differences between Hindu and Christian views of God. Ewing knew Vavra liked to teach and argue complex ethical positions, weighing philosophical views and differing religious perspectives, and he guessed, correctly, that Vavra would not look anxious or shocked at the mention of illegal abortion. Ewing had many questions he wanted the physician to answer: Exactly how was an abortion done? What equipment was used? Did Vavra know anyone who could be relied upon to do an abortion well?

"I said, 'Look, we're not going to be able to *prevent* women from going to abortionists,' " Ewing recalls. " 'But I just want to talk with you and see if there's any way we could arrange a system to incorporate both some counseling and safe medical practice.' "

Vavra said he believed there was, and over a period of weeks during the fall and winter of 1965, the Reverend James Ewing and Dr. John Vavra set out together to begin breaking Missouri abortion law—for Vavra did know of reputable abortionists; he knew about the Vinyards, and he knew there was a physician in town who performed the occasional abortion in his office, and he knew also, as did Ewing, that Missouri criminal law made the very act of passing on this information a felony punishable by up to five years in prison. It was therefore a contained and extremely careful system that the minister and doctor established as they got underway: women making inquiries around the Washington University campus were directed first to Ewing, who talked to them long enough both to counsel the women and to satisfy himself that they had genuinely made up their minds. Then Ewing would explain that what he was about to do was illegal, that others were taking serious risks for them, and he would send the women to Vavra. Then Vavra sent them on to the abortionists, keeping discreet records as he did so, tiny handwritten markings in small datebooks he carried in his breast pocket. He used the conventional doctor's notation for Last Menstrual Period, as though he were entering notes into a medical chart. *Sherry, LMP Mar. 17. Set up with Mrs. V. $350.*

THE WORD "underground" did occur to both of them, the medical professor and the minister of God; from their offices on the Washington University campus they were running an underground, and as the college semesters progressed, Ewing and Vavra began hearing reports of similar clandestine alliances in other cities—all of them involving ministers who had volunteered to defy their own states' abortion laws. At the University of Chicago a Baptist minister named E. Spencer Parsons had since 1965 been directing pregnant students to illegal abortionists whose work Parsons trusted. In East Rutherford, New Jersey, an answering machine in a United Methodist parsonage took messages from women who wanted the name of a competent abortionist. At Southern Methodist University in Dallas, a Methodist chaplain named J. Claude Evans kept in his home a kimono sent to him by the grateful parents of a student who had flown to Japan for an abortion after Evans helped make the arrangements.

In Wisconsin, during the late 1960s, a pregnant woman could walk into the office of certain Baptist or Methodist or United Church of Christ ministers and walk out an hour later with the name of an illegal abortionist in Chicago or Juarez. In Pennsylvania, a call to a certain telephone number rang an answering machine in a Baptist nursery school in the town of Wayne; the recorded message left instructions on finding the nearest minister who could provide "counseling" on abortion—and, it was understood, a referral to an abortionist. In California, fifty ministers and rabbis up and down the state passed around memoranda like this one, typed on an old manual typewriter and dated August 16, 1968:

Dr. Madrid is on again. He and his assistants moved all equipment from Juarez to Nogales, Sonora, since we last communicated with you. The new arrangement is as follows: Fly to Tucson, Ariz. Take bus to Nogales, Ariz., a 1 hr. 10 min. ride. From the U.S. side of Nogales call 21 70 5 on the Mexican side. Say "this is Mary" and receive the reply "this is Pete," as before. An English-speaking man will be on duty from 9AM to 7PM daily. He will instruct the caller as to meeting place and the driver will identify himself with the green card with our stamp on it, as usual. New prices: $250 to 8 wks., $350 8–13 wks.

By 1968, there were in fact approximately one thousand ministers working in this semishadowed fashion around the American states, and in a few cities a good distance from Missouri these ministers had already printed letterhead stationery or obtained listings in the local telephone directory: the Clergy Counseling Service for Problem Pregnancies; the Illinois Clergy Consultation Service; the New Jersey Clergy Consultation Service on Abortion. A single-minded prosecutor might have studied the names and discerned before him a ragged multistate enterprise to help large numbers of people break the law, and he would have been right. A new form of civil disobedience had started up—in churches and ministers' homes and the sparsely furnished offices of college campus ministries— and its spiritual and organizational center was lodged in a nineteenth-century brick Baptist church on the south side of Washington Square Park in Greenwich Village, New York.

The church was called Judson Memorial, and the man who assumed its pulpit every Sunday was a tall Texas-born minister with a big chest and ears that stuck out and a haircut like a Marine. A lot of unusual activity went on at Judson under the Reverend Howard Moody's supervision; he had arrived in 1956, after graduating from the Yale Divinity School and serving five years as chaplain at Ohio State University, and over the next decade the new minister and his associate pastors filled the old church halls with dance and experimental theater and noisy political meetings. Civil rights marchers, Vietnam War protestors, local Democratic Club organizers all mingled and argued at Moody's church, which by the mid-1960s was regarded less as a temple of God than a stewpot for liberal Democratic politics.

Such was the reputation of Judson Memorial that back in 1965, when a young Jewish woman named Arlene Carmen began asking around Democratic circles about possible sources of fund-raising help for some black Southern sharecroppers about to be evicted from their land, she had been urged repeatedly to go see Howard Moody. "I'd never had anything to do with ministers," Carmen recalled in an interview before her death in 1994. "But the community leaders I talked to all pointed their fingers at Judson Church." When she finally overcame her uneasiness and went to the preacher's office, he agreed at once to help her raise money, and by 1967 Carmen was on the Judson Memorial payroll. She was hired as a church administrator, but from the outset Howard Moody asked for her help with one project in particular: he had the idea that women looking for abortionists should be able to receive detailed information from a minister who could counsel them first.

Such a minister, were he to conclude his counseling by writing down the name of a competent abortionist, would almost certainly be acting in violation of his own state's abortion laws. Moody and Carmen both knew that. "Though we never admitted it publicly," Moody recalls. Publicly the stance was to be at once truculent and gravely ministerial. When Howard Moody and the New York pastors he had pressed into partnership organized in May 1967 the first formal clergymen's counseling service, they placed the word "abortion" in the name, declining euphemisms, and called *The New York Times* to announce their debut. The article appeared on the front page, along with the telephone number to call, OR 5-5000, and a picture of Moody in his clerical collar, which was an accessory he put on only when he was obliged to look ecclesiastical and dignified. "Twenty-one Protestant ministers and rabbis have announced the establishment of a Clergymen's Consultation Service on Abortion to assist women seeking abortions," began the *Times* story. And some paragraphs down: "The clergymen cited 'higher laws and moral obligations transcending legal codes,' and said that it was their 'pastoral responsibility and religious duty to give aid and assistance to all women with problem pregnancies.' "

Any district attorney who wanted to advance upon that kind of territory clearly had his work before him—pastoral responsibility, religious duty, private confidences between women and their ministers. Howard Moody believed in the practical value of civil disobedience and public arrest; he had marched in Alabama with Dr. Martin Luther King, Jr., and had been arrested himself in Brooklyn when Moody and hundreds of others lay down in front of advancing cement mixers to protest segregation in the construction industry. But he had no desire to go to jail for passing illegal information to pregnant women. "I wasn't the least bit interested in that," Moody recalls. "I wanted to help the *women*. If we got arrested, we weren't going to be able to do anything."

To that end, the ministers set out to make prosecution as unpalatable as possible. Lawrence Lader, a New York journalist whose 1966 book *Abortion* had inspired enough urgent requests for help from pregnant women to turn Lader into a vigorous public critic of American abortion laws, had first suggested to Moody that ministers might make ideal conduits for the sort of referral information Lader was trying to pass along himself; nobody was eager to arrest a man of the cloth for counseling his flock. Ephraim London, an attorney on the board of directors of the New York Civil Liberties Union, advised that the ministers refer only to licensed physicians, and always across state lines: if New York clergy sent women to New Jersey and New Jersey clergy sent women to Pennsylvania, the jurisdictional confusion might work to their advantage.

"The legal advice which gave us the most assurance was Mr. London's recommendation that we never either assume or admit that we were breaking the law," Carmen and Moody wrote in *Abortion Counseling and Social Change*, their 1973 account of the Clergy Consultation Service. "At all times we were to behave as though we were acting within the laws of

New York State and that as clergy we were bound to follow a higher moral law."

The telephone began to ring the day the *New York Times* story appeared, and within a week distraught female voices had backed up on the answering-machine tape and Howard Moody was spending seven hours a day ushering women in and out of his office at Judson. The sheer volume astonished him, but more startling were the women who had telephoned Manhattan and then gotten on airplanes in places hundreds of miles away. "They would come all the way from the Midwest, or drive all the way from Pittsburgh, just to talk to you," Moody recalls. "Not to get the abortion. Just to talk."

Sometimes they brought their boyfriends or their husbands or their mothers, and in the halls of the church Arlene Carmen would step around the women and men leaning stony-faced against the wall or sitting up in straight-backed chairs with their purses in their laps. It looked to Carmen like human misery on parade, and there were so many of them that in short order the clergymen's abbreviated lists of abortion doctors were inadequate to the task. They had a doctor in Puerto Rico, a polite little man whose patients came back from San Juan with reports of his scrubbed tile floors and attentive nurses; they had a doctor in Louisiana, a Tulane-educated gynecologist who worked all alone out of a New Orleans hotel and went to the trouble of putting potholders on the examining table stirrups to keep the metal from chilling the women's feet. But they needed many more, and after a while Carmen began making regular trips out of Manhattan to visit abortionists herself.

She took notes. *"Code: Nina." "Call & ask for Dr. SOS." "On probation for three years; slum area; filthy; bad procedure." "Danger of infection: Avoid." "Butcher; avoid at all costs."* There were times when she never proceeded past the first telephone call. "You'd be told, 'Meet me at such-and-such a parking lot, at such-and-such a time, at such-and-such a place,' " Carmen recalled. "You'd say, 'Who is the doctor? Where is the office?' " The telephone contact would tell Carmen not to worry and she would say thank you, she certainly wouldn't, and she would scratch that name from the list. Still the list kept growing; Howard Moody was traveling to places like Cleveland and Chicago, working social webs laid out by the civil rights movement and Vietnam War protest, calling meetings together in church halls or ministers' offices or bare-floored college rooms with folding chairs and Stop the Draft posters on the walls. He was passionate in his delivery and could drawl when it suited him and he never grew out the crewcut, so that when he got up to speak he created the startling effect of a Southern drill sergeant expounding on the rights of women.

"Howard started telling us about the ravages of illegal abortion, and what was happening to women, and what the death rate was like," recalls Eleanor Yeo, a United Church of Christ minister who was working a

University of Wisconsin campus ministry in Milwaukee when she first heard Moody speak. "Maybe it was lucky it was during the Vietnam War, because I think a lot of people at that point were questioning authority in ways they never had before. People who ordinarily would not think of doing anything illegal were all of a sudden thinking of things they had to do."

Moody would instruct the gathered ministers that they must understand this to be a counseling service, not a hand-over-the-address service; some of the women who showed up at Judson and the other New York churches were not certain that abortion was what they wanted, and the ministers were supposed to remain alert to this ambivalence, to encourage women to keep their babies or give them up for adoption if the idea of having an abortion disturbed them too much. Even the more adamant women needed a pastoral person to talk to, Moody would say, although there was a famous Clergy Service story about the woman who called Judson Memorial in the early months, when Moody assumed all their telephones were tapped. "Is this where you can get an abortion?" the woman asked. Moody blanched and explained that she had called a church, but was welcome to come in for abortion *counseling*. "I don't need the Sermon on the Mount, Reverend," the woman said. "I need an abortion." And she hung up.

There were a few critics who chafed even then at the assumption that adult pregnant women required the sage advice of ministers, particularly since most of the ministers were men. "Hominy Dominy Counseling Service," read one barbed cartoon drawn by Patricia Maginnis, a fiercely anticlerical California activist who used to hand out lists of Mexican abortionists to anybody who asked for them; in her drawing, which she reprinted onto postcards, Maginnis put a balding little priestly man with his hands over his face and a second caption that read, "Women's Counseling for Problem Clergymen."

But California's clergy service was one of the busiest, churning out advisories on clinic closures, answering letters from Mexican doctors who wished to offer their services, and instructing the women on travel across the border: don't drink the water, put your money in your brassiere, rehearse your story of kidney or bladder infection in case the customs agent asks why you have antibiotics in your purse. ("If a woman is asked whether she went to Mexico for an abortion, she should be prepared to look totally surprised.") A file drawer at Judson Memorial began to back up with dispatches from California and Illinois and Michigan and New Jersey, and within a year of its *New York Times* debut, the New York Clergy Service office was also ferrying occasional information to and from the quiet regional abortion network headquartered on the campus of Washington University in St. Louis, Missouri.

By comparison to some of its more ambitious counterparts in other states, the St. Louis operation in 1968 was still both modest in scope and discreet in its public presence. John Vavra had enlisted a few medical students to help examine and interview pregnant women before and after their abortions, but there was no letterhead stationery or telephone listing. The service did not even have a name. It was simply known around campus —and within widening circles of friends, faculty, Washington University graduates, and *their* friends—that Dr. Vavra and the Reverend Ewing helped women find abortionists.

And when Gwyndolyn Harvey set out to learn about abortionists in St. Louis, she too was directed to John Vavra. Gwyn felt that she ought to visit the medical professor in person, that this was not a subject one spoke about over the telephone, and when Vavra received her in his office she found that she was taken aback by his appearance: gray-haired, bow-tied, surrounded by his academic books, Vavra was soft-spoken, scholarly, and entirely calm—"unflappable," Gwyn Harvey recalls. "He looked as far from anybody doing anything radical as you could possibly imagine."

Just as he had three years earlier with the Reverend Ewing, Vavra explained to Gwyn Harvey some of the medical details of induced abortion —how physicians generally used surgical instruments to scrape away the lining of the uterus, but skilled lay abortionists could also bring on miscarriage by packing sterile gauze into a woman's cervix. Gwyn saw that Vavra was being careful about precisely what he would and would not say; he offered up no names or telephone numbers, during that first meeting in his university office, but when Gwyn told him about the Suicide Prevention callers, he was interested at once. Gwyn wondered if they could perhaps work together, Vavra making available his abortionist referrals and Suicide Prevention providing volunteer counselors. Vavra said he thought they surely could. There was no conversation at all, either that day or any time after that, about *why* such a service might be justifiable, about what it was in their backgrounds or general philosophies that might be prompting two well-educated professionals to sit in a university office and make arrangements for the ongoing violation of state law. "We didn't have to discuss it," Gwyn Harvey recalls. "We were almost like kindred spirits. The issue was: there were people getting hurt."

FROM HER CIRCLE of colleagues and the Suicide Prevention volunteer roster, Gwyn selected her likeliest initial recruits: a Washington University psychiatric social worker, and Judy Widdicombe.

"I remember it was a Sunday, and Gwyn Harvey came in, and we were going to make gumbo," Judy Widdicombe recalls. "Chicken gumbo. We had bought fresh okra. We did this, oh, several times a year. And she bounded in, I can see her coming through the door, and she said, 'Well! Guess what. We're going to start doing Problem Pregnancy Counseling.'

"And I said"—a low whistle, and a slow intake of breath—

" 'Oooh-kay.' And she said, 'Yep. I'm going to handle the counseling. You're going to do the medical.' "

The model for what she had in mind already existed in some other states, Gwyn Harvey explained: counselors, ministers, abortionists, and doctors, all linked together by a telephone network that would serve as each woman's initial point of contact. John Vavra would lead them to the abortionists; as the service expanded, Vavra himself would continue to make the personal referral to the doctor or midwife who was to perform the illegal abortion, but Gwyn and Judy were going to assemble a medical and pastoral corps to attend to the women beforehand and afterward. They needed clergymen, Gwyn said, and they needed sympathetic doctors. Judy was in hospitals five days a week; she must know a few gynecologists who might be approached as volunteers, who might be willing to conduct preparatory and follow-up medical examinations even though the act of doing so was probably in itself a violation of the law.

Judy thought about what she was going to say to these doctors, and how best to say it, and within a matter of weeks she had approached Dr. Melvin Schwartz and Dr. Robert Duemler and Dr. George Wulff, who taught obstetrics and gynecology at the hospital where Judy had trained as a nurse, and who had worked the Infected OB ward at Kansas City General Hospital in 1933, and who had learned as an intern how to recognize the caustic blackened stain of a self-administered potassium permanganate douche, and who answered Judy, like the others, without hesitation: Yes.

TWO

The Wedge

1969

On February 10, 1969, under the splendid limestone and filigree rotunda of the Missouri state capitol, State Senate Bill 206 was read for the first time into the legislative record. The words "termination of human pregnancy" appeared in the sixteenth paragraph, in boldface type.

334.310. 1. A physician licensed by the state board of registration for the healing arts may terminate a human pregnancy or aid or assist or attempt a termination of a human pregnancy upon the written request of the patient requesting the termination of human pregnancy or upon the written request of the parent or legal guardian of a minor or incompetent patient requesting the termination of human pregnancy if the physician finds that one or more of the following conditions exist:

(1) Continuation of the pregnancy is likely to result in the death of the mother;

(2) There is a substantial risk that continuation of the pregnancy would gravely impair the physical health of the mother;

(3) There is substantial risk that the child would be born with a grave permanent physical defect;

(4) The pregnancy resulted from rape, incest or other felonious intercourse. Any woman requesting a termination of human pregnancy

based on the contention of rape must have reported the rape to the police within seventy-two hours after the alleged occurrence and must have filed an affidavit or complaint to that effect with the police.   .

The bill was printed into a handout approximately the size of a theater program, which is how proposed legislation looks when it begins its passage through the capitol of an American state. The state printing office duplicated one thousand reprints of the original bill, and the reprints were copied on shiny-paper 1960s copying machines, and a misaligned glued-and-stapled copy of Senate Bill 206 arrived one day on the desk of a St. Louis obstetrician named Matthias H. Backer, Jr., who believed that he could see it for precisely what it was: a wedge.

In 1969 Matt Backer was forty-two years old, a compact man with thin brown hair and a trim mustache and a meticulous shoes-just-polished look that was honed from many years' service in the Naval Reserve. If he had ever encountered Judy Widdicombe directly, perhaps brushing past her in some hospital hallway, it had made no strong impression on him; Backer shared with one partner a bustling obstetrics and gynecology practice, one of the busiest in St. Louis, and nobody had ever suggested to him that an organized underground might be spreading among licensed physicians and illegal abortionists. He knew illicit referrals took place from time to time, the occasional doctor willing to nod his head in the direction of an abortionist. He knew it was not unheard of for a respectable physician to slip into an operating room himself and perform an abortion that was probably illegal under the strictest reading of the criminal code. Backer didn't particularly like knowing this; he thought his colleagues were wrong to do it. But it happened infrequently, it seemed to him, and more to the point it happened quietly, discreetly, unsanctioned by law, a long way from the public eye.

Matt Backer believed he could live with that.

But if Senate Bill 206 were to be approved by the legislature and signed by the governor, it would rewrite state law with nearly a century and a half of tradition behind it. The word "abortion," which since 1825 had remained in the manslaughter felony section of the criminal code as a grave reminder that a dreadful matter was being addressed, was to be replaced in print by the coolly clinical "termination of human pregnancy." The state of Missouri was being asked to declare formally, for the first time in its history, that these terminations of human pregnancy might be acceptable even when no life-threatening emergency confronted the physician. A careful reading of the language made it evident that a minimal number of women would actually qualify for legal abortions under the terms of Senate Bill 206, but for three years Matt Backer had been alert to a grander agenda that was pushing its way toward the state of Missouri; he knew, because he had been reading the newspapers and talking to his colleagues and studying season by season the disturbing reports from other parts of the country, that Senate Bill 206 was not where it was going to stop.

"Catholic obstetrician and gynecologist, father of 12, and one Mongoloid child": that was how one of the local newspapers would summarize Dr. Matt H. Backer, Jr., when the reporters began calling him up for quotes, as though those details were all a person needed to know about what he believed and why. And this was not inaccurate information (although offensive, perhaps, in its suggestion that Edward, the Down Syndrome boy, was somehow not one of the thirteen children), but Backer did not see how it was especially instructive. He made a point of explaining to newspaper reporters that his interest in Senate Bill 206 had nothing to do with his religious faith or the size of his family; Backer had no objection to contraception, for couples who wished to use it themselves, and at the Jesuit-run St. Louis University Hospital, where Backer taught obstetrics in addition to running his private practice, he had developed a close friendship with the obstetrics director Denis Cavanagh, a good-humored Scotsman who was regarded as a suspiciously liberal Catholic by some of the mustier Jesuits around the hospital. Cavanagh was not troubled by contraception either, and in fact had told Backer many times that he had once believed loosening Missouri's abortion law might be a reasonable idea.

By 1969 Cavanagh had emphatically changed his mind, and at the end of their hospital workday he and Backer would take off their white coats and climb into Backer's two-tone Cadillac and drive the two hours west to Jefferson City, where the legislators had scheduled public testimony on Senate Bill 206. The drive to Jefferson City was plain, rural, and virtually without landmark until the pale dome of the capitol suddenly lifted above the flat horizon, and often as they drove Backer and Cavanagh would try to predict the course of the evening hearing before them, wondering which fictive statistics their opposition intended to repeat into microphones this time. Cavanagh had a ruddy face and a rust-colored mustache and still spoke in a brogue that thickened when he was agitated; after a while Backer could do a fair imitation of the Scottish, rolling his R's the way Cavanagh would when he was exclaiming over some new report out of Jefferson City: "Hey, Matt! Look what they're trying to pull!"

Cavanagh had his own war stories about illegal abortion complications; in Miami, where the Scotsman had spent seven years teaching obstetrics and gynecology before his relocation to St. Louis, he had written professional papers and a textbook chapter about the treatment of infection-induced shock. Many of the patients Cavanagh had treated for shock had been brought to the hospital after criminal abortions; a few of these women had died in the hospital, and some had lived but presented particularly gruesome detail for their medical charts (there was one Havana abortionist who had evidently been trying to set off an abortion with phenol, a caustic disinfectant, but had missed the uterus and run the phenol into the bladder instead), and Cavanagh told Backer how sorry and angry these cases had made him feel, the women lying there so sick, sometimes with the police at the bedside pressing them for details about the abortionist.

Indeed, Cavanagh used to say, it would have been a great convenience, both for him and for those women who evidently wanted so badly to end their pregnancies, if a developing baby could be redefined somehow as part of a woman's body—removable at will. But Cavanagh did not believe that was true. He did not believe writing it into law would magically make it be true. When he had first arrived in Misssouri, Cavanagh had listened to other doctors complain that the nineteenth-century state law was so rigid that rape victims and women with severely deformed fetuses could not legally obtain a clean hospital abortion, and when he began to read the language of what in the 1960s were nicknamed "reform" bills—legislation that proposed for the first time to legalize certain limited categories of abortions—Cavanagh, like Matt Backer, believed for a while that his colleagues were well intentioned, that it might be tolerable to make legal room for women in these unusual situations while still condemning abortion as a matter of state policy.

But Senate Bill 206 was Missouri's second feint at one of these laws, and both Backer and Cavanagh had come to believe that "reform" was the wrong word. The supporters of the abortion bills were inventing statistics and piling one tragic story atop another, as far as Backer and Cavanagh could see; everybody understood that legalizing the very occasional rape and fetal deformity abortion was not really what the "reform" proponents wanted at all. Both Backer and Cavanagh knew there was a larger plan, and that this was not the imagining of some overwrought Catholic academics; there *was*, one had only to pick up a newspaper to see it. The state of California had changed its law. The state of Colorado had changed its law. North Carolina, New Mexico, Oregon, Arkansas—to the east and west and south of Jefferson City were legislatures that had been persuaded for the first time in American history to write into the statute books formal approval for certain kinds of abortions, and Backer and Cavanagh knew that in some of those states the very legislators who had introduced the idea in the first place were now denouncing their own legislation as insufficiently accommodating. What they wanted looked plain enough, to the doctors from St. Louis, and Backer and Cavanagh had a shorthand they used as they readied their public testimony: abortion as birth control. Abortion—this was a strident way to phrase it, but seemed to Backer a suitable expression both of attitude and long-range goal—on demand.

No one wrote these words on placards, in the state of Missouri in 1969; there were other state legislatures in which abortion adversaries had been photographed shouting or pounding tabletops or brandishing small sealed jars containing preserved human fetuses, but in Missouri the proceedings were still civil and thoughtful and Backer occasionally found himself looking forward to them, the way he might anticipate a good teaching rounds at the hospital. He and Cavanagh would get out of the Cadillac in Jefferson City with packets of slides under their arms, sober slides in black and white, with phrases like FETAL CONDITIONS and MOTIVES FOR ABORTION typed in outline fashion so the words could be projected overhead. They carried newspaper articles and academic reprints, and after they disappeared into the capitol building the streets outside grew quiet and very dark, because Jefferson City was a place of little movement at night, when most of the government had gone home for supper, and even the gleam off the river was hidden from view.

Some hours later one of the first-floor doors would push open and Backer and Cavanagh would come out again, stretching and rehashing and buoyed by the night air. They would tally their Assembly votes. They would argue about which opposition doctor had made the worst impression. Somebody's emotion-wracked story would make them roll their eyes and grimace as they remembered it, and then they would climb into the Cadillac and Cavanagh would fall asleep and Backer would drive in the moonlight, south and east, down to Linn, around the curve at Rosebud, up through Union and Pacific and the darkened barns and farmhouses along either side of the highway. He knew the shapes of the towns, and the agricultural fields at night, and the bridge that crossed the Meramec River before the lights of the St. Louis suburbs began to come up; in the fifth decade of his life, Matt Backer had learned by heart the route between his home and his statehouse. Abortion had made him learn it.

THE LAW that defined and set punishment for the crime of abortion was first placed into the Missouri penal code in 1825, four years after the territory became a state. It was the second abortion law passed in the United States—Connecticut's was the first, in 1821—and in its initial form, in what the historian James Mohr has suggested was a reaction to popular midwife-administered and folk abortifacients like the juniper oil called savin, the law prohibited only abortions induced by poisoning. In 1835 the wording was revised to prohibit instrumental abortions as well, and by the late 1800s every American state included in its criminal code an abortion law with provisions comparable to Missouri's.

This is what the Missouri abortion statute said:

> 559.100. Any person who, with intent to produce or promote a miscarriage or abortion, advises, gives, sells, or administers to a woman (whether actually pregnant or not), or who, with such intent, procures or causes her to take, any drug, medicine or article, or uses upon her, or advises to or for her the use of, any instrument or other method or device

to produce a miscarriage or abortion (unless the same is necessary to preserve her life or that of an unborn child, or if such person is not a duly licensed physician, unless the said act has been advised by a duly licensed physician to be necessary for such a purpose), shall, in the event of the death of said woman, or any quick child, whereof she may be pregnant, being thereby occasioned, upon conviction be adjudged guilty of manslaughter, and punished accordingly; and in case no such death ensue, such person shall be guilty of the felony of abortion, and upon conviction be punished by imprisonment in the penitentiary not less than three years nor more than five years, or by imprisonment in jail not exceeding one year, or by fine not exceeding one thousand dollars, or by both fine and imprisonment; and any practitioner of medicine or surgery, upon conviction of any such offense, as is above defined, shall be subject to have his license or authority to practice his profession as physician or surgeon in the State of Missouri revoked by the state board of medical examiners for its discretion.

When condensed into ordinary English, Section 559.100 was not a remarkably complicated part of the criminal code. It said, as did every other American abortion statute during the first half of the twentieth century, that a person could be imprisoned and stripped of his medical license, if he possessed one, for performing or helping with an abortion. It said these penalties were inapplicable only if the abortion was necessary to preserve the life of the woman.* It offered no definitions of either of the crucial words: "preserve" and "life."

"Therapeutic abortion" was what doctors were trained to call these legal-exception procedures—hospital abortions, mandated as medical therapy, and protected by law. A pregnant woman with failing kidneys might live if she had an abortion and die if she did not. A pregnant woman with critical heart disease might be viewed by all prevailing medical opinion as unable to survive the stress of carrying the baby to term. There were other medical emergencies in which it was obvious to everyone consulted that only an induced abortion would keep the patient alive—but by the 1950s,

---

* It is unclear why the phrase "or that of an unborn child" was included in this line of the Missouri law, since performing a "miscarriage or abortion" to "preserve" the life of an unborn child would appear a contradiction in terms. James Mohr, now history department chair at the University of Oregon, speculates that physicians may have been closely involved in the bill's drafting during the early 1800s, and may have wanted to make certain that the new law was not invoked against physicians who had performed certain emergency obstetric procedures like cesarean sections or other instrumental deliveries. A detailed account of Mohr's examination of American abortion laws—which Mohr argues were written largely in response to pressure by organized nineteenth-century physicians, and reflected a change from earlier, more tolerant attitudes about abortion—can be found in his book *Abortion in America: The Origins and Evolution of National Policy.*

with rapid advances in medical diagnosis and treatment, there were not very many of them. Mary Calderone of Planned Parenthood, writing in 1960 for the *American Journal of Public Health*, dismissed outright the role that genuinely life-threatening medical crises had come to play in American abortion procedures. "Medically speaking," Calderone wrote, "that is, from the point of view of diseases of the various systems, cardiac, genitourinary, and so on, it is hardly ever necessary today to consider the life of a mother as threatened by a pregnancy."

And since no state legislature had ever written out an explanation of exactly what was meant by preserving a life, a gray area of considerable proportions was suggested by the language of abortion statutes like Missouri's. Did the woman have to be threatened with imminent and certain death? What about possible death, or a shortened life expectancy? What if her physical life was not threatened by pregnancy, but her emotional life was? Could a doctor argue that he was "preserving a life" by performing an abortion on a woman who insisted that she would commit suicide, abandon her family, or suffer a nervous breakdown if she were forced to bear another child?

In a very few states the word "health" had been written into the abortion statute; "safety" also appeared in Maryland's statute, and "serious and permanent bodily injury" in New Mexico's. But the law offered no guidance on the questions those words might inspire, either, and for many years individual physicians answered them one by one in the relative half-light of their own medical offices. By the 1950s some hospitals had convened formal committees to examine each case, but even the therapeutic abortion committee system operated without what might be described as consistent or mutually accepted principles. A published 1959 survey of California hospitals found that what one hospital considered an acceptable therapeutic, another rejected as impermissible under the law. The hospitals were asked to consider hypothetical case histories and came back with contradictory responses: a fifteen-year-old minister's daughter, raped by a mental defective who had escaped from a state institution? Yes from fifteen hospitals, no from seven. A married woman with Hodgkin's disease whose X-ray treatment would endanger the fetus? Yes from ten hospitals, no from eleven. A depressive mother of three who took an overdose of pills after learning she was pregnant again? Yes from seventeen hospitals, no from four.*

---

* Critics of the therapeutic abortion committee system observed at the time that in many hospitals the onset of these case-by-case review boards raised new barriers that had not existed before the committees got underway—because some board members were deeply opposed to abortion, because some members felt obliged to hold down their hospital's abortion numbers regardless of the merits of the particular case before them, and because physicians grew discouraged about submitting case requests and thus subjecting already-distraught patients to group interrogation by committee members. Some hospital records showed significant drops in the number of medically approved therapeutic abortions after the committee review process was introduced.

Under the strictest reading of the law, none of these abortions would have physically preserved the mother's life. The closest call was the woman who attempted suicide, but it was widely understood that Will Kill Herself If Denied Abortion was a vast catchall into which a very great deal could be tossed if individual doctors were so inclined. In the printed studies pregnant women appeared to have an exceptionally low suicide rate, but those were statistics, and a doctor could study them carefully and still not know what the lone anxious woman before him might do if the hospital told her no. *It is therapeutic because we say it is* had become by the late 1950s the defining principle for the small number of abortions performed in American hospitals: if we want this abortion done, because the woman is seriously ill or was sleeping with someone important or knows the doctor well enough to get herself signed in, then Therapeutic will appear on the record and the odds are that no district attorney will come making inquiries.

In the spring of 1959, at a gathering of legal academics at the May-flower Hotel in Washington, D.C., a University of Pennsylvania law professor named Louis B. Schwartz made what in retrospect reads as the first significant formal and detailed public proposal as to how this system might begin to be dismantled. "Inevitably there will be strong feelings about it," Schwartz said by way of introduction, in what proved to be thundering understatement, as he described to the assembled members of the American Law Institute (ALI) a suggested revision to the penal codes of the American states. For seven years the ALI, a half-century-old organization of influential law professors, judges, and practicing attorneys, had been at work on a massive project to modernize state criminal codes and give them a coherence that America's state-by-state statutory development had so far worked to discourage. The Model Penal Code, as the ALI called its work-in-progress, was meant to offer every state a sample set of statutes that could define and set out appropriate punishment for crimes ranging from homicide to bigamy to obstructions of governmental operations.

Article 207 of the Model Penal Code, Sexual Offenses and Offenses Against the Family, included in its subsections the crime of abortion. The ALI attorney directed to supervise the rewriting of state abortion statutes was Schwartz, a former Justice Department criminal attorney who had never had occasion to give much thought to abortion before Herbert Wechsler, the Columbia University law professor in charge of the Model Penal Code project, asked Schwartz to assume responsibility for much of the specific language of the code. Abortion law was only one of dozens of subheadings Schwartz was expected to address, but he and his younger associates could see at the outset that this was an area in which proposals for "modernization" would have to be presented with exceptional delicacy.

"We knew this was very controversial—as Schwartz describes it, the notion of trying to disguise change," recalls Curtis Reitz, a University of Pennsylvania law professor who was a student at the university law school when Schwartz asked him to help with the abortion law research. Schwartz was inclined as a matter of general belief toward concepts of individual liberty and minimal state intervention, and he knew the idea of rethinking abortion law had been publicly broached before. Revision of state criminal abortion statutes had been advocated in a few published books, articles, and broadsheets, and as Schwartz examined the principles at stake in abortion law he was particularly impressed by the writings of Glanville Williams, a Welsh legal scholar whose popular 1956 lecture series at Columbia University—the topics ranged from abortion to euthanasia and suicide—had been worked into a book called *The Sanctity of Life and the Criminal Law.*

Williams was something of a philosophical bomb-thrower in his time, examining abortion and contraception in unusual detail and arguing in his book that most criminal abortion law was as impractical as Prohibition— an evil that, as Williams put it, was "greater than the alleged evil of abortion itself, without in fact preventing abortions." Schwartz recalls that at heart he thought Williams was somewhat extreme, but fundamentally sound; it seemed to Schwartz that women were going to go on looking for abortionists no matter what the law prohibited, and that this put them at terrible risk for their own safety. Indeed, as Curtis Reitz conducted his research he became fascinated by the dramatically publicized 1955 newspaper accounts of a socially prominent young Philadelphia woman who had died after her own mother brought her to an illegal abortionist. (The mother, charged as accessory to a homicide, pleaded *nolo contendere* and received a suspended sentence.)

But Schwartz knew that proposing even incremental change in such an emotionally loaded body of law might set off passionate protest both in the ALI and in the state legislatures where the Institute hoped to present its finished work. "Glanville Williams was—how can I put this?—on our left," Schwartz recalls. With Williams' ideas as a kind of specter of the drastic possibilities they were too sensible to endorse, Schwartz and his colleagues began in 1959 offering up for the ALI a sample statute that would revise abortion law on a scale at once modest in its practical application and sweeping in its philosophy. "Abortion, at least in early pregnancy, and with consent of the persons affected, involves considerations so different from the killing of a living human being as to warrant consideration not only of the health of the mother but also of certain extremely adverse social consequences to her or the child," Schwartz wrote in his proposal preface, thus upending—in the view of one segment of the legal and medical community, at least—a cherished if imperfectly applied American principle that was more than one hundred years old.

The abortion statutes in the American states, Schwartz proposed, should be rewritten to permit "a policy of cautious expansion of the categories of lawful justification of abortion." "Cautious" was meant to be the operative term. Abortion might be "different from the killing of a living human being," but it was an act of very great gravity nonetheless, and society as a whole, in the form of its doctors and lawyers and lawmakers, had the right and obligation to make rules detailing when a woman could and could not do it. Schwartz tried to buttress each of his categories with research from the academic and popular press, and laid them out as follows: Abortion to Prevent Gravely Defective Offspring. Abortion of Pregnancy Resulting from Rape and Other Criminal Intercourse. Abortion to Preserve the Health of the Mother.

This last category, which on the face of it contained perhaps the least controversial language—in a few states the concept was already written into the statute—had the makings of a legislative Trojan horse, and Schwartz knew it. For what he wanted in the Model Penal Code was a definition of health that included "mental health," which both critics and supporters of the abortion proposals recognized as a term of wondrous elasticity. Schwartz himself had worried in his 1959 proposals about all the difficult cases their categories were leaving out:

> The moderate enlargement of justifiable abortion discussed in the preceding comments removes some of the anomalies in existing penal laws against abortion. But it does not legalize abortion in a variety of cases where individual hardship might seem to call for a dispensation from the general rule, e.g. pregnancy of a deserted wife, pregnancy of a woman who is the working member of a family supporting a dependent husband or other children, pregnancy of a woman inmate of a penal or other institution, pregnancy where the child will be one of multiple illegitimate offspring of a woman who has already demonstrated her incapacity to rear children decently.
>
> Moreover, experience has shown that hundreds of thousands of women, married as well as unmarried, will continue to procure abortions not sanctioned under proposed Section 207.11, in ways that endanger their lives and subject them to exploitation and degradation. We cannot regard with equanimity a legal pattern which condemns thousands of women to needless death at the hands of criminal abortionists. This is a stiff price to pay for the effort to repress abortion, notwithstanding that the purpose of the law has the religious support of very many of our citizens and commands the highest consideration of others, since it comprehends a reverent attitude toward the life process itself and a concern for the spiritual implications of terminating pregnancy.

By including "physical or mental health" in the final version of the Code, with no explicit language about what that ought to mean, Schwartz hoped he was offering state legislatures an opening to make some of those "hardship" abortions legal without actually saying that was what they were doing. "We have often turned to psychiatrists for an out when we were solving social problems," Schwartz recalls. "I would have been in favor of a very flexible reading of the mental health part of it."

When Schwartz finally presented his written proposals at the ALI meeting in Washington, a heated discussion began the moment he had stopped speaking. There were references to the Holocaust, to infanticide in China, to the killing of human beings in the name of progress. "The state cannot give the authority to perform an abortion, because it does not have the authority itself," protested one Illinois attorney at the conference. "Those lives are human lives, and not the property of the state. . . . We are giving the right to impose a life or death sentence."

Schwartz fielded some dissent from the opposite direction as well. A Texas attorney worried that the model statute was too restrictive because it required not one but two doctors to certify in writing that the planned abortion was covered by one of the categories. (Maybe so, Schwartz replied, but to write emergency exceptions into the law leaves "a great big loophole" everywhere else.) Another attorney wondered whether the model statute shouldn't permit abortions of *all* pregnancies outside marriage. (Too offensive to people's sensibilities, came a Maryland attorney's reply; any girl who "goes out and gets herself into trouble" would be able to "get out of it" with an abortion.) The question of statutory rape was raised—if the ALI was going to start down this controversial path in the first place, shouldn't abortions be legal for teenagers who were by law too young even to consent to sexual intercourse?

That motion carried, and was incorporated into the final draft of the Code. "All illicit intercourse with a girl below the age of 16 shall be deemed felonious for the purposes of this subsection," the abortion passage read when the Model Penal Code was finally approved, in its entirety, at the American Law Institute's annual meeting in Washington, D.C., in May 1962. Except for the statutory rape addition, the language was very nearly what Schwartz and his colleagues had worked up three years earlier: the statute, if states chose to adopt it, would create for the first time the formal legal concept of Justifiable Abortion. Not Therapeutic, with its implications of urgent medical necessity, but Justifiable: in the phrasing of Matt Backer or Denis Cavanagh or the protesting ALI lawyer from Illinois, a woman and her physician were to be permitted to end another human life, legally, in a hospital, with the codified approval of the society around her, because *her situation justified it.*

Across the United States, as attorneys and judges of the American Law Institute carried the product of their Model Penal Code efforts back to law schools and law offices and the capitol workrooms of freshman legislators, the abortion passage under Article 207 began insinuating itself into the machinery that produces the nation's state laws. The wording was changed a little from one stapled Senate bill to another, the categories revised, the terminology altered to suit a particular legislator's tastes. But the passage, and the one extraordinary proposition from which it resonated, was the American Law Institute's.

Even most of the Institute members thought it didn't stand a chance.

• • •

In California, the Model Penal Code language turned up in 1961, before the final Code was approved, in the form of an Assembly bill to permit abortions for rape, incest, fetal deformity, and threats to mental or physical health. The bill was introduced by a freshman Democrat named John Knox and hustled swiftly into Interim Study purgatory without ever coming to a vote.

"It was, at that time, something which was not talked about by polite people or written about," the Democratic California congressman and former state senator Anthony Beilenson, who in 1961 was still a year away from his entry into the California legislature, recalled in an oral history project interview some years later. "It was very much a sort of under-the-rug kind of subject. So it was very difficult. People thought it was going to take us ten or twelve years, you know, at the earliest, before we would be able to liberalize the laws."

Two American dramas, unrelated except by the grim similarity of the newspaper headlines they inspired, worked together to help prove Beilenson wrong.

The first began and ended in a single two-month period, the summer of 1962, when a Phoenix children's television hostess named Sherri Finkbine took about three dozen tranquilizers during the early weeks of pregnancy with what was supposed to be her fifth child. The pills had been carried home from Europe by Finkbine's husband, who was a high school teacher and had been chaperoning students on a trip to Europe when he bought medication to help him sleep. It was not until Sherri Finkbine read an alarming article about tranquilizers that she asked her doctor to check the contents of the pills; he had to wire London for the information, Finkbine recalled in later years, and when London wired back, the doctor called Mrs. Finkbine and asked her to come into his office on a Saturday so he could tell her what he had learned.

The medication was Thalidomide, a German-made pharmaceutical that was enormously popular in Europe until women who had taken the pills while pregnant began giving birth to children with profound deformities: flipperlike stumps instead of arms and legs. The research and warnings of a single diligent American Food and Drug Administration official had kept Thalidomide off the U.S. market, but approximately five thousand Thalidomide-affected infants had been born in Europe by the time Mrs. Finkbine swallowed her pills; one of them, an infant born in Belgium in 1961, had died shortly after birth when his distraught mother put a lethal dose of sedative into the baby bottle. Medical journals were already running photographs of the Thalidomide babies, and Sherri Finkbine's doctor told her in the office that she might or might not wish to examine the pictures, with the babies propped up, and black strips over their eyes for privacy, and no legs and no arms.

"And the doctor looked at me," Finkbine recalls. "And he said, 'You know, if you were my wife I'd tell you the same thing.' He said, 'We have four children too. If you really want to have a fifth baby, let's terminate this pregnancy and start again next month under better odds.' "

In the medical idiom of the times this was warm-up, commonly understood, for therapeutic abortion. Technically it was outside the confines of the law. Arizona's abortion law was as specific as Missouri's; it permitted abortion only to preserve a woman's life, and technically an abortion was not going to preserve Sherri Finkbine's physical life, since carrying a deformed baby was not medically dangerous, and she had four other children to dissuade her from suicide. Still the system was ready to receive her, in its muffled and guarded way. Finkbine's doctor had made that plain, and if Sherri Finkbine had been a different sort of person or the telephone lines had been busy that night or some other glitch of history had kept everyone's voices as low as they were supposed to be, then she and her husband would almost certainly have passed without incident through Good Samaritan Hospital and gone home to recuperate.

But Sherri Finkbine worried about the families of local Air National Guardsmen, many of whom had been been shipped to the Berlin Wall crisis in Europe and might have come home with similarly ill-labeled Thalidomide. On the night before she was supposed to have her abortion, Finkbine called an *Arizona Republic* editor, who urged her to talk directly to a reporter, and with the first edition of the morning newspaper the abortion was public business. The *Republic* ran a large, urgent headline: PILL CAUSING DEFORMED INFANTS MAY COST WOMAN HER BABY HERE. The hospital attorneys balked. Judicial orders were sought to determine how Arizona statutory law might apply to Mrs. Finkbine's therapeutic abortion.

For the next thirty years Sherri Finkbine would insist that she had talked to the newspaper only to warn women about the pills and only after a promise of anonymity, but her name appeared in court documents that were part of the public record, and reporters from the papers and wire services began to call the house and walk around on the lawn and wait for Mrs. Finkbine to come out and say something. She was refused an abortion in Arizona. Japanese officials discouraged her from seeking a visa to Tokyo. Strangers began writing letters to Sherri Finkbine, suggesting ammonia or roller coasters or two quarts of gin or a certain physician in Juarez, Mexico; a couple in San Francisco offered to adopt the deformed baby if Sherri Finkbine would carry it to term. A letter in cursive expressed the hope that someone would strangle Sherri Finkbine's other four children, "because it is all the same thing," and someone in McKeesport, Pennsylvania, sent the Finkbines a cartoon of an embryo recoiling from a large knife. Attached to the cartoon was a card that read: "I curse you and curse you and curse you and everything you do for the rest of your filthy life."

Sherri Finkbine and her husband took an airplane to Sweden, where they waited thirteen days for the Royal Swedish Medical Board to approve a therapeutic abortion at a hospital in Stockholm. Reporters saw them off in Los Angeles, reporters gathered in Copenhagen for their transfer of airplanes, and at the Stockholm airport a Swedish reporter swept them up in his little car and offered to pay them money for an exclusive. *Life* magazine sent photographers to Arizona to record her living room and her children and her opened suitcase, and by the middle of August it was possible to stand in a supermarket line almost anywhere in the United States and study the pictures of pretty Sherri Finkbine taking a brush from the hand of her wispy-haired youngest child. In the pictures she was pregnant, preparing to have her abortion. There was also in the magazine an explanation of *phocomelia*, which was the medical name for the Thalidomide deformity; the literal translation was "seal limbs," and when Sherri Finkbine was asked afterward what the Swedish doctors had said about her abortion, she said they had told her that deformities were present and that "baby" was not the best description for what she had been carrying. It was better to think of it, Mrs. Finkbine reported the doctors having explained, as a growth.

So all at once, during the summer and autumn of 1962, induced abortion in the United States was something that was "talked about by polite people," as Anthony Beilenson has put it, and written about and argued about and thrown for good measure into one of that year's Gallup polls, which found that just over half those surveyed believed Sherri Finkbine had done the right thing. Thirty-two percent did not, and in both public and private forums the sorrows of a single married lady gave instant animation to the questions implicit in the efforts of the American Law Institute. Should the law have bent for Sherri Finkbine? Should the law have changed for Sherri Finkbine? Was there any occasion upon which society might declare, by the wording of its printed rules, that a woman could go to her doctor with a growing embryo inside her and instruct the doctor to pull it out?

Could the multiplying cells that would become a human infant, if they were left intact inside a woman's body, *ever* be thought of as a "growth"?

In 1963 the American rubella epidemic began, and the argument rose in pitch.

German measles, which is the common name for rubella, presents itself in adults with symptoms ranging from dangerous fever to a scarcely noticeable rash. But when the virus infects pregnant women, it can move through the bloodstream, cross the placenta, and wreck the development of embryonic cells. The statistical risk factor has been raised and lowered repeatedly as additional medical studies have appeared, but in the early 1960s, twenty years before the distribution of an effective vaccine, it was widely believed that women who contracted rubella during the early weeks of their pregnancies ran about a fifty percent chance of bearing babies with congenital damage that could include blindness, hearing loss, mental retardation, and a variety of heart and blood and bone malformations that might or might not prove fatal.

In the United States, rubella infections multiplied in epidemic numbers from late 1963 through 1965, pushing west as the months went by, and by the time the epidemic reached California and Hawaii there were hospitals on both coasts in which the local medical community understood that a pregnant woman who had confirmed an early pregnancy rubella exposure could obtain a "therapeutic abortion"—technically illegal—and go home the next day. At the University of California at San Francisco (UCSF) hospital, the gynecologist in charge was of the increasing conviction that he didn't care how the law was worded; women were obliged only to call up and say, Well, I had the rash, I think it might have been rubella. "Anybody who had a possibility of having a deformed baby could have an abortion, it was just that simple," recalls Alan Margolis, the hospital's chief of gynecology from 1965 to 1972. "Why do you want a kid born blind and deaf and all that? It just didn't enter our thinking to question it."

Then in May 1966 a San Francisco obstetrician named Paul Shively was formally charged by the California State Board of Medical Examiners with "unprofessional conduct" for performing and assisting in abortions for women who had been exposed to rubella. A second physician was also charged, and then a third and fourth and fifth; by the end of the summer, nine San Francisco doctors had been threatened with the loss of their medical licenses for taking improper liberties with the state abortion law, and spokesmen for the state attorney general's office had announced investigations of more than forty in all.

Alan Margolis and many of his medical colleagues were astonished when they heard about the investigators marching into Shively's office for records. Margolis himself was doing dozens more abortions at UCSF than any of the physicians facing state charges, and the Board of Medical Examiners seemed to have picked a remarkably odd target in Paul Shively, who was the widely respected chief of obstetrics at an Episcopalian-run hospital set in the midst of the most Catholic neighborhood in San Francisco. That Shively had done the abortions was never in dispute; San Francisco Medical Society president George Herzog, who was also an obstetrician, said immediately and publicly that he was only sorry *he* hadn't been doing rubella abortions too. "This is a distinguished company of doctors," Herzog was quoted as saying in the following morning's newspaper. "They are humanitarians."

Not since Sherri Finkbine's trip to Sweden had a single running news story worked so well to suggest that the act of abortion might be openly and approvingly described as a deliberate medical decision to keep a particular baby from coming into the world. Of the doctors singled out for punishment in 1966 by the state of California, not one fit the tabloid profile of Abortionist; these were solidly established physicians performing inpatient operations on women who appeared—when the newspapermen wrote about them, at least—to be married, earnest about their families, and generally above moral reproach. "A group of San Francisco's most distinguished physicians asserted yesterday that threats of abortion charges against them are based on a political-religious vendetta using an 'archaic statute' as a pretext," one of the early *San Francisco Chronicle* accounts began:

> All the abortions were performed with the full approval of physicians' committees in the top Bay Area hospitals. . . . All the women suffered a grave risk of giving birth to severely deformed babies; all wanted children, but preferred to start their pregnancies anew so the resulting children would have the best possible chance for normalcy.

The story spread to New York, and to the medical press, and out across the webs of gossip and information that professional colleagues pass back and forth. In the summer of 1966 it reached the St. Louis medical office of Dr. Matt Backer, who registered briefly its salient details and turned his attention back to work. Backer was in Missouri. The rubella abortions were in California. Half a nation stretched between St. Louis and a place where a newspaper writer might be permitted to compose a phrase like "start their pregnancies anew," and in any case Backer was too busy to brood about a procedure everybody understood was against the law. He ran a two-hospital obstetrics practice, he taught at St. Louis University, he worked in spare hours for the Navy, he went to Mass in the mornings, and supper at his home was served at a refectory-style table long enough to accommodate a dozen people at once; Matt Backer was not a man with social controversy much on his mind, and the California doctors sounded to him like decent men, a long way away, doing the wrong thing for motives both honorable and profound.

"I had the conviction in my own mind," Backer recalls, in the cautious way he has of selecting his words, "that if abortion were only for indications like this, it would not be a preoccupation of mine to try to prevent it."

Which was not to say he thought it was all right. Matt Backer had cared for rubella patients himself—not many, because rubella passed lightly through Missouri during the epidemic years. But on three occasions he had been obliged to look at women in his office and deliver this somber news, that the rash was German measles and that German measles had been found to increase very greatly the chances of giving birth to a deformed baby. When the women asked him what they could do, whether there was some way to get rid of it and try again, Backer had a standard reply: There are other doctors who will do that for you, but I will not.

His patients stayed, and bore their babies. Two of them gave birth to

infants who showed no sign of any abnormality. The third woman gave
birth, about the time rubella reached its peak in California, to a baby girl
who was nearly blind and nearly deaf and hobbled for life by a congenital
heart defect that would keep her from any active childhood play. Backer
had complicated feelings about that baby from the moment she was born;
thirty years later he would still remember the child's first name, and
the way the mother had accepted without argument his explanation that
abortion was a procedure he would not perform, and also this, to keep the
weight from settling too lightly upon his memory: long after the birth,
when the child was growing older, Backer saw the mother again and she
talked about how hard it was, how fierce her daughter's disabilities. The
mother looked at Backer and said: If I had known what was coming, what
it would really be like, I would probably have gotten an abortion.

It was not a simple business. Backer knew that, even in 1966, when
this rubella-damaged baby girl was still tiny; Matt and Laverne Backer's
seventh child had an IQ of 35, and Backer was not among those parents
who profess to see disabled children as special gifts from God. Eddie was
an exuberant boy and the family loved him, but sometimes Backer felt as
though he had been honoring God about as energetically as a mortal fellow
could, welcoming each of these new lives into his rapidly multiplying
family, and that there was some deep injustice in adding to this load a child
who would never be able to dial a telephone or prepare his own supper or
spend a night without adult supervision. Raising disabled children was
work, formidable work, and grave: Backer once brought one of his patients
over to the house, after the woman had given birth to a Down Syndrome
baby and was deciding whether to raise the child at home, so she could
watch the daily life of a family with a retarded child. The woman never
did tell him exactly what she saw that afternoon, but her reaction to his
household was not lost on Backer. She put her baby in an institution.

Thus Matt Backer could follow, to a point, the logic of the doctors
willing to describe abortion as an act of kindness and a promise of relief.
The problem, as Backer saw it, was that the logic didn't gel. If they genuinely believed the most compassionate remedy for damaged infants was
putting them out of their misery, then why do it at a stage when no
one could be sure the infant was damaged at all? Even if fully half the
rubella-exposed pregnant women were carrying babies whose development had been affected, why abort the other half who were healthy? Why
not wait until the babies were born, and then do a sort of Aldous Huxley–
like triage, disposing of infants who didn't meet the national standard?
You don't kill people because they're imperfect; that was how Laverne
Backer would phrase it, watching Eddie lumber merrily beside the backyard swimming pool, and when Matt Backer picked up the St. Louis
County Medical Society Bulletin that October, he saw finally that he
was going to have to say something publicly—that somebody needed to
speak up.

On the tenth page of the *Bulletin,* in capital letters, Backer read the headline: DECISION . . . IN THE WRONG HANDS.

And below the headline, in italic type: *"An Editorial—Read Carefully."*

So Backer did.

"The decision in Missouri to perform an abortion is not in the hands of the members of the medical profession but is lodged in the law written by a group of elected public officials (some farmers, some lawyers, some businessmen, some professional politicians)," the editorial began. "That is wrong."

Backer went on reading: ". . . antiquated laws written by nonmedical zealots . . . must resort to the quack, the untrained, or the village midwife . . . the old bromide, 'It was good enough for mother, and it's good enough for me.' . . ."

There was a statistic about illegal abortion deaths in Italy, and a reference to Dark Ages theologians. There was a suggestion that modern and civilized man see fit to rise above obsolete hundred-year-old laws. The last paragraph of the editorial ended in capitals, like the title. "Missouri could be the first State in the Union to move into the twentieth century," it said. "WHY DOESN'T IT?"

MATT H. BACKER, JR., which was how he would sign his name when he finally put his thoughts to paper, spent many weeks considering the dismal implications of the eight paragraphs he had read in the Medical Society *Bulletin.* "A medical decision," the editorial had said—like repairing a heart valve or setting a broken leg. Not a word about the second human life. Not a word about Only for Deformities. This was where it led, Backer saw, once the initial threshold had been crossed, once society decided in its sympathetic way that women could legally rid themselves of certain babies because those babies would be so terribly difficult to raise.

Backer had a study in his basement, a little windowless room that made him feel sometimes as though he had descended into the hold of a submarine. He kept a lot of books in the room, and a walnut desk that had belonged to his father. A big Smith-Corona manual sat square on top of the desk.

"I am well aware of the widespread professional sentiment for liberalization of abortion laws and even more aware that doctors favoring this are highly ethical humanitarians," Backer typed, rapidly:

Within the proponent group, however, is a well-organized, vocal minority which hopes that the new laws will be loose enough to permit the use of abortion as a form of population control. Little is said about this because these astute extremists realize that the best way to convince a majority of the American people that the law has to be changed is by emphasizing the need for abortion in cases associated with rape, incest, possible fetal abnormality or danger to maternal health.

Blueprints made a lousy metaphor for human embryos, Backer wrote; it had become quite the vogue, among the doctors and politicians trying to tinker with the laws, to suggest that tearing up an architectural drawing was philosophically very different from blowing up the completed building. "No 'blueprint' left alone in its natural habitat develops into a building or machine," Backer wrote, "much less a living, breathing organism. On the other hand, the human fetus has never been known to develop into anything but a human being, albeit occasionally an imperfect one."

How could they call a human fetus a blueprint one day, Backer wrote, and the next day a citizen of the United States? How could they justify making abortion legal for the "severely anxious" unmarried girl without acknowledging that *every* pregnant unmarried girl was severely anxious? How could they push from their minds the Nuremberg Trials, and the good upstanding doctors who followed state directives about who was and was not entitled to live? "If these laws are changed to permit the destruction of intrauterine life, except where it constitutes a truly grave danger to the life of the mother," Backer wrote, "what will remain in principle to prevent the eventual destruction of life in the infirm, crippled, feeble-minded, aged or even, perhaps, politically unpopular ethnic groups?"

He sent his letter to *Missouri Medicine*, the glossy-paged journal of the Missouri Medical Association, where doctors across the state might pay attention to what Backer had to say. He wrote a rebuttal headline that seemed to him a good slap in the face. "THE DECISION TO KILL," the headline read. "WHO WANTS IT?"

But by the time Backer's letter was published, in June 1967, two states had already toppled, and the third was on its way.

On April 25, the governor of Colorado had signed into law the first working state legislation based on the abortion language of the American Law Institute. A young Denver-based assemblyman named Richard Lamm (who was in fact worried about overpopulation, but had also believed for a long time that pregnant women ought to be able to assert their own interest in not carrying a baby to term) had introduced the bill only three months earlier and watched, with increasing astonishment, as both the state House and Senate approved it by two-thirds majorities. "It caught everybody by surprise," Lamm recalls. "People would just say, 'That sounds like a good idea! I'll buy that! Okay!' And we just started down this list—and by God, by the time I introduced the bill, I had enough co-sponsors in each house to pass it."

On May 8, the North Carolina House and Senate had voted into law (until 1996 the North Carolina governor had no veto power) the second working legislation based on the Model Penal Code abortion language. Arthur Jones, the state representative who introduced the North Carolina law, was a retired banker who told people he had been interested in birth control for half a century, ever since he and some similarly enterprising fellow students at Ohio's Oberlin College had sold contraceptive jelly on campus—at cost, Jones added, and only to married couples. "I was writing papers about family problems for social studies courses," Jones recalled, for a newspaper writer in Charlotte, "and I realized then how cruel, inhuman, and obsolete the abortion laws were. I made a mental note at the time to do something about it if I ever had a chance."

The Indiana state legislature passed a Model Penal Code abortion bill, but the governor vetoed it. The Oklahoma state legislature considered a Model Penal Code abortion bill and put it off for a year. Abortion bills rose and were argued over in New Mexico and Minnesota and Georgia and New York, where during one legislative hearing a politician and district attorney interrupted each other with angry questions about euthanasia, atheism, the Ten Commandments, and the morality of justifiable homicide in wartime. The weekly *National Catholic Reporter* began grouping its abortion dispatches in brief urgent bulletins, the datelines ticking off state capitols as the numbers grew: Hartford, Trenton, Santa Fe, Phoenix. "ABORTION BILLS WIN, LOSE," the *Reporter* headlined in the spring of 1967, and in subhead, lest the tenor of the season be missed: "Fight Is On from Maine to Colorado."

Then in California, where the legislator Anthony Beilenson had for four years been championing his version of the Model Penal Code statute, the state Assembly and Senate weathered two last grueling abortion hearings (the Senate's was eight hours long, ended at two in the morning, and was described by one state senator as one of "the greatest speaking marathons I have ever witnessed") and passed Beilenson's Humane Abortion Act in June. Two days later the abortion bill was signed into law by Ronald Reagan, then in his sixth month as governor of California. Reagan had maneuvered unsuccessfully to kill the new abortion bill before it reached the gubernatorial desk; he let it be known that he didn't much like the Humane Abortion Act, and that what troubled him most was the prospect of abortions for birth defects—the very issue that had helped haul the argument into 1960s newspapers. "I cannot justify the taking of an unborn life," Reagan announced to reporters during the last weeks before signing the bill, "simply on the supposition that the baby may be born less than a perfect human being."

Anthony Beilenson, a young New York transplant who liked to use the adjectives "barbaric" and "hypocritical" when he talked about the nineteenth-century statutes then serving as abortion law in California and nearly every other state, had rigorously worked the legislature to accommodate and ultimately outmaneuver the governor; the final version of Beilenson's Humane Abortion Act dropped fetal deformity as a legal justification for abortion. But the other Model Penal Code categories stood essentially intact: rape, statutory rape, incest, and danger to health—physical *or* mental. The most populous state in the country had turned Louis Schwartz's American Law Institute proposals into law, and it was plain to a lot of people (including Reagan himself, who expressed his misgivings aloud but signed anyway) that "mental health" could encompass birth defects very handily if the presiding doctors were of the right sympathies. All they had to do was announce that the woman was severely distraught at the prospect of bearing a deformed child, and even Matt Backer knew that was not much of a stretch. A massive shift was in the making, eight years after Louis Schwartz had stood before his colleagues in Washington with the typewritten paragraphs that Schwartz worried would elicit strong feelings: by the fall of 1967 more than half the legislatures of the American states had examined and debated and in some instances taken votes on the abortion language of the Model Penal Code, and so pervasive was the "reform" label—suggesting, as one contemporary dictionary definition put it, "the correction of evils or abuses"—that Backer's own diocesan newspaper had itself taken up the loaded shorthand.

For the Model Penal Code had reached Missouri, too. "Group Circulating Petitions Opposing Abortion Reform," read the headline in a spring 1967 issue of the *St. Louis Review,* the weekly publication of the Catholic Archdiocese of St. Louis, after a Republican state senator named Robert Prange introduced into the Missouri legislature his version of the American Law Institute statute. (Prange's secretary was said by family legend to have looked aghast when Prange handed her the text of his bill. "She said, 'That's political suicide,' " recalls Mildred Prange, the senator's widow. "And he said, 'Yes, I know. But it's something that needs doing.' ")

Prange's earliest abortion bill was modified in committee until its provisions were even narrower than the standard Model Penal Code version; the bill dropped the rape and mental health categories and would have legalized abortion only in cases of fetal deformity or a threat to the mother's physical health. But the inaugural appearance of *any* proposal for legislative abortion "reform" was in itself enough to inspire an organized committee of opposition. "Immoral actions cannot be made humane by legislation," read the petitions of the newly assembled Missourians Opposed to Liberalization of Abortion Laws. "We hold that the arresting of innocent human life, the deliberate destruction of innocent human beings, is wrong, regardless of the state of development or of the circumstances under which that human came into being."

A modest list of male Missouri names formed the makeshift letterhead for Missourians Opposed: a pathologist, an obstetrician, three engineers, two editors for Catholic publications. Matt Backer's name was not on the list. He signed a petition and read what reports he could find in the newspapers and his weekly *St. Louis Reviews,* but he really had no interest in attaching himself to some purely theoretical antiliberalization campaign; the 1967 bill died quickly and without much publicity in the state House of Representatives, and because the Missouri legislature of that era convened only every other year, Backer assumed that a St. Louis physician might now consider the matter closed for many months to come.

But as he read his newspapers Backer began to see that something profound had given way and that the public conversation was no longer even nominally about Thalidomide or rubella or rape. "Abortion: Once a Whispered Problem, Now a Public Debate," read the broad top-of-the-page headline on a long and prominently displayed *New York Times* article in early 1968, and the opening paragraphs in that article were not the customary heartrending account of a pseudonymous *in extremis* Model Penal Code category patient. Instead they described, quite matter-of-factly, a modern South Dakota abortionist whose illegal business had thrived for thirty years—an "elderly widow," wrote the *Times* reporter, Jane E. Brody, "a one time nurses' aide . . . assured herself of future business by giving each customer a small, white calling card . . . counts among her friends some of Rapid City's most prestigious citizens."

The *Times* article went on to suggest, in careful but unmistakably sympathetic prose, how very many illegal abortions were undertaken for what Backer had always thought of as "social" reasons. ("Joan R., a 21-year-old New York girl who has been struggling along as a theatrical technician, said that in her job, 'I could not even afford to be pregnant, much less have a baby.' Joan was unable to turn to her parents for help, and the abortion she obtained from a careless Pennsylvania doctor left her sterile for life.") More disturbing still was the implication that some physicians were now publicly willing to urge that state laws be changed to legalize these purely elective abortions—to accommodate the Joans as well as the Mrs. Finkbines. Here was one physician declaring that everybody knew most of New York's in-hospital Therapeutics were performed on "suicidal" women who were not truly suicidal at all; here was another complaining that many would-be patients at his Midwestern hospital had been, as the *Times* writer paraphrased it, "shortchanged" by the arbitrary refusals of the hospital's Therapeutic Abortion committee.

Shortchanged. Backer found this an ominous choice of words, and it was around this time that Denis Cavanagh began stopping by Backer's hospital office with occasional bulletins and ominous news items of his own. Did Backer know the American Civil Liberties Union (ACLU) had pronounced abortion during the first twenty weeks of pregnancy to be a "civil right"? Did Backer know abortion proponents were advancing the ludicrous claim that eight to ten thousand women per year died from criminal abortions? Had Backer seen the item about the Colorado state representative, Richard Lamm, complaining that even under his liberalized abortion bill the Denver hospitals were turning too many women away? Was Backer watching the offhand way in which abortion-sympathetic doctors in California appeared to be defining "mental health"?

By the time the Missouri legislature reconvened for its 1969 session, ten states had adopted or were in the process of adopting some version of the Model Penal Code abortion legislation. In most instances the "mental health" loophole was written right into the new legislation, and when Senator Prange's 1969 Missouri version was introduced in February in its earliest form, Backer took some small comfort in the fact that his home state, at least, appeared still to be exploring the more conservative route: Senate Bill 206, Missouri's second-round adaptation of the Model Penal Code language, legalized "health" abortions only for those pregnancies that would "gravely impair the physical health of the mother."

But even this reassurance faded quickly, for Prange then attached to his bill an amendment adding "mental health" to the criteria for legal abortion. And that, finally, was what pushed Matthias Backer to begin driving to the state capitol to address the legislature in person. As he began his Jefferson City visits, Backer knew that two separate antiliberalization organizations were simultaneously trying to draw attention and volunteers in St. Louis, but Backer still wanted no formal relationship with either of them; there was the Missourians Opposed group, and there was another entity called Fight for Life, and Backer knew from his newspapers that in nearly every American state in which the legislature had seriously considered some version of a Model Penal Code abortion statute, some comparably named regional organization had sprung up to mount opposition to the new legislation—to urge to their statehouses men and women like himself, ordinary citizens roused to public protest, often for the first time in their lives, by the prospect of any change in the traditional understanding of abortion under state law.

Backer was not sure what it was about these new groups that made him uneasy. The lead organizer for Missourians Opposed to the Liberalization of Abortion Laws appeared to be the editor Harvey Johnson, who worked out of St. Louis for a passionately conservative anti-Communist Catholic monthly called the *Social Justice Review;* and although Backer and Johnson were friendly enough when they talked by telephone (Johnson had reprinted into the *Review* Backer's entire *Missouri Medicine* essay, displaying it nicely and including the "DECISION TO KILL" headline Backer liked so much), Backer had never envisioned himself a religious or political crusader and was not at all interested in attaching himself in print to some of the right-wing causes Johnson championed with such enthusiasm. It unsettled Backer to see his essay appearing alongside editorial copy assailing "the weeds of modern paganism" and "the hot irons of Communist slavemasters," and as he collected newspaper references to the various new antiliberalization groups now announcing their presence at statehouses around the country, it struck Backer that the organizational names alone suggested a reach somewhat grander than he had ever had in mind for himself. Fight for Life, Right to Life of Southern California, New York Right to Life, Minnesota Citizens Concerned for Life, Mobilization for the Unborn—a certain pattern was emerging in the titular vocabulary, the suggestion that the central mission was not simply the protection of traditional abortion statutes but rather the *defense of the right to life,* the *championing of life itself.*

And it was not that Backer disagreed with this, exactly: it *was* life they were talking about; his own *Missouri Medicine* essay had included a warning about "the cheapening of unborn life." But Backer had written that essay for doctors. He had taken some care to word it respectfully, aware that "the widespread sentiment for liberalization of abortion laws," as Backer had put it, extended to men he knew and personally liked. He had imagined that the argument over rewriting state abortion laws would remain at its core a medical matter, a collegial exchange, an earnest debate among physicians worrying together about the best and most ethical way to practice medicine.

He was only beginning to understand how wrong he had been.

THREE

Heaven and Earth

1970

WHILE THE MEN of the legislature pushed pieces of paper around hearing-room tables—she knew there were women in the capitol, but a great hush appeared to have fallen over their small and far-flung voices—Judy Widdicombe took pregnant women into her house and brought them to the back bedroom to finish out their abortions.

It was 1970, the very early part of the year, and the law had changed a little in places that appeared eager to discourage as many potential abortion patients as possible. The North Carolina law required women to prove they were residents of the state. The Colorado hospitals were turning down far more women than they accepted. Kansas had approved its Model Penal Code legislation, but the law was not yet in effect. A woman could fly to California if she had the plane fare *and* the hotel fare *and* the abortion money *and* the names of two doctors who would interview her to decide whether she was mentally ill, or poised to become mentally ill, or willing to pay them to declare for the record that "continuance of the pregnancy," as the California law put it, "would gravely impair the physical or mental health of the mother." If she had a tremendous amount of money and a current passport a woman could sit on airplanes for many hours and then be ferried by a foreign taxi driver to a legal abortion doctor in London or Tokyo, an option that appeared from St. Louis to be about as practical as reporting to a clinic on the moon; the women who called the

Clergy Consultation Service of Missouri, during the early months when the telephone number was passed by sympathetic doctors and college campus word of mouth, were not going to apply for passports in order to end their pregnancies.

The number rang at the Widdicombe house, on the line in the back, on the porch. "Clergy Consultation Service," Judy would say, in her nurse's voice, musical and brisk at the same time, waiting. In the evenings Judy was away working labor and delivery at the hospital, and during those hours Art Widdicombe would answer the porch telephone for her, his opening lines rehearsed until he could repeat them comfortably and with some measure of Judy's assurance: Yes, we can direct you to someone who will counsel you about abortion. No, we cannot give you the name of an abortionist.

That was the rule, the Howard Moody model: counsel only in person, minister-to-parishioner, where the district attorneys will be least enthusiastic about wading in to make trouble. Here is the minister on duty in your area, Judy or Art Widdicombe would say, and then a while later the telephone would ring at the chapel office of the Reverend Ken Gottman in Kirkwood or the Reverend Tom Raber in Springfield or one of the other ministers and counselors Judy and Gwyn Harvey had coaxed into service. The ministers had the names of the abortionists.

It was a short list at first, and not entirely satisfactory; Martha Vinyard did reliable instrumental dilation and curettage, which meant that she supervised the full procedure, but she charged a minimum of $600. There were abortionists in town who did a reasonably sterile job of "packing," the insertion of a gauze irritant into the cervix, but a woman who had been packed had to leave and wait for the miscarriage somewhere else. It was cheaper and quicker to send the women to Mexico, where *clinicas* in the border towns and the capital stayed open by bribing the appropriate authorities; or to Chicago, where Judy was developing a businesslike relationship with a young abortionist she came to think of as the Corpsman.

The Corpsman's abortions were illegal, but so were nearly everybody else's, and he had built his reputation doing abortions for the undergrounds in Chicago—first "the Service," the clandestine abortion network run by women who all used the pseudonym Jane, and then the Chicago branch of the Clergy Consultation Service. Chicago's clergy service was managed by the Reverend E. Spencer Parsons, the American Baptist minister who was serving as dean of Rockefeller Memorial Chapel at the University of Chicago, and after he met the Corpsman, Parsons confidently added the man's name to the national network's roster of competent Chicago abortionists. "Very trim, fine-looking guy," Parsons recalls of his first encounter with the Corpsman, in 1967, when a local physician brought the abortionist to campus to present him personally to Parsons and vouch for his qualifications. "I could see why people were impressed. The question was, how come this guy's doing this, what background does he have in this? He gave out that he was an army medical corpsman, and he'd been doing these in Korea."

None of this was true, as it turned out—many months later the Corpsman told Parsons that he had never served in the medical corps at all and that he had learned his technique from the physician who accompanied him to Parsons' chapel office. But it looked to Parsons as though he had learned it very well, and by the time the man changed his story, Parsons no longer cared what his history was; women came back from Corpsman abortions with high commendations. He ran an efficient operation, along the lines of a tidy bookmaking outfit, with a woman to answer the telephone and a tough little driver who told people his name was Jimmy. As a rule Judy Widdicombe did not pay much attention to automobiles, but when she made her trip to Chicago to survey the Corpsman's procedure it struck her that Jimmy was chauffeuring what looked like a Mafia car, some big dark American model that he drove her around in for a long time until she was thoroughly lost, which was plainly Jimmy's intent. He had picked Judy up in a prearranged Chicago hotel, and when he finally dropped her off they were somewhere in the big city, Judy had no idea where, ascending the elevator in a massive apartment complex to the unmarked door of a two-bedroom apartment. The living room had been rearranged enough to look like a waiting area, and in the bedrooms the Corpsman kept medical tables, foldable for sudden rapid relocation.

Judy watched him for an afternoon, the women knocking one by one on the plain apartment door, the female assistant showing them where to undress and lie down on their backs, the Corpsman bending over them in his white jacket with his surgical instruments clicking quietly against each other. He used no anesthetic and the women would grimace or blanch in pain while he talked to them, gently, Judy thought, and actually rather oddly for a medical man; she had attended at many hospital D&C procedures, but the patients were usually draped and knocked out by drugs, so that even the kindest doctors sometimes told jokes or chatted about their weekends over the women's limp forms. It was interesting to watch a doctory person work alone on a woman he had to *listen* to, Judy thought, and when the Corpsman's day was finished, she was satisfied with what she had seen: he was good, he was courteous, he used sterile technique. He knew what he was doing.

It was the local abortionists Judy had to back up, the ones who wedged gauze into the women's vaginas and then sent them away to abort. Most of these women went home or were taken in by friends while they waited for the bleeding to start, but some would call the back porch telephone number and say, Please, come get me, I can't go home now. So Art or Judy would climb into the station wagon and go out to find them, pale young women standing out on street corners with their purses under their arms and the cotton gauze pushed up inside them. When Art tried to remember how these women looked, a long time later, he could not quite bring their faces into focus; they seemed to him very young, really girls more than women, and what came to his mind was the back bedroom where the women would lie in the Widdicombes' spare bed and abort. He remembered that their hands would reach up behind them over their heads and that they would grab hold of the headboard, crying out while it was happening, and that if Judy was away at the hospital, Art would try to think of what he might say or do that would be of comfort. Most of the time he would tell them it was going to be all right; he would just stand there saying this over and over until it seemed as though things were coming to a peak. Then the bleeding might surge, or the woman might have a pushing feeling, and Art would show her the spare toilet, where he and Judy had fixed a plastic basin inside the bowl. If they had timed it right the basin would catch the abortion, the blood and the fetus together, and then Art would put some newspaper over it, so the woman did not have to look at it, and lift the basin into the bathtub for Judy to examine when she got home. After a while Art learned to study the contents of the basin himself, Judy showed him how important this was, to make sure nothing had been left inside to set off infection, and when the women were back in bed he would go over to the bathtub and raise the newspaper and look, knowing what he was supposed to find: two arms, two legs, a spine.

He was relieved when he found it all in the pan. Sometimes he wondered why it didn't bother him very much, the blood and the feet the size of rice grains and the fragile pale splinters he could see were the beginnings of bones. Art could throw up just like that if he happened to see someone else get sick nearby, but sifting the contents of a stranger's abortion pan never made his stomach heave; the way he looked at it he was helping, he was a lifeline: a girl was reclaiming her future and he was one of the people holding her hand while she did it. He would call Judy out at the hospital and give her the report, It's complete, or I can't find it all, and then Judy would tell him over the telephone what else he might have to do: monitor her temperature, massage the abdomen, see if anything else comes out.

Judy took these calls at the nurses' station, keeping her voice low and turning her back to the other nurses, and when she believed Art had the situation under control she would hang up the telephone and go back to work. At the hospital only the obstetrician Robert Duemler and a few others knew what she was doing, working labor and delivery in a Catholic obstetrical ward while shuttling women to and from abortionists, and when Judy stood in the hospital delivery room and watched a doctor place one large gloved hand around the emerging head of a baby, she would check herself sometimes, to see if she could still summon the throat-tightening exhilaration that had made her love obstetrics in nursing school. *We are helping a woman push a human being into the world*, this was the work she had wanted to do, and when she thought about it now she found that "potential" was a word she was repeating to herself: what landed in the plastic basin in the Widdicombes' back toilet was a potential human being, the premonition of a human being, the first rough arrangement of eyelids and fingerprints and human skin. The actual human being was the woman making her way out of the bathroom to lie back down on the bed. There was a story attached to every one of those women, and it seemed to Judy that the words they mouthed as they lay on their backs were the same words she heard at the bedsides on the hospital obstetrics floor, It hurts, You're a nurse, Help me through with this, I can't do it by myself.

The words were the same and the work was the same: Judy scrubbed, she soothed, she laid a palm on the forehead, she checked for complications. When she was a teenager and working as a hospital volunteer, she had watched a woman hemorrhage to death after an illegal abortion. In later years it would make a suitable story, the shouting ambulance crew, the hospital doctors pounding at the woman's chest, the damp blond hair splayed out across the gurney—but the truth was that Judy didn't walk around afterward brooding about abortion and the law and the nature of medical treatment. It was part of the hospital drama and she put it in the back of her mind. In nursing school she never saw an abortion complication, unless they had disguised one and labeled it as something else; Judy's mother never talked about it, her supervisors never talked about it, she believed in God but could see no evidence in or out of church that God in His specific plan ever worked out precise rules for women who got pregnant by mistake. In St. Louis the 1962 Sherri Finkbine articles ran in both newspapers, the *Post-Dispatch* and the *Globe-Democrat*, and Judy read them in her dining room with her first son baby-lurching around the legs of the table. She loved that boy in ways she had not understood the human heart could manage, but when she read the Finkbine articles she had a hard time even grasping the terms of the debate: of *course* Sherri Finkbine should be able to have her abortion. They were two different things, a fetus and a baby, and it was a terrible story because a terrible thing had happened to Sherri Finkbine, to the living person, to the woman, to the mother of children already born. This was women's business, and if you were a nurse who took care of women—for that was how Judy thought of her work in labor and delivery, the care of women—you did what they needed you to do, you helped them have the baby or not have the baby, they came to you in crisis and you eased them to the next place. Either way it was a kind of delivery, and her own minister was helping arrange the abortions. From the porch phone to the reverend to the abortionist, with an obligatory pause for contemplation: that was as unsullied a moral passage as Judy Widdicombe could imagine, and on Sundays she stood in the church choir, her conscience at peace, and let her good voice soar.

•   •   •

"My object would not be to rush, but to reassure," recalls the Reverend Ken Gottman, who was third pastor down at the Kirkwood United Methodist Church when Judy Widdicombe recruited him as a Clergy Consultation Service abortion counselor. "I would say to the woman, 'What do *you* think? How do *you* feel about it?' I would say, 'I don't know how God feels about it. I can tell you what the Bible says about it, which is nothing.' I would say, 'The churches are in various stages of agreement or disagreement about the isssue. It's very convoluted. We're all just sort of feeling our way.' I might say, 'I can cite you a biblical passage to support whichever point of view you're leaning toward.' "

Ken Gottman was a young minister when Judy Widdicombe met him, twenty-five years old, and chafing. He had straight brown hair that came a little over his ears, and he would have grown a beard if his chin hair hadn't been so spotty; the Kirkwood United Methodist Church had hired him in 1968 into a proper-sounding position, Director of Program Development and Delivery, but Gottman liked to think that somewhere inside his inoffensive presence lurked the kind of bearded Christian radical who could mobilize rent strikes downtown. He worried about his pastoral obligations to the poor. He marched against the war in Vietnam. The world seemed a place of furious turmoil at the close of the 1960s, and when Judy Widdicombe asked Gottman to join the Clergy Service, he said yes immediately; this was direct, it was vigorous, it had the feel of engagement.

It was also a violation of the Missouri criminal code, by the most prudent legal analyses, and Gottman grew frightened as Judy and Gwyn Harvey added his name to the ministers' lists. He believed in civil disobedience, when circumstances called for defiance of the unjust law, but he was not certain that his congregation would believe it at his side, and anyway Gottman was not being asked to go limp in the arms of an arresting officer or provoke a confrontation by performing an abortion in public. All he was supposed to do, in the phrasing of the Missouri statute, was "advise"; once a week or so, when the calls came over from Art and Judy Widdicombe's house, Gottman was to walk down to the Unitarian chapel in Kirkwood during the late afternoon, let himself into the deserted offices, and wait for the women who showed up to talk. He was to listen, to counsel, to describe the difference between a D&C and a packing procedure. If they wanted to go through with it he was to give them an abortionist's name, and the instructions: telephone here, drive here, meet him here, clean yourself with Betadine antiseptic, buy sanitary pads before you go. And when they asked—some of them didn't, as it turned out, some of them appeared never to have tripped over an ethical quandary in their lives, but some would look at him in a certain way, so that he could see the uneasiness gathering around their eyes—he was to assume his collar, as best he could, and talk to them about induced abortion and God.

So Gottman had to work out what he thought about this, about induced abortion and God. Among the Clergy Service ministers around the country an improvised theology had arisen, during the last three years of the 1960s, suggesting the ways a woman's visit to an abortionist might be reconciled with Christian and Jewish teachings: Life is a continuum. We are called upon to cherish it, but not to understand the precise moment of its beginnings. The Bible offers us guidance but no instructions; all is contextual; the word "abortion" does not appear in the Old Testament or the New Testament. Instead we are given in Scripture a wondrous tapestry in which faith, human freedom, and moral responsibility are interwoven in patterns that God helps each of us to understand, together and alone.

"For those making their own decision to seek an abortion, I would do whatever I could to assist them, and I would do it in the name of Jesus Christ who came that men might have the promise of life, life abundant," preached the Reverend J. Claude Evans, the Southern Methodist University chaplain who ran the Clergy Consultation Service in Dallas. From the university chapel pulpit, on a Sunday morning, this is what the Reverend Evans said:

Unwanted pregnancies happen to the good girls, not to the bad girls. The promiscuous girls, the ones who sleep around with all the campus studs, they are worldly-wise; they know what to do. Even when they get pregnant, they don't show up in my office. They know what to do. It is the innocent I'm concerned about. The girls who didn't plan to go to bed, and who therefore took no precautions, who get in trouble. Countless recent studies verify this. So I had to ask myself, Claude, what would you have done if that one daughter of yours, that apple of your eye, had she gotten pregnant?

One day I really thought this through, not only rationally and abstractly, but emotionally. I pictured Peachie coming to me with her tears and her agony. I saw her through the streams of tears on other young faces I have counseled. I saw myself through many a parent of real, live SMU students—some who acquitted themselves magnificently when confronted with a daughter's pregnancy, and others who failed their daughters, at least at first, until they had worked through their own puritan hang-up. At last I began to see why the life of the girl is more important than the potential life in a bit of protoplasm. I felt the weight of an agonizing choice between evils, but I saw clearly what I would have done. Had my daughter found herself pregnant, had she concluded that a marriage was not indicated, that there was not enough mature love between the two to support the demands of parenthood, and had she come to me early enough, then I would have moved heaven and earth to find a respectable physician who took such work as his obedience to the promise of God that man is meant for a community of love, in the family and in the world, and this act would be an act of obedience to a higher law under the promise.

Evans used to type his sermons before he read them, and underlined to guide his expression: *demands of parenthood, act of obedience, promise of God.* Every pastor had his favored turns of phrase. In Greenwich Village, where Howard Moody ran his ministry in the manner of a keyed-up Democratic community organizer, the Judson Memorial Church minister used language like "man's vengeance on woman," and "a most odious violation of constitutional liberties," and "a matter of individual conscience and free choice." In the southern Missouri city of Springfield, where the Reverend Tom Raber was working as a Southwestern Missouri State University teacher and campus minister when the Clergy Service recruited him, Raber studied the Bible and considered his own ministry for some weeks before he said yes; Christian studies scholars sometimes make use of biblical concordances, volumes that serve as alphabetical indexes to the Bible, and Raber examined his concordances for references that might be of help. He found "womb" in the concordances, and "miscarriage" and "conception" and "life-giving" and "life," but Raber could not find a single Bible passage in which the words spoke directly about deliberately induced abortion.

"And the conclusion I came to," Raber recalls, "is that the Bible is at least ambivalent, if not *open* to abortion." Raber's parents had worked for the Salvation Army, and human despair had been part of the landscape around him for nearly as long as he could remember; as a teenager he lived for a while in a North Dakota Salvation Army building in which thirty homeless men slept every night in the basement. It was not a romantic vista either of poverty or of the "unwanted child," a phrase that was making its way into every public abortion debate, and when Raber said yes to the Clergy Service work he began to find that he was talking quite a bit about a compassionate God, a forgiving God, a God whose mercy could surely extend to the woman making a decision to abort rather than bear a child she was not prepared to raise. "That my first understanding of the nature of God is that God is a God of love," Raber recalls. "And compassion. And that from *that* you begin to make the rest of your decisions about things. I looked at it and said, What would be the appropriate response for a Christian person dealing with another person who is just faced with an unbearable dilemma? I guess the answer I got is, first of all you have to help that person sort it out for themselves, and let them know they're valuable people, regardless of what decision they've made."

The ministers had no texts, not on this subject, and in St. Louis Ken Gottman felt sometimes as though he was operating mostly on instinct, with the hundred-year-old abortion laws and the unhappy women in the chapel office and all around them the wild exploding late-1960s mess of free love and birth control pills and posters that said, If It Feels Good, Do It. The women were intensely frightened, he saw that very quickly, and it seemed to relieve them to know that a certified man of God did not think they were going to burn forever for what they were planning to do. In later years Gottman would have his scriptural references lined up and he could speak confidently about God creating human beings with minds and choices of their own—"God has so constituted life," Gottman would say, "that we are granted the awesome privilege and responsibility of *choosing* how we shall live." But when he started with the Clergy Service he had no worked-out oratory to unroll: the abortion laws were just old and wrongheaded and rendered obsolete, Gottman thought, by the very women who sat before him and wished to know exactly what they needed to do to break them.

They had the right to know it. It was wrong for them *not* to know it. Howard Moody had wondered in print what certain practices of the era might look like to an observer two centuries hence: "Such a futuristic explorer could only surmise from the data he discovered that either these 'medieval' 20th century people were ignorant and superstitious," Moody wrote in a particularly heated 1967 magazine article, "or women in this society were the victims of an unforgivably cruel punishment stemming from some inexplicable hostility on the part of men." This was more flamboyant language than Gottman was inclined to use, but fundamentally not so different from what he believed; the problem was not free love or convulsing morals, although certainly these made life at the personal level look a great deal more complicated than it had a decade earlier. The problem, if Gottman had to identify the single meanest hurdle for the women the Widdicombes sent down to the chapel to meet him, was the laws.

The laws were walls, massive walls, with little doors that were kept locked all the time, and moreover were hidden from view so that ordinary women couldn't find them unless men with important titles could be bothered to come around and point out their locations. Then the men with important titles might on very special occasions pull out their keys and make a grand throat-clearing show of unlocking or not unlocking, perhaps we will today, perhaps we won't. The ministers might all have been tunneling, there in the Clergy Service chapel offices, clearing out their small conduits for the women to crawl through: keep your head down, keep quiet, call us when you get to the other side. And as the laws began to change, in the months between 1967 and the beginning of 1970, Howard Moody shook his head impatiently and exhorted them to keep on tunneling: "Eighty percent of our cases," Moody declared in a 1969 *New York* magazine interview, when the writer asked him what he thought of the efforts around the country to work the Model Penal Code guidelines into law, "wouldn't qualify under the proposed model abortion codes."

The shorthand for the Model Penal Code–inspired abortion law change was "liberalization," a word that was supposed to suggest modern thinking and magnanimous flexibility. But it sounded faintly soiled when Howard Moody said it. "Liberalization doesn't mean nuthin'," Moody drawled for the *New York* magazine writer, whose anecdotal stories about the Clergy Service described a clientele of women grateful for referrals not to hospital review committees but to men like the Corpsman, with efficient drivers in unmarked cars. Under liberalization some abortions were permissible and some were not. Under liberalization a woman had to sit before doctors and plead her case. Liberalization was a fancier set of locked wall doors: a woman could see the doors now, she didn't have to be wealthy or well-connected or near death to have them displayed to her, but she couldn't open them herself and still there were those men, with the keys, looking her over to decide for themselves what might happen next.

It had taken Howard Moody and his Clergy Service ministers only a few months to see this, Moody would say later, to understand that the fiercely debated new laws were proving to be of minimal use to the women who called their referral numbers for help. In Colorado, where the governor had signed the Model Penal Code law with a small flourish of first-state pride, uneasy hospital review committees worked so diligently to fend off "abortion mill" accusations that statewide the number of abortions approved for medical reasons rose to fewer than three hundred the year after the law took effect—and that number included patients accepted from other states. In North Carolina, where the new law formally closed off the state to nonresident women who wanted abortions, most hospitals reported their 1968 abortion totals in numbers so low that it might have been some unusual and remarkable disease they were tabulating: five in one hospital, eight in another. In California the numbers looked more encouraging on paper—a sevenfold increase in therapeutic abortions the first year after liberalization, and tenfold the year after that—but every abortion still demanded the two-doctor approval, for a fee, and the hospitalization, for a bigger fee.

A writer for *The New York Times Magazine* addressed the California situation at some length, in December 1968, and the title of his article generated considerable head-nodding among the Clergy Service ministers: "How California's Abortion Law Isn't Working." Here were accounts of women denied abortions by hospital committees—a married woman who reported having been raped (turned down, the article said, on the grounds that the pregnancy might be from her husband); a seventeen-year-old seduced by the man whose children she had been baby-sitting (she leaped from her family's garage roof after the abortion was denied, the article said, and died in the fall). The California clergy service had its own collection of comparable stories. "For every therapeutic abortion performed at the County Hospital, two are turned away," the two California clergy service directors wrote in a summer 1968 memorandum from Los Angeles to their statewide ministers, most of whom were directing steady traffic to the illegal Mexican clinics a full year after the liberalization law had gone into effect. "About six or seven are done per week. We sometimes see up to ten pregnant women per *day.*"

The liberalization laws were failing, as Moody and most of his referral service ministers saw it, in the crudest and most practical terms: women who wanted abortions were being told no. And they were failing in philosophical terms, too, even for some of the women who were being told yes. There was no procedure known to American medicine that required *men* to come before hospital committees and beg. No other variety of hospital patient was expected, as a matter of routine, to feign mental illness in order to receive the treatment she came for. When Moody and Spencer Parsons wrote and successfully lobbied into passage a resolution at the 1968 gathering of the American Baptist Convention, the denomination into which both ministers had been ordained, the word "liberalization" never appeared in the text: early abortion, the resolution declared, should instead be "at the request of the individuals concerned, and be regarded as an elective medical procedure governed by the laws regulating medical practice and licensure."

This was muted language, carefully arranged, with a radical notion at its core. The words "governed by the laws regulating medical practice and licensure" had a particular meaning in abortion debates: what they meant, as Glanville Williams had proposed more than a decade earlier, was Not Singled Out in the Legal Code. *This procedure is no more the state's business than the suturing of a wound:* that was the message of the American Baptist Convention's resolution, which in 1968 made for so startling a proposition that even Howard Moody felt compelled to temper it for his fellow Baptists' approval. As finally adopted, the resolution called for the abolishing of all laws restricting abortion during the first twelve weeks of pregnancy only. Abortions after the twelfth week, the resolution declared, should be permitted solely in hospitals and solely for pregnancies following rape or incest, pregnancies likely to result in babies with serious birth defects, and pregnancies posing a physical or mental health danger —all the categories, in other words, of the Model Penal Code.

"I think we had a feeling that that was what we could get," Howard Moody recalls. "It was a very early resolution on abortion, and it was unusually liberal for that time, for the Church to be doing. We were kind of shocked when it passed. But we had already pretty much decided by then, in the Clergy Consultation Service, that repeal was the only way to go."

AMONG THE ACTIVISTS—and they were multiplying now, mailing out mimeographed newsletters from New York and Oregon, California and Illinois, buoyed by reports of victory or promising legislative votes—a debate with a name like a prizefight began to spread around the states: Reform vs. Repeal. It was an argument suited to the contentious 1960s, shall we smash the system or transform it gradually from within, and in letters, broadsheets, and long-distance telephone calls, partisans trod some of the same ground as rebellious university students as they made their impassioned pitches about philosophy and compromise and the political realities of introducing change.

What California and Colorado and North Carolina had done to their abortion laws, if one believed enough in this approach to use the vocabulary without irony, was Reform. The Model Penal Code language was Reform. Reform was every bill that encompassed within its provisions the basic principle of century-old American abortion law, the idea that the decision to perform an abortion must remain the business of the state. Acid-tongued activists in certain parts of the country had written about what seemed to them the mealiness of the Model Penal Code language even before the earliest reform bills were signed into law. One of the very earliest, an indefatigable California woman named Lana Clarke Phelan, wrote this letter to the Oklahoma abortionist W. J. Bryan Henrie just as the California abortion bill was nearing the governor's desk in 1967:

I am sure you have heard prior to now that Tony Beilenson's pathetic bill got through the Senate and has passed on to the California Assembly, and that Ronald the First, Boy Idiot Governor, has stated he will pass it if it comes to his desk for signature now that it has been revised to refuse abortion to women bearing deformed fetuses. . . .

On the bright side of the ledger, the law must go, in its entire damned savagery, and every thinking person knows it, and so if California will at least speak the dirty word "abortion" right out loud from the hallowed halls of Sacramento, perhaps the smaller states will grab the much-needed ball and run with it by way of further amendments which will lead to more sanity. I note Beilenson has his medical requirements limited to consent of two doctors before 13 weeks, and 3 after that until just 20 weeks, after that—absolutely nothing can be done! If we could just delete the hospital requirement, perhaps it might be workable, but only for a few privileged people. I cannot imagine any decent woman subjecting herself to this humiliation. I would go to Mexico in a flash.

Lana Phelan, who held up the Southern California end of a noisy little West Coast organization called the Society for Humane Abortion, was an extremely respectable-looking woman, employed as a legal secretary and married to a parole officer and grandmother to three children; when newspapers ran her photograph she was usually smiling, with a choker of pearls around her neck and the kind of precise curled hairdo that suggested regular visits to the beauty parlor. She was a prolific correspondent, sitting up late at night at the electric typewriter in the den of her small house in Long Beach, and as far back as 1965, amid the nation's earliest overtures to abortion law change, Phelan had formed a merry alliance with a younger Northern California medical technician who often managed to look outrageous in print. "Politicians insist they have to have their noses up our skirts," the outrageous Patricia Maginnis once remarked at a speaking engagement in Rapid City, South Dakota, prompting the local newsman to lead his write-up the following morning with what must have been a real conversation starter of a quote. Maginnis was an Oklahoma-born Women's Army Corps veteran, unmarried and attractive in a delicate, distracted sort of way (when they traveled together, Phelan would remind her to comb her hair and make herself look a little "prissier," as Phelan put it, before the photographers arrived), and in California the Phelan-Maginnis letters flew from one end of the state to the other, trading breaking news and dispatches from the Mexican clinics and venomous tidbits about this or that misguided male politician. To Rowena Gurner, the woman who shared with Maginnis the running of the Society for Humane Abortion's cramped little office in San Francisco, Phelan wrote around 1966:

> The lecture at Long Beach State Police Administration (vice division) went beautifully, and for three hours I ate 50 cops for breakfast. . . . You would have loved hearing me tell those cops how brave and professional they must feel questioning a bleeding and terrorized woman and threatening her with death if she didn't talk before her family and lawyer got there. One man denied it vehemently, said he'd never heard such nonsense, and another stood up and said, "Shut up. You've done it a hundred times, and the lady's telling the truth. You feel like a louse, and it's a hell of a job and we'd be better off without the laws on the books."

Maginnis and Gurner, a rebellious New Yorker who had relocated to the West Coast by bicycle in 1963 (Gurner spent four months riding a three-speed alone across the country, arriving in California with a sleeping bag and a flyswatter and a packful of clothing), had joined forces as the Society for Humane Abortion in 1964, renaming an older and meeker California group called the Citizens' Committee for Humane Abortion Laws. It was not incidental that the word "Laws" disappeared from the new group's letterhead: in 1965, even as early versions of the Model Penal Code bill were being shoved into desk drawers by legislators who could not imagine such controversial language surviving a vote, Pat Maginnis was posting herself on San Francisco street corners to hand out pamphlets that demanded the repeal of *all* abortion laws. It was not possible to fix laws that gave men the right to tell women what to do with their bodies, the Society declared; such laws must be abolished, and if they could not be abolished, then they ought to be defied.

"ARE YOU PREGNANT?" one of the early leaflets began, in the Maginnis-Gurner sledgehammer style. They printed detailed, annotated lists of illegal Mexican abortionists, indicating who spoke English and who used anesthetic and who might be "on vacation" to avoid the police. They offered precise instructions on negotiating price, managing hostile taxi drivers, and watching the abortionist for cleanliness. ("Tell the specialist that you want to *see* him scrub his own hands and arms with Phisohex. . . . If the specialist carelessly rests his gloved hand on *anything* or *any* part of the operating table, his hand is no longer sterile. Tell him to re-scrub.") And they advertised free classes—"AN ACCELERATED COURSE IN ABORTION," one of the flyers read—in which women gathered in a private home or rented hall could learn details not commonly available during the mid-1960s:

CURRICULUM

1. Anatomy of the female reproductive organs
2. Sterile technique
3. After-abortion care and methods of contraception
4. Methods of abortion
    A. Dilation and Curettage
    B. Vacuum Aspirator
    C. Intra-Amniotic Saline Injection
    D. Self-Induced abortions (only for women who have given birth previously by normal delivery, not Caesarian section)
        a. Digital Method
        b. Saline Lavage
5. Dangers involved
6. How to deal with police questioning
7. How to locate abortion specialists in foreign countries

Lana Phelan and Pat Maginnis began running these classes together in 1966, usually with social workers and newspaper writers and undercover police officers scattered throughout the audience. ("The third time I was in the San Diego area to do the class," Maginnis recalled in a 1975 oral history, "this poor plainclothesman—I felt so sorry for him. He had heard this lecture four times.") They started in California and moved rapidly north and east, to Oregon and New Mexico and Wisconsin and Ohio and a packed-to-capacity apartment in Washington, D.C., and between them they worked a defiant road show that in every locale promised to be quite unlike anything that had ever come through town before. Lana Phelan had a small background speech she liked to offer as warm-up: the first contraception and abortion laws were European canon laws, Phelan would explain, the laws of the Catholic Church, which wanted women to produce as many little worshippers as possible. She would tell her audience that Western governments had embraced these archaic laws, which dated from a time when women had neither vote nor voice in the running of the land, and used them to try to compel motherhood whether women wanted it or not. She would tell them that people are inventive in spite of Church and government authority, that a woman would drag herself like a wounded animal to an illegal abortionist rather than bear a child she could not raise. Then she would look at her audience and say: I have to tell you how I know this. It happened to me.

When Lana Phelan had her abortion she was seventeen years old, married to a much older man, and already the mother of one baby—she would always begin the story this way. It was 1938, high Depression times, and Lana Kitchen, that was her name then, lived in the central Florida town of Lakeland, where she worked behind the counter for Walgreen's. She had married at fifteen and given birth at sixteen, in a labor that went on for three days and left her so sick that the nurses tiptoed in and out of her hospital room. Her doctor told her she would never survive another labor (but did not, Phelan would always point out, tell her how to keep from getting pregnant). Within a year, she was pregnant again.

The doctor confirmed it for her in his office, Phelan recalls, after she missed a menstrual period:

> He said, "I told you not to get pregnant." And I said, "But you didn't tell me how not to." And he said, "Well, I'll see if I can keep you alive." That was what I had to go on. "I'll see if I can keep you alive."
>
> And all I knew was—I was not quite seventeen years old, I had a baby that was in such bad shape, and my husband, dear and sweet as he was, was a man that things happened to. He never made things happen. And I thought, "I'm going to die." I was worried sick about it. I talked to a couple of girls I knew, and a girl over in cosmetics said, "Well, I've heard of a woman in Ybor City, outside of Tampa, I heard that she would do this."
>
> And I went to see her. Great, big, stout, black woman. Whether she was Cuban or black, I don't know. But she was—oh, she was kind. She didn't question me. She just said, "Well, honey, I'll do it, but it'll cost you fifty dollars."
>
> *Fifty dollars.* I went back to work. Fifty dollars, it was a lot of money then, and I couldn't tell anybody what I was doing. I was still sick, I was all to pieces, and I thought, What am I going to do?
>
> So I started saving. It took me nearly four months to get that fifty dollars. I finally got forty, and I sold everything I had, that I could get away with, I lied and said I had gotten paid less than I had. I didn't tell my husband. At that time he was going to school. He didn't know I was pregnant. This is not a thing you could tell. Nobody suspected. So I had forty dollars, I needed ten more dollars, and in desperation I borrowed it from a customer, a man that ran a local automobile agency. To this day I

think that man knew what I wanted that money for. It was secret money, so secret, you can't imagine, I'm sure I had it in a little bag or a little purse—because I didn't want anybody to stop me. I knew what I had to do. I knew I had to stay alive, because I had to raise that child.

So I took the streetcar as far as I could, and I walked, it was about eight or ten miles from my house to where she was. She had this little gurney in the back, it was clean, rough but clean. It had white, white linen, sheeting or something, and a little pillow. It was covered with white. And it had pans on it, I guess surgical pans, enamel, and they had instruments and things, and they were covered. I'm sure there was not a sterilizing device or anything there. I was so frightened. All I knew was I had to have this thing done—I didn't even know what it *was.*

She was very kind to me. She put the stuff in, slippery elm, I think. And she said, "Now, go home, and in four or five days your period will come, and then you will expel this. Everything will fall out of you." That's the way she put it. Everything will fall out. So I went home, and two or three days later I was running a fever, and I knew. But I went to work. I worked with this fever. I couldn't tell anybody. It was two or three days I worked with that fever. And this was an absolute horror, I'll tell you because maybe someday other women won't go through it quite so bad, I went to my sister-in-law's for dinner, with my husband. My sister-in-law was the matriarch of the family, considerably older than both of us, and dinner was a command performance. And I was at the table when I felt these awful pains.

I excused myself and went to her bathroom, and I looked down at myself. I sat down on the john and looked at myself. And my God, I nearly collapsed. I was bleeding. There was blood all over me. And the worst thing—the worst thing—there was a little tiny limb of some kind, I don't know whether it was a foot, or an arm, protruding from my vagina. All I could see was that. It was so tiny, it was like, oh God, a pipe stem thing. But all I knew was, what do I *do* now? How do I fix this? I hadn't told anybody anything.

So I quick took the toilet tissue and made gobs of toilet tissue, and I poked everything back up inside of me, just pushed, pushed, until everything went back inside, and I wiped up all the blood, and cleaned myself up, because it wasn't on my dress, and I went back to the table and I said, "I have such a headache, I can't even hold my head up, I've got to go home." And my husband said, "I'll take you home." And I said, "No, no, stay and have dinner." I got out of there and I took a cab as far as I had money for, and I walked the rest of the way, it must have been a mile or so, because I didn't have any more money. When I got there the house was dark. And I thought, Oh, my God. She had said not to come back. But I didn't have any choice.

I went around the house, and it was so black, and dark. I knocked on her side door, and I guess she had gone to bed, because she came to the door and she said, "I told you not to come back here." And I said, "I had to, I had to," and I was crying, and she let me in. She put me up on the gurney, and she said, "Oh, my, you're almost finished, I have to clean you up." I felt really comforted when I got back to her, because it was somebody to share my secret with. And she cleaned me up, and I was laying on that gurney just sobbing my heart out, and I'll never forget that woman, she was wonderful. She came around, big black lady, she put her arms around me on the gurney, and she put her face down near mine, and she kind of put her cheek up next to mine.

And she said, "Honey, did you think it was so easy to be a woman?"

That was the story Lana Phelan told.

She did not cry when she told it, because it had happened a long time ago, and she had made a kind of decision about that when she saw that she was going to begin talking about it before public audiences. But in the audience women would cry, usually without making any noise; an awful stillness would come over the room when Phelan began her abortion story, and when she looked at the faces of the women who were listening, she could see them sitting there with their cheeks wet and their hands up to their mouths, and in that moment she believed sometimes that she could tell, by watching, which among them had done the same thing.

She would introduce Pat Maginnis to talk about medicine, mechanics, and practical matters that must not be forgotten. ("Have the people write down the phrase 'No hablo espanol' for use over the phone if necessary," read one of the Maginnis class outlines. "Taxi drivers sometimes know the only doctors operating, or the quacks. Sophisticated women MAY know how to differentiate. e.g.—girl and father were beaten up in a motel, story of the two women, on separate occasions, who paid over 600 and 400 respectively, the 'doctor' pulled out catheter from pocket and shoved it in vagina—no infection, luckily, no abortion either.")

And when the class was moving toward its close, and both Maginnis and Phelan had delivered many warnings about what they were about to say, Maginnis would talk about her abortions, too—the illegal one in Mexico, from a Tijuana doctor who never said a word to her, and the two that she told people she had managed to bring on herself.

Pat Maginnis was capable of telling her abortion stories with a manic black humor; she had been raised in a strict Catholic family in Oklahoma and left to join the Army, abandoning her religious upbringing with such fervent distaste that when she talked about either Church or family, she tended to use phrases like "crippling Roman Catholic dogma" and "I wouldn't give a person a dime for marriage." She led what by her own account was a monogamous but spirited heterosexual lovelife, and on three separate occasions during the late 1950s, as Maginnis put it in an oral history some years later, "my great Catholic uterus, with all its Catholic eggs, managed to put out another pregnancy."

The first she disposed of, angrily but without lingering physical complications, in Mexico. The second two she decided to handle herself—that was how she would tell the story—by running catheters into her own uterus and then driving to hospitals once the miscarriages had begun. "All of these hospital people were just horrified that when I went in I said, 'I have an induced abortion in process,' because you didn't say this, no woman ever said this, you had to lie and say you were spontaneously aborting," Maginnis recalled. "So they left me alone in the examining room, and I just thought my whole bottom was going to come off, the pain, it was just dreadful . . . finally I did manage to produce about a six-month fetus, which got up off the table and cursed my Catholic uterus."

Pat Maginnis used language like this with interviewers, not abortion-class audiences. But even on a public stage she was direct and unrepentant when she described the catheters and the Lysol solution and the finger pushed deep and repeatedly into the vagina to irritate the mouth of the uterus into miscarriage. These were dangerous techniques, Maginnis would warn, requiring great patience and scrupulous cleanliness; the only thing immoral about them was the state laws that made them, for some women, the method of last resort.

It is probably safe to assume that no one in the United States had ever spoken in quite this way before a public audience. Rowena Gurner, the office administrator and third leg of the Society for Humane Abortion tripod, would describe with minimal prompting the details of *her* illegal abortion: she had flown to Puerto Rico, back when she lived in Manhattan, and when she talked about the experience her only audible emotion was angry impatience, as though she had been exiled to foreign territory for what ought to have been a medical procedure as straightforward as appendectomy. Gurner was exasperated by hand-wringers who wanted to carry on about the Tragic Decision; in 1967 a caller suggested the Society check into a certain plant's abortifacient properties, and Gurner cheerfully told a newspaperman that she planned to become pregnant so she could try swallowing some. "If it doesn't work, I'll get an abortion anyway," Gurner said. "In any case, it won't come to term."

The plan never materialized, but the public point was made: these were women who would shout the word from any available parapet because they had *been* to abortionists. They knew what it *felt* like. They were not the first women in America to run Lysol into their bodies or come weeping to midwives in the middle of the night, but they were the first women sufficiently inflamed by resentment, and unencumbered by timidity or moral ambiguity, to hold forth about it in lecture halls.* And

* Some studies of abortion in the nineteenth- and early twentieth-century United States have offered evidence that for many generations before legalization, abortion was discussed openly but *euphemistically*, in elliptical language that was nevertheless widely understood. James Mohr's *Abortion in America* cites numerous nineteenth-century examples, from newspaper advertisements and home medical guides, of references to treatments for such complaints as "obstructed menses" and "female

as the names of new abortion organizations began materializing one by one in different parts of the country, Pat Maginnis and her comrades formed the rowdy Western radical wing for a movement working hard to present its controversial proposals in somewhat less flamboyant style. The Wisconsin Committee to Legalize Abortion, the California Committee for Therapeutic Abortion, the Organization for Abortion Law Reform, the Association for the Study of Abortion—it was not coincidental that almost none of the new organizations contained in their capitalized titles the word Repeal.

"Repeal sounds radical," recalls Lonny Myers, a Chicago anesthesiologist who joined the minister Spencer Parsons in 1966 to help set up a statewide group that operated under the relatively restrained-sounding name of Illinois Citizens for Medical Control of Abortion. Myers and Parsons both believed in repeal; the Colorado liberalization sounded to both of them like "a bunch of men deciding for a woman whether or not this should happen to her," Myers said in an oral history, "as though it were any of their goddamn business." But they thought it strategically adept to leave a certain ambiguity in their new organization's name:

I wanted it to present the most radical idea *and* something where the name would express the individual or conservative viewpoint . . . I wanted to get the medical profession behind me. It is the perfect example of government interference in the practice of medicine. And I wanted to appeal to doctors, saying, "Look, the government is interfering. Your whole philosophy at AMA is that a government shouldn't interfere with the practice of medicine."

It didn't work, at least not in the early years; pushing any change in the wording of abortion laws was controversy enough, and when polls were taken they tended to show physicians, like the general public, leaning steadily toward the Model Penal Code principle—legal abortion only for those cases that physicians and society have collectively ruled appropriate.* When the Harvard Divinity School and the Joseph P. Kennedy Jr. Founda-

---

irregularity." Leslie Reagan's *When Abortion Was a Crime* also lists a variety of early twentieth-century euphemisms for abortion: to "bring my curses on," to "bring her around," to be "fixed up," etc. But the Society for Humane Abortion leaders appear to have been the first women in the country to make a public crusade of describing abortions—including their own—in deliberately blunt and specific language.

* A 1967 survey of the thirteen thousand members of the American Psychiatric Association, for example, produced overwhelming Yes responses when the psychiatrists were asked whether they believed pregnancy "should be interrupted" if the pregnant woman met any of the Model Penal Code categories—rape, incest, fetal deformity, or endangerment to life or physical or mental health. But more than three quarters of the respondents answered No to the final category: "Whenever the woman requests it."

tion co-sponsored a Washington, D.C., abortion symposium in 1967, physicians, ethicists, lawyers, and theologians spent three days wrestling over anguished questions about whether the law ought properly to be tampered with at all. "What proof should be required of rape to justify abortions?" a published summary of the conference proceedings read, highlighting some of the starting-point questions for lengthy discussion among the participants. "What proof of defectiveness in the fetus will be required? Should a pregnant unwed girl be allowed an abortion?"

For the duration of that symposium, Lana Phelan and Pat Maginnis paraded outside the conference hotel with large cardboard signs that younger Society for Humane Abortion volunteers had laid out for them. On one side, above the words NO CONTRACEPTIVE IS 100% SAFE, was a small array of birth control devices, including real diaphragms and condoms; during one windy moment, the condom on Phelan's sign tore partway loose and unrolled into a startling little banner that set off enthusiastic hooting and waving from the passing traffic. The other side displayed a knitting needle and a wire coat hanger (it is possible, Lana Phelan has suggested, that this marked the American public debut of the coat hanger as a visual symbol for illegal abortion), and the words END BUTCHERY along the bottom of Phelan's sign, in thick black letters, so they would show up when the newsmen came to take her picture on the sidewalk.

Neither Phelan nor any of the other vocal repeal advocates had been invited to participate inside, and the views of groups like the Society for Humane Abortion were examined during the proceedings only as the farthest end of a very long spectrum. From the outside, snubbed repeal advocates snapped back. "This conference, despite the trappings of objectivity, seemed to those of us who had been trying to build a movement to be primarily aimed at smothering recent progress," Lawrence Lader recalled in his 1973 memoir, *Abortion II: Making the Revolution.* "The speakers and panelists were largely Catholics or antiabortion in their views, and the 'repeal' position was virtually ignored."

There was some exaggeration to this swipe; the panelists' list included Protestants, Jews, and at least one law professor (who was in fact a Jesuit) willing to make a convoluted case for full repeal.* But it did not include Larry Lader, who in the wake of his 1966 book *Abortion* was becoming an insistent and widely published advocate for the proposition that abortion was a national scandal precisely because of the laws that prohibited it. "In the name of observing the law, shameful injustice is committed," Lader had written in *The New York Times Magazine* in 1965, as he was complet-

---

* The Reverend Robert F. Drinan, then dean of the Boston College School of Law, suggested that removing abortion entirely from the criminal code might prove more moral than enacting statutes specifically approving certain types of abortions—even for the person who finds abortion morally unacceptable. It was preferable for the law to remain silent on the subject, Drinan argued, than to venture in the perilous direction of "eliminating the right of *some* fetuses to be born."

ing what was to become the first of his three books on abortion. "Women suffer needless grief and pain. The medical profession is intimidated and confused. . . . The law is a heritage of ecclesiastical history, based on religious dogma rejected by most of our population."

In the modest roster of men and women willing to attack American abortion law in public, Larry Lader was something of an oddity. He was not a minister with memories of parishioners' private troubles; he was not a doctor with war stories from the hospital wards. His own conversion, the way he told the story, included no desperate friends or family members knocking on abortionists' doors. Lader was a writer, he would explain; he made his living writing magazine articles and books. For some years he had also volunteered in New York liberal Democratic organizations—before 1950, he had taken on an unpaid district organizing job for a local Democratic congressman whose leftist politics attracted accusations of Communist sympathizing—and when Lader set about looking for a subject suitable for full-length biography, what finally attracted him was what seemed to him a life of uncompromising political and social defiance. His book *The Margaret Sanger Story* was published in 1955, after two years of research and interviews.

It was a lively book, because Lader was an entertaining writer, but the Margaret Sanger it described was so saintly a crusader, in her fifty-year campaign for birth control in the United States and abroad, that some critics and subsequent biographers dismissed Lader's research as a rehash of Sanger's own sanitized version of her complex and controversial life. Lader remembers that he did let his surpassing admiration for Sanger rather vividly color the text ("It isn't what you'd call an objective book," he recalls), but in any case objectivity had always interested him less than militancy. His second book, in keeping with this deepening interest, was about militant New England abolitionists. His third book—because at first no one would touch it as a magazine article, he recalls, even after he sent written proposals to dozens of editors around the country—was *Abortion*.

It was birth control that led him to abortion, Lader told interviewers afterward: the two subjects followed each other by logic, historically and socially, because even when contraceptive methods are legal and readily obtainable (as they assuredly were not during the years of Sanger's famous campaigns), they do not always work. Margaret Sanger herself had assailed abortion as barbaric; but Lader assumed that Sanger based her opposition not on the procedure itself, but rather on the danger to which turn-of-the-century New York Lower East Side women exposed themselves by frequenting illegal abortionists.* To Lader, the underlying principle was what

---

* The modern Sanger biographer Ellen Chesler, whose *Woman of Valor* examines in detail Sanger's public and private writings, would concur; Sanger's birth control clinic in New York, Chesler writes, referred some pregnant patients to hospitals for therapeutic abortions as early as the 1930s.

mattered: Margaret Sanger had devoted her public life to vehement prose and oratory about a woman's right to the control of her own body. If women had the right to prevent conception, then they must also have the right to prevent birth. The logic of this proposition appeared unassailable to Lader, and when his publisher held a press conference to announce the release of the book, Lader explained that since the publication the previous year of his *New York Times Magazine* article, hundreds of women had written or telephoned him to ask how they might find one of the abortion doctors he had written about. "And I decided, in a flash of a second, All right, I've shot my mouth off, *I will become an activist*," he recalls. "I said, 'I will help women. If they write or phone me, I will get them to medical help.' "

By 1966, in Larry Lader's Upper West Side apartment, the letters began to accumulate, just as they were accumulating in the little Society for Humane Abortion office in San Francisco, with their postmarks from many states away: "Please call me collect in Houston." "Time is of the essence." "I am pregnant and my husband has disappeared." "It is *very urgent* that I talk with you." A few physicians made contact with Lader themselves, saying, Well, I know of an organized ring that will do a clean job, or, Look, I can only do eight or nine a week. This was what Lader had written about in his abortion book, a world and a practice that ran on its own without any perceivable regard for the wording of the law, except that the law made it ridiculous, as Lader came to see it: demeaning, dangerous, profoundly hypocritical. He began giving lectures, offering himself for interviews, seeking publicity where he could find it. He announced, on the radio and at speaking engagements, that he would refer women to abortionists. He urged on the launching of the Clergy Consultation Service. He joined the Association for the Study of Abortion (ASA), which in 1966 was New York's genteel counterpart to the Society for Humane Abortion.

The ASA was a resolutely scholarly organization, its letterhead heavy with M.D.'s and Ph.D.'s and Esqs. Nobody from the ASA talked about self-induced abortions or politicians sticking their noses up women's skirts. After its charter meeting in 1964, as the Model Penal Code abortion language was receiving an early airing in New York, the ASA put out a newsletter that was meant to look both academic and appropriately prudent for a nonprofit group trying to maintain its tax status ("to ascertain and evaluate the attitudes of lay and professional groups toward existing abortion practices," the statement of objectives began), reprinting persuasive articles and surveys rather than direct denunciations of the abortion laws. Many of the members were veterans of the birth control organizations of the 1950s and 1960s, Planned Parenthood and the Human Betterment Association for Voluntary Sterilization; for some years now, Margaret Sanger's women's emancipation speeches notwithstanding, their arguments had focused far less on women's rights than on what was described as the evolving public and social good. The world's population was exploding and must be contained; families in poverty must be given the means to keep their numbers manageable; the human race, as one Human Betterment Association press release phrased it in 1964, must look both to birth control and to voluntary sterilization "to protect itself against the invading horde of the yet unborn."

At the Association for the Study of Abortion, the Model Penal Code was described as a laudable goal, at once prudent and enlightened. A society modern enough to contain its population by birth control, the argument suggested, ought to be able to reason its way to legal abortion next—as humane medical practice, geared toward the social welfare. And just as birth control was advancing gradually into American society, pushing step by step past barriers of social disapproval, religious condemnation, and the law, now legal abortion might be poised to make the same delicate passage.

The ASA bulletins were careful and upbeat, reporting promising developments from around the states. And by 1968, both Lader and the Illinois abortion activist Lonny Myers were part of an impatient ASA contingent—"the militants," as Lader would always refer to them later —who had lost interest in the slow pace and philosophical timidity of Reform. When Myers and Lader helped organize a 1969 conference in Chicago to launch an entirely new abortion group, they put the challenge in block letters on the cover of the conference flyers: FIRST NATIONAL CONFERENCE ON ABORTION LAWS: MODIFICATION OR REPEAL?

The argument was stacked from the outset. "I've already told Lonny I agree that groups favoring moderate reform (ALI, etc.) should be invited and represented," Lader wrote in a memo to Myers and other conference organizers, "but that we must prove decisively that this approach is pointless and bound to fail." In the circles Lader was traveling in, this was not hard to prove; it took the conference planners so long to find a speaker to defend "The Case for Modification" that they had to print "to be announced" on the programs. How to Start a Repeal Group was a conference workshop, with panelists from New York and California and Nevada and Colorado; there was no workshop on How to Start a Reform Group, and on the conference's final day a long and tumultuous meeting produced an organizational acronym that left no uncertainty when it was spelled out. NARAL: the National Association for *Repeal* of Abortion Laws.

ONE LIFE: that was the essence of Repeal. If it were two lives, it would be society's business; society has something to say about what we may and may not do to another person. But it was not two lives, it was one. A woman who wanted an abortion was one person standing there with her own insides, the tissue of her uterus, an internal organ, a matter for medical people to manage without being forced one way or the other by the dictates of the law. In New York an assemblywoman named Constance Cook, an upstate Republican of mild good humor and famously calm demeanor, listened over and over to her colleagues in the legislature as they shouted and argued and rewrote and compromised and still could not manage the votes for a Model Penal Code bill that had been reintroduced repeatedly since 1965. "What Al kept doing was compromising, trying to negotiate," Cook recalled in an oral history many years later, remembering the efforts of Manhattan assemblyman Albert Blumenthal to make his Model Penal Code bill palatable enough to get through the legislature:

> He kept adding things, like consent of the father, consultation with a social worker, that sort of thing, which made it even more impossible. It became clear to me, every time he added one of those things, he was losing, not gaining. . . .
> And the other thing that bothered me about the reform debates was that the speeches were so outrageous, and so male-oriented, that I decided, Oh, I'd love to show them what a *real* bill is. If they think *this* is bad, they should really get the bill. And that did set my mind, I must say, to introduce outright repeal.

There was no Model Repeal bill to work from, as Cook and a few other state legislators around the country began writing the texts of their blunt new proposals, but in principle there was not very much to say. "An abortional act is justifiable when performed by a licensed physician." That was the essential language of what came to be called the Cook-Leichter bill, hyphenated to include the name of Manhattan Democratic assemblyman Franz Leichter; in its original form, the bill was more a proposal to remove certain subheadings of the statute books than to replace them with anything else. "I checked the health law," Cook recalled. "And I reached the conclusion that this was a medical procedure, and the health law required its being done by a doctor. So we didn't need a statute that said so."

Constance Cook was fifty years old when her abortion bill, widely (and accurately) described as one of the most far-reaching proposals in the United States, began finding its way to the front pages of the newspapers in New York. She and Leichter introduced it in 1969, watched it wither in committee, and reintroduced it for the 1970 session, when her name and photograph began appearing alongside the abortion headlines. She had schoolmarmish glasses, an all-business face, and a resume that suggested nothing but the diligent career of a widely respected legislator: law degree,

Fulbright scholar, legal work for the governor and legislative committees. When she ran for office, she toured Ithaca and the countryside in a 1924 Packard, her husband at the wheel; that was generally reported as her only feint at showiness, and in the legislature she specialized in education law and budget formulas for local school districts, not traditionally an attention-grabbing corner of the Albany arena. "A tailored, matronly manner," a *New York Times* reporter wrote in 1970, "rarely raising her voice and never doubting ultimate success."

Cook had no stories to tell about unwanted pregnancy in her own life; she had suffered a miscarriage at thirty-five, when she very much wanted to start a family, and then went on to give birth to a daughter and a son. But she was a woman at work beside long rows of men, one of four women in the 150-member assembly, the only woman Republican. She had been invited one afternoon in 1967 to join a gathering of women at an apartment in the Dakota, a splendid old Manhattan apartment building with soaring ceilings and views out onto Central Park. "The women had decided that they had to move politically, they had to get some legislation, and they wanted to know how to do it," Cook recalled.

> For many years I had been the only woman Republican legislator. At least this is why I thought I was being invited there—and because they were interested in women's issues, and I've always been interested in women's issues, and there had never been any kind of political support for them. I thought, well, all right, I'll go. So I went down, and it was just fascinating to me. I felt like I was sort of in another world. The women that were there have since become quite well-known, but just at that time, they weren't. Except for Betty. Her book was just out.

"Betty" was Betty Friedan, the president of a memorably acronymed organization then marking its first year of existence: the National Organization for Women (NOW). By 1967, Friedan's book had actually been out for four years, but so formidable was its wake that in 1966, when NOW was created by women from two tables at a Washington, D.C., conference of state commissions on the status of women, *The Feminine Mystique* had sold millions of copies and was widely regarded as a founding document. The chapter titles alone had entered the American vernacular, for a certain breed of woman left profoundly dissatisfied by the expectations of postwar society: "The Problem That Has No Name." "The Crisis in Women's Identity." "The Happy Housewife Heroine." "Progressive Dehumanization: The Comfortable Concentration Camp."

The Feminine Mystique, which began as a fifteen-years-after-graduation survey of Friedan's Smith College classmates, was a grand-scale lament about educated women of the 1950s middle class—in particular, "suburban housewives," a phrase Friedan, a regularly published magazine writer, often used to describe herself. "Gradually, without seeing it clearly for quite a while, I came to realize that something is very wrong with the way American women are trying to live their lives today," read the opening lines in the preface to Friedan's book, and for the next four hundred pages Friedan went on to describe a world of women made desperate and deeply depressed by an insistent modern mythology that women by nature are genuinely fulfilled only as wives and mothers. The word "abortion" never appeared in *The Feminine Mystique*. Friedan exhorted women to continue their educations and find their way back into the world outside their households, and the National Organization for Women was launched with opening statements that spoke in sweeping terms about "full participation in the mainstream of American society" and "truly equal partnership with men."

The political meeting to which Constance Cook had been invited, that afternoon in 1967, took place in Betty Friedan's apartment. And the women gathered in the apartment had much on their minds besides the eventual fate of the Model Penal Code bill. "They asked about labor law changes, jury duty, and then somebody at the meeting said something about abortion," Cook recalled. "Everyone agreed it was a terrible situation and ought to be changed. . . . And they said, 'How would you do it?' I said, 'Well, the first thing you have to do, you have to have a bill.' "

From the outset there was discord about placing legalized abortion on the NOW agenda—for that was precisely what was taking shape in gatherings like this, an ambitious social and political agenda that proposed to alter permanently the role of women in American society. When three hundred women and men met in Washington in November 1967, for the contentious NOW conference that was to produce an eight-item WE DEMAND manifesto, items one through seven ran through child care, job training, poverty programs, educational opportunity, the tax code, the Equal Opportunity Employment Commission, and the United States Constitution. "Abortion" was the very last word in item eight, The Right of Women to Control Their Reproductive Lives, and at the conference some women spoke ardently against including it—too controversial, they argued, and too likely to alienate potential support.

But the yes votes carried, and when Right VIII was printed into the NOW Bill of Rights, it displayed no uncertainty about Reform vs. Repeal: "The right of women to control their own reproductive lives by removing from the penal code laws limiting access to contraceptive information and devices, and by repealing penal laws governing abortion," read the full text of NOW Right VIII, announcing with a vengeance the arrival of an indecorous new presence in the abortion debates.

"Don't talk to me about abortion reform," declared Betty Friedan, from a public stage, at the 1969 meeting that produced the National Association for Repeal of Abortion Laws:

Abortion reform is something dreamed up by men, maybe good-hearted men, but they can only think from their male point of view. For them, women are passive objects that somehow must be regulated; let them only have abortions for thalidomide, rape, incest. What right have they to say? What right has any man to say to any woman—you must bear this child? What right has any state to say? This is a woman's right, not a technical question needing the sanction of the state, or to be debated in terms of technicalities—they are all irrelevant.

In her speech Friedan described NOW's Bill of Rights, and the decision to list abortion among them:

At that time, New York State was having a constitutional convention, and Larry Lader invited me to a meeting of all the different groups —church groups, medical groups, Planned Parenthood, and the rest—who were working on abortion reform. I said, we're going into the New York State constitutional convention demanding a Bill of Rights for women, and we are going to demand that it be written into the Constitution that the right of a woman to control her reproductive process must be established as a civil right, a right not to be denied or abridged by the state.

Most of the people at that table, people working on abortion reform, were men. They looked at me in absolute horror, as if I was out of my mind. They said, you don't know what you are talking about, you're not an expert on this, you women have never done anything like this. You're just going to rock the boat—this isn't the way to go about it—you listen to *us:* A.D.A., A.C.L.U., the clergymen, the medical people.

It was not an entirely new concept that Friedan was introducing, the description of abortion as a woman's civil right. But no respectable abortion group had suggested aloud this tip-of-the-iceberg view of the work before them—that legal abortion be sought not as a public health measure or a compassionate moral compromise, but instead as part of a massive change in traditional assumptions about women in American society. "Without the full capacity to limit her own reproduction, a woman's other 'freedoms' are tantalizing mockeries that cannot be exercised," wrote Lucinda Cisler, an architectural student who became intensely involved in the New York abortion repeal effort, in the 1970 collection *Sisterhood Is Powerful: An Anthology of the Women's Liberation Movement.* "With it, the others cannot long be denied, since the chief rationale for denial disappears. . . . Demand that your legislators repeal all laws limiting access to contraception, sterilization, or abortion, and that they actively oppose any measures that restrict the access in *any way.*"

The National Organization for Women was the earliest and most widely publicized of the scores of organizations that came to be grouped as the "modern women's movement" (social historians sometimes use the term "second wave feminism," to distinguish this from the suffragists of the nineteenth century who began their campaign for the vote for women). Legal abortion was deliberately left off the lists of some of these organizations' public goals; the Women's Equity Action League, for example, was founded in 1968 by Midwestern women who believed NOW was inviting trouble and offending otherwise-interested women by including abortion in its WE DEMAND manifesto. But much of the new movement was adamant about the linkage between legal abortion and women's equality, and during the closing years of the 1960s, as Constance Cook was commencing her effort to repeal the New York abortion law, "second wave" feminists— particularly defiant left-wing women applying certain organizing and political techniques from the civil rights and student antiwar movements— became some of the most insistent, inventive abortion repeal advocates in New York. At Judson Memorial, Howard Moody's church in Greenwich Village, women gathered in the big basement room to take turns speaking publicly about their illegal abortions. Entire sessions of "consciousness-raising," the women-only private discussion groups that formed such an integral part of the movement's early years, were given over to stories about illegal abortion. A New York legislative committee discussing abortion law reform was broken up, its members finally fleeing to closed-door session, when a dozen women stood in the audience to shout that almost all the speakers were men.

And by the spring of 1970, as Cook and Franz Leichter began preparing for a second try at a bill to repeal the New York abortion law, a small but vocal and strongly feminist group called New Yorkers for Abortion Law Repeal was publicly denouncing the Cook-Leichter bill on the grounds that the bill did not go far *enough*. Legislative subcommittee changes had modified the bill in certain philosophically important ways: no longer an unadorned repeal of the state abortion statute, the Cook-Leichter bill now contained language requiring physicians —not nurses or paramedics, whose work might come more cheaply—to do the abortions. It set a time limit on elective abortions—twenty-four weeks into pregnancy, unless the mother's life was at stake. "At a certain stage, your body suddenly belongs to the state," wrote an acerbic Lucinda Cisler, one of the leaders of the group denouncing the bill, in an attack that was published both in the women's journal *Notes (from the Second Year): Radical Feminism* and in the more widely read *Ramparts*. If one believed in feminist principles, Cisler argued, then one must understand a woman's body always to be her own; compromise, by its very nature, missed the essential point: "*She belongs to herself and not to the state. A chain of aluminum does feel lighter around our necks than one made of iron, but it's still a chain, and our task is still to burst entirely free.*"

There was some alarm about compromise at the fledgling National Association for Repeal of Abortion Laws, too, particularly with regard to the twenty-four-week limit. The word "Repeal" was supposed to mean precisely that; this was what the assembled activists had debated so heatedly at the founding conference in Chicago. "It plunged the Executive Committee into a two-day crisis," wrote Larry Lader, who by 1970 was completing his first year as NARAL's executive director, recalling the New York legislative campaigns in *Abortion II: Making the Revolution*. But Constance Cook and her state senate allies were resolute when NARAL board members telephoned to press for full and unqualified repeal. The abortion bill was already a tough sell, Cook declared, what with the Catholic Archdiocese denouncing it and legislators making speeches about the murder of children; and without the twenty-four-week limit and the physicians-only provision, it was lost.

"I said, 'Look, it's this or nothing,' " Constance Cook recalls. "I said, 'We haven't sacrificed our basic principle, which is that it's a matter of right. We've only cut the time, and most people realize that isn't terribly significant.' Nobody really wants to have an abortion in the second trimester." Cook herself did not much like the compromises either, but in later years she acknowledged that she would have given away far more if she had felt it were necessary. What she really wanted was legal, no-questions-asked abortions during the first few months of pregnancy, which Cook believed was when most women went to abortionists anyway—or would go to abortionists, if the procedure were legal and available without the indignity of the hospital therapeutic abortion committee.

And when the NARAL board did vote to support her bill, splitting what had been the New York chapter into two warring camps, it seemed to Cook that it was not so bad to have the word "compromise" flung like an epithet in her direction. She had always believed that what would finally bring her bill to passage was the guilty conscience of a roomful of male legislators. Cook knew those men, she knew they had daughters and sisters and girlfriends, she had listened to a few of them finish their eloquent antiabortion speeches and then approach her privately to tell her that of course she was right. (" 'I hope you win,' " Cook recalls one of her Senate colleagues telling her. " 'I hope and pray you win. . . . But I'm going to vote against you.' ") She knew also that politicians will vote with more enthusiasm for something that looks like compromise, and by the early weeks of April 1970, when the Cook-Leichter bill arrived before the New York Assembly with the potential to change twentieth-century abortion practice far beyond the boundaries of New York State, Cook was not especially sorry to be branded a compromiser as the legislators prepared to cast their vote.

On paper, what the members had before them was not altogether unprecedented. A fifth of the nation's legislatures, from Colorado on, had already approved the principle of qualified legal abortion by adopting some version of the Model Penal Code. And earlier that spring, after intensely emotional statehouse and community debates, legislators in two other states—first Hawaii, then more recently Alaska—had approved laws that came far closer to actual repeal: legal abortion, for any reason, as long as the fetus was "nonviable" and the abortion was performed by a physician.*

But Hawaii and Alaska were lightly populated states, a daunting distance from the mainland, and both states' new legislation contained language limiting abortions to residents only. A matter of very different magnitude confronted the New York legislature: the Cook-Leichter bill contained no residency requirement, and if the bill were to pass into law, a vast network of buses, trains, freeways, and airlines would suddenly link the entire nation to a state where abortion was legal and available to any woman with the means to pay for the procedure and her transport to New York. No screening committees would pass judgment on abortion patients; no psychiatrists would inquire into their mental health; no penalty-of-perjury statement would be placed before them to ensure that they lived within the boundaries of the state in which they were about to abort. In practical terms, the Cook-Leichter bill proposed to legalize abortion not only for New York women but also for women in Chicago and Philadelphia and Boston and Detroit—and St. Louis, which by airplane was two and a half hours away and had in place an efficient referral system, the Clergy Consultation Service, to help Missouri women travel safely to New York for their abortions.

The crucial vote came on April 9.

The New York state Senate had voted preliminary approval of the abortion bill already, after five hours of standing-room-only debate. The state Assembly had defeated the abortion bill once, narrowly, after *eight* hours of debate. Governor Nelson Rockefeller had all but guaranteed his signature, should a bill survive both Senate and Assembly vote—"I continue to believe very strongly that the state's archaic abortion law must be changed," he declared in his January address to the legislature—and now some parliamentary maneuvering had brought the bill back to the Assembly for reconsideration, for one last vote that would kill it or send it back for certain Senate reapproval and then the governor's signature.

Once more the speeches flowed, the arguments banged around the great Assembly chamber, the galleries filled with women leaning down to make out the sound of the assemblymen declaiming into the microphones at their desks. "At one point during the height of the debate, an elderly

---

* The Alaska state Senate had approved on April 2 a bill permitting physicians to perform abortions for any reason during the first twenty weeks of pregnancy. On April 10, one day after the key Assembly vote in New York, the Alaska abortion bill—which contained a thirty-day residency requirement—was given final approval by the Alaska state Assembly. The legislation was vetoed by Alaska governor Keith Miller, but the state legislature overrode Miller's veto at the end of April, and the new Alaska law went into effect on July 29, 1970.

woman in the galleries had to be stilled by the capitol police," wrote the *New York Times* reporter at the Albany statehouse. " 'Murderer,' she called out, her thin voice quivering in the still chamber. 'You are murderers, that's what you are. God will punish you. You are murderers.' "

Seventy-five votes: that was what they needed, in the Cook-Leichter camp, to bring legal abortion to the state of New York. Technically, the required number of votes was seventy-six, but because the New York Assembly Speaker Perry Duryea had promised in the event of a tie to cast a vote in favor of the bill, despite the fact that by standard state practice the Speaker does not vote, all present understood that the crucial number was seventy-five. The earlier Assembly vote had given them seventy-three, almost enough, two votes short; the ensuing week's urgent last-minute lobbying had pushed a few votes this way and a few votes that way.

And now in the Assembly chamber an extraordinary thing was about to happen, a thing that began in such confusion that at first only a few people saw, and sucked in their breaths, and the television cameras swiveled and bumped into each other on the Assembly floor. It was late in the afternoon. The speeches had exhausted themselves. The clerk began once more his slow tabulation, ayes down one column, nos down the other. One by one the assemblymen cast their votes, pronouncing the word aloud into their microphones, as is the custom in close votes or matters of intense public interest: Yea. No. No. Yea.

The roll clerk added up, and Constance Cook added up, and the newsmen and the secretaries and the legislative aides and the women in the galleries all added up. The roll clerk stood and prepared to announce that the yea votes added up to seventy-four. The abortion bill had failed again, this time by a margin of a single vote.

Then out in the chamber, on the Democratic side of the floor, directly across from the Republican desk of Constance Cook, a lone assemblyman rose from his chair. He stood up very straight, and although the news accounts reported that his hands were trembling and his face was drained of color, it seemed to Cook that he kept his voice steady as he called for the floor. What he said at first was lost to the melee in the chamber and the sudden scramble of the cameras, but the critical words of Assemblyman George Michaels appeared the next morning in the pages of nearly every newspaper in New York, along with the photograph of his graying head, bent down toward his desk, after he had begun to cry.

Twice so far he had voted against the abortion bill, Michaels said, and his family was assailing him for it. "My own son called me a whore for voting against this bill," Michaels said, and the people who were present to hear it remember that this was when his voice began to break. "I realize, Mr. Speaker, that I am terminating my political career," Michaels said. "But I cannot in good conscience sit here and allow my vote to be the one that defeats this bill. I ask that my vote be changed from No to Yes."

In the furor that erupted around the desk of George Michaels, the legislative aides weeping and the NARAL people embracing and the assemblymen shouting at Michaels and thumping him on the back in congratulations, Constance Cook watched from across the chamber and knew that what he had said was true. Michaels' district was just north of hers, in upstate New York, a traditionally Republican area that Michaels had landed largely by culling support from the city of Auburn's Democrats, many of whom were Roman Catholic. As a Democrat amid Republicans, Michaels had nobody if he lost the Catholics. His own family was Jewish, a point he managed to stress there on the Assembly floor. "I can't go back to a family seder and say I defeated the bill," Michaels said, referring to the ceremonial Jewish meal held during the spring holidays of Passover. Cook had known before the vote that Michaels intended to switch if a single vote proved to be all she needed—he had promised it to her, she says, and it never occurred to her to doubt Michaels' promise—but she understood at once that what he was promising was political suicide for the sake of the biggest abortion bill in the United States.

"I was deeply grateful," recalls Constance Cook. "And as far as I was concerned, it was worth it."

ON APRIL 11, 1970, in his office at the statehouse in Albany, Governor Nelson Rockefeller signed into law Chapter 127 of the Laws of 1970, adding to the New York penal code the following passage:

> "Justifiable abortional act." An abortional act is justifiable when committed upon a female *with her consent* by a duly licensed physician acting (a) under the reasonable belief that such is necessary to preserve her life, *or, within twenty-four weeks from the commencement of her pregnancy.* A pregnant female's commission of an abortional act upon herself is justifiable when she acts upon the advice of a duly licensed physician (1) that such *act* is necessary to preserve her life, *or, within twenty-four weeks from the commencement of her pregnancy.* [emphasis in original]

A brief New York–datelined wire service article appeared that same afternoon in the *St. Louis Post-Dispatch.* Many more immediately eye-catching stories ran that day in the newspaper: Apollo 13 was preparing to lift off from Cape Canaveral, carrying three astronauts toward the moon; the Cardinals had opened baseball season the night before by beating the New York Mets 7–3 before a record first-day crowd at Busch Memorial Stadium; President Nixon was reported to be considering for the U.S.

Supreme Court a sixty-one-year-old Eighth Circuit Court of Appeals judge named Harry A. Blackmun.

But Judy Widdicombe read the *Post-Dispatch*'s New York story first.

NEW YORK, April 11—Liberalized abortion laws were on their way today to the desks of New York Governor Nelson A. Rockefeller and Alaska Governor Keith Miller.

Rockefeller was in New York City last night when the measure received final legislative approval, 31 to 26, in the Senate. The Governor said that he would return to Albany today to sign the measure.

The New York law will take effect July 1. Unlike liberalized abortion laws in other states, the law does not require a woman to be a resident of New York State, nor does the operation have to be performed in a hospital.

After 24 weeks of pregnancy, abortion could be performed only to save the mother's life.

There was a small photograph of George Michaels, his eyes downcast, his shoulders slumped, one hand clasped to his forehead. "Changed Vote," the caption read. And below: "He changes his vote to 'yes,' brings passage of New York's abortion law." And Judy Widdicombe began calling airline companies to get the fares and daily flight schedules from St. Louis to New York.

FOUR

The Writing in the Heart
1971

AFTER HE SPOKE before civic groups, or argued onstage in campus abortion debates, or showed his small collection of slides in darkened auditoriums, the mail cheered Matt Backer and warmly urged him on.

Dear Dr. Backer,
    I'm writing this short letter to tell you how good it is to know someone like you is fighting against the upcoming abortion law in Missouri. How anyone can rationalize the killing of an embryo or fetus I'll never understand.

Dear Dr. Backer,
    We would like to express our agreement with you on the stand that you have taken in regard to the unborn fetus, and praise your involvement. We do not feel that society should condone abortion by liberalizing abortion laws. We hold the belief that society does have an obligation to protect the rights of its citizens—the right to live being one of those. Our support is with you.

Dear Dr. Backer:
    As one of your patients, I feel that I must congratulate and commend you on your firm stand against legalized abortion in Missouri. In our

highly technological society, where the process of dehumanization often takes place, it is really great to know there are men like yourself who still care about life and human dignity. You are speaking in behalf of a vast public who still feels the same way as you do and supports you whole-heartedly. Thank you for being that kind of man and doctor who still believes in the sacred moral code of life which still dwells in the minds of many. Good Luck.

Dear Doctor Backer:
    May God be with you in your fight for the unborn children.

He kept the letters, sliding them into a manila folder with the clippings and photographs people had sent him. "Enclosed is the picture of our son," a St. Louis woman wrote. "He was born June 26, 1954 after a 26 weeks pregnancy." The son was smiling, his shaggy hair combed artlessly sideways for the picture, and his jacket bore the logo of the best Catholic high school in the city, St. Louis University High, where Backer himself had gone to school. If the woman had counted her pregnancy dates correctly, her son had been born two weeks after what the state of New York now defined as the cutoff for elective abortion: up to twenty-four weeks of pregnancy, under the terms of the astonishing legislation Governor Rockefeller had signed into law, a woman could legally report to a New York hospital and have her baby aborted. The imagery was so vivid, the premie nurseries in one hospital and the second-trimester abortionists in another, that Backer's correspondent had evidently felt no need to describe it. She and her husband, she wrote, simply wanted to thank him.

Backer paper-clipped the picture to the handwritten note and stored it, carefully, as though archiving small bits of evidence. He was glad the lady had written. But he was infinitely sorry that he and Denis Cavanagh had been right. It was *not* Tragic Cases Only the abortion people were after; it was abortion without justification, abortion on demand, abortion for any woman who did not want to give birth to the baby she was carrying. The New York law had not even included the usual half-hearted language about protecting women's "mental or physical health." The only prerequisite to a New York abortion appeared to Matt Backer to be cash up front, and there was plainly quite a lot of that to be had; the January 1971 *Medical Economics*, in a lengthy article entitled "The Mad Scramble for Abortion Money," told story after story of the sudden entrepreneurial fervor of New York businessmen, doctors, and for-profit referral agencies. Real estate agents were scouting vacant medical suites as potential clinic sites. Abortion facilities were offering over-the-counter stocks for investment purposes. The Clergy Consultation Service had set up its own New York City abortion clinic, called Women's Services; within months of opening, in 1970, it was taking in a hundred women a day and paying its physicians $75 per abortion, which meant that a swift and efficient doctor could make $1,000 during a single eight-hour shift.

And the women were traveling to New York by airplane, by train, by automobile, by Greyhound bus—the trip from St. Louis took twenty-four hours by bus, crossing five states, but Backer had no doubt that Missouri women were making it. Every record-keeping effort by the New York health authorities indicated that the legislature in Albany—as both supporters and opponents of the new abortion law had predicted—had succeeded, with its April vote, in legalizing abortion far beyond the borders of the state of New York. The new law went into effect on July 1, 1970, and over the following year more than half the abortion patients in the city of New York were identifying themselves as residents of another state: 55,347 residents to 83,975 nonresidents, according to city health department surveys.

The competition for those abortion patients was inventive and fierce. Taxi drivers under kickback arrangements worked the airport and train stations, trolling for anxious young women who might without much protest be deposited at the wrong clinic. Newspaper advertisements and direct-mail solicitations to physicians spread the telephone numbers of agencies that could charge $100 simply to act as intermediaries for women who wanted New York abortions. An Ohio businessman set up an enterprise called Aeromedic, which promised to use light planes to fly women round-trip from Cincinnati to an abortion clinic in western New York. One hired airplane, its trailing banner announcing the number of a private referral agency, startled December vacationers by buzzing Miami Beach.

New York was not the only option, either; Backer's newspapers and medical newsletters had begun to carry unsettling reports from other parts of the country. In Washington, D.C., where a federal court ruling in 1969 had overturned the conviction of an abortion doctor by finding the District of Columbia's abortion law unconstitutional, a new clinic called Preterm was preparing to accept any patient who could manage the trip to Washington. The California screening process, according to which "mental health" abortions were supposed to be approved only after two-doctor psychiatric review, had been stripped to a formality that appeared to accommodate nearly anybody who had enough money for the abortion. Alaska, Hawaii, and Washington State now offered abortion on demand to in-state residents, and even those residency requirements proved somewhat flexible in practice; in Hawaii, for example, a woman seeking a hospital abortion had only to sign a piece of paper promising that she had lived within the state for the thirty days before she intended to abort.

Even the good Midwestern state of Kansas, up against Missouri's western border, was relaxing its abortion barriers by the spring of 1971; at the Kansas University Medical Center, a five-minute drive across the state line from the Missouri side of Kansas City, any woman could obtain an abortion if the doctors in charge decided that her case suited the state's new Model Penal Code categories. Up and down both coasts, with tentative forays into the heartland: a kind of sickness was encircling the United States, and sometimes it bewildered Backer to think that he had to do this in the first place, that he was actually being called upon to go into public auditoriums and explain *why* it was a bad idea to permit the legal killing of the unborn. It was as though someone had pulled him aside one day and said, *Remember how it used to be illegal to murder somebody else? We're rethinking that concept. Perhaps you have something to say on the subject.*

In his basement study, where Backer kept his books and his journals and his chronicles of the battles around the states, the abortion file grew. *Linacre Quarterly,* the medical ethics journal of the National Federation of Catholic Physicians' Guilds, ran regular, voluminous essays by physicians and clergy: "Proposed Abortion Laws: 'Slaughter of the Innocents,' " "Who Speaks for the Fetus?," "The Physician and the Rights of the Unborn," "Man Plays God." A Georgetown University philosophy professor named Germain Grisez wrote a scholarly book, five hundred pages long, using Christian doctrine, legal history, and philosophical argument to attack point by point every justification for legalized abortion. "A new name is needed for prejudice against the unborn," Grisez wrote at the conclusion of *Abortion: The Myths, the Realities, and the Arguments.* "I suggest it be called 'prenatalism,' since it is based on the fact that *we* are already born, while *they* are unborn."

Backer took some from one article, some from another, culling for his talks. Here was a careful look at abortion in Japan; here an editorial reminding people what the Nazi doctors had done; here a dissection of "therapeutic" abortions, which were hardly ever truly mandated by medical necessity.

Here was a psychiatrist explaining that *every* distressed pregnant woman faces challenges to her "mental health," and that the physician's duty is to help such patients manage their feelings without destroying another human life. And here was a critical view of the post-liberalization situation in New York—the public health officials might be crowing over low complication rates, but they maintained dubious follow-up on out-of-state patients, and in any case they always failed to point out that every successful abortion produced at least one death.

Backer made lists of his discussion points. He typed them in outline form for his slide presentation:

UNBORN HAVE LEGAL RIGHTS
    To Recover for Prenatal Damage
    To Inherit
    To Due Process
    Not to be killed by unconstitutional authority

MODERN HISTORY
Social Pressures
Medical Controversy
ALI Model Penal Code
Rubella Epidemic
Group Pressures

He made a slide of a magazine cartoon, to stir up his audiences a little, especially the Catholics: husband saying to pregnant wife, as they stroll down the sidewalk with their nine children, "Notice that people don't smile at us the way they used to?"

And he selected, with some uneasiness, his abortion pictures.

For Matt Backer understood this, just as every regular voice on the new American right-to-life circuit was beginning to understand it: you could bury your audience in history, you could repeat fact after fact of medical science and biology, you could appeal to moral reason until your breath gave out, but nothing you could say in any public forum had the power of the pictures. In later years a line would emerge among right-to-life speakers, passed along with amusement and the conviction that it was true: the pro-abortion people had three rules for debate, it was said, and they went like this:

1. Don't let them show pictures.
2. Don't let them show pictures.
3. Don't let them show pictures.

For the men and women who showed them—usually as a slow-building climax to their urgent appeals in living rooms or church meeting halls or Catholic school auditoriums—the pictures did a single, terrible, unimpeachable thing: they made people see. The abortion law repeal activists could dodge and feint with their talk about "zygotes" and "fetal tissue" and "reproductive rights," but the pictures showed *legs*. The pictures showed feet with the toes all formed, miniature toes, the size that entrances adults when they look at a newborn baby. The pictures showed small fingers, stretched and separate as if just learning to flex. The pictures showed crooked elbows and bent knees and tiny rounded heels, but all of these were in pieces, literal pieces, with jagged ground-meat colored flesh where the joints should have been. Sometimes the pieces were tumbled and still bloody, the way they looked when they were first dumped into a surgical pan, and sometimes they were laid out more neatly for the pictures, as though some laboratory technician was lining them up to see what might still be missing: a spine, perhaps, or a severed right arm.

The first abortion pictures had surfaced during the very early years of the legislative wars; Right to Life of Southern California had a few, during the late 1960s, and New York Right to Life distributed pictures in the statehouse during the last urgent weeks of the campaign to keep the state's old criminal abortion law intact. But by 1971, three years into the legislative wars, the chief national stewardship of the abortion pictures was being ceded to a Cincinnati couple named John and Barbara Willke.

The Willkes were "lecturers," as their biographical handouts put it; he was a physician and she was a nurse, and during the 1960s they had developed what they liked to think of as a sensible Christian-morals-in-the-real-world roadshow to talk about sexual behavior and individual responsibility. Both Willkes were Catholic, but they tailored their talks for more general religious audiences, and in the conservative circles in which they tended to travel—church meeting halls and parochial schools were often the sites of their lectures—the Willkes perceived that they raised a few eyebrows when they insisted that married men and women ought to enjoy sex as a God-given expression of intimacy and revel in pleasing each other in bed. But they were no advocates of the sexual experimentation that had attracted such attention during the late 1960s, either; their lectures explained that premarital sex was always wrong, and that teaching young people about contraception made roughly as much sense as showing drivers-education students how to break the law without getting caught.

The remarks of a colleague, as the Willkes would tell the story later, had prodded them to add abortion to their lecture agenda. ("He said, 'What are you doing on right to life?' " John Willke recalls. "And we said, 'We aren't going to get involved in that. If we do, it'll swallow us up.' And it did.")

They were natural and implacable recruits. In their 1964 guide on teaching sex to children, the Willkes had advised that children inquiring about abortion be told that "some mothers don't want a new baby after it starts growing in their bodies, so they have a small operation performed that kills the baby in their womb." When the child asked why, the book continued, the parent might answer that while it was unwise to judge other people's motives, "it is usually based on selfish reasons—'what I want—' 'my convenience—' 'my comfort.' " Now that message became the centerpiece of their talks, and around regional right-to-life groups in the Midwest and beyond, they made rapid entry into the ranks of the busiest speakers. John Willke was a second-generation Ohio general practitioner, a tall lean man with an air of collected certainty that impressed some people and irritated others; when he spoke in public, with his cool hazel eyes and his stolid three-piece "country doctor" suits, his voice had a precision and smoothness that made some of his debate opponents feel as though they had been patted on the head and sent off to their rooms. His wife was small and vigorous and played off him rather well, the caring nurse to his more clinical doctor, and when she remembered what it was like to work with the pictures, she would tell a particular story about the early days, when the Ohio Assembly was considering an abortion legalization bill and the Willkes first volunteered to join the bill's testifying opponents.

They had taken a vacation in Florida, Barbara Willke recalls. They were giving a public talk as their holiday commenced. Afterward a doctor approached them, and urged upon them a small present for use in their work: a fetus, preserved in formaldehyde, and set inside a plastic bucket.

"He was well-meaning," Barbara Willke recalls. "And I remember saying, *oh, you poor little thing,* to the baby, and it bothered me the whole week. All I could think was, *oh, if somebody had just loved you.* It was sad. A perfect little baby. And I decided that if I was going to stay in the movement, I had to develop a thicker skin."

Politics, and one swift lesson in the shock of the visual, thickened up Barbara Willke's skin. Early in their abortion talks, the Willkes, like Backer and many of their other counterparts around the states, had begun bringing along the celebrated Lennart Nilsson photographs that had appeared in *Life* magazine in April 1965. The pictures had been hailed around the world as a revelation of the wonder of life's beginnings: taken over seven years both inside and outside the uterus, they showed in extraordinary color and detail the week-by-week swelling of the developing human embryo—from the magnified bulge of a blood-filled four-week heart to a membrane-shrouded face at twenty-eight weeks of pregnancy.

That a society could know this about human life and still approve abortion—could gaze directly at the week-by-week creation of a living human person, and exclaim over its beauty, and still open the door to anyone who wished to kill in an operating room a life at exactly the same stage of its growth—this was a mystery to the Willkes, and they were adept at conveying in lecture halls their own sense of horror and sorrow. When Barbara Willke held up the Nilsson pictures before an audience, she could feel a visceral reaction in the room; the magazine cover shot, the most famous of the series, showed a shimmering amniotic sac, and floating inside, its thumb in its mouth and its almost-formed eyes staring whitely out, something that Barbara Willke imagined anybody would have called a baby unless they intended to get rid of it. It was eighteen weeks along, six weeks short of the New York cutoff date. All a person had to do was call it a "fetus" and it was *legal* to get rid of it, once the laws began to change: it was legal to cut a woman's uterus open and remove the baby before it could survive on its own, which was the method for the operation called hysterotomy; or to inject a woman's uterus with strong saline solution and wait for her to go into labor, which was the preferred method for abortions in the second trimester. "Expel the fetus" was the technical term for what happened after a saline infusion, as though sufficiently chilly terminology would distract attention from the act itself, as though any woman grieving her way through a spontanous miscarriage would be told she was losing her "fetus." The word "fetus" began to make Barbara Willke feel sick when she heard it. There was a kind of Third Reich insistence to the way the abortion people clung to that word; repeat it often enough and it's not a baby anymore, it's just a tumor or a kidney or some other body part you can toss in a bucket after you've snipped it out.

The first abortion picture Barbara Willke was forced to study did in fact have a bucket in it: a metal bucket, with deep shiny sides. In the file drawers of right-to-life speakers the bucket shot was famous, and passed in duplicate from state to state; Matt Backer had a copy of the bucket shot, and in Ohio the right-to-life group working the state legislature in 1971 held vigorous private debate about whether to spread copies of the bucket shot through church groups and other audiences that might be roused to mobilize. Barbara Willke remembers being against it, at first. "I just thought it was too awful," she recalls. "But they said, No. If they're doing it, people ought to know what they're doing."

In the center of the bucket shot, at the bottom of the bucket, curled against the metal as though in nasty parody of a newborn tucked inside its cradle, lay what Barbara Willke could see was the product of a late abortion, probably a hysterotomy. The baby's knees were tucked toward its chest. The outlines of an ear were clearly etched against the head. A mess of placenta lay beside the small round belly, still funneling toward the baby's navel. No detailed documentation came with the bucket shot, as the picture was passed from hand to hand; the pathologist who later took credit for the picture was a St. Louis physician named William Drake, and according to Drake's own account some minor rearrangement had gone into giving the bucket shot its full effect.

The infant's body had been delivered to his hospital laboratory in 1969, Drake recalls, in a plastic container. A supervisory nurse told Drake that the baby had been taken from its mother in an induced abortion. Drake recalls that he never learned specifics about who had aborted this baby and why (in Missouri in 1969 any induced hospital abortion would have carried the famously inexact label "medically indicated"). But he put the body and some of the bloody gauze into a metal bucket—"It just looked like it would be a better picture," Drake recalls—and later gave the photograph to a Missourians Opposed to the Liberalization of Abortion Laws member for use in petitions and lobbying in Jefferson City.

Drake's photographic instincts were clearly very good. Everything about that picture suggested cold and steel and casual discard, as precise a metaphor for abortion as a right-to-life speaker might imagine, and in the Willkes' view the documentation on that particular baby was less important than the story the photograph worked to illustrate: physicians were performing hysterotomies, nobody could pretend they were not, and when a successful hysterotomy had been completed, this was what the medical crew left behind in the operating room or the pathology lab: a dead baby in a plain hospital pan.

And it was true that hysterotomy was by the close of the 1960s a relatively uncommon procedure; the preferred method for abortion in the second trimester was saline infusion. Saline infusion pictures were worse. When a doctor completed a hysterotomy abortion he pulled out what looked like an intact but underdeveloped baby, but when a woman had a saline abortion, the "uterine contents"—that was the vocabulary of the clinical reports—could emerge with skin as red and shiny as if someone had scalded it. In bleak moments John Willke took to calling the saline pictures "candy apple babies." It turned the stomach to look at them, the tight scarlet scalps, the small arms corroded and bent, but at least as bodies they were intact. It was the first-trimester abortions that left pieces, cut or ripped; the standard procedure for first trimester was either suction, which pulled everything through a narrow vacuum tube, or dilation and curettage, which required scraping the inside of the uterus with a sharp instrument and then using a forceps to ease the fragments out. In either case, once magnifying lenses were used to aid in identification of the pieces, separate and recognizable body parts were discernible from about nine weeks of pregnancy on. These pictures required good lighting and a skilled photographer, but when they were taken from the proper angle, an attentive viewer could sometimes make out every toe.

Around the country, as abortion bills surfaced one state legislature at a time, certain sympathetic pathologists and other physicians began making available photographs that were quickly reproduced and spread—there was one energetic obstetrician in Rockville, Maryland, who sent the Willkes both an excellent D&C-scraped-parts picture and what the obstetrician described as a midoperation photograph of a surgeon cutting the umbilical cord during a hysterotomy. And it was at the 1971 Ohio State Fair, Barbara Willke recalls, that a single memorable episode taught her how crucial the photographs might become. Ohio Right to Life had set up a booth at the fair, with literature and bumper stickers and copies of *Handbook on Abortion*, the slim paperback question-and-answer volume the Willkes had completed earlier that spring. The volunteers sat at the booth and waited, prepared to answer questions. Nobody stopped to look.

"Then the volunteers opened up a couple of the *Handbook*s to the pictures," Barbara Willke recalls. "And all of a sudden they had people ten feet deep. They were just mobbing the booth."

Only four of the color abortion pictures had appeared inside the first edition of the *Handbook on Abortion*, but the Willkes set to work. Pictures, they now understood, were going to remake forever the terms of debate. Nobody could look directly at these pictures and launch into a speech about the right of women to do what they wished with their bodies. The Willkes began collecting pictures in earnest: a six-week embryo, caught in a golden light and blown big enough to make visible the budding fingers and the black pinpoints of eyes; an eight-week embryo, floating in its sac and bowing its head as though in prayer; a grown man's hand, the magnification so high that the whorls of his fingerprint looked wide as stripes, holding gingerly between thumb and forefinger two delicate, pale, perfectly formed human feet.

Jack Willke called that photograph Tiny Feet, and he was always interested by people's response to it. A pathologist from Oregon had handed him the picture, at a meeting in California, and told him the feet showed development at ten weeks. Because they were perfect feet they suggested to the human imagination a perfect body, a person minutely proportioned but exquisitely made from top to bottom, and it was also apparent to Jack Willke that women had some mysterious fascination with babies' feet. When the Willkes began sending out copies of their earliest slide collection, which was to become the single most popular visual teaching tool for right-to-life organizers around the country, they made certain that Tiny Feet was among the pictures; the embryos were included too, and photographs of living babies who had been born prematurely. You want to teach them before you shock them, the Willkes urged other speakers, let them understand step by step what is genuinely at stake. Their written suggestions were explicit:

> In our presentation, we would show the eighteen-week LIFE Magazine cover, and ask, is this being human? Subsequently, we will show very human-looking babies at sixteen, fourteen, twelve, and eleven weeks. In each case, we repeat the question, "Is this one human?" Backing gently down this age ladder, we would hope to lead our audience down with us, never giving them any visual or intellectual reason for changing the initial mindset they started with, i.e. "It's a baby." By eleven weeks, if due to size alone, some will begin to question, so we stop at that point and discuss in some detail the perfection at eleven weeks. We will detail all of the functions of breathing, swallowing, urinating, tasting, sensitivity to pain, capability of learning, etc. Hopefully, this will reaffirm in those who were still wavering, a firm conviction that this is "still human life." The next photo shows the very, very tiny human feet at 10 weeks. We next drop to eight weeks, noting that "this little one doesn't look quite so human any more." At this point we will challenge our audience to enter the scientific age and to extend their eyes with a microscope and their perception with instruments of precision. . . .
>
> We will then show them one more visual at six weeks, and then *we'll show no visuals under six weeks. . . .* We will not show visuals under six weeks because we feel that if we do, the audience may change their minds from their conviction that this is human life. Not that this would make any sense, as the biologic facts and intellectual facts compel us to judge for humanity right down to conception. A viewing audience, however, is not always that rational, and some will make a visual judgment even though we ask them not to. [emphasis in original]

When that lesson in human development was complete, the Willkes advised, any speaker using their slides should proceed directly to the war pictures. That was what they called them in private (in their written guidelines, "pictures of aborted babies" was the more careful phrase), and as the Cincinnati Right to Life office began mailing the Willke slides to addresses all over the country, the Willkes were firm in their view that the war pictures must always be part of the presentation. "If this be shock, so be it," they wrote:

> The pictures at Mai Lai [sic] were also shock, but taught us something about war that words never could. . . . These pictures are powerful, but they are absolutely authentic, and unless you tell this part of the story, you have only half the story. We, therefore, feel strongly that pictures of aborted babies should always be used.
>
> The dynamic thing to do with aborted babies is to start at the youngest age, in contradiction to telling the story of life as above where we start with live-born babies. The reason we start at a ten week level is that these are the least ghastly of a very upsetting set of pictures. One can often feel the growing tension in an audience as we move up the ladder of age from the scraps of arms and legs to the final terribly upsetting, "cut the cord, drop the baby in the bucket and let her die." It is important here again that pictures of fresh specimens be used, such as the ten week suction scraps of torn away arms and legs in the two sets mentioned above. It is equally important here not to use pickled specimens of limbs, because of their whitened, blackened, shrunken, and less human-looking appearance. In describing the types of abortion, we would suggest that you stay fairly close to a simple clinical description of what goes on. We believe that too much added editorialization or emotion-tugging comments are not needed as the facts and pictures speak powerfully enough for themselves. We would progress, then, up the age ladder from suction, to D&C, to salt, to hysterotomy, and cap it with the full garbage bag of tiny corpses.

The garbage bag shot was the Willke slides' answer to the bucket shot: stark, black-and-white, sweeping in its message. The Willkes said the picture had come from what they called "the undeveloped roll," a single roll of film mailed to them by a Winnipeg doctor too career-conscious to risk public identification, and when they saw the film's pictures and chose the one shot that was to climax their slide show, the Willkes composed the label that was always to serve as the caption: "The result of one morning's work at a Canadian teaching hospital."

The garbage bag shot worked like the Dachau concentration camp photographs from 1945: bodies upon bodies, apparently ready for disposal, piled into what appeared to be a large plastic-lined trash can. In the picture the infants looked nearly the size of newborns; a hospital producing that many smooth-fleshed late-term abortion bodies in a single morning would have to have been doing a remarkably brisk business in hysterotomy. It was a death picture without visible dismemberment or blood, so it could make people angry and sick without actually making them queasy, and over the years it was to be reprinted hundreds of thousands of times in books and flyers and slide shows and teaching collections. As propaganda there was extraordinary resilience to the garbage can shot, but Matt Backer was appalled by it; he thought it pushed a long way past the limits of acceptable taste. Even if the trash bag was real there seemed to Backer something sordid about this kind of imagery, and anyway late-term abortion was still a rare enough practice that Backer had misgivings about making extensive use of those pictures in his own presentations. The shock factor made him uncomfortable, and when he selected his visual lineup he made sure that in his pictures the D&C-cut body parts were laid out above a six-inch ruler so that no one could mistake their actual size.

Backer ignored the Willkes' admonition about visuals before six weeks; one of Backer's pictures showed a pale bulging column that he would tell people was the human neural tube two to three weeks after conception. He had graphs, too, slides that showed nothing but numbers and thin black lines going up or down. Maybe his show didn't jolt people out of their chairs quite as sharply as one of the Willke talks, but Backer didn't fly from one city to another presenting himself as a "lecturer," either; he was a St. Louis obstetrician, quite a busy one at that, and if he was going to talk to people about this subject he intended to do it in language that suited him.

"Abortion destroys a normal human life," Matt Backer would say. "The child is not the unjust aggressor," he would say. "I don't think we should institutionalize a system in which we have to take a physical every morning to stay alive," Matt Backer would say, which was his way of swatting back the audience members and debate opponents who wanted to abort children like his own retarded son. He was not an especially polished public speaker, but he was calm and hard to ruffle, and he learned to sense as if by instinct when someone in the audience was going to raise a hand and start in on the Church. *Isn't it a fact*—they were always grouped, these questions, and usually phrased in the manner of a withering attorney on cross examination—Isn't it a fact that Catholicism condemns abortion? Isn't it a fact that you yourself are Catholic? Isn't it a fact that Catholics make up most of the membership of the groups that call themselves Right to Life? Isn't this whole issue simply a case of one religious group imposing its view on all the rest of us?

And Backer would wait, his face impassive, his hands on the lectern —for always there was that final Catholic question, the dismissive question, the one without which the others had no point. Then Backer would nod, and know that he was not supposed to look angry or even faintly irritated, and answer the questions in order, one by one: Yes. Yes. Yes. *No.*

•   •   •

THIS WAS TRUE: the Roman Catholic Church, the church in which Matthias H. Backer, Jr., had been baptized and confirmed and formally educated from kindergarten through his final year of medical school, instructed its parishioners that abortion was a mortal sin—an "unspeakable crime," in the 1965 language of the Second Vatican Council, which used the word "abortion" as though it were half of a hyphenated phrase, grammatically and philosophically inseparable from the word "infanticide." "Life must be protected with the utmost care from the moment of conception; abortion and infanticide are abominable crimes," read the Council's Pastoral Constitution on the Church in the Modern World. The original document was printed in Latin, and the adjective describing the crimes, *nefanda*, varies from one English translation to another—"abominable," "nefarious," "horrible," "unspeakable."

And this, indisputably, was also true: from the earliest years of the Model Penal Code abortion proposals, it was Catholics who had taken up the opposition—who drove to the statehouses and mimeographed position papers and stood before lecture audiences to plead the case for abortion as a legal and social wrong. Denis Cavanagh was Catholic. The Willkes were Catholic. The founder of Missourians Opposed to the Liberalization of Abortion Laws was Catholic, and the pathologist William Drake was Catholic, and William Hogan, the Maryland obstetrician who sent the Willkes some of their first slides, was Catholic.

The head of New York Right to Life was Catholic. The head of Right to Life of Southern California was Catholic. The leading testimony against Colorado's 1967 Model Penal Code law had come from the Catholic Lawyers Guild and the Colorado Catholic Conference. A Catholic lawyer headed Voice for the Unborn, the Washington State group formed to fight the 1970 abortion referendum; a Catholic research physician headed the Value of Life Committee, the Boston-based group formed to fend off changes in Massachusetts abortion law; a Catholic cardiologist and a Catholic attorney helped set up Illinois Right to Life. The National Right to Life Committee, in the earliest years not much more than a two-man effort to staff an office in Washington, D.C., was headquartered inside the United States Catholic Conference building in the capital; the director was a priest, Monsignor James McHugh, and the first multistate National Right to Life conference, in January 1971, took place in a Chicago Catholic college named after a French saint.

Everybody cultivated exceptions, the person who could be pointed to, with some indignation, when the subject came up: in Chicago there was Victor Rosenblum, a Jewish law professor who helped found the legal organization called Americans United for Life. In Arizona there was Carolyn Gerster, an internist who had been raised without religious training and was attending an Episcopalian church by the time she began testifying against changes to the state abortion law. There were Lutherans working in the Minnesota group and a Methodist doing public speaking for the Massachusetts group; the governor of Alaska was a Methodist too, and had vetoed his state's permissive abortion bill before the legislature over-rode him.

That these points of information were known, in the gradually over-lapping circles of state right-to-life organizations, was a measure of the nature of the problem. When Carolyn Gerster was recruited into action in 1970 in Arizona, the surgeon who contacted her expressed his alarm about possible changes in the state abortion law and added, as Gerster recalls, "You're Protestant. And we're trying to be sure we have fifty percent non-Catholic." Victor Rosenblum once fended off a newspaperman who said with evident bewilderment, "Rosenblum, that's not a Catholic name, mind telling me your wife's maiden name?" Her name had been Rann, Rosenblum said, and the newsman nodded: "Oh, Ryan. That explains it." Actually no, Rosenblum said, now beginning to be angry, it was Rann, shortened from Ranovsky. The Catholic label lay over all of them, and with it a logic that men like Matt Backer found maddening to try to penetrate: if your church teaches that the law must not permit abortion, and you believe that the law must not permit abortion, then by definition you are mouthing the teachings of your church.

And if you are mouthing the teachings of a church—here was the logic, Backer could see, at its tidily reasoned conclusion—then you have no business interfering in the secular matters of government. The National Association for Repeal of Abortion Laws had from its first months of organization been describing regional right-to-life groups as clumsily dis-guised arms of the Roman Catholic Church; in the 1970 annual NARAL meeting minutes, typed underlining emphasized the strategy suggestion aimed directly at the Church. "Expose the tax-deductible lobbying efforts of Catholics," the NARAL recording secretary wrote. "Point out the fact that hospitals refusing to sterilize people or perform abortions are practic-ing religion on public tax money!"

Practicing religion: plainly that was the essence of it, that every time Dr. Matthias Backer drove down to Jefferson City and talked about the incidence of rubella damage or due process for the unborn, he was, like his Catholic counterparts around the states, "practicing religion"—dispatched by the pope, as it were, to foist papist ideas upon a secular democracy. Was it practicing religion when he gave money to charity, or volunteered his medical services at the local clinic? His church taught the deep importance of charity and community, too. Thou Shalt Not Kill was a basic proposi-tion, in the teachings of the Catholic Church; was it practicing religion when Matt Backer refrained from shooting people in the street? *Just be-cause my church teaches it doesn't make it a religious teaching:* in legisla-tive hearings and debate halls all over the country, Matt Backer and many of his Catholic colleagues tried over and over to make people see that the principle they were defending, the sanctity of human life, was not a Catholic principle but an *ethical* principle, a moral bedrock, solid under all of us.

This conviction, that a church can champion a moral principle so universal that it ought to apply to people of any faith or philosophy, was the most profoundly important driving force for the gathering American movement that called itself Right to Life.

In the mouths of their opponents it was made to sound like nonsense: roomfuls of Catholics, letterheads full of Catholics, all of them insisting they are not promoting a Catholic idea. But it was true. That was what they insisted. In another decade religious people would join the campaigns against abortion with unabashed reliance on ministers' teachings and readings of Scripture, but in the years before 1973, when the groundwork was being laid, men and women who built right-to-life organizations in cities all over the country described themselves the same way Matt Backer did: born to Catholicism, raised in Catholicism, and guided to Right to Life by the plainest possible intersection of medical science and common moral sense.

Backer had learned these things in Catholic school—both medical science, as he would understand and practice it, and common moral sense. He learned long division in Catholic school, too, and the proper use of pronouns, and the history of the American Revolution; when he thought about it afterward it seemed to Backer that there was not very much he *hadn't* learned as a Catholic. Being Catholic was as indelibly a part of him as being a Missourian, it was what he was born to, who he *was,* and when he surveyed the terrain of his childhood neighborhood, the parish of St. Anthony of Padua spread about him as a social and geographic organizing principle, defining school and playground and even the bowling alley where Matt Backer got a summer job setting pins. The bowling alley was ten lanes wide, on St. Anthony's property, in the basement of the girls' school. Backer's two sisters went to the girls' school—he was the middle child, the only boy—and in the mornings Backer walked the same route they did, toward the brick girls' school building and into his own, in his dark trousers and light shirt, the uniform of the St. Anthony's boys.

He lived in South St. Louis, the most thickly Catholic neighborhood in a city already so Catholic that men and women joked about the standard St. Louis getting-to-know-you greeting: What parish are you from? Certainly there were Protestants and Jews who had settled over the generations in St. Louis; since the mid-1800s the city's population had swelled rapidly with the arrival of immigrant European workers and freed slaves from the South. But the city was founded by French Catholic traders, the first schools and university were Catholic, and many of the immigrant Europeans were German Catholics like Backer's grandparents—"Scrubby Dutch," in the parlance of the town, supposedly a reference to diligent wives out washing their white front steps clean once a week.

Backer's mother was not out front every week, but as a boy he understood that he was purebred Scrubby Dutch; all four of his grandparents had immigrated from northern Germany, and Backer's parents still broke into German sometimes when they wanted to talk over the children's heads. His father was a real estate and insurance salesman, a genial man of modest ambition. His mother had worked as a seamstress once, but stayed home with the children while Matt was a boy. Their street was plain and straight, the narrow brick houses lined up with their small patches of grass front and back, and Backer knew the religion of each neighbor family as surely and casually as he knew their last names; for some reason Dakota Street had attracted more non-Catholics than was common around South St. Louis, and on schooldays Matt watched certain neighbor children walk east to the public school while he joined the uniformed crowd heading north toward St. Anthony's. On weekends he and the Protestant children sometimes played together in their fenced backyards or the little grass field at the end of the block, but his closest friends were Catholic, the boys from his class; by the time Matt was old enough to be an altar boy, he had spent so many years with the other altar boys, both in church and in school, that they could launch without warning into the priest-and-response game. They might be lounging around somebody's living room, with their gangly legs and their bad teenage complexions, and somebody would grin and intone *Introibo ad altare Dei,* I will go unto the altar of God. Those were the opening words of a priest at Mass. *Ad Deum qui laetificat juventutem meam,* one of the other boys would answer back, as fast as he could: To God, who gives joy to my youth. The game was in the instant back-and-forth, the priest's recitation and the altar boy's response, without missing a single syllable of the Latin.

Backer was good at it. He could rattle off the syllables one after another while he stood under the shower, or between panting breaths while he was running up the street to church; fifty years later, when he was a grandfather, he would still be able to repeat without prompting the Latin, *Confiteor Deo omnipotenti,* I confess to Almighty God. In parochial school he learned the words and the instructions that accompanied them the way his schoolmates did, in religion class, the first class of the day. The Sisters of St. Joseph taught, in their long black habits, and once a week a priest came in to lecture and answer their questions. They had begun with the Baltimore Catechism, the beginnings of religion class: Lesson One.

> The Purpose of Man's Existence
> 1. Who made you?
>   God made me.
> 2. Did God make all things?
>   Yes, God made all things.
> 3. Why did God make you?
>   God made me to show His goodness and to make me happy with Him in heaven.

Matt Backer made his first communion when he was seven years old, his hair slicked back and his shoes shined black, and throughout his boyhood the nuns and the priests and the immediate world around him taught him the lessons of an observant Catholic boy. A Catholic boy did not eat meat on Fridays. A Catholic boy examined his conscience before making confession to the priest. A Catholic boy raised his hat when passing a Catholic church, and bowed his head reverently whenever the Sacred Name of Jesus was mentioned, and genuflected in the presence of the Blessed Sacrament, the body and blood of Christ in the bread and wine at the altar. Sometimes Matt managed these obligations and sometimes he did not, but by the time he was in high school he understood that these were Church rules, Catholic rules, duties imposed upon *him*.

His neighbors, the Lutheran sign painter next door and the Baptist minister up the street, lived by different sets of rules. His neighbors were permitted to eat meat on Fridays. If he had been interrogated about it Matt Backer supposed he would have said it was a pity that his neighbors were forsaking the comfort and salvation of the Church, but there was no proselytizing in Backer's immediate world; the neighbors had their own religion, and nobody in Backer's house was going to march next door and insist that God meant the sign painter's family to be eating fish or praying in Latin or otherwise attending to the intricacies of what Matt had been taught all his life were the Laws of the Catholic Church.

But Matt Backer had also been taught that there were other laws too, broader laws, the laws of human decency, the universally recognized laws of God. At St. Louis University High School, the Jesuits taught their students ethics, honor, codes of conduct—principles meant for any moral man, as Matt understood it. In his first year of high school one of his teachers began the term by announcing that at the top of every paper, each student was to write out one definition from the teacher's code of conduct, which was written out and passed around the class to ensure that it be properly understood. Integrity, Loyalty, Enthusiasm, Honesty, Manliness, Courage—for the rest of his life Backer would remember the categories, and remember that the definition of "Integrity" included the words "making decisions based on principle." The notion was not Catholic integrity but human integrity, and in Catholic schools some of the strongest principles to be upheld were taught not as religion but as *ethics*, the study of right and wrong in human conduct, also a required course.

One of the most basic of human rights is the claim to life. This is not a positive but a natural right. It is not granted by any state or society, but flows from the nature of a free living being who is entitled to the means necessary to work for the attainment of his ultimate end. Life is the most fundamental natural means to this end. Ultimately, this right to life comes from God, as does everything else which man has.

This paragraph, under the heading "The Right to Life," introduced the eleventh chapter in an ethics textbook written by Vernon J. Bourke, a philosophy professor at St. Louis University during the 1940s, when Matt Backer was an undergraduate. By the time the book was published Backer had left St. Louis University to enlist in the Navy, but it was based on already-taught ethics courses and almost certainly reflects the tenor of discussion that would have swirled around young Catholic college students like Matt. These were complex arguments, drawing not on Scripture but on the kind of point-by-point reasoning used by St. Thomas Aquinas, the greatest theologian in the history of the Church. The chapter subheadings suggested the weight of debate: "Is It Unjust to Kill an Innocent Man?" "Is It Wrong to Kill an Unjust Aggressor?" "Is Capital Punishment Justifiable Morally?" "Is It Just to Kill an Unborn Child?"

Classroom argument and many long written treatises offered lengthy responses to questions like these, but in the textbook they were dispensed with efficiently. It is always wrong to knowingly kill an innocent man because every person has a right to life, and others have a duty to do everything morally possible to promote that right. It may be moral to kill an unjust aggressor because the right of self-defense is part of the right to life, but one must be certain that no other less serious means of defense will suffice. Capital punishment is morally justifiable, as it is justifiable to remove a diseased part of the human body, when a criminal has seriously offended the common good of the community. It is always wrong to kill an unborn child, because the child is an innocent person and thus by definition cannot be thought of as an unjust aggressor.

The word "abortion" appeared only three times in Vernon Bourke's textbook, and if it was uttered aloud in Matt Backer's classrooms it sailed right past him. The United States had entered World War II, and the arguments were about more pressing matters, the definition of just war, the protection of adult human life, the rational defense of one's own religious and philosophical beliefs. Long ethical discussions had really never held Backer's interest anyway; he was a focused student, impatient to graduate and get on with his work, and it was not until his last years of medical school that he was forced to think seriously about the elaborate but rigid moral lessons of the men who had taught him and written his textbooks. *Direct killing of innocent human life is never permissible,* that was one of the essential guiding principles for a Catholic obstetrician, with its special reference to innocence: a life blessed by God, as all human lives are blessed by God, but still too new to have committed any offense in the world.

And as a young physician, understanding the guiding principles but still working at how to apply them, Backer sometimes felt himself pulled between conscience and instinct as he worried about what to do for the women in his care. A patient had cancer of the cervix, when Backer was a resident, but she was three months pregnant. Backer sought out the hospital priest. What were his options? The cancer was confined to a single site, and a hysterectomy would probably get it all out, but wasn't a hysterectomy direct killing of her unborn child? The priest urged Backer to explore every possible alternative, but Backer was as certain as a resident could be; without the operation she would die, leaving her other children motherless.

The verdict, from both priest and Backer's department chairman: perform the hysterectomy and baptize the semiformed baby. "Double effect" was invoked, a principle taught in philosophy classes and immediately applicable here: a person may in good conscience commit a moral evil *if* it is the unavoidable by-product of committing a moral right. The person must not *intend* the moral evil, that was important to double effect, and the reasons for taking this perilous route must be profound indeed.

They were difficult discussions, but for the rest of his life Backer would have no trouble remembering what they were about, and more significantly what they were not about. They were not about what that woman was carrying inside her uterus. Backer *knew* what she was carrying inside her uterus, just as the priest knew it and the department chairman knew it and everyone else appeared to know it amid the surroundings in which Matt Backer had come of age: she was carrying a baby, a second human life, and when Backer did the hysterectomy, it was the principle of double effect that comforted him in the operating room. The direct act he had performed was the removal of a diseased organ. The unintended by-product, the moral evil made unavoidable by this lifesaving operation, was the killing of an unborn child.

Where had he learned this? Where had he *not* learned it? Backer had taken Human Biology during his first year of college. The classes were held in an old laboratory and lecture room deep inside the Basic Sciences building, and there a Father Carroll made his students see that human biology was a wondrous thing: the egg and the sperm, the moment of fertilization, the cells dividing one after another, the small arrested life suspended mutely inside the lined-up shelf jars the color of iced tea. This was not an atmosphere in which a professor of science might stroke his chin and say, Well, life begins at conception, but what do we really *mean* by "life"? Look at the *pictures*, Backer would want to cry later, when the unfriendly men and women in his lecture audiences would start in on the Deep Uncertainties of Life's Beginnings; you can *see* it there, the egg and *sperm*; they *join*, it *begins*.

It would always seem to Backer that if there was a single venue for the source of his understanding about the nature of uterine life, it must have been that classroom. He thought himself an ordinary Catholic, paying respectable if not particularly scholarly attention to the teachings of his faith; he knew from some of his Catholic journals that the Church itself had over the centuries demonstrated some confusion as to the stage in pregnancy at which abortion ought to be considered the equivalent of homicide. The first authoritative collections of canon law, compiled during the twelfth and thirteenth centuries, defined abortion as homicide only after the fetus had been "animated," or infused with a soul. The point at which this ensoulment took place had been the subject of philosophical debate for centuries before that (many early theologians were influenced by the Aristotelian view that the embryo's soul progressed through various stages of development before acquiring the divine nature that distinguished it as human), and well into the 1800s, most papal decrees to address the subject repeated the principle that although abortion was always a moral wrong, it was a sin less grave than murder if it was performed before the fetus had been animated.

It was not until 1869, during the papacy of Pius IX, that official Church doctrine was changed permanently to the teaching Matt Backer had absorbed as young Catholic man: that abortion must be regarded as murder regardless of the stage of pregnancy at which it occurs. Since the scientific understanding of human conception advanced so dramatically during the nineteenth cetury (the mammalian ovum was first identified in 1827, and the fertilization of egg by sperm first observed microscopically and accurately explained during the second half of the century), the Church's evolving position seemed to Backer both sensible and enlightened, further confirmation that "life begins at conception" was a declaration not of religious faith but rather of the demonstrable science he had learned in Father Carroll's biology class. Four times a year Backer received his subscription copy of the Catholic physicians' ethical journal *Linacre Quarterly*, and in the pages of *Linacre* "from conception on" resonated between the lines as a kind of first principle, a starting point so universally understood that there was no need to waste ink articulating it.

How to act upon this principle, as a Catholic physician and an ethical human being—that *Linacre's* theologians and medical writers addressed with vigor. Was it acceptable for a Catholic doctor to abort an unborn child in order to save the child's critically ill mother? No, *Linacre* reported, it was not; the doctor's duty was to do all he could to save both patients without a direct mortal assault on either. Was it permissible to rescue a woman from ectopic pregnancy, the accidental implanting of a fertilized egg inside a fallopian tube? Yes, one *Linacre* article advised, but only with scrupulous care; a ruptured tube was potentially fatal if an ectopic was left alone, and *Linacre* instructed that the tube could be removed with the pregnancy inside, that this was the excising of a diseased organ, a moral act with an unfortunate side effect. But it was not permissible to make an incision and pull the pregnancy from a still-unruptured tube, even if the doctor's intent was to save the tube so that the woman might more easily become pregnant again. That was a direct assault upon the "living, inviable fetus"—the intended destruction of innocent human life.

Catholic hospital physicians are expected to abide by the official Ethical and Religious Directives for Catholic Health Facilities, and Backer knew these intricacies of moral distinction as well as any of his colleagues, although in truth some of them gave him a headache. There were days when he struggled with what he was supposed to do; he could see what made sense to him and what did not. During his first months of practice he was asked to attend temporarily to another doctor's patient, a woman with six children and rheumatic heart disease so severe that it looked to Backer as though she could never survive another pregnancy. This was a lady who needed a tubal ligation desperately, as far as Backer could see, but tubal ligation was sterilization and thus prohibited by the laws of the Church. If the woman had been Backer's patient and come to him in time, he would have sent her to a non-Catholic hospital for a ligation—maybe an immoral act on Backer's part, one step removed, he didn't much care. But she was not his patient. She was devoutly Catholic. She did not have a tubal ligation, and she got pregnant again, and she lived until the baby was six weeks old. Then her heart gave out and she died.

Where was the justice in this? Where was the moral rectitude? Where was God's hand, leaving seven children without their mother because her fallopian tubes were not supposed to be tied off? For a long time Backer felt angry and a little guilty about the heart patient who died; it began to seem to him that Catholic hospitals were obliged by tradition to include in their staffs some self-appointed moralist, the Keeper of the Tubes, whose principal mission was poking into other people's business to make sure everybody else was as holy as he was. By the early 1960s the entire Catholic posture on birth control had begun to trouble Backer, in fact; ever since he was old enough to understand it, he had been taught that artificial contraception was gravely wrong: that it frustrated nature, that it thwarted cooperation with God in His plan for the creation of life, that it blocked the natural and God-given purpose of sexual intimacy, which was the begetting of children by a husband and wife. These were part of the rules for being Catholic, and it was obvious to anybody watching the Backer household that Matt and Laverne Backer had followed the rules. They had never intended to have thirteen children—when they were courting, Backer would say later, they had talked about five or six, a nice crowded Catholic doctor's family.

But they knew the Church teachings about marital relations and the ready welcoming of new life, and Backer's wife, whose own brother was a Jesuit priest, was even more deeply committed to the faith than he was. The two of them—Backer would say this wryly later, as though it were amusing to imagine that it needed saying in the first place—proved to be a spectacularly fertile couple. And Backer was not a man to complain about birth control being philosophically unavailable to him: they were his children, he loved them, he helped bring them into the world, and he and his wife were to put every one except Edward through school and on toward college and families of their own. They shopped at the supermarket four carts at a time, took their vacations in seven- or eight-children shifts, drove very large automobiles, handed down a lot of clothing, and bought loafers instead of tie-ons (fewer laces to help with in the morning)—Matt Backer once calculated that his wife had prepared more than a hundred thousand individual meals, sometimes forty-five in a single day.

Backer had built up a successful obstetrical practice, and he could afford amenities to make this somewhat easier: the housekeeper; the second housekeeper hired solely to iron parochial school uniforms; the custom-built suburban house, its bathrooms equipped with two sinks apiece and its small bedrooms lined up dormitory-style along the long hallways. Still it was staggering work for Laverne, and it grated on Backer when he read theologians explaining how artificial contraception violated the "natural law." He had a cloudy understanding of the notion of natural law, anyway: the eternal law, Backer had been taught this was supposed to mean, the law upon which all else is based, the law that need not be written on paper or tablet to be instinctively and universally understood as truth. "Written in the heart," someone had once said to Backer, trying to explain it, quoting the apostle Paul in the Bible.

But no matter how diligently Backer looked, he could not find written in his heart the image of a married couple sleeping in the same bed without intimate relations if they wanted to wait before having more children. That was "rhythm," as the Catholic theologians explained it, and that was supposed to be morally acceptable. That was a natural use of God's gifts—refraining from sex during a woman's fertile periods, and engaging in sex during what was presumably the sterile period, so as to delay, if a couple felt pressing and ethically defensible needs for doing so, the arrival of new children.

Nothing about this seemed natural to Backer; disciplined, perhaps, but not natural. Birth control pills made the dissonance worse. The first pills arrived on the United States market in 1960, carrying with them the promise of the simplest and most reliable artificial contraception in history, and in 1963 a Boston gynecologist named John Rock wrote an influential little book arguing that birth control pills ought to present Catholics with no greater moral qualms than rhythm. Rock was Catholic, a point he stressed in his opening pages, and his own research in infertility studies had helped prove oral contraceptives' effectiveness in suppressing ovulation. "They provide a natural means of fertility control," Rock wrote, "such as nature uses after ovulation and during pregnancy."

Rock called his book *The Time Has Come,* and Matt Backer agreed that it had: Catholic theologians themselves were arguing over contraception, and by the mid-1960s Backer knew it had become standard practice for a faithful Catholic husband and wife to priest-shop until they found a cleric willing to give them formal dispensation to "follow their own consciences," as the Catholic newspapers delicately phrased it, on the use of birth control pills and other artificial contraception. As a social phenomenon the birth control pill looked less than wonderful to Backer, who could see that women might now imagine themselves free to behave as stupidly about sex as men sometimes did; Backer knew it was old-fashioned, amid the era erupting jubilantly around him, but he believed sex outside marriage was a bad idea made considerably more inviting by the celebrated new pills. Still here they were, he couldn't do anything about that, and women—married women, respectable women, women clearly trying to attend to the welfare of the families they already had—would show up in his office asking for them.

Backer would say: I'm sorry, I can't do that, it's against the teachings of the Church.

The women would say: Are you practicing medicine here, or are you practicing religion?

And he began, finally, to write prescriptions for birth control pills.

It felt odd to Backer at first, because he was unaccustomed to such brazen defiance of the Church, but the women were right: he didn't want to practice religion in his medical office. He could see their objection. It made sense to him. Obliging every one of his patients to bow to the Catholic Church's teaching on contraception—*that* was practicing religion, and a particularly soggy bit of religion it was, too, with the Vatican dithering about and taking years and years to gather itself up for a definitive statement on the morality of birth control pills. When the definitive statement was finally released in July 1968, in the intently anticipated papal encyclical entitled *Humanae Vitae,* Dr. Matthias Backer was one of the many Catholic parishioners who read it with a disappointment bordering on disgust. A husband and wife must understand their mission of "responsible parenthood," Pope Paul VI declared in the lengthy text, but "they must conform their activity to the creative intention of God, expressed in the very nature of marriage and of its acts, and manifested by the constant teaching of the Church."

The creative intention of God, the pope had discerned, permitted rhythm but not the birth control pill. "There are essential differences between the two cases," the encyclical declared. "In the former, the married couple makes legitimate use of a natural disposition; in the latter, they impede the development of natural processes." There were some lengthy passages about the rhythm method giving "proof of a truly and integrally honest love" while artificial contraception might cause the man to "lose respect for the woman," and Backer studied the encyclical from beginning to end in his archdiocesan newspaper, looking for the logic that would convince him the pope was right. He couldn't find it. The ship had *sailed,* that was how Backer would think of it, and all the labored reasoning of *Humanae Vitae* was not going to call it back; there were Catholic women all over the world taking birth control pills in the morning before they went to Mass.

In his office, when his married patients asked for them, Backer went on writing prescriptions for birth control pills. Maybe he was a bad Catholic. Clearly he was ignoring the explicit instructions of the magisterium, the official interpreters of Roman Catholic doctrine. In the confessional Backer generally found a variety of moral failings to unload from his conscience, but helping dispense birth control pills was not one of them; perhaps it was rationalization, perhaps the conscience was more elastic than it should have been, but when he looked at what he was doing he could find no genuine sin to confess.

It was none of his business, that was all. His patients could follow the magisterium or not follow the magisterium. They could go make their own confessions, if their consciences were troubling them; he was not going to lecture them about artificial contraception, he was not going to debate artificial contraception on a public stage, and he was certainly not going to drive to the capital city of his home state and demand that his church's directives on artificial contraception be incorporated into the statute books. The statute books were for everybody, Catholics and non-Catholics alike, and you didn't spend two hours negotiating the highways to Jefferson City, or lose your Thursday evening to the Creve Coeur Township Republican Organization, unless everybody was involved—unless the men in charge of the statute books were tampering with an idea as basic as the protection of human life.

And here, finally, was something Matt Backer found written in the human heart. Safeguarding unborn life was not Catholic business, in any way that Backer understood the special comforts and rigors of Catholicism; it was not denominational, it was not religious at all except in the sweeping sense that all of us are God's creations and made, born or unborn, in His image. Matt Backer happened to believe that, but he was not going to get up and argue about it before an audience. Science was what he would argue, and science was always how he began: "I would like to present some medical evidence," Backer would say, and the house lights would darken, and he would proceed to the first abortion slide. *Making decisions based on principle:* the Jesuits had taught him how to act but not what to say, that was Backer's way of thinking about it, and if debate audiences heard him and Jefferson City legislators heard him it was because his medical evidence was sensible, it was rational, it was universal, it was *right.* It wasn't his fault that Catholics had happened to see it first.

HIS PICTURE ran in the *National Catholic Register,* Backer white-jacketed, looking grim, with a stethoscope around his neck, and behind him an anatomical drawing of an infant in the womb. More letters came in the mail.

From Yardley, Pennsylvania: "We want to encourage you in your fight to prevent liberalization of abortion law in Missouri. We too are fighting it in Pennsylvania."

From Flushing, New York: "Enclosed is my check for $5.00 to aid you in this fight. I'm a New Yorker who has been horrified by the law our State Assembly has passed."

From Klamath Falls, Oregon: "We are very much concerned here with the new change made in our state abortion law where an unborn child can be aborted up to five months and for almost any reason. 'Right to Life,' an organization protecting the right to life of the unborn as well as other people, is fighting this type of legislation."

He wasn't famous, not really; by 1971 there were Matt Backers all over the United States, each with the lecture notes and the photograph collection, traveling from statehouse to debate stage and back again. Big-city newspaper writers knew whom to call when they needed a ready voice for No on Liberalization: some of the No leaders were physicians, some were lawyers, some were nurses or homemakers or businessmen (Edward Golden, the head of the rapidly growing New York Right to Life Committee, was a building contractor), and some of them—too few, in Backer's view, and too meek when they did make a showing—were clerics. In St. Louis, Cardinal John Carberry read for a news conference a joint pastoral letter co-signed by all seven Missouri bishops and intended for Sunday dissemination at every Catholic parish: "A society which legalizes the killing of the unborn is not only violating God's law, but is opening the door to more hideous abuses of life," the letter read.

If the pastoral letter was meant as a solemn declaration to put the matter to rest, it failed; the pace was quickening, and Backer knew it. In March 1971 a repeal bill was introduced in Missouri—a no-categories, no-committees, abortion-through-the-second-trimester bill, like the one that had passed in New York. A citizens' committee had drafted its own version of a repeal bill and pleaded publicly for a legislator to carry it; the representative they found, DeVerne Calloway, was a fifty-five-year-old St. Louis Democrat, and one of the only black women in the Missouri legislature. "Social hypocrisy," DeVerne Calloway wrote, in a letter to the *St. Louis Globe-Democrat*, describing the century-old abortion law she was now proposing to dismantle. "Despite these laws, crimes continue to rise and women increasingly are having abortions."

Backer packed up his notes and got back on the highway to the capitol. At least now he need waste no more breath with his explanations that Model Penal Code–based abortion bills were an open door to abortion on demand; DeVerne Calloway's bill, as Backer read the wording in the folded state handout, *was* abortion on demand. They might as well have written it into the legislative language. Be It Enacted By the General Assembly of the State of Missouri: Abortion on Demand.

No abortion shall be performed in this State unless: 1) Such abortion is performed by a licensed physician, and 2) Such abortion is performed in a licensed hospital. That was the critical sentence in H.B. 650, relegating abortion to the status of a gallbladder operation, and on the evening of the bill's first public hearing, the crowd at the statehouse was so large that the meeting was moved from the modestly sized legislative hearing room into the full House of Representatives chamber. Calloway and her co-sponsors had brought in a lineup of witnesses to defend the bill, and as the witnesses took their turns one by one at the microphone, a Missouri Catholic Conference employee listened and made notes: "She indicated that she was brought up morally oppposed to abortion but does not think that one group has the right to legislate for all." "He claimed that the poor do not want children but that they have neither the funds nor the education to keep from having them." "She claimed that historically there were only practical reasons why abortion statutes were written—i.e. surgical conditions were so poorly developed at one time as to make abortion extremely hazardous. Thus laws were written to prohibit them."

These were familiar arguments already, for anyone who had spent the last three years culling useful material from the debates around the country. They were beginning to be old arguments, in fact, and it worried Backer to hear them again in the halls of his own state legislature; it was unimaginable to him that the cautious state of Missouri might accept into its statute books an idea as radical as abortion on demand, but he knew that if you hammer at anything long enough it begins to show cracks. Perhaps the legislators would hear them all out, the witnesses for and the witnesses against, and then retreat to their offices to settle on something that looked to them like compromise: *some* permissible killing of innocent human life.

And Backer knew this as well: the legislative chamber of a capitol building, at least in theory the democratic center of every American state, was no longer the only place where the conflict might finally be won or lost. In the first months of winter, five months before the introduction of the Calloway repeal bill, the attorney general for the state of Missouri had called Backer's office to enlist the doctor's defense against an entirely different sort of attack: a lawsuit, in St. Louis County Circuit Court, contending that the abortion law of the state of Missouri violated the individual protections guaranteed by the United States Constitution.

The suit was called *Samuel U. Rodgers, M.D.. v. John C. Danforth, Attorney General of the State of Missouri.* Its full formal name was longer than that: four physicians, three clergymen, and three women "of childbearing age," in the language of *Rodgers v. Danforth,* were bringing a complaint against Danforth and St. Louis County prosecuting attorney Gene McNary. There was no personal reference in the lawsuit to either Danforth or McNary; they were named solely as individuals charged with enforcing state law—just as Henry Wade, the district attorney for the county of Dallas, had been named the defendant in a similar suit in Texas. And if the courts approved it, Danforth had explained to Backer, a separate third party might be permitted to enter such a lawsuit too, as a *"guardian ad litem,"* the court-appointed adult representative for an infant whose own interests might be at issue in a legal dispute.

They needed both Backer's testimony, Danforth said, and the use of his name. Would he enter *Rodgers* v. *Danforth* on behalf of a class of infants—"children presently in existence but unborn," as the legal papers would phrase it—who could be understood by the judicial system to be arguing, through Backer, that they were human persons who possessed a constitutional right to life?

Certainly, Backer said. Of course he would.

Thus M. H. Backer, Jr., M.D., temporary legal guardian for a fictitious American fetus whose official name in the Missouri court papers was Intervenor Defendant Infant Doe, joined in early 1971 the great funnel of abortion litigation aimed at the United States Supreme Court.

FIVE

Exhibit A
1970

ONE EVENING Judy and Art Widdicombe, who had wondered whether
police listened in on the conversations that took place over the back porch
telephone, decided to look around for wiretaps. They felt beneath the
telephone and looked under the desk and studied the interior baseboards
and corners of the screened-in porch, and then they found flashlights and
went outside and shone their lights up and down the telephone wires,
wondering whether a tap was something a person could see while squinting up from the backyard lawn.

They never found anything. But they knew they were being observed.
At occasional intervals an unmarked car cruised slowly up and down their
street, its windows rolled shut, its occupant making no effort to identify
himself. There was one two-day scare that frightened Judy terribly at
first, a message passed to her one day at the hospital: urgent, call your
mother-in-law, and Judy's mother-in-law whispering into the telephone,
"The sheriff's office called, they have something they want to serve you,"
and Judy frantically calling Gwyn Harvey, and Gwyn Harvey calling the
county coroner, because the county coroner had access to arrest warrants,
and the county coroner telling Gwyn he would find out what he could but
that in the meantime the Widdicombes must get out of the county at once.
Judy and Art secured their sons at a friend's house and retreated across
the county line, to spend the night at Gwyn's—"trying to figure out,"
Judy recalls, "who was the sonofabitch that turned us in."

Finally the Widdicombes went home, seeing that there was no point lurking beyond the county boundaries, and waited for the call from the sheriff's office. When the call came Art answered the telephone, and as he handed the receiver to Judy she was startled to see that he was smiling. "It's okay," Art said. "The guy wants your *services*. His girlfriend's daughter is pregnant."

The Clergy Service accommodated the young woman, collectively violating the Missouri criminal code one more time—collectively, according to the defining paragraph of Section 559.100, risking up to five years' imprisonment for offering advice "with intent to produce or promote a miscarriage or abortion." The word "advises" appeared twice in the manslaughter-abortion statute, alongside "gives," "sells," "administers," and "procures," and a separate state law listed items that might be seized by police upon issuance of a warrant:

Any of the following articles, kept for the purpose of being sold, given away or otherwise distributed or circulated, contrary to law, *viz.*, pills, powders, medicines, drugs or nostrums, or instruments or other articles or devices for producing or procuring abortion or miscarriage, or other indecent or immoral use, or any letters, handbills, cards, circulars, books, pamphlets, advertisements or notices of any kind describing or purporting to describe any of such articles, or giving information, directly or indirectly, when, where, how, or of whom any of such things can be obtained.

The wording of the statutes had not been changed to account for the fact that abortion was now legal under some circumstances in certain states outside Missouri; inside Missouri it was still a felony, and although nothing in Missouri law prohibited women from crossing the border to engage in an act that was lawful elsewhere, handing over the telephone number of an out-of-state abortion doctor could surely qualify as advising with intent to promote. By mid-1970 three hundred women per month were coming through the Clergy Service referral system, and even after New York opened up, the counselors continued to make occasional referrals to efficient illegal abortionists who were closer or cheaper than the physicians so busily opening up shop in Manhattan. Women drove up the highway to the Corpsman in Chicago, or flew down to Mexico City with Dr. P——'s telephone number penciled discreetly, as per Clergy Service instruction, amid the other numbers in their address books: "Some of the women have been asked at the border when coming back if they have been to Mexico to see Dr. P——," Gwyn Harvey warned the Clergy Service counselors in a December 1970 memo. "Needless to say, the girls 'didn't know who he was.' "

Lapel ribbons, for the Mexico abortion patients; that was one of the details Missouri Clergy Service counselors were expected to remember, that every woman traveling to Mexico for an illegal abortion was to pin on a small white ribbon so the driver could pick her out at the Mexico City airport. Also the New York clinics had to be paid by traveler's check, money order, or cashier's check, no cash, please; also the California doctors would not accept personal checks, please ignore the rumor that the Los Angeles doctor took Mastercharge; also be sure to remind the women how to handle an airplane ticket, ask for a boarding pass, write down the number of the departure gate. Sometimes the women were casual and competent at matters like these, but more often it seemed to Judy that they were not; they were frightened and young and had never been on an airplane before, and they were leaving the state alone because it cost too much to have somebody come along to hold their hands. In the college town of Columbia, where the Clergy Service counselors worked out of the University of Missouri ecumenical center, the local coordinator had prepared herself for students and was startled to see all the others who came in too, the professional women in tailored clothes and the girls from the farming country in cars that bumped awkwardly onto campus with unsmiling relatives squeezed into the back seats.

"I have an image of sort of a big car that has rust on it, and the car is full," recalls the coordinator, Georgie Gatch. "I remember one where there literally was an uncle and a grandfather and a grandmother. I mean, the whole family came up. And they did not have a lot of the defenses a more educated person does. They would say she was pregnant, or she was in the family way—they usually pretty much knew the timing—and that there was a problem, and we want to get rid of the pregnancy. There was very little hesitancy about an abortion. They knew what they wanted to do. *How* to do it was very difficult."

There was a humorless little counselors' joke, Georgie Gatch recalls: Give them the information and have them swallow the piece of paper. As a layperson, Gatch probably had no claim even to the uncertain protection of priest-penitent confidentiality when she volunteered illegal information in a Clergy Service session, but it was not the threat of arrest that troubled her most. The entire scenario was wrong, Gatch thought, especially when the farm families sat before her; these were girls for whom the drive to Columbia was an excursion to the big city, and here was a stranger on a college campus handing them a telephone number for Chicago or New York or a place where people answered the telephone in Spanish. Once a month Judy Widdicombe would buy an airline ticket for herself and board Flight 468, the first TWA jet of the day from St. Louis to LaGuardia Airport in New York, to link up with the morning's collection of abortion-bound women—walking the beat, as Judy saw it, making sure the connections were working and the clinics were managing their pickups correctly. And even as Judy talked to the uneasy women at the boarding gates, she knew that at least these women were waiting for airplanes, that they were not sitting alone in Greyhound stations in the midst of an all-night bus trip. That was what the clergy-counseled women did when they could not afford the airplane fare: they climbed aboard interstate buses or they cadged car rides or they scrabbled together gasoline money and drove, west or northeast or south toward Mexico, fifteen hours at a stretch if they had to, as long as it took to reach an abortionist the Clergy Service ministers had promised them was reliable and safe.

That the routes were now marked, that they were no longer so secret, that some of them led to real clinics with physicians' names on the front doors and expensive medical equipment inside—this was a kind of progress, as Judy saw it, and the newsletters from the National Association for Repeal of Abortion Laws were beginning to fill column after column with roundups of promising statehouse action around the country. "*Arizona's* bill lifting all restrictions against abortion performed by licensed physicians won House approval (32–37), cleared the Senate Judiciary Committee (6–3), but was bottled up in the Senate Rules Committee," one NARAL summary reported during the summer of 1970. "An ALI-type reform law in *Kansas,* enacted more than a year ago, went into effect July 1 . . . *Pennsylvania* has a repeal bill filed. . . . Rep. John Galbraith of *Ohio* has filed a beautiful repealer that simply strikes out all reference to abortion from the criminal code."

But when NARAL grouped its brisk roundups under subheadings, "Near-repeal Laws" and "New ALI-type Laws" and "Legislative Defeats," the Defeats column was three times the length of the others. Among the modest number of states that had succeeded in some modification of their abortion laws, only New York and California were opening their doors to significant numbers of out-of-state women. Missouri showed no sign of wavering at all. And forward-thinking legislation was a form of government action that could always be undone; the New York Assembly was already facing intense lobbying from right-to-life groups demanding that the state's new abortion law be repealed at once. Some ground had been gained, since the era of medical articles with titles like "Foreign Bodies Lost in the Pelvis During Attempted Abortion," but from Judy's watch the gain sometimes appeared modest indeed; every time she came home on one of those TWA flights from New York, she spent some portion of the flight in the back of the plane, pointing out for the stewardesses the seating location of the passengers who were recovering from abortions.

"Do you know what to do if one of them starts to hemorrhage?" Judy would ask, to the stewardesses' alarm. Then Judy would balance herself back there, beside the metal cupboards full of Coca-Cola and miniature liquor bottles, and show them: lay the woman on her back, pack clean towels in to stanch the bleeding, press both palms against the pelvis to massage the uterus from above. As far as Judy knew there were no actual calls for this form of first aid on any of the New York–St. Louis flights, but the point was in the lesson itself, alerting the stewardesses, making them more solicitous of the pale young women who were flying over the eastern half of the United States just to return home from the doctor.

Lobbying, political organizing, the redrafting of criminal codes—this was fine democratic work, for a person who believed that nineteenth-century state abortion laws no longer made any sense, but it was not enough. The city of St. Louis was still a place in which a man from the sheriff's office had to seek abortion help from volunteers who were breaking the law to oblige him. From Judy's perspective all the vigorous state-house effort had in fact produced only two good states, New York and California, and beyond that a mess of state-by-state fine print that required a fifty-item checklist just to remember which states were saying yes for rape but no for mental health, and which were saying yes for mental health and forcible rape but no for statutory rape, and which demanded one consultant or two consultants or a three-member "therapeutic abortion board." Judy was a nurse, not a lawyer or a lobbyist, but the checklist itself was enough to convince almost anybody that the campaign to change the country's abortion statutes was going to have to be run by separate assault teams: one to work the legislatures, one to commit civil disobedience, and one—this was how the civil rights activists appeared to have managed *their* successes, a generation earlier—to challenge the statutes in court.

In Missouri the lawsuit was filed in May 1970, in United States District Court, and again in October 1970, in the Circuit Court of St. Louis County. Gwyndolyn Harvey, identified in the legal papers as vice-president of the Clergy Consultation Service of Missouri and "a married woman of child-bearing age," was named as one of the fourteen plaintiffs.

"Plaintiffs Harvey, Halverstadt and Steinbach are disturbed and distressed over the possibility of increasing their family size at an unplanned rate," read the opening plaintiffs' brief in *Rodgers* v. *Danforth*. "They are keenly aware of the frequency of contraceptive failure and the inability to obtain abortion in the State of Missouri. . . . Plaintiffs Harvey, Halverstadt and Steinbach have an inalienable right to decide for themselves whether or not to bear children and this decision is one in which the state has no legitimate interest."

There were many arguments in the plaintiffs' brief, and multiple affidavits, lengthy legal footnotes, and references to the First, Fourth, Fifth, Eighth, Ninth, and Fourteenth Amendments to the U.S. Constitution—it was a long brief, 120 pages, and by the time it was filed, so much abortion litigation was already under way that new cases like *Rodgers* were taking on a cut-and-paste quality: one of these arguments, one of those, two of those. But there resonated in this one small passage of *Rodgers* v. *Danforth* a particular urgency for Judy Widdicombe, who was, during those opening months of Missouri litigation, a married woman of childbearing age who made regular, careful use of a contraceptive diaphragm. The diaphragm had failed. She was pregnant.

• • •

IN 1970 THE BUSINESS of the St. Louis County courts was conducted inside a severe-looking six-story brick-and-concrete building that appeared to spread in many directions at once through the commercial center of Clayton, which is one of the suburbs abutting the boundaries of the city of St. Louis. The courts complex was less than a year old at that time, and each floor overhung the slightly smaller floor below it, a design that was probably meant to look modern but instead gave the building a mass at once hulking and unbalanced, as though the entire structure had been lowered upside down onto a bare city block. A narrow escalator carried the citizenry up from the lobby to a small bank of elevators, and up this courthouse escalator, one morning in early October, a twenty-nine-year-old attorney named Frank Susman carried the opening *Rodgers* v. *Danforth* papers toward the fifth-floor reception counter of the Clerk of the Circuit Court.

Susman was a trim young man, taller than average, with a mustache and wire glasses and thick wavy hair that was somehow going to make him look even younger and more earnest when the newspapers began running photographs of his serious face. He was married, not happily (the marriage was about to end in divorce), to a woman he had met while he was an undergraduate at Brandeis. He had an infant son. During the early years of his marriage Susman had served for a while as an Army Reserves drill sergeant, teaching bayonet and hand-to-hand combat, and when he happened to mention this to people afterward he was amused to hear them say that they found it either very difficult or very easy to imagine; he had a way of addressing people abruptly, as though he was still snapping off orders to nervous recruits, but amid the more experienced lawyers at his father's law firm, Susman still spent most of his working hours as the low-man junior apprentice, taking busywork assignments to earn his place.

He did divorces, corporate work, research on cases he was too much the novice to argue himself. The firm was in downtown St. Louis, and when Susman had to pull from casebooks he walked over to the civil courts building library, where the ceilings were thirty feet high and the long tables lit from the ends by low glass-shaded lamps. All the codebooks were here, on the shelves of the library, the criminal statutes and the taxation rules and the dense, dry guidelines that are meant to direct and proscribe the life of citizens of an American state. Here too were the reported opinions of every appellate court in the country, and in *those* volumes, the hardbound somber-colored collections with names as plain as monograms, U.S., Fed. Rep., Cal. App., S.W.2d—within those pages a person might find the most exhilarating commotion of conflict and drama and clashing ideas. That was supposed to be the purpose of legal argument, the resolution of conflict in courts of law, and recorded amid the small conflicts were the written proceedings of the great ones. What are the protections of the United States Constitution? What is the relationship between legislatures and the courts? How far may a state go in controlling its citizens' behavior? What is the meaning of "liberty," of "due process," of "equal protection under the laws"?

Frank Susman, four years out of Washington University law school and eight months shy of his thirtieth birthday, was delivering to the St. Louis County Circuit Court in October 1970 a stapled set of typewritten pages on which every one of these questions, implicitly or explicitly, had been addressed. The formal name for the papers was PETITION FOR DECLARATORY JUDGMENT AND INJUNCTIVE RELIEF, and that was how it was written, in capital letters, with Frank Susman's name right below that, also in capitals. There were three other lawyers' names on the cover sheet with Susman's, but his was the first; in oral argument and the newspaper clippings it was going to be Frank Susman's case, and his western Missouri co-counsel, a more seasoned Kansas City lawyer named Charlotte Thayer, had concluded when she met Susman that this was a young man who was not much interested in splitting either the work or the public credit. "He didn't take to suggestions kindly," Thayer recalls. "He obviously was out to make his reputation on this."

For more than a year before the filing of the first Missouri abortion suit, Frank Susman had been volunteering during some of his off-hours as legal adviser, a title that sounded somewhat more official than the work itself proved to warrant, to the Clergy Consultation Service of Missouri. Susman had met Gwyn Harvey and Judy Widdicombe as they were planning the Clergy Service, and when he examined for them the wording of the state abortion statute, it looked to Susman as though for legal purposes the women, and all their pastoral co-conspirators, were essentially flying blind.

"If we got caught, we were going to worry about it then," Susman recalls. "I'm convinced that we did not have a legal defense." There was something darkly appealing about this, for a lawyer just starting out; Susman had never before been prompted to think one way or the other about abortion, but the statute seemed to him so barbaric and stupid, forcing a woman to be pregnant when she didn't want to be, that he liked the idea of helping women set out to defy it.

His parents—his mother had died when he was a teenager—had been political liberals, Jewish, only remotely religious. His father had joined in the preparation of *Shelley v. Kraemer*, a famous 1948 U.S. Supreme Court case that struck down as unenforceable in state courts the written covenants meant to ensure that only white people could buy homes in certain neighborhoods. Frank Susman had studied that Supreme Court ruling in his own constitutional law texts, and like most law students he had learned case by case and cause by cause the potential power of the planned and politically focused lawsuit. A lawsuit could function as a counterweight, in a democratic society; under the right circumstances a lawsuit could work as a written reminder that what the voting majority and its elected politicians want is not necessarily right or just. A properly argued lawsuit could kill an entire state statute if a judge could be convinced that the statute was unconstitutional—that it denied to individual citizens some basic and profoundly important right that was written into the U.S. Constitution.

Because the Constitution was composed in 1787, and amended only occasionally during the two centuries that followed, many generations of contentious legal minds had been put to work trying to deduce exactly what the language of the document itself might mean—how the words as written might be applied to more modern problems, and also what ideas might be understood to be implicit in those words. And as Frank Susman grew increasingly interested in his work with the Clergy Service, attending regional conferences and studying legal news from around the country, he learned that attorneys in more than a dozen other states were trying to find in the language of the Constitution a *right* to legal abortion, a right important enough to make judges strike down abortion statutes no matter what state legislators thought about the merits of the laws.

In New York, in the midst of the heated statehouse debate on the proposed new abortion statute, four separate teams of lawyers had brought lawsuits in federal court to attack the existing law as unconstitutional. In Georgia, where the legislature had voted in a limited-categories Model Penal Code law, the local ACLU director had tapped an Atlanta lawyer (who was herself on pregnancy leave, preparing for the birth of her first child) to lead a federal court challenge arguing that even the revised law was unconstitutional. In Wisconsin, two lawyers were using a constitutional challenge to defend a Milwaukee doctor who had been arrested for performing an illegal abortion; in South Dakota, a physician was challenging on constitutional grounds his arrest for performing an illegal abortion; and in Austin, where University of Texas graduate students at the local alternative newspaper had helped organize a clandestine abortion referral network, two attorneys even younger than Susman had volunteered to prepare a constitutional case against the Texas abortion statute.

As a national effort this experimental new litigation was haphazard at first, with no central coordination more formal than long-distance telephone calls and back-and-forth mailings of clippings and sample legal complaints. The American Civil Liberties Union had taken its strong position on abortion law two years earlier (after more than a decade of internal argument, the national ACLU board had finally approved in late 1967 a resolution proclaiming women's right "to have a termination of pregnancy prior to the viability of the fetus"), and when Susman began his research in early 1970, he was directed across the state to Charlotte Thayer's Kansas City law office. Thayer was fifteen years Susman's senior and had co-founded the ACLU's Western Missouri chapter, but she had no personal experience with abortion either; she had been asked some months back to host at her home a retired judge and ACLU board member who wanted to talk about the principle of a right to an abortion, and twenty-five years later Thayer would still remember the way the women in her living room had sat silently after the judge's talk, afraid to look at each other or say anything aloud. Then one woman had put her hands together, slowly, beginning to clap, and a second woman had joined her, and then all the women in the room were clapping so vigorously that it seemed to Thayer that relief had washed over the room, that the judge had said things all of them believed without even realizing they believed them. After that Thayer had studied the background papers and commenced some public speaking of her own: the abortion decision is properly made by a woman and her doctor, Thayer would say; the abortion decision is a matter of personal privacy; the abortion decision is not the business of the state.

Frank Susman knew this argument reasonably well by 1970, and he happened to agree with it. But by itself any argument could be dazzling in its eloquence and useless before a court of law. Judges were supposed to defer to the wisdom of legislators, unless the Constitution dictated otherwise; that was central to the constitutional system, to ensure, at least in theory, that judges kept to themselves their private opinions of laws written and adopted by the people's elected representatives. A judge could be personally appalled by the Missouri abortion statute, could think it ridiculous that women should be risking post-abortion hemorrhage on the TWA flight home from New York, and still declare that he had no authority to second-guess the legislators who were keeping that statute on the books. The authority, if it existed at all, was going to have to come from the Constitution, from those eighteenth- and nineteenth-century words and the centuries' subsequent interpretations of them.

So Frank Susman—like Margie Pitts Hames in Atlanta, and Sarah Weddington and Linda Coffee in Austin and Dallas, and a score of attorneys in separate law offices across the United States—set about constructing a brief to help a judge find that authority.

He began, as lawyers do, with the work that had gone before him. Had any court in the nation explicitly decided that language in the Constitution could be read to promise women a right to legal abortion?

The courts of California gave Susman his first answer. The case was called *People* v. *Belous,* indicating by its title that someone had been accused of a crime and was fighting the People, the state of California. The facts in the *Belous* case were laid out at the beginning of the opinion: Leon Phillip Belous, a Los Angeles obstetrician who had occasionally held forth on television and in newspaper letters-to-the-editor with his views that criminal abortion laws were too restrictive, had been telephoned in his office in 1966 by a distraught young couple who had seen Belous on television. The young man and woman were unmarried, she was pregnant, and when Belous told them there was nothing he could do for them, they told him they were desperate and would go for a risky Tijuana abortion if he refused to help. Belous finally examined the young woman at his office, and when he declined either to perform an abortion or to give them an abortionist's name, they pleaded with him again, insisting, in the words of the ruling, that the young woman would have her abortion "one way or the other."

Finally Belous gave them a telephone number for an abortionist, a physician licensed in Mexico but unlicensed in California, where the physician had recently taken up residence and begun what Belous had testified appeared to be a competent and relatively safe practice in illegal abortion. The couple made their arrangements and went to the doctor's makeshift apartment office, and while the young woman was recovering from her abortion, police raided the apartment. The abortionist was arrested, Belous's name was found in a notebook, and after the abortionist pleaded guilty, Belous was convicted by a jury of violating the state abortion law —the then-applicable law, which defined as felonies both abortion and conspiracy to commit abortion. He was fined $5,000, placed on probation for two years, and stripped of his medical license.

From those unlikely circumstances—a physician convicted for passing along a telephone number—had emerged a statewide legal battle that climaxed in 1969 in the California Supreme Court. Procedurally it was a complicated case, because by the time *People* v. *Belous* went up on appeal, the California legislature had already revised the state law under which Belous was convicted; by 1969 the Model Penal Code language was in place in California and already being interpreted loosely enough to attract abortion patients from other states. But Belous had links to the Committee for Therapeutic Abortion, the first California professionals' organization to agitate for abortion law change, and as his conviction was appealed, a powerful lineup of legal and medical talent pitched in for a defense intended not only to exonerate Belous by finding the old California law unconstitutional, but also to argue that the new Model Penal Code–based "reform" law had proved an inadequate remedy.

One brief alone, a friend-of-the-court argument prepared for the California Supreme Court, was signed by 178 medical school deans, including department chairmen and the heads of every medical school in the state. A second brief was signed by four law school professors and more than a dozen other attorneys, including the president of the California State Bar. There were opposing briefs, to be sure—"The Right and Duty of the State to Protect Fetal Life," began one friend-of-the-court brief joined by a long list of clerics and physicians—and by the time the *Belous* case reached the California Supreme Court in 1969 it had escalated into a full-scale duel over the legal and philosophical merits of restricting abortion in any way under the language of a state criminal code.

Because the California Supreme Court of the 1960s was regarded by many legal observers as the most influential state supreme court in the nation, its rulings frequently looked to as examples of how other courts might proceed on a given issue, the outcome of the *Belous* appeal was destined to resonate a long way beyond the boundaries of California. And when the Supreme Court ruled in favor of Belous—by a four to three vote, the state justices reversed his conviction—the writing of the opinion was assigned to a liberal justice named Donald Peters, who happened to have among his legal clerks that year a thirty-two-year-old woman with very strong ideas about abortion and women's rights.

Here was detail Frank Susman could never have learned from the printed language there in a law library in St. Louis, but it left a deep enough imprint on the shaping of the law that twenty years later, from her Superior Court judgeship in Los Angeles, that onetime law clerk could say without hesitation, in response to an inquiry, "I don't need you to refresh my memory—it was the most important thing I ever did." Her name was Abby Soven, and she was a Brooklyn native who had graduated near the top of her University of Southern California law school class and gone on to this prestigious clerkship with the state supreme court. She had watched the *Belous* case with extremely keen interest—the women on the court staff that year had all paid close attention to the appeal—and when the opinion was assigned to her justice, who was known for delegating to his clerks a great deal of the actual writing, Soven asked outright for the assignment. She had never undergone an abortion herself, but she had a vivid memory of having asked her physician what would happen if she were ever to want one; the physician had told her she would have to go to Mexico, and even then, in the late 1950s, Soven thought this outrageous. Some years later she had lent her Southern California apartment to a friend who had arranged to have an illegal abortion inside, and Soven remembered too the disquieting hours of waiting outside the apartment for the abortionist to finish his work. "I remember being very upset that things should be done that way," she recalls. "It struck me as being degrading and risky. I don't think I really thought it through that there ought to be a right."

Now, with the *Belous* opinion assigned to her, Abby Soven could think through the question of rights. When she examined the facts of the case, it was obvious at once that the statute they were considering was no longer valid, but Soven dismissed any worries about that; there was a larger question at stake, and if the opinion was worded right she could have a shot at answering it. "This is the *opportunity*," she recalls of her own excitement at the case. "I have just graduated from law school. I'm a woman. I'm a liberal. And *this is the issue*. This is the opportunity to say, coming from the most respected state supreme court in the United States, bar none, that the woman has a fundamental right and the state can't impinge on it. And no court had *done* it."

The legal briefs on behalf of Leon Belous—particularly the friend-of-the-court briefs written with an eye toward the larger issues at stake—had articulated the argument that Soven imagined would prove central in the resolution of *People* v. *Belous:* the old abortion law violated both the state and federal constitutions because, as one of the briefs put it, the statute

> drastically interferes with a woman's right to control the use of her own body. . . . There may be a thousand medical reasons why the fertilized ovum should not be allowed to develop into a child, and why the woman's very strong choice might be against allowing such development, but the wisdom of the medical profession and the woman's own right of decision are nullified by the prohibitory California statute. As a result the woman is required, whether she likes it or not, to use her body as a baby factory from the moment of conception until the moment of childbirth.

But a court clerk is not a justice, and there are rules to be observed: the marching orders, the principles on which the opinion is to be founded, are agreed upon by the majority's judges. And when Soven reported to Justice Peters for the details of her assignment, he gave her perplexing news. "He said, 'We have four votes for vagueness,' " she recalls. "I said, 'You must be kidding. *Vagueness?*' And he said, 'That's what we got four votes for. Now go write it.' "

"Vagueness" was a doctrine courts had developed as part of the general constitutional principle that citizens are entitled to fair treatment under the laws, which requires, at a minimum, understanding what it is that a given law prohibits. A statute could be overturned on grounds of vagueness, the U.S. Supreme Court had ruled many years earlier, if persons of "common intelligence must necessarily guess at its meaning and differ as to its application." Now Soven was being told to write an opinion explaining why the old California abortion law was unconstitutionally vague—which seemed to her both preposterous and ultimately self-defeating, since all it promised to produce was a more carefully written law that kept abortion illegal.

She had her orders. She had done her research. She had once read a local newspaper columnist define legal precedent as one court taking in another court's laundry, which struck Soven as a useful metaphor, each court examining what the previous court had placed into the bundle. "And that was when I decided to pack it," Soven recalls. "I was really resolved to give them a *big* bunch of laundry."

Into the theory that Soven didn't believe—it had never seemed to her an exceptionally vague law, it said abortion was illegal "unless the same is necessary to preserve her life," and moreover the law that contained that line wasn't even in the statute books any more—the law clerk packed the theory she did believe. We have no standard definition of "necessary," we have no enlightening definition of "preserve," we have no clarification on how immediate is death supposed to be, et cetera—all this went into the opinion, explaining vagueness, why the law was impermissibly unclear. Then, startlingly, twenty-four paragraphs into the text:

> Moreover, a definition requiring certainty of death would work an invalid abridgment of the woman's constitutional rights. The rights involved in the instant case are the woman's rights to life and to choose whether to bear children. The woman's right to life is involved because childbirth involves risks of death.
>
> The fundamental right of the woman to choose whether to bear children follows from the Supreme Court's and this court's repeated acknowledgment of a "right to privacy" or "liberty" in matters related to marriage, family, and sex. That such a right is not enumerated in either the United States or California Constitutions is no impediment to the existence of the right. It is not surprising that none of the parties who have filed briefs in this case have disputed the existence of this fundamental right.

The language kept appearing, crimson shirts amid the plain white sheets, all the way through the *People* v. *Belous* Void for Vagueness opinion. *The great and direct infringements of constitutional rights . . . an invalid infringement upon the woman's constitutional rights . . . the pregnant woman's right to life takes precedence over any interest the state may have in the unborn.* At first Soven wrote it into footnotes, but her senior clerk at the court had taught her, no, either put it in the opinion or take it out entirely. So she put it in, and even when her drafts were reworked by others, the constitutional rights language stayed. "I knew *exactly* what we were doing, and what I wanted this to be used for," Soven recalls. And when it was done and published she was quiet about it, watching to see what would happen. "Just sit back," Soven remembers, "and wait for the apples to fall."

The ruling was issued in September 1969, and the apples began falling almost at once. "This is the first State Supreme Court decision in the history of the nation which declares unconstitutional any anti-abortion statute," wrote Keith P. Russell, a Los Angeles obstetrician who had helped found the California Committee for Therapeutic Abortion, in an elated letter to his fellow signers of the medical school deans' friend-of-the-court brief. "It is a landmark decision. . . . In voiding the pre-1967 anti-abortion law, the California Supreme Court has pointed the way for approximately 40 states with virtually identical laws, allowing abortion *only* to save the life of the woman, to have their laws declared unconstitutional, or to reform such laws by legislative action. Legal attacks have already begun in other states."

The legal attacks were gathering both in state courts and in federal courts, and on November 10 a U.S. District Court judge in Washington, D.C., delivered the next surprise: because it was impermissibly vague, the judge had declared unconstitutional the felony abortion law of the District of Columbia, which had been challenged by an indicted physician and permitted abortion only when "necessary for the preservation of the mother's life or health." The ruling, pending a Supreme Court appeal, made Washington, D.C., at least temporarily the only jurisdiction in the United States with no statutory restriction on abortions by physicians. And in St. Louis, during the early months of 1970, Frank Susman studied the printed opinion in *Belous* and led off his complaint, the causes of action under which the Missouri abortion law might be found unconstitutional, with vagueness.

1. *Said sections 542.380, 559.100 and 563.300 of the Missouri Revised Statutes, 1959, as amended, challenged herein, are unconstitutionally vague and uncertain on their face.* But that was not nearly enough. If a judge was going to be asked to stretch, to read between the lines and fit together new ideas, it was important to offer him considerable stretching room. Missouri judges were not like California judges, to understate the situation almost comically, and Susman assumed he was going to have to mention, at least in passing, every argument that had so far been raised in the abortion briefs prepared by his colleagues in other states.

2. *Said statutes, challenged herein, on their face and as applied, deprive plaintiffs . . . of the right to privacy in the physician-patient relationship, which is protected by the First, Fourth, Fifth, Ninth and Fourteenth Amendments to the United States Constitution.* Privacy: that was number two. The *Belous* opinion had talked about privacy, even as it was supposed to be talking about vagueness. The privacy cases cited in *Belous*, most prominently the U.S. Supreme Court decision called *Griswold v. Connecticut*, were by the close of the 1960s being examined by abortion lawyers as another promising route to the overturning of abortion laws. A former Supreme Court justice named Tom C. Clark, who had resigned from the Court in 1967, wrote a 1969 law journal article exploring the developing doctrine of the right of privacy and its possible application to abortion law challenges. And although Clark's article was not exactly a call to arms (he concluded by arguing that trying to change abortion law case by case in the courts would be "slow, expensive, and possibly disastrous"), it was read in some quarters as a signal from an extremely well-placed source that the Supreme Court might consider this an approach worth pursuing: state prohibitions on abortion as impermissible encroachments on the personal privacy guaranteed by the Constitution.

The very idea that the Constitution contained such a guarantee—that constitutional protections encompassed, as the *Belous* opinion had put it, "a 'right to privacy' or 'liberty' in matters related to marriage, family, and sex"—was in itself less than a decade old. The phrase "right to privacy" had been introduced into legal discourse more than eighty years earlier, but for many years it was invoked principally to describe a person's right to fend off unwanted intrusion by other persons—overly aggressive newspaper reporters, for example, or advertisers trying to use a personal photograph in an advertisement without the subject's permission. It was not until 1965, in a case involving the sale and use of contraceptives, that a Supreme Court ruling first gave the word "privacy" the meaning that now appeared to hold such interesting potential for abortion law challenges like the one Frank Susman was trying to assemble. "Would we allow the police to search the sacred precincts of marital bedrooms for telltale signs of the use of contraceptives?" Justice William O. Douglas wrote for the majority in that 1965 case, *Griswold* v. *Connecticut:*

> The very idea is repulsive to the notions of privacy surrounding the marriage relationship.
> We deal with a right of privacy older than the Bill of Rights—older than our political parties, older than our school system. Marriage is a coming together for better or for worse, hopefully enduring, and intimate to the degree of being sacred. It is an association that promotes a way of life, not causes; a harmony in living, not political faiths; a bilateral loyalty, not commercial or social projects. Yet it is an association for as noble a purpose as any involved in our prior decisions.

The Supreme Court's ruling in *Griswold* v. *Connecticut* was the culmination of a long campaign to permit the use of birth control in a state whose criminal code prohibited it. Before 1965 the Connecticut birth control law was among the most restrictive in the United States; it banned not only the distribution and purchase of contraceptive devices, but also their use: any person who used a contraceptive in Connecticut was guilty of a felony. It was, as one of the *Griswold* Supreme Court opinions was to observe, an "uncommonly silly" law; condoms could be purchased legally under the fiction that they were to be used for the prevention of disease, and it was widely understood that women could sometimes obtain contraceptives surreptitiously from private physicians.

But the Planned Parenthood League of Connecticut could not by law perform what was in those years the principal function of any Planned Parenthood clinic, which was to provide women with birth control devices.* For some years before 1965 Planned Parenthood had worked in

---

* During the late 1930s Planned Parenthood did dispense contraceptives at a few Connecticut clinics, but in 1940, when the Connecticut Supreme Court upheld the state anticontraception law, the contraceptives service ended for the next twenty-five years. A detailed account of the early Connecticut legal battles and their impact on subsequent cases appears in David J. Garrow's *Liberty and Sexuality: The Right to Privacy and the Making of Roe v. Wade.*

Connecticut as an odd combination of lobbying organization and jitney service, making futile feints at the legislature and ferrying station wagons full of women across state lines to be fitted for diaphragms. The shuttle runs had been started up by Estelle Griswold, the League's energetic executive director, who had seen both the practical and public relations merits in driving carloads of grown women into New York and Rhode Island, where artificial contraception was legal. "You could really publicize that in your pamphlets," Griswold recalled in an oral history many years later. "I had a drawing made of a group of women in a car, waving flags, and going toward Port Chester [New York]."

It was Estelle Griswold, according to legal legend, who first brought together two Yale professors with a common distaste for the Connecticut birth control law: Lee Buxton, the recently appointed obstetrics and gynecology chair at the university's medical school, and Fowler Harper, a popular law school professor who was known as an outspoken liberal. Catherine Roraback, a Connecticut attorney who worked on the case with the Planned Parenthood League, recalls that the collaboration was said to have begun over Griswold's martinis. "She was notorious for her martinis," Roraback recalls. "She put Lee Buxton and Fowler Harper in touch with each other—gave them each a drink, as the story went, and left them to their conversation."

The product of that 1957 cocktail party conversation was a constitutional attack seven years in the making. The Supreme Court had never before addressed what appeared to be the central question presented by the situation in Connecticut: did a state legislature, presumably expressing democratically the will of the community at large, have the right to criminalize the use of birth control by married couples? Or did the U.S. Constitution contain some overarching protection that prevented states from enforcing laws which intruded so directly on the privacy of married couples? *Was* there a personal "right of privacy" that in certain situations outweighed the state's right to control the behavior of its citizens?

It took the Connecticut attorneys two separate tries to bring these questions directly to the attention of the U.S. Supreme Court. Their first lawsuit, brought on behalf of Dr. Buxton and of two married couples complaining that the anticontraception law kept them from receiving proper medical treatment, was dismissed by the Court on the grounds that since the law was almost never enforced, there existed, in legal terminology, no "controversy"—no actual circumstances warranting Supreme Court intervention. But the ruling was a close one, five to four, and in their dissenting opinions Justices William O. Douglas and John Marshall Harlan argued that the Court should have taken on the married couples' complaints and that the Constitution did protect what Harlan called "their right to enjoy the privacy of their marital relations free of the enquiry of the criminal law."

Even though the framers of the Constitution never wrote the word "privacy" into their written guarantees of individual protection, Harlan argued in his long and influential dissent in the first Connecticut birth control case, *Poe* v. *Ullman,* the *idea* of a right to personal privacy in intimate decisions involving marriage and family was implicit in various Supreme Court rulings dating back to the early part of the century. In the 1920s, for example, the Court had ruled that states could not intrude into parents' relations with their children by forcing the children to attend public rather than private or parochial schools. In the 1940s, the Court had ruled that the right to marry and procreate was so "very basic" that states could not require the sterilization of convicted felons. These and other rulings could be drawn together, Harlan argued, to prove that marital privacy was indeed a right protected by the Constitution—a qualified right, to be sure ("I would not suggest that adultery, homosexuality, fornication and incest are immune from criminal enquiry, however privately practiced," Harlan wrote), but a right nonetheless.

Five years after that encouraging but ultimately fruitless set of Supreme Court opinions, the Connecticut birth control advocates came back to the Court for their second try. As setup for the new challenge, Estelle Griswold and Lee Buxton had opened a well-publicized contraceptive-dispensing clinic in New Haven and cooperated eagerly in their arrest ten days after opening day. Now there was an indisputable legal controversy, two individuals arrested and convicted of violating the Connecticut law (Griswold and Buxton were fined $100 each), and in June 1965 the Supreme Court handed down the ruling that formally gave new constitutional meaning to the term "right of privacy." The Connecticut law was unconstitutional, Justice Douglas wrote for the majority, because it intruded upon

> the zone of privacy created by several fundamental constitutional guarantees. . . . We do not sit as a super-legislature to determine the wisdom, need, and propriety of laws that touch economic problems, business affairs, or social conditions. This law, however, operates directly on an intimate relation of husband and wife and their physician's role in one aspect of that relation.

Douglas' majority opinion in *Griswold* was to become famous among many generations of law students chiefly for the elegant and somewhat mysterious vocabulary it used to identify the places in which this "zone of privacy" could be located in the Constitution. The Constitution's Bill of Rights, Douglas wrote, contained certain "penumbras, formed by emanations from those guarantees that help give them life and substance." And what exactly was a penumbra? Apparently it was a semishadow, one law review author wrote after the ruling, having consulted his *Oxford English Dictionary:* the "partially shaded region around the shadow of an opaque body, where only a part of the light from the luminous body is cut off." Like a shadow, in other words, Douglas' "zone of privacy" was a kind of shimmer reflected off brighter and more directly visible portions of the Bill of Rights: the First Amendment, for example, which guaranteed freedom of association and privacy from government intrusion.

Douglas concluded his brief lead opinion with a rousing tribute to marriage (". . . a coming together for better or for worse, hopefully enduring, and intimate to the degree of being sacred"); and in a more businesslike concurring opinion, Justice Arthur Goldberg elaborated on the applicable protections of the Ninth Amendment, which declares that the people retain rights *besides* those explicitly and specifically named in the Constitution. "Although the Constitution does not speak in so many words of the right of privacy in marriage, I cannot believe that it offers these fundamental rights no protection," Goldberg wrote. "The fact that no particular provision of the Constitution explicitly forbids the state from disrupting the traditional relation of the family—a relation as old and as fundamental as our entire civilization—surely does not show that the Government was meant to have the power to do so."

Legal scholars had exceedingly mixed reactions to the *Griswold* ruling during the first years after its release (Douglas' "penumbra" metaphor was subject to particularly biting attack), but the principle was now set: the Constitution protected a "zone of privacy" that included the right of married people to use birth control. *Married* was a crucial part of the *Griswold* ruling; the opinions made repeated reference to the sanctity of the husband-wife relationship and said nothing about the state's power to keep unmarried people from buying or using contraceptives.

But among those interested both in birth control and abortion, the potential next steps were compelling to imagine. In June 1968, the *North Carolina Law Review* published a long article about possible constitutional routes to the dismantling of state abortion laws, and privacy led the list:

> The values implicit in the Bill of Rights suggest that the decision to bear or not bear a child is a fundamental individual right not subject to legislative abridgment—particularly in light of *Griswold v. Connecticut.* . . . It is an anomaly that a woman has absolute control over her personal reproductive capacities so long as she can successfully utilize contraceptives but that she forfeits this right when contraception fails. Clearly no government is permitted to compel the coming together of the egg and spermatozoon. Why then should the state sanctify the two cells after they have come together and accord them, over the woman's objection, all the rights of a human being *in esse?*

The author of that law review article was a twenty-seven-year-old University of Alabama law professor named Roy Lucas, who in 1968 was working hard to establish a name for himself as one of the nation's leading scholars on abortion law. Lucas was an ambitious young man, possessed in those years of such an earnest passion that in both demeanor and physical appearance he reminded Uta Landy, the German-born woman who became his first wife, of pictures she had seen of the young Robert Kennedy. They had met while Lucas was traveling in Scotland, on a fellowship break from his studies at New York University Law School, and Landy remembers that even during the early months of their courtship they talked at length about social issues and the pressing importance of changing abortion laws.

"He had known a woman who needed an abortion during his previous student years," Landy recalls. "I believe he accompanied her to Puerto Rico, and felt that it was absolute travesty that women had to go through this. He believed women should be able to have an abortion like any other surgical procedure, that first of all, it should not be illegal, and secondly, it shouldn't be a big deal. He would say to me that human beings are meant to control their lives." *

Landy herself was uneasy about abortion as they argued; she had been raised in an intellectual German Catholicism, and she still had a feeling that a person had no right to interfere with a pregnancy that had apparently been ordained by fate. But Lucas would prod her, teasing her when she went to Mass, quoting the great philosophers and calling the Catholic Church the "thought police." By the time they came back to New York, so that Lucas could finish his final year at NYU before accepting a teaching job in Alabama, he was fired with excitement about his coming career, Landy recalls, and had ideas of one day becoming a Supreme Court justice. He began making contact with some of the older attorneys and physicians who had formed the Association for the Study of Abortion. And he had begun to envision his *North Carolina Law Review* article, which Lucas had written in rougher form as a student essay in law school, as the first offering in a sequence of materials that would bring a major abortion law test case before the U.S. Supreme Court.

In subsequent years Roy Lucas was to become a controversial figure in abortion law circles, alienating many of his colleagues during increasingly bitter struggles over money and public recognition; and the earliest unpleasantness appears to have begun with that article. Some abortion lawyers insisted afterward that Lucas' long and carefully annotated study was the principal early roadmap to constitutional attack on state abortion laws, while others, particularly some of the women, would tell interviewers

* Roy Lucas left the field of abortion law by the mid-1980s and subsequently lost touch with his former legal colleagues; even his ex-wife Uta Landy, at the time of this interview, had not heard from him in years and said she did not know where he was. But David Garrow, who located Lucas and interviewed him for *Liberty and Sexuality*, reports that the young woman in Lucas' abortion story was Lucas' girlfriend, and that the relationship, in Lucas' words, "ended because of this. . . . [I] felt very bad about it" —*Liberty and Sexuality*, p. 336.

that they never even read Lucas' piece as they were working up their cases. But Frank Susman certainly read it. By the end of 1969 Roy Lucas had moved back to New York and was hustling foundation money to start up a full-time abortion law concern, and with a weighty institutional name, the James Madison Constitutional Law Institute, he had begun sending sample briefs to attorneys all over the country. The reactions to Lucas' offerings were mixed; Margie Pitts Hames, who had begun working in Atlanta on the Georgia case, was referred to the James Madison Law Institute (rather brusquely, she recalled in an interview before her death in 1994) when she asked the ACLU national office for help. So she contacted Lucas, and what she received in the mail was some medical articles and a copy of Lucas' law review piece.

"And a bill for twenty-five dollars," Hames recalled. "And I said, Well, screw this. It wasn't any help at all."

Frank Susman had a different feeling about the Lucas material; when he studied it, at his St. Louis law office, the suggestions in Lucas' writings looked to Susman like perfectly respectable arguments. Here was an attorney who had already looked up scores of cases, read a lot of law review articles, and thought through a variety of different ideas that might be offered to the federal courts, starting with Privacy and moving on to Vagueness, Equal Protection, Discrimination Against the Poor. "In all candor, I'm not sure there was that much research in our briefs, as opposed to a great deal of plagiarism," Susman recalls. "Roy Lucas had already *done* it. There was no point in reinventing the wheel."

Into Susman and Thayer's opening complaint went Causes of Action Three through Thirteen—"throwing dice," as Susman recalls, trying anything that might make sense to a judge. The abortion law deprived doctors of their right to practice proper medicine. The abortion law deprived women of Eighth Amendment guarantees against cruel and unusual punishment by forcing them to bear every child conceived. The abortion law violated guarantees to equal protection by discriminating, with no compelling justification, against different classes of people. The abortion law infringed upon freedom of speech, freedom of religious belief, due process of law, and—a nod here to the insistent law clerk at the California Supreme Court—"the fundamental right of a woman to determine for herself whether to bear children."

But plausible arguments by themselves are not enough to bring a legal case, as the *Griswold* lawyers had learned; the case needs plausible people, too, the parties who are said—in what is sometimes a fiction created for the purposes of a lawsuit—to be approaching the court with a problem that affects them directly and immediately. Finding the right people had for some time been a sticking point in abortion litigation. Arrested abortionists were one source of lawsuit; state statutes could be challenged as unconstitutional when they had been invoked to arrest doctors like Leon Belous in California, or Milan Vuitch, the Washington,

D.C., gynecologist whose indictment had led at least temporarily to the overturning of the District of Columbia's abortion statute. For some months abortion activists like Larry Lader had been casting about, to no avail, for a physician willing to invite arrest deliberately by publicly arranging an in-hospital abortion that would defy the law even while meeting the highest possible medical standards. Paul Shively, the San Francisco obstetrician whose legal difficulties following 1966 rubella abortions had galvanized medical support around the country, had been approached early on by a wealthy Missouri industrialist who had become a passionate convert to the causes of legal abortion and population control; Shively recalls that the industrialist offered Shively and his lawyer complete financial support—every penny their lengthy defense might cost—if Shively would demand a criminal trial and thus escalate the state medical board's charges against him into a public full-scale legal challenge of state abortion laws.

Shively and his San Francisco colleagues, who were eventually cleared of the charges against them, all declined the offer; a life's career was too vast a gamble. And in these early years only one other American doctor came forward in a sacrificial public flouting of the law: Jane Hodgson, a fifty-five-year-old obstetrician from St. Paul, Minnesota, who deliberately notified legal authorities before perfoming a hospital rubella abortion, technically illegal under the Minnesota abortion statute, on a woman who already had three young children at home. Hodgson, a lean and plain-spoken former Mayo Clinic doctor who had been the first woman elected president of the Minnesota Society of Obstetricians and Gynecologists, had grown weary of what seemed to her the damnable hypocrisy of respectable doctors—herself included—who turned down abortion patients and then shrugged away responsibility when the same women turned up infected by street abortionists. She was indicted in May 1970, with her criminal trial set for that fall; from New York, at his Madison Law Institute, Roy Lucas was helping prepare her defense.

In the wake of Hodgson's indictment, no other physician rushed to join her in some fraternal open defiance of *his* state's abortion law. Hodgson, should she be found guilty of violating the Minnesota abortion law, faced possible jail time and the loss of her medical license. But there was a second option that did not require arrest: an individual or group of individuals could approach a judge and ask for a "declaratory judgment," a pronouncement on the validity of a particular statute. Or they could ask for "declaratory judgment and injunctive relief"—pronounce the statute unconstitutional, Your Honor, and then issue an injunction, a judicial order to prevent the state from enforcing it.

Plainly this was the simpler and safer route for any lawyer who wanted to test the law without imperiling some doctor's future. In Texas and Georgia, where the lawyers Sarah Weddington, Linda Coffee, and Margie Pitts Hames were working up their own lawsuits, women had seemed the obvious choice as lead names in the declaratory judgment suits; the first hurdle in bringing a declaratory judgment is convincing the judge that the plaintiff, the person bringing the lawsuit, has what the law calls "standing," which means the person must be directly and seriously burdened by the statute in the first place. And Weddington and Hames had been drawn to the abortion issue by women's rights activists; it was not physicians' rights that interested them most. If a lawyer wanted to do what Abby Soven had done, to find in the promises of the Constitution that "fundamental right of a woman," then the plaintiff ought logically to be a woman, preferably a pregnant woman, a woman who could seriously contend that this very statute, at this very moment, was depriving her of the right to obtain an abortion within her own home state.

In Texas, Weddington and Coffee were introduced to a pregnant young waitress named Norma McCorvey, unmarried and already the mother of one child, who brought them a convincing account of the economic and emotional hardship a second child would cause; they promised her anonymity, gave her the pseudonym Jane Roe, and prepared simultaneous complaints describing the dilemmas both of Jane Roe and of a second woman who was married and suffered from a neurochemical disorder that had led her doctor to advise against both pregnancy and birth control pills.* In Georgia, Hames and her Atlanta legal colleagues found a pregnant young woman named Sandra Bensing, also unmarried, and rejected for abortion by a hospital committee that was following the Model Penal Code restrictions of the new Georgia statute; the Atlanta lawyers gave their plaintiff the pseudonym Mary Doe and joined this name to a list of individuals and organizations challenging Georgia's abortion law. The New York and Ohio and Kentucky cases used pregnant women too, along with physicians and ministers and a social worker; the Illinois case used heads of medical school obstetrics departments as well as women seeking abortions; and New Jersey's was to combine both doctors and women affiliated with women's organizations, so that the consolidated name of the case, *Y.W.C.A.* v. *Kugler*, would convey at first glance the interesting if slightly misleading impression that the Young Women's Christian Association (Princeton branch) was suing the New Jersey attorney general for the right to legal abortion.

Frank Susman and Charlotte Thayer led their case with doctors. They wanted to bring the case in U.S. District Court, where most of the abortion cases were headed; at the close of the 1960s federally appointed judges were widely perceived as far more sympathetic to individual rights than most state judges, and the abortion challenges were going principally to three-judge District Court panels, each one formed specifically to hear a

---

* Three weeks after filing their original complaints in the case that would come to be called *Roe* v. *Wade*, the Texas attorneys requested and were granted permission to add to their list of plaintiffs a Dallas-area physician named James Hubert Hallford, who was facing prosecution for performing abortions in his medical office.

particular case. Now summoned only rarely, these three-judge panels had for many years been convened, in a salute to the gravity of the matter at hand, when a state statute was being challenged under the U.S. Constitution. Each panel held two District Court judges and a U.S. Circuit Court of Appeals judge, and Susman was familiar enough with the St. Louis regional lineup to know they would have a far better shot at a sympathetic hearing if they filed their suit in Kansas City. So their lead doctor was Samuel U. Rodgers, a veteran Kansas City obstetrician with teaching appointments at the universities of both Kansas and Missouri.

Rodgers was black, and deeply respected in the Kansas City black community; he and a handful of other black doctors had pressured the city during pre-integration days into including obstetrics and gynecology training at the city's Negro hospital, and many years later a major health clinic was to be named after him. He had handled his share of medical complications from criminal or self-induced abortions, and he was sympathetic to the aims of Susman's lawsuit. But Rodgers was under no illusions about why he had been asked to step to the head of the plaintiffs' list. "I wouldn't have been asked to be on it if it had been easier to find somebody," he recalls. "It's as if the big-time doctors didn't want any part of it."

Following Rodgers on the complaint were three other Missouri physicians, three representative Missouri women, one rabbi, and one Protestant minister. Under the title *Samuel U. Rodgers, et al., v. John C. Danforth, Attorney General of the State of Missouri,* Susman and Thayer filed in May 1970 a formal complaint asking the U.S. District Court for the Western District of Missouri for a declaratory judgment and a permanent injunction against the enforcement of the Missouri abortion law.

Four months later, the District Court threw it out.

Interpretation of state law belonged in state courts first, the *Rodgers* three-judge panel ruled: Susman and Thayer would have to bring their case through the state system. It was gloomy news. Other federal panels had assessed the judicial mandate differently; *Roe* v. *Wade,* the Texas case, had already won in U.S. District Court, and *Doe* v. *Bolton,* the Georgia case, had emerged with a largely favorable ruling from its District Court round too, which meant—since these three-judge panel rulings could be appealed straight to the top, bypassing the Court of Appeals step—that both cases were probably headed directly for the U.S. Supreme Court. But *Rodgers* v. *Danforth* was going to have to start again at the local level, so Susman brought the case back to the big St. Louis County Court building in Clayton—the only St. Louis–area jurisdiction, Susman knew, where he was likely to draw the right kind of judge.

He got him. *Rodgers* v. *Danforth* was assigned in October 1970 to Herbert Lasky, circuit judge for the county of St. Louis. "A co-religionist," Susman recalls, with a small grin: Lasky was the only Jewish judge on the bench, and looked to Susman like a possible liberal abortion vote. Now the legal papers began to stack up in Lasky's chambers, and copies were delivered to Susman's office one by one. Motion from the District Attorney's office that the case be dismissed outright: motion denied. Motion for emergency judicial order, because of the urgent situation of all women needing legal abortions, temporarily voiding the state abortion law while the case was being heard: motion denied. Motion to let into the case another party entirely, one M. H. Backer, Jr., M.D., on behalf of Infant Doe "and all other unborn children": motion granted.

Susman was nonplussed at the introduction of Backer into his abortion case. The notion of asking for *guardians ad litem* had spread around the country among attorneys who were trying to defend their states' abortion laws, and for public relations purposes alone it was a clever maneuver, suggesting that the fetus required the same kind of grownup protection the court might afford an abandoned child. The phrase "to defend the fetus," with its hints of threat to a helpless little person, appeared in every newspaper story about Backer's *guardian ad litem* appointment. And in May 1971, as *Rodgers* v. *Danforth* was being readied for its hearing before Judge Lasky, attorneys for the state presented the court with Exhibit A: Matt Backer, in sixteen pages of text with a sixty-item medical bibliography, expounding for the Circuit Court on the nature of life before birth.

From conception on the child is a rapidly growing organism, Exhibit A began. *By the end of the seventh week, we see a well-proportioned small scale baby.* Susman read with mounting uneasiness: what was he supposed to *do* with this?

When one views the present state of medical science, we find that the artificial distinction between born and unborn has vanished. Was he expected to argue these assertions in a court of law? *The ears are formed by seven weeks and are specific, and may resemble a family pattern. . . . By the beginning of the ninth week, the baby moves spontaneously without being touched. . . . The whole thrust of medicine is in support of the notion that the child in its mother is a distinct individual in need of the most diligent study and care.*

Strategically, their big abortion case only a week away from the scheduled day of argument, Susman and Thayer were trapped. What they wanted from the court was "summary judgment," a ruling based solely on the information in the briefs already delivered to the judge. Summary judgment would mean no actual trial need take place, no witnesses, no examination and cross-examination, no arguing over important facts. If they wanted to attack the veracity of Backer's exhibit they would probably end up going to trial, and if they went to trial Susman would have to come up with witnesses to rebut propositions like *from conception on the child is a rapidly growing organism.* How the hell was he supposed to do that? Bring in a scientist to argue about brain waves? Bring in a metaphysician to argue about what brain waves *meant,* or the conflicting definitions of individualism? It might take months just to figure out what a lineup like that was supposed to look like, much less how to make useful sense of each witness' testimony. Frank Susman had been practicing law for three years, but he had yet to try his first case, and he was not about to start with this. Courtroom trials were supposed to examine who was right and who was wrong, Susman thought, not some philosophy-class quagmire like the moment when human life begins.

So Susman and Thayer, pushing for their summary judgment, took a small wild gamble in the county circuit court: they stipulated as to the truth of Exhibit A.

A stipulation is an agreement, for legal purposes: okay, accepted, now let's get on with the matter at hand. In legal disputes stipulating can save time and effort; the stipulation is a way of declaring to the court that both sides concede to a particular fact so that nobody needs to argue about it. Stipulating was the surest way to avoid a trial, as far as Susman could see: it *was* a gamble, but so was the entire enterprise, marching into court with a lot of constitutional theories that some imaginative lawyers had worked up in their heads. "Come now plaintiffs," Susman dictated, in the weighty Shakespearean language of the court, "and for the purpose of the pending motion . . . and for those purposes alone hereby stipulate as to the truth and accuracy of the Affidavit and attached Exhibit A."

On May 20, 1971, in his courtroom in the Clayton county building, with no jurors present and a court reporter transcribing the proceedings, the Honorable Herbert Lasky conducted his hearing in *Rodgers v. Danforth*.

"May it please the Court," Frank Susman began:

> Miss Charlotte Thayer and myself appear here this morning not only, really, on behalf of the ten plaintiffs in this cause, who are four physicians, three college men, three individual women of child bearing age and married, all residents of the State of Missouri. But in many ways I think it can honestly be said we appear here this morning in behalf of thousands of others, not the least of which, for example, are the three hundred fifty some-odd women who have every month . . . and who continue every month, to obtain abortions, through the local services of a social welfare agency in the State of Missouri known as the Clergy Consultation Service of Missouri, plus the hundreds of other women who obviously go through different channels. . . .

He strode right into the Exhibit A stipulation, and delivered a nice rhetorical flourish or two:

> There have been some suggestions that there are facts in dispute as to when life begins. In effect, there is no dispute. Medical and scientific

knowledge do not differ. Plaintiffs are more than willing to concede that the fetus is a living organism. Plants are living organisms. Other animals, egg and sperm in their own right, prior to conception, are living organisms. . . . The problem of what is in dispute is merely a question of nomenclature. What do you call life? Are you talking about spiritual life? No evidentiary hearing before this court could ever help answer the question of when spiritual life begins.

No other court in the abortion cases around the country had entertained serious legal argument about when life does or does not begin, Susman argued, and surely this court did not wish to become the first. "I have serious worries that in many cases it would resemble what is now an infamous case, that of the Scopes trial," Susman said. "And what this Court and what the participants would do would be to inherit, unfortunately, more than the wind."

Then Susman expounded on what he hoped were some of his real arguments: the statute was vague, the statute denied freedom of speech to those who wished to advise on abortion, a fetus possessed no assertable rights under law. He listened to the other side's attorneys: one for the county prosecutor's office, one for the attorney general's office, and two—one of them worked as legal counsel to the Missouri Catholic Conference—arguing before the court, at some length and with repeated references to "unborn children" and "the right to live," on behalf of the fictitious Infant Doe.

When they were finished, Charlotte Thayer got up for a heated last word. "Surely none of us are supposed to be so absurd that we would claim this fetus does not have the *potentiality* of being born," she said, sounding angrier as she warmed up:

> I would point to the defendant intervenors' brief and mention, on page twelve thereof, the statement that this is a tiny human being, independent, as though he were lying in a crib with a blanket wrapped around him, instead of his mother. How absolutely absurd! This is no independent existing person. This is a human life that could *mature* into an individual nine months after conception. We haven't removed the issue as to the right of a mother to determine whether this life that lies within her should mature into a human being. We have not destroyed the issue of the question of abortion by agreeing to this. We think we have simplified the issue, and gotten out sociological arguments, and gotten down to the true facts.

It was the Constitution that the court should be considering, Thayer argued, not philosophy or sociology or spiritual supposition. "And I will tell you," said Charlotte Thayer, who was that morning the only lawyer in Judge Lasky's courtroom capable of becoming pregnant, "that I am pretty bothered when we have people in the Legislature sitting down there drawing statutes, and our doctors are supposed to be bound by nonprofessionals telling them how to practice medicine, and by them telling a woman how she is going to make use of her own body."

The judge thanked her, pronounced the hearing finished, and took *Rodgers* v. *Danforth* under submission. Now all they could do was wait.

WHILE THESE arguments had begun to take shape, these large and important debates about implicit rights and evidentiary procedure and the shaping of constitutional law, Judy Widdicombe was for a brief and dismaying time a pregnant woman in the city of St. Louis. "The universe," she would say many years later, remembering, "works in strange ways."

They had not wanted another child—not then, and probably not ever, although Judy had thought about this sometimes, what it might be like to have one more baby. Art had talked some months earlier about having a vasectomy, but he had never gotten around to scheduling the procedure, and when Judy's pregnancy test came back positive, she smoldered for a while: if he had *made* the damn appointment they would not be sitting in the kitchen with their heads in their hands trying to make up their minds what to do. Judy loved raising children; she truly believed she was good at it, the nurturing, the holding, and the people around her did too. Some months back the Widdicombes had befriended a neighborhood teenager who was having trouble with his own divorced parents, and the boy ended up moving into the middle bedroom at the Widdicombes', an informal foster child. Then a pregnant girl had found her way through the Clergy Service to Judy and Art's house; after the girl decided against abortion and put the baby up for adoption, she moved into the Widdicombes' back bedroom for a while, a second informal foster child.

So now there were six of them living together in the suburban St. Louis house, Judy and Art and the Widdicombe boys and the two young boarders, and it was lively and loud, with the good-natured din of teenagers slamming in and out. But they were stretched, Art working his newspaper route and Judy working what amounted to one shift for the hospital and an unpaid second shift for the Clergy Service.

And they had a medical problem that was likely to complicate any future childbirth: an A-O blood incompatibility between Judy and Art had forced both their sons to undergo total blood transfusions shortly after birth. Judy's doctor had advised her, after her second son was delivered three weeks early by induced labor and transfused almost immediately, that if she got pregnant again they might have to induce labor even earlier and take on the medical risks of a seriously premature baby.

"So we looked at all of that," Judy Widdicombe recalls. "We looked at where we were. We looked at what the quality of our life was like. It was good. And we decided to go to Chicago."

Thus on a summer morning in 1970, one day after Art Widdicombe had his vasectomy, so that all the way up he sat on a small inflatable cushion and winced with every jostle of the Ford, Judy and Art drove three hundred miles to Chicago to break the law in the office of the Corpsman. The midwife Martha Vinyard could have accommodated Judy locally, but her procedures cost too much and took three days start to finish; a New York doctor might have worked Judy into his schedule, but the New York law was just cranking up amid a chaos of overbooking. Judy knew the Corpsman would take care of her quickly. She knew precisely what the Corpsman would do, because she had watched him do it, and when they arrived in Chicago Judy took out the telephone number and repeated step by step the coded signals of her own underground: call the number, ask for Jimmy, wait for the callback.

She felt like a fool. She was pregnant by accident. She was a *nurse,* she *ran* this illegal system, she sat there every week dispensing sage medical advice about birth control and responsible sexuality and how to avoid unplanned pregnancy in the future blah blah blah, and here was the Mafia car to drive her zigzag through the streets of Chicago so as to disorient her before delivering her to an abortionist. It was a medical office this time, not an apartment building; the Corpsman appeared to have come up in the world, some physician allowing him into professional quarters off-hours, and when Art and Judy rang the doorbell they were let in by a woman who smiled, unself-consciously, as though she was about to ask for the insurance forms. Judy followed her into a procedure room. A methodical calm had come over her, the narrowed nurse's eye, *I am observing this and can explain medically its every step,* and when the Corpsman came in they greeted each other cordially and quickly so that he could get down to business. Judy undressed and lay on the table. Art stood by her head. She asked him how his vasectomy was feeling. A fierce deep pain shot up inside her and she pressed her hands to her abdomen: *I knew this, I knew this, he doesn't use cervical anesthetic.*

"Breathe," the Corpsman said.

Judy breathed. When he scraped she could hear the sound of the curette, grating, as though it was sawing at something gritty. The pain was terrible, tighter than labor, sucked to a searing center that moved with the scraping curette, right horn of the uterus, left horn, Judy could envision the anatomical drawing as though she were moving a pointer back and forth for her nursing students. At least labor pain came and went. Every woman she sent up here had to lie back and endure this, all those women, their feet in the steel stirrups and this amazing pain, what must it feel like when an incompetent person did it?

Breathe.

When he was finished the pain stopped roaring and Judy sat up on the table. The Corpsman left the room. She climbed off the table to steady herself, put her clothes back on, and followed the Corpsman into the other procedure room. She felt obliged for some reason to remain calm and wholly lucid, to keep behaving in the manner expected of a medical professional. She assisted, nurse at the procedure table, at the next two abortions.

Afterward they went out to dinner in a restaurant and Judy fainted in the bathroom. It was not much of a faint; she just felt an urgent need to lie down and when she opened her eyes she saw that she was on her back looking up at the ceiling. When she went back to the dinner table she told Art that she felt better. They spent the night in Chicago and drove to St. Louis the next day.

The Corpsman never charged them for the abortion.

Art would make a mild joke about that, a long time afterward: that was the only Clergy Service payoff I ever got, a $250 abortion for free. He and Judy talked to each other during the drive home, small conversation, careful, Judy trying to see how she felt. What was it that she had lost? Already the word *potential* had carried her through some tremulous moments and she found she was holding hard to it now, the death of a certain potential, a fantasy erased. She and her vasectomied husband would have no more children. For a while Judy would slump into something like grief, a mourning not for a person but for an idea, a shadow, a not-yet person. But she never wavered when she talked about it. This was a decision we made: that was how she would describe her abortion, her voice almost always firm, and as the months went by that word too became essential to the lexicon, *decision*, responsible decision making, private decisions. It was right to have made the decision, Judy would say; what was wrong was the way they were compelled to behave once they had made it. Code names, unmarked station wagons, anonymous telephone numbers—it was as though Judy Widdicombe had stepped from the pages of *Rodgers* v. *Danforth* or *Roe* v. *Wade*, skulking around the Midwestern states in the exercise of a woman's Fundamental Right to Determine for Herself Whether to Bear Children.

Judy didn't need a constitutional law professor to write some learned definition of "fundamental" for her, either. You could look it up in the dictionary. *Basic. Underlying. Indispensable. Essential part.*

ON JUNE 7, 1971, three weeks after taking *Rodgers* v. *Danforth* under submission, Judge Herbert Lasky made his ruling. It was three pages long. It gave everything, the entire case, to Susman and Thayer.

"Broadest abortion ruling in the United States—*ever*," Susman recalled two decades later, with evident pleasure. He was in two-week Army summer camp when the news came over the radio: without elaboration or any detailed explanation of his reasoning, Lasky had ruled that the Missouri abortion law was unconstitutional for every reason Susman and Thayer had offered up. The law was vague, the law advanced no compelling state interest, the law deprived women of their rights, the law violated five different amendments of the Constitution—it was as though Lasky had decided to dispense with an opinion and to hand down instead a telegram that read *Yes.*

The appeal would go straight to the state supreme court, a prospect so dispiriting that Susman could see in the conservative seven-judge lineup perhaps one potential sympathetic vote. But there was bigger excitement outside Missouri, where U.S. District Court cases were churning their way through the system as the list of abortion statute challenges began to grow. From one judicial panel to another, the resolution of the cases varied as sharply as personal conviction on the morality of abortion: in Wisconsin a court found the state statute unconstitutional, an impermissible violation of "a woman's right to refuse to carry an embryo during the early months of pregnancy." In Louisiana a three-judge panel upheld the state abortion statute, declaring in the majority opinion that contraception and abortion were wholly different matters—"the first contemplates the creation of a new human organism," the opinion read, "but the latter contemplates the destruction of such an organism already created." Contradictory opinions issued from courts in Illinois, Kentucky, North Carolina, Ohio. The U.S. Supreme Court rattled both sides at once by delivering in April 1971 its ruling on the District of Columbia's abortion statute: the statute was *not* unconstitutionally vague and thus must still be enforced, the Court ruled, but prosecutors should remember that the statute permitted abortions for "the preservation of the mother's life or health."

And "health," the Supreme Court ruled in its District of Columbia case, *United States* v. *Vuitch,* was a term of considerable flexibility. "The general usage and modern understanding of the word . . . includes *psychological as well as physical* [emphasis added] well-being," read the majority opinion by Justice Hugo Black. By general consensus the Court therefore managed to uphold abortion practice in Washington, D.C., already an active business in local hospitals and doctors' offices, while preserving in theory the statutory law that restricted it. The justices signaled, furthermore, that they did not intend to stop there: in May 1971, exactly one month after *Vuitch* was handed down, the Supreme Court announced that it would consider the Georgia and Texas abortion cases, *Doe* v. *Bolton* and *Roe* v. *Wade.*

Now the stakes were very high indeed: together, with their small array of provisions and exceptions, the Texas and Georgia statutes encompassed all of what Judy Widdicombe and her compatriots around the country saw as bad abortion law. The Texas statutes read somewhat like Missouri's, exempting only life-of-the-mother abortions: the state penal code made it a crime to administer, procure, or furnish "the means for procuring an abortion, knowing the purpose intended." Georgia's was a far more modern statute, adopted in 1968, as the Model Penal Code language spread around the country: the Georgia law made abortion permissible after rape (forcible or statutory, and formally reported to some law enforcement agency), or to prevent "serious and permanent" damage to a woman's health, or to prevent the birth of a baby who would "very likely" carry some "grave, permanent, and irremediable mental or physical defect."

The Georgia abortion law included an assortment of procedural requirements, all of which made the statute stricter in practical terms than the now loosely interpreted older law of the District of Columbia. A prospective abortion patient had to swear under oath that she was a legal resident of the state. The abortion must be performed in a full-sized accredited hospital, not a clinic or a physician's office. The abortion required the written approval of three physicians *and* a medical committee from the hospital in which the operation was to take place. Both in concept and in practice, the lawyers bringing the Georgia case were arguing, all these obligations added up to "a procedure so cumbersome, time-consuming, and restrictive as to manipulate out of existence the rights of an abortion applicant."

Everything that had disturbed abortion activists about the increasingly popular "reform" laws—the categories, the bureaucratic gauntlet, the written rules that worked to accommodate the rich and shoo away the poor—appeared to be enshrined in the modern Georgia abortion statute. And by taking both cases at once—the oral arguments were scheduled back-to-back for the same day, December 13—the Supreme Court alerted attorneys around the country that the justices intended to examine more than simple *prohibitions* on abortion. Could a state ban some abortions while permitting others? Could a state ban abortion outright? Could a state adopt laws that made abortion legal but set up a half dozen formidable barriers between every patient and her abortionist?

The legal work accelerated. *Roe* v. *Wade's* Sarah Weddington, stunned by the news that her first contested legal case was about to be heard by the Supreme Court, packed up and took an airplane to New York to spend the summer preparing at Roy Lucas' law offices; at night, with the windows opened to stale summer air and the riotous noises of Manhattan street life keeping her awake, Weddington bedded down on an old chaise lounge in the switchboard room of a downtown abortion counseling center. "The first taxi ride of my life was in a car with a bulletproof partition between the passenger and the driver, a sign that said the driver did not carry more than five dollars in change, and an enclosed money bucket in the floor of the front seat, welded to the transmission," Weddington wrote in her 1992 memoir *A Question of Choice,* recalling her disconcerting arrival in New York City. "My first night I stayed with an acquaintance; the day before, a stewardess had been murdered three blocks away."

In the summer of 1971 Sarah Weddington was twenty-four years old, a Texas Methodist minister's daughter who captured the fancy of newspaper photographers by showing up for work with her thick strawberry blond hair combed long and full over her shoulders. She had taken a job some months earlier as a Fort Worth assistant city attorney, the first woman ever hired in that position, and had abandoned it, by her own account, when her supervising lawyer wrote Weddington a seven-word note apparently meant as an ultimatum: "No more women's lib. No more abortion."

Weddington had weighed her alternatives, Fort Worth job or "women's lib" and Supreme Court case, and quit. Now she made her permanent home in Austin and shared a law practice with her husband, who had accompanied her, four years earlier, when she drove to Mexico for an illegal abortion.

Nothing gruesome had happened to Sarah Weddington during her Mexican abortion; the office was clean, the doctor was pleasant, and she came out of the anesthesia frightened and dizzy but physically unscathed. For a long time she was intensely private about her personal history; even in abortion law circles nobody knew Weddington had obtained an abortion herself until she wrote about it twenty-five years later in her book. When she was asked afterward about her long reticence she would tell people that the abortion had proved an episode of such comparatively low drama —she and Ron Weddington were not married yet, he did not want children, she wanted to finish law school—that she had always believed she should stick to her lawyer role and keep her personal history to herself. But as a political cause abortion activism had appealed to her the first time she learned about it: she was introduced after law school to the feminist branch of the lively Austin counterculture, and by 1969 was donating legal research time to local abortion referral volunteers. In the University of Texas law library Weddington made her way through the same collection of cases that was beckoning her counterparts around the country, *Griswold, Belous, United States* v. *Vuitch*. When the local abortion referral organizer Judy Smith finally asked her to sue, to bring a formal legal challenge to the Texas abortion statute, Weddington describes having cringed at first:

"No, you need someone older and with more experience," I told Judy. "You need somebody in a firm, with research and secretarial backup." My mind whirred with all the reasons I was not the right person. . . . After all, my total legal experience consisted of a few uncontested divorces for friends, ten or twelve uncomplicated wills for people with little property, one adoption for relatives, and a few miscellaneous matters. I had never been involved in a contested case. The idea of challenging the Texas abortion law in federal court was overwhelming.

She began preparing the case, despite her misgivings, and recruited her former law school classmate Linda Coffee to help. Coffee was three years older than Weddington, which still made her one of the youngest associates at the Dallas bankruptcy firm where Coffee was working; together, the two lawyers' accumulated years added up to less than the age of any *one* of the black-robed justices they were now preparing to face. In Atlanta, Margie Pitts Hames was more seasoned, thirty-eight years old and a veteran of both private practice and volunteer ACLU cases, but she had never argued before the U.S. Supreme Court, either; Hames had done her civil rights work in local police brutality and racial discrimination cases. It had not gone unnoticed in Washington that all three were women —Supreme Court personnel, particularly after Georgia officials announced that an assistant attorney general named Dorothy Beasley would argue for the state, began referring to December 13 as Ladies' Day—and for older attorneys like Connecticut's Catherine Roraback, it was exhilarating to count up the sheer numbers of women whose names turned up on the legal papers now circulating with new urgency around the states.

Certainly there were men among the attorneys bringing abortion cases—Missouri had Frank Susman, for example, and one line of influential research had been explored by a New York Law School professor named Cyril Means, who wrote at voluminous length about the ecclesiastical and political history of abortion law. Roy Lucas, for that matter, was antagonizing some of his own colleagues by spreading himself so widely around the abortion circuit: he was simultaneously handling cases in Minnesota, North Carolina, New Jersey, and Ohio; he had hired himself out as private counsel to individual abortion doctors; and he had tried, without success, to take over Supreme Court argument of the *U.S.* v. *Vuitch* challenge to the Washington, D.C., abortion law.

Lucas had tried to move in on Margie Pitts Hames' Georgia case as well, deeply irritating Hames, who declined Lucas' suggestion that he replace her as lead attorney in the Georgia appeal. Lucas had been somewhat more successful forging a role for himself in *Roe* v. *Wade;* both Sarah Weddington and Linda Coffee had accepted Lucas' early long-distance offer of help, and Sarah Weddington, who temporarily relocated to New York chiefly because she worried that Lucas appeared to be making no progress on the *Roe* Supreme Court appeal, had agreed to a temporary part-time staff job at Lucas' Madison Law Institute, so that in a formal sense she was spending her summer preparing *Roe* v. *Wade* as Roy Lucas' employee.

By the final weeks of summer, as Sarah Weddington and her husband were preparing to return home to Austin, Roy Lucas had begun insisting that as the more experienced abortion attorney, he ought to present the Supreme Court oral argument in *Roe*—that he would not even consider Sarah Weddington's suggestion that the two of them divide evenly their allotted time before the Court. And when Weddington hesitated about

demanding that she argue her own case, it was the don't-back-down telephone calls from other women attorneys that steeled her resolve; from
their work in civil rights, the student left, and the burgeoning women's
movement, a network of ardent young women lawyers, many of them
working out of ill-funded little offices with macrame plant hangers and
Sojourner Truth posters on the walls, had made what seemed to them the
natural leap to abortion cases. Catherine Roraback herself had an abortion
law challenge underway, a Connecticut version of *Roe v. Wade*, and because she had helped argue *Griswold*, she assumed a venerated elder status
among some of the younger attorneys, who looked to her for guidance
as they composed and passed around their angry and grand-scale legal
complaints.

"Humility was not a strong suit for any of us in those days," recalls
the former Center for Constitutional Rights attorney Nancy Stearns, who
was barely out of New York University law school when she helped prepare an ambitious class-action abortion case, one of the four lawsuits filed
simultaneously in 1969 to challenge the existing New York abortion statute. Their approach was shaped more by political organizing strategies
than by fine points of the law, Stearns recalls—she and some of her
colleagues deliberately sought out large numbers of women plaintiffs, both
to rouse public interest and to impress upon the court the vast reach of
illegal abortion as a women's plight. ("Remember, the judges were almost
one hundred percent male," Stearns recalls. "Here you were, trying to
explain to these basically male, middle-aged, middle-class white men what
it *was* to have an unwanted pregnancy. So just bringing in a single pregnant woman might give you legal standing—but it would not teach very
much.") *

In Chicago, the lead attorneys were similarly inexperienced as they
plunged in—"really novices of the first order, nobody looking over our
shoulder," recalls Sybille Fritzsche, who had barely made the transition
from the University of Chicago law school to her new job as counsel to
the local ACLU chapter when she and a young Chicago Legal Aid Bureau
attorney named Susan Grossman filed a challenge to the Illinois abortion
statute. The Illinois case was constructed from the ground up, like New
York's, and Fritzsche and Grossman loaded their plaintiffs' list with prominent Chicago obstetricians and pored over every written abortion study
they could find, anxious not to appear the brash young revolutionaries
before conservative Illinois judges.

By telephone and by mail, the fat manila envelopes shipped halfway
across the country from one law office to another, the lawyers tracked each
other's progress and kept a running score: this argument is working in

---

* The New York cases were rendered moot by the adoption of New York's 1970
abortion liberalization, but many of the attorneys involved in their filing quickly
shifted their attention to similar lawsuits in other states.

some courts, this one is being tossed aside, this one is politically effective but will never convince the Supreme Court. The Association for the Study of Abortion began leading its decorous newsletter with a lengthy roundup called "Cases in Court," and from the ASA's rosters an older generation of stalwarts now offered up their years of research and statistics, like aging scholars hauled out of the campus library to join the raucous demonstration outside.

Christopher Tietze, the internationally respected Vienna-born medical statistician who had spent twenty years with Planned Parenthood and now directed the World Population Council, had volumes filled with his documentation on contraception and induced abortion. Planned Parenthood Federation of America president Alan Guttmacher, who had been writing books about birth control since the 1930s, had elaborate analyses of the medical arguments for dismantling state abortion laws. And in the offices of Harriet Pilpel, a New York attorney who had been part of the founding core of the ASA, telephone advisories and multiply shipped sample briefs had created an improvised lawyers' clearinghouse, with the remarkable Pilpel dispensing legal wisdom and war stories as the Supreme Court arguments approached.

Harriet Pilpel was a *Griswold* veteran herself, a longtime American Civil Liberties Union board member, and more to the point a direct link to Margaret Sanger and the contraception battles of the first half of the century. She had finished Columbia Law School in 1936, the second-highest ranked student in her class, and was hired straight into the New York law firm of Morris Ernst, whose client list had since the early 1920s included Margaret Sanger. For more than twenty years now Pilpel had handled the office's Planned Parenthood work, and although Planned Parenthood itself was a nervous and sometimes standoffish ally in the early abortion battles (some regional chapters worked vigorously to keep from sullying themselves with the issue of abortion law repeal), Harriet Pilpel had used both the lay and legal press to agitate for changes in the laws. "We continue to maintain strict anti-abortion laws on the books of at least four fifths of our states," she wrote in *The Atlantic Monthly* in 1969:

> denying freedom of choice to women and physicians and compelling the "unwilling to bear the unwanted." . . . If we really want to cut our population growth rate on a voluntary basis, we should make abortion available on a voluntary basis, at least in the early stages of pregnancy. When Japan liberalized its abortion laws some years back, it halved its rate of population growth in a decade.

Pilpel was a striking and fiercely intelligent woman, with a reputation both for startling vanity—professional women around her were sometimes disconcerted by watching Pilpel attend to her makeup and elaborately arranged hair in the midst of public meetings—and for a breakneck work pace. It took two secretaries to keep up with her dictation, and Eve Paul, a younger associate who was later to replace Pilpel as general counsel to Planned Parenthood, remembers Pilpel hurrying from office to office with so many papers and law books that she pulled them along in a wheeled cart. "You could meet her at a party, she was out almost every night, or the theater, or a concert, and there would be Harriet, with her little cart," Paul recalls.

And now Pilpel began, from her office in New York, to help choreograph some of the preparation for the abortion cases' grand debut at the Supreme Court. She visited Atlanta. She consulted with Sarah Weddington. She composed, with other attorneys, friend-of-the-court briefs to add to the principal arguments being directed at the Court. She arranged "moot courts," the staged question-and-answer sessions in which attorneys help each other prepare for oral argument, so that abortion lawyers could gather in one room and fling at the twenty-five-year-old Sarah Weddington every rancorous query a Supreme Court justice might be likely to think up.

It was a long way from St. Louis, all of it. Frank Susman was on the mailing lists and the moot court invitations were sent out to him, too, but mostly he stayed away and waited, glumly, for his *Rodgers* ruling from the Missouri Supreme Court. Hayseeds, that was the word he would use: the courts were abuzz with abortion cases, here was the federal court out of Illinois with a wonderful ruling that voided the state statute for vagueness *and* denial of right to privacy, here were state courts as far away as Florida and Vermont issuing opinions that would have resounded as victories if they had come in Susman's case. *Violative of our state constitution*, wrote the Florida Supreme Court. *Deprives a woman of medical aid, even though she may be afflicted in body or mind, or both*, wrote the Vermont Supreme Court. A federal court in Kansas took on the Kansas Model Penal Code law, which had already set off a modest but brisk abortion business inside the state, and loosened the new statute even further: no more three-doctor review committees, the Kansas court declared, and no more requirement that abortions be performed only in certain formally accredited hospitals. "This Court is convinced of the existence of a fundamental right to individual and marital privacy," read the opinion from directly across Missouri's western border, "which includes within its scope the right to procure an abortion."

But *Rodgers* v. *Danforth* was headed for Jefferson City, the capital of the Show Me state, the state that emblazoned on its license plates a motto that many Missourians translated wryly for outsiders as No, I'm Not Ready to Try That, You Go First. Hayseeds ran the capital city and hayseeds ran the Supreme Court, that was Susman's considered opinion; even a rigid Model Penal Code reform law could not seem to find its way out of the Missouri General Assembly, and now Susman was depositing his abortion case before judges who worked half a block from the capitol building and probably split lunch checks every week with the politicians who kept voting No.

He gave a speech, not an upbeat one, in the spring of 1972. Among the audience members were physicians interested in the Clergy Service, and Susman advised them that there was not much risk of prosecution for the referring doctors; most of the women were being directed to legal clinics in New York, Susman said, and any prosecution-minded authorities would probably come after the Service organizers first. But technically, Susman told his audience, the Missouri statute did still prohibit "advising" for abortion unless the mother's life was in danger.

"I could sit here and not recall the last time that's ever really occurred, where in order to save the life of the mother, based on her physical well-being, you had to perform an abortion," Susman said in his speech.

> They get around it and say, well, they classify her as suicidal, and therefore a woman who knows the right doctors and has the right amount of money goes to the right two psychiatrists. Each one writes her a letter indicating she is suicidal. She then takes these two psychiatric letters to her favorite hospital, and each hospital has an abortion committee, excluding the Catholic hospitals, and this abortion review board is usually composed of three ob-gyns. And she walks in with one letter in each hand and waves them, and they all sit there and nod their heads, they assent, and then she goes into the hospital and has an abortion, which is necessary to preserve her life because otherwise she would take it—and this way, avoids the statute. . . . But the amount [local hospitals] do is just a drop in the bucket compared to the number of people that we see. It's just nowhere near meeting the demand. And it's a very expensive proposition.

Judy Widdicombe could see that the Clergy Service was not meeting the demand, either. Even during periods of relative calm, when her underground appeared to be running as smoothly as a commuter airline, Judy knew she was missing women too poor or too timid to seek out directions on how to leave the state for an abortion. By 1972, the Service had gone sufficiently public to make "underground" the wrong term, anyway; it was the Pregnancy Consultation Service now, with letterhead stationery and business cards and papers of incorporation that Susman had drawn up. They had telephone numbers in five cities and a small office in St. Louis. The newspapers had written about them, occasionally in a tone that bordered on frank admiration; one headline in a county suburban paper read "For Problems with Pregnancy Call 388–0310," and in the women's pages of the *Post-Dispatch* an advice columnist ran the telephone number in the matter-of-fact voice the columnist often used to list information about charities or civic organizations. Under the column's headline a newspaper artist had drawn a small illustration, for emphasis: an airborne jet, the New York skyline looming below, and every passenger seat occupied by a crossed-out stork.

And it was plain by 1972 that unless some zealous new attention-grabber came to work at one of the local prosecutors' offices, the legal authorities intended to leave the Service alone. Newspaper writers used Judy's telephone number whenever they needed comment on abortion news; in their articles she was Mrs. Judith Widdicombe, Director of the Pregnancy Consultation Service, and when the public hearings were called, Judy now made the drive down to Jefferson City so that she could take her turn at the microphone, too. She gave the legislators numbers, sometimes with dollar signs attached: five thousand women a year referred out by the Service, husbands or companions sometimes traveling alongside, at least a million dollars on airline tickets alone, another million on abortion fees paid to doctors outside the state of Missouri. "Abortion Causing Outflow of Money from State," one St. Louis headline ran, crassly, but it was true, all one had to do was sit down with a calculator and the fee lists from California and New York. If it were *legal* in Missouri—if that vicious, winking, we-don't-really-mean-this language could just be erased one day from the pages of the statute books—

FROM WASHINGTON, where the Supreme Court was due to release its opinions in *Roe* v. *Wade* and *Doe* v. *Bolton*, there was silence. Then in June 1972, a brief announcement, delivered without elaboration: both cases had been held over for the next term, with reargument scheduled for the fall. Justices Hugo Black and John Marshall Harlan had retired in 1971, leaving only a seven-man court to hear the first oral arguments; now the two vacancies had been filled, with President Richard Nixon's appointees William Rehnquist and Lewis Powell. And although the Court never made explicit its reasons for holding the cases over, all the *Roe* and *Doe* attorneys were to come back to Washington and present their arguments again, in October 1972, to the full Supreme Court.

Once again the briefs were filed, the moot courts staged, the arguments rehearsed in anticipation of the justices' questions from the bench. Then one week before the *Roe* and *Doe* rearguments, the Missouri Supreme Court handed down its ruling in *Rodgers* v. *Danforth.* It was worse than Susman had imagined.

The opinion was short and blunt. The argument that the statute was unconstitutionally vague, the Missouri court ruled, was "without merit." The rulings of other federal courts, finding similar statutes unconstitutional elsewhere, were irrelevant. As to the many other arguments Susman and Thayer had proffered, the claims of privacy rights and equal protection and due process and so on, the Missouri court dispensed with them all in two brief paragraphs. "The issues in this case," the *Rodgers* opinion read, "are sharply and significantly narrowed by the following facts stipulated to by the parties. . . ."

The *stipulation.* " 'Human life is a continuum from conception to death' "—they were skewering the case on the stipulation, on that sixteen-page embryos-are-fetuses-are-children treatise that Susman hadn't known what to do with. "The United States Supreme Court has expressed itself on the taking of 'human life' in the case of *Furman* v. *Georgia,*" the Missouri Supreme Court went on, invoking what seemed to Susman the spectacular non sequitur of a recently decided landmark capital punishment case. "As we read the opinions in *Furman,* the Court generally expressed its disapproval of the practice of putting to death persons who, some would argue, had forfeited their right to live. We believe we must anticipate at least equal solicitude for the lives of innocents."

Well, there was a lot of interesting jurisprudential reasoning emanating from the abortion cases around the states, but Susman had never seen this argument before—that if states are forced to let criminals live, then they must also be allowed to prohibit abortion. The *Rodgers* ruling included one frosty dissent, twice the length of the majority opinion: "There is a great difference between finding that the death penalty as it has been imposed in this country is unconstitutionally cruel and unusual punishment and finding that an abortion in the early stages of pregnancy is thereby an unconstitutional taking of a 'human life,' " wrote Justice Robert Seiler, the one judge Susman had suspected might lean his way. "If the parties were to stipulate that two plus two equal five we would not consider ourselves bound by it, and I do not consider myself bound by a stipulation which purports to establish as an immutable fact that a month old embryo is the same, except for age and maturity, as an adult."

No other justice joined in the dissent. *Hayseeds.* "The reasoning of this decision has been criticized in many quarters," Susman wrote to his plaintiffs the following week, phrasing very sparely his assessment of the *Furman* v. *Georgia* logic. "We believe this legal precedent is not analogous."

The practical effects of the *Rodgers* ruling, Susman wrote, were slight. An appeal would of course be forthcoming, but the U.S. Supreme Court already had on its docket two cases whose outcome could be expected to address all of the constitutional questions raised by *Rodgers* v. *Danforth.* The Texas law before the Supreme Court was nearly identical to Missouri's, Susman wrote, and there was a strong feeling that the Court was prepared to take on directly the question of state control over abortion. "It is my considered opinion," Susman wrote, "that the United States Supreme Court will hold that a woman has a constitutional right to have an abortion, at least during the first trimester."

But what the Court *did* hold, three months later, on the third Monday of 1973, astonished even him.

SIX

The Ruling
1973

FOR THE REST of his life Bill Cox would remember that he first heard the news on the radio, KWOS or KLIK, one of the Jefferson City stations that mixed music with hog price updates and dispatches from the national wire services. In the basement office of the Missouri Catholic Conference a table radio sat on one of the cluttered desks, and it seemed to Cox afterward that for some moments he had stood over this radio, his face turned toward the plain wood-paneled wall, and wondered whether he could possibly have heard right.

WASHINGTON, Jan. 22—The Supreme Court barred the states today from interfering with the decision of a woman and her doctor to end pregnancies within the first three months.

In a 7–2 decision striking down the Texas abortion law, Justice Harry A. Blackmun said medical data indicated abortion in the first three months "although not without its risk, is now relatively safe."

Therefore, he said, "any interest of the state in protecting the woman from an inherently hazardous procedure . . . has largely disappeared."

The ruling was issued after two years of deliberations by the Justices. It was based predominantly on what Blackmun called a right of privacy.

He said the right "is broad enough to encompass a woman's decision whether or not to terminate her pregnancy."

The bulletins changed as the day went on, and the headlines were confusing in their hurried reports. "Court Bars States from Banning Abortions in First 3 Months." "High Court Rules Abortions Are Legal Up to Seventh Month." It was a crowded news day because Lyndon Baines Johnson died of a heart attack that afternoon; the papers and radio news were full of tributes to the former president, and late into the night, as Bill Cox and his Catholic Conference colleagues scrambled for details on the Supreme Court abortion cases, one central piece of information was repeated again and again until they finally came to see that this, at least, must against all instinct and common sense be true: *every state.* The Supreme Court of the United States had made the practice of abortion legal in every state. No matter what the statute books said, no matter what the legislators believed, no matter how convincing the photographs of small dismembered human feet—for the last three years Bill Cox had watched no-holds-barred abortion work its way into New York and California, but those were distant places, wild and ethnic, and famously unfettered by Midwestern restraint. Cox had grown up in Missouri, the son of a Jefferson City lobbyist, and even Kansas seemed to him a culture so different from his own, Westward Frontier and Every Man for Himself and so on, that he had never imagined the newly legal Kansas abortionists spreading their practice east across the border. "It was not *conceivable* that in Missouri, at that time, the law would change," Cox recalls. "It was just not conceivable."

Now the Supreme Court had changed it, in a single morning, from a thousand miles away.

Passage by passage, as Cox and his Catholic Conference coworkers examined the second-day articles and finally the Supreme Court opinions themselves, the vast sweep of what the Court had done began to take shape before them. In the Texas case, *Roe* v. *Wade,* the Court had struck down the Texas abortion statute and, by extension, the thirty other state laws like it—including Missouri's. In the Georgia case, *Doe* v. *Bolton,* the Court had struck every provision of the Model Penal Code law that actually hampered the practice of abortion—the residency requirement, the hospital requirement, the three-doctor approval, the mandatory hospital abortion committees. They were lengthy opinions, more than a hundred pages of print, but if one studied them both together, as Blackmun himself had indicated the reader was meant to do, then neither the First 3 Months nor the Legal Up to Seventh Month headline had done an adequate job of suggesting just what it was the Supreme Court was now demanding of the American states. As long as doctors approved them—as long as attending physicians could be said to be exercising their "best medical judgment"—

then abortions were to be legal, as Cox read the rulings, in every state and *all the way into the ninth month of pregnancy.*

Nobody seemed to understand that at first, but it was written there, in plain English, in the text.

Bill Cox was not a lawyer; politics was his passion, bred in his own childhood, so that once he had thought he might run for public office himself. He was no abortion expert, either; the Missouri Catholic Conference had hired him straight out of Notre Dame as a legislative "policy analyst," emphasis on education, and what Cox knew about abortion practice he had learned on the job, taking his turn as an on-request speaker with the Willke abortion pictures. It was the pictures that had radicalized him, it would seem to Cox later, but only in the way that famine photographs can rouse a complacent person to write letters or send money: no one needs a picture of a starving infant to understand that famine is bad. "Human rights" was becoming a term of some currency, the human right to speak freely and to live without political repression, and every time Cox put before an audience one of the dilation and curettage slides, listening in the darkened auditorium for the silence and then the swift sucking in of breath that meant someone had taken an instant to understand exactly what the picture was showing, it seemed to him that everyone in the room must surely see how wrong it was to pigeonhole abortion as a Catholic issue or even as a women's issue. How could the term "human rights" have any meaning at all if it did not include the right to keep from being dismembered before birth?

ROE et al. v. *WADE, DISTRICT ATTORNEY OF DALLAS COUNTY* was the first Supreme Court ruling Bill Cox ever read from beginning to end, every line of every opinion, concurrence and dissent, including the footnotes.

1. A state criminal abortion statute of the current Texas type, that exempts from criminality only a *life-saving* procedure on behalf of the mother, without regard to pregnancy stage and without recognition of other interests involved, is violative of the Due Process Clause of the Fourteenth Amendment.

(a) For the stage prior to approximately the end of the first trimester, the abortion decision and its effectuation must be left to the medical judgment of the pregnant woman's attending physician.

(b) For the stage subsequent to approximately the end of the first trimester, the State, in promoting its interest in the health of the mother, may, if it chooses, regulate the abortion procedure in ways that are reasonably related to maternal health.

(c) For the stage subsequent to viability, the State in promoting its interest in the potentiality of human life may, if it chooses, regulate, and even proscribe, abortion except where it is necessary, in appropriate medical judgment, for the preservation of the life or health of the mother.

In other words there were supposed to be dividing lines now, markers the United States Supreme Court appeared to have just invented for the states, one rule for first-trimester abortions, another for second-trimester, another for those "subsequent to viability." Even if you put aside the chorus of questions those directives quickly inspired (Why trimesters? Who defines viability? Why didn't the Court just go ahead and write the obligatory legislation itself?), you had to read *Doe* v. *Bolton*, the second case, to see the very large loophole in *Roe's* option (c). The *Roe* opinion had discoursed with apparent great feeling about "the potentiality of human life," and how important it was to remember that the state *did* have an interest in this "potential" life, and that this interest became forceful—"compelling" is the legal term—once the fetus was able, at least in theory, to live on its own outside the mother's womb.

Thus a state government, the Court had ruled, could if it wished prohibit third-trimester abortion—*except where it is necessary.* And what was necessary? *Preservation of the life or health of the mother.* The Supreme Court had already written, in *U.S.* v. *Vuitch*, a definition of "health" so freewheeling that it allowed the District of Columbia some of the East Coast's brisker abortion business. Now the Court repeated that definition, as it proceeded through its reasoning in *Doe* v. *Bolton:*

> . . . The medical judgment may be exercised in light of all factors—physical, emotional, psychological, familial, and the woman's age—relevant to the well-being of the patient. All these factors may relate to health. This allows the attending physician the room he needs to make his best medical judgment. And it is room that operates for the benefit, not the disadvantage, of the pregnant woman.

Or in blunter language: if "appropriate medical judgment" pronounces the abortion necessary, for "familial" factors or "psychological" health or a woman's "emotional" well-being, then the state may not interfere to save the fetus.

Had the state been left with anything at all?

In the months leading up to January 1973 three men worked full time in Jefferson City for the Missouri Catholic Conference: Bill Cox; a former seminarian and city councilman named Anthony Hiesberger; and a former Missouri state attorney named Louis DeFeo. DeFeo came from Kansas City, which in the mythology of Missouri geography is supposed to provide the relaxed Western counterbalance to the more Eastern and formal St. Louis, but he was a man of sober demeanor and great reserve, his desk often stacked with yellow legal pads that he filled with tight handwritten notes. To Cox and Hiesberger, DeFeo was the slow-talking conservative senior man, methodical and lawyerly, the one who wrote language like "shall be construed" and "pursuant to Section 211" into Conference-drafted legislation. When Matt Backer had agreed to let his name be used as an intervenor in *Rodgers* v. *Danforth,* thus assuming guardianship of the symbolic Infant Doe, it was DeFeo and a second attorney who had researched and written the Backer briefs. Now the earliest copies of the *Roe* and *Doe* opinions came to Lou DeFeo first, and after he read them and reread them an uncharacteristic fury filled his voice as he told the nonlaw-yers in his office what the Court majority had done: it was biologically *illiterate,* there was *absolutely no precedent* for it, it was *wrong.*

They had five weeks left before the March 1 state capitol deadline for introducing new legislation for 1973. *Now,* Lou DeFeo said, we start draft-ing *now.* We hit the floor at a *run.* Bill Cox flew to Washington, where a dozen state Catholic Conference representatives gathered at the U.S. Cath-olic Conference headquarters for a grim comparing of notes; in the face of such grand-scale restrictions on the states, what kind of legislation could possibly stand up in court? Was there any way to restore to the states some supervisory power, some authority to ensure at least that the taking of human life not become as commonplace and morally unconstrained as minor elective surgery?

Cox brought a few ideas back to Jefferson City, and DeFeo telephoned the state Conference offices in New York and other strong Catholic states. Rapidly, a rough list began to form. Perhaps they could require written approval of the baby's father. Parental consent could be made a prerequisite for abortions on underage girls. Live-birth language might be conspicu-ously written in, a mandate that late-stage abortions be performed using a method more likely to deliver a live infant. The legislation itself could include directives on rushing medical aid to an infant that managed to survive a saline or hysterotomy; probably the reporters would pick up on wording of this nature, with its immediate dreadful imagery, and the ensuing articles might help people understand what the Court was actually requiring of their own state laws.

Abortion: the destruction of the life of an unborn child in its mother's womb or the termination of the pregnancy of a mother with an intention other than to produce a live birth or to remove a dead unborn child. Even a legislative definition could be phrased to wave before the public a certain way of thinking about what the Supreme Court had done, and DeFeo's legal pads filled with possibilities for a last-minute omnibus everything-we-can-think-of abortion bill. He felt a certain queasiness, as the draft legislation neared completion, and his discomfort deepened when he heard a few Jefferson City legislators express the same concern: was the very act of writing this legislation turning them into collaborators? Had DeFeo spent thirteen years as a working attorney and a lifetime as a practicing Catholic so that he could bend over a basement desk in Jefferson City and write *regulations* on the taking of human life?

There was not much besides abortion on DeFeo's mind, those first few weeks after the January 22 news, and as he talked to men and women in Jefferson City and around the country, some repeated aloud the argument that had been troubling him privately: we must not be accomplices. When Hitler passes a law saying all Jews must wear yellow stars, do we rush to our offices and compose regulations for the yellow star distribution centers? Is there no point at which a person must stop, refuse to budge, and declare, *I will not participate in evil?*

DeFeo didn't know. It was moral triage. A general sends his soldiers into war, knowing some of them will be killed, because he believes that is how he can limit suffering on a larger scale. If no one were willing to write legalistic paragraphs about the procedures to be followed before taking the life of an unborn child in its mother's womb, then the limits of the Supreme Court's ruling would never be tested and abortionists would work without any restraints at all. And so DeFeo went on writing his subsections. *Only after the woman certifies in writing her consent . . . only after obtaining the written consent of the father . . . in the event that a child shall manifest a heartbeat, pulse, or other vital sign subsequent to leaving the body of the mother. . . .*

He was going to be the general, and send the soldiers in. He had no right *not* to, that was how it seemed to DeFeo; he was not going to keep himself holy while watching his companions lose the war. "If I take the idealistic approach, the babies die," DeFeo recalled many years later, remembering his own time of doubt. "If I go into the fray, then maybe it's something like this: No babies die post-viability, while we fight for the rest of them."

AMID ALL the written pages in *Roe* v. *Wade* and *Doe* v. *Bolton*, all the historical background and constitutional discourse and citations from previous cases, there were three passages in particular that most inflamed ordinary men and women in state right-to-life organizations—and that helped make activists of abortion opponents who had never before thought to sign a petition or join a committee.

One had to read through thirty-seven pages of *Roe* v. *Wade* to reach the first of these passages, and in those pages themselves was a very great deal to trouble a person of Lou DeFeo's beliefs. DeFeo had a passing acquaintanceship with the Supreme Court justice who wrote both the *Roe* and *Doe* majority opinions; as a novice lawyer DeFeo had clerked for an Eighth Circuit Court of Appeals judge in 1961, when Harry Blackmun was also serving on the Eighth Circuit and would fly down regularly from his home in Minnesota to St. Louis for weeklong convocations of the multistate court. DeFeo thought Blackmun a likable man, affable and kind, and he knew the soft-voiced Minnesotan had a reputation for working so slowly and painstakingly on his assigned opinions that the final products sometimes generated some eye-rolling around the halls: they were long, scholarly, and given to so much discursive information that lawyers preparing law journal articles—which often required extensive research into case rulings and legal history—cheered up instantly when they saw that Blackmun had already addressed their subject in a judicial opinion.

It was President Richard Nixon who had elevated Harry Blackmun to the Supreme Court, in April 1970, and until now nothing the relatively new justice had said or written had offered any strong signal as to how he might approach the question of abortion law. Nixon was opposed to loosening abortion laws, or so he had maintained in public, and at the time of Blackmun's Supreme Court nomination it was supposed, based on Blackmun's own record and Nixon's declaration that he wanted to send the Court another "strict constructionist," that the new justice from Minnesota would cast moderately conservative votes, with a cautious inclination toward civil liberties. Only one item of biographical data might have stirred any speculation about Blackmun's views on matters of medical practice: for ten years, from 1950 until his Court of Appeals appointment in 1959, Blackmun had served as resident counsel to Minnesota's famous Mayo Clinic.

In post-*Roe* years several published accounts of the *Roe* and *Doe* deliberations would provide background detail on the lengthy Supreme Court discussion that followed the first oral arguments: how Blackmun worked especially hard at first on the *Doe* v. *Bolton* ruling, grappling with his own reluctance to dispense with the idea of hospital approval committees ("I have seen them operate," Blackmun wrote in an early memo to his colleagues, "[and] they serve a high purpose in maintaining standards and in keeping the overzealous surgeon's knife sheathed"); how one justice after another added insertions and revisions as the opinions evolved; how Blackmun flew back to Minnesota the summer before the reargument and spent two weeks alone in the Mayo Clinic library, studying medical volumes and researching the history of the Hippocratic Oath.

None of this detail was available in January 1973 to a lobbyist in Jefferson City, Missouri, but Lou DeFeo was familiar enough with Supreme Court procedure to know that any important seven-to-two ruling reflects more than the thoughts of a single justice and his clerks. DeFeo also remembered that Blackmun had spent a decade on the premises of one of the nation's most prestigious hospitals, and this information only deepened DeFeo's anger at the justice whose name was most prominently attached to the majority opinion. *Blackmun* should have known better, DeFeo declared aloud, as he read with his colleagues the opinion's especially offensive paragraphs; Blackmun should have used his medical background to correct whatever misguided collaboration had taken place among the justices. How could a learned man, steeped for ten years in the evolving dramas of biological science, make these extraordinary conclusions in the name of the Supreme Court?

There were twelve separate sections to *Roe* v. *Wade* alone, each headed by a Roman numeral. The very first was preceded—uncharacteristically, for the court that was supposed to act as the unassailable arbitrator of American legal disputes—by a preface that came tantalizingly close to apology for the reaction the justices were likely to provoke:

> We forthwith acknowledge our awareness of the sensitive and emotional nature of the abortion controversy, of the vigorous opposing views, even among physicians, and some of the deep and seemingly absolute convictions that the subject inspires. One's philosophy, one's experiences, one's exposure to the raw edges of human existence, one's religious training, one's attitudes toward life and family and their values, and the moral standards one establishes and seeks to observe, are all likely to influence and color one's thinking and conclusions about abortion.

Then the judicial work of the opinion commenced in earnest. The facts of the case were outlined, and various procedural questions dispensed with. Yes, Jane Roe did have standing to bring this suit, even though she herself was no longer pregnant ("If that termination makes a case moot, pregnancy litigation seldom will survive much beyond the trial stage, and appellate review will be effectively denied," the opinion read). No, the other Texas plaintiffs, the physician convicted of performing abortions and the married couple who had been advised not to use birth control pills, did not. There was a lengthy discourse (very Blackmunian, DeFeo thought with some exasperation) on the history of abortion, going all the way back to ancient Greece and the wording of the Hippocratic Oath. The oath includes a pledge never to induce abortion, and although none of the principals in either case had thought to bring it up, Blackmun was clearly troubled by the apparently inflexible words in "the famous Oath that has stood so long as the ethical guide of the medical profession."

A published 1943 history of the Hippocratic Oath, the opinion read, provided a satisfactory explanation: when the oath was composed, it reflected the beliefs of ancient Greece's Pythagorean philosophers, who condemned abortion as the destruction of a living being, although theirs was at the time a minority view within their own culture. That this view advanced in popularity over the centuries was undisputed, Blackmun had written, but Church and criminal laws had contained their own ambiguities even as they condemned the act of abortion. Here the work of the New York NARAL-affiliated law professor Cyril Means appeared to have paid off nicely, for the opinion's references to early Christian theology and the development of the common law included lengthy citations to Means' law journal articles and concurred in much of their argument.

It is undisputed that at common law, abortion performed *before* "quickening"—the first recognizable movement of the fetus *in utero*, appearing usually from the 16th to the 18th week of pregnancy—was not an indictable offense. The absence of a common-law crime for pre-quickening abortion appears to have developed from a confluence of earlier philosophical, theological, and civil and canon law concepts of when life begins. . . . Although Christian theology and the canon law came to fix the point of animation at 40 days for a male and 80 days for a female, a view that persisted until the 19th century, there was otherwise little agreement about the precise time of formation or animation. There was agreement, however, that prior to this point the fetus was to be regarded as part of the mother, and its destruction, therefore, was not homicide.

The law of England was detailed, from the nineteenth century on, as were the first American statutes, which like the early English law distinguished in severity between abortion before quickening—not a crime at all, in some of the earliest state laws—and abortion after quickening. Even the early laws that did criminalize early abortions established milder penalties for abortions performed before a woman could be said to feel the movement of the fetus—a pre-quickening abortion under the early New York law was a misdemeanor, for example, while abortion of a quick fetus was second-degree manslaughter. It was not until the late nineteenth century that the country's laws began to show uniformity in condemning abortion at any time during pregnancy, Blackmun had written:

It is thus apparent that at common law, at the time of the adoption of our Constitution, and throughout the major portion of the 19th century, abortion was viewed with less disfavor than under most American statutes currently in effect. Phrasing it another way, a woman enjoyed a substantially broader right to terminate a pregnancy than she does in most States today.

Why, then, had the laws of every state been rewritten before the turn of the century to criminalize abortion throughout pregnancy? Several principal explanations had been advanced, Blackmun had written: first, that abortion laws were "the product of a Victorian social concern to discourage illicit sexual conduct." But the state attorneys were not even trying to raise those arguments in defense of the Texas abortion law, the opinion observed, and no court had taken seriously the proposition that states may try to prevent sexual immorality through their abortion laws.

Second, Blackmun had written, citing a 1940s history of medicine, abortion during the nineteenth century was dangerous—the procedure itself, that is, even when conducted under what for that era were medically sound conditions:

> When most criminal abortion laws were first enacted, the procedure was a hazardous one for the woman. This was particularly true prior to the development of antisepsis. Antiseptic techniques, of course, were based on discoveries by Lister, Pasteur, and others first announced in 1867, but were not generally accepted and employed until about the turn of the century. Abortion mortality was high. . . . Thus, it has been argued that a State's real concern in enacting a criminal abortion law was to protect the pregnant woman, that is, to restrain her from submitting to a procedure that placed her life in serious jeopardy.

But modern medical techniques had made abortion far safer for women, Blackmun had written, citing a half-dozen medical studies—even safer, when conducted during early pregnancy, than full-term pregnancy and childbirth itself. So the state's rationale for criminalizing the procedure because it was inherently dangerous had "largely disappeared," the opinion read.

Thus far, for a person up-to-date on the essays and scholarly writings circulating among the more active state right-to-life organizations, this *Roe* opinion had already struck out in a dubious direction. It was worrisome to see Blackmun going to such lengths to explain away the abortion reference in the Hippocratic Oath, or citing as historical authority articles by the legal abortion advocate Cyril Means. But then the Blackmun opinion listed the next argument that had been advanced to explain and justify the enactment of criminal abortion laws:

> The third reason is the State's interest—some phrase it in terms of duty—in protecting prenatal life. Some of the argument for this justification rests on the theory that a new human life is present from the moment of conception. The State's interest and general obligation to protect life then extends, it is argued, to prenatal life.

Here was the first of the *Roe* v. *Wade* passages whose special offensiveness would become a subject of animated discussion in right-to-life circles. The very choice of vocabulary jolted the right-to-life reader and warned of what lay ahead: ". . . rests on the theory that a new human life is present from the moment of conception." This was a *theory?* The principle that life begins at conception was to be regarded as a kind of educated guess, a plausible hypothesis? Had Blackmun and his brethren, in all their voluminous research, declined to pick up an obstetrics textbook?

The opinion proceeded cautiously, a step at a time. One did not need to accept this belief in life at the moment of conception, Blackmun had written, in order to defend a state interest in *potential* human life. It was a less "rigid" claim, but certainly a valid one. If one recognized this interest —this justification, by the state, for laws that dictate what a person may and may not do—then the federal courts could step in only if there was some other and stronger conflicting interest involved. And there was, the opinion read: privacy.

Even though the word was nowhere in the Constitution's text, Blackmun had written, the rulings commonly grouped together as privacy holdings—including, most famously, *Griswold* v. *Connecticut*—established within the protections of the Constitution "a guarantee of certain areas or zones of privacy." Courts might find the roots of that guarantee in different parts of the Constitution, the opinion read; the U.S. District Court *Roe* v. *Wade* ruling that preceded this one had based the privacy guarantee in the Ninth Amendment, which declares that the people may not be denied rights simply because they are not among those specifically named in the Constitution. Another defensible source for the privacy right was the Fourteenth Amendment, which was ratified after the Civil War and wrote protections for the former slaves into the Constitution; the applicable passage, which for more than a century had been invoked and debated in a wide variety of appellate cases, declared that no state could "deprive any person of life, liberty, or property, without due process of law."

The Supreme Court majority in *Roe* v. *Wade* now found the privacy right, Blackmun had written, in this phrase of the Fourteenth Amendment. And regardless of where its constitutional roots might lie—in the Ninth or in the Fourteenth Amendment—the right of privacy was

broad enough to encompass a woman's decision whether or not to terminate her pregnancy. The detriment that the State would impose upon the pregnant woman by denying this choice altogether is apparent. Specific and direct harm medically diagnosable even in early pregnancy may be involved. Maternity, or additional offspring, may force upon the woman a distressful life and future. Psychological harm may be imminent. Mental and physical health may be taxed by child care. There is also the distress, for all concerned, associated with the unwanted child, and there is the problem of bringing a child into a family already unable, psychologically and otherwise, to care for it. In other cases, as in this one, the additional difficulties and continuing stigma of unwed motherhood may be involved.

None of these arguments would support a right to abortion, the opinion read, if the fetus were directly protected by the Fourteenth Amendment, as the state of Texas and some friend-of-the-court briefs had argued. If the fetus itself could be defined as a "person," the opinion acknowledged, then the amendment would seem to guarantee the fetus' *own* right to "life" and "liberty." But elsewhere in the Constitution and in other appellate cases, the opinion went on, "person" was used to mean a person already born. "All this," the opinion read,

> together with our observation ... that throughout the major portion of the 19th century prevailing legal abortion practices were far freer than they are today, persuades us that the word "person," as used in the Fourteenth Amendment, does not include the unborn.*

Here was the second particularly disturbing passage, for a reader of right-to-life inclinations. Nearly anyone who had encountered the Fourteenth Amendment in a history class understood it as a measure of protection—as a set of paragraphs meant to ensure greater security for the "life, liberty, and property" of individual Americans. That those very words should be cited to permit the *taking* of fetal life was in itself hard to grasp. But more viscerally upsetting, especially for the nonlawyers who made up most of the right-to-life volunteer forces around the states, was the spectacle of seven justices announcing that for the purposes of the Fourteenth Amendment—and by extension, logic would seem to suggest, the entire American legal system—a fetus was not a person. To a man or a woman who had seen what Jack Willke intended in the shining magnification of the perfect tiny feet, the Supreme Court of the United States had just defined out of existence one branch of the human race.

But the worst was still to come. Even apart from questions of Fourteenth Amendment interpretation, the opinion read, the attorneys representing Texas had argued that life began at conception and that protecting human life was one of the duties of a state. And now the Court made good on its ominous earlier suggestion about the "theory" of new life's beginnings. "We need not resolve the difficult question of when life begins," began what in right-to-life ranks were soon to become the most

---

* In a footnote, Blackmun observed that Texas faced a logical dilemma in urging that fetuses were entitled to Fourteenth Amendment protection: "Neither in Texas nor in any other State are all abortions prohibited. Despite broad proscription, an exception always exists. The exception contained in Art. 1196, for an abortion procured or attempted by medical advice for the purpose of saving the life of the mother, is typical. But if the fetus is a person who is not to be deprived of life without due process of law, and if the mother's condition is the sole determinant, does not the Texas exception appear to be out of line with the Amendment's command?"

famous two sentences in the history of the Supreme Court. "When those trained in the respective disciplines of medicine, philosophy, and theology are unable to arrive at any consensus, the judiciary, at this point in the development of man's knowledge, is not in a position to speculate as to the answer."

Two and a half pages of elaboration followed that passage, with examples of religious argument and philosophical debate, and the uncertainties raised by new biological developments like "morning-after" pills and embryo implantation. And the opinion went quickly on to say that this did *not* mean the state must always let pregnant women do anything they wished with the potential life they were carrying. "The State does have an important and legitimate interest," the opinion read,

> in preserving and protecting the health of the pregnant woman, whether she be a resident of the State or a nonresident who seeks medical consultation and treatment there, and . . . it has still *another* important and legitimate interest in protecting the potentiality of human life. These interests are separate and distinct. Each grows in substantiality as the woman approaches term and, at a point during pregnancy, each becomes "compelling." [emphasis in original]

"Compelling" is a crucial word in constitutional argument. Even when a deeply important personal right is found to be protected by the Constitution, that right can be taken away if the state proves it has a reason that qualifies as "compelling."* And with its complex three-part *Roe* v. *Wade* mandate, the Court now explained that the state had dual compelling interests, which appeared at different points in pregnancy.

The barriers were to be set at the trimesters, which were the divisions obstetrical studies sometimes used to distinguish among different periods of pregnancy. In a normal gestation each trimester would last about twelve weeks, and what the state could require of a given woman—how much control it could exert through its abortion laws—would from now on depend upon which trimester her pregnancy had advanced to on the day she showed up for an abortion. During the first trimester the state had *no* compelling interest that could logically be invoked, and could do nothing to interfere; the abortion decision was solely up to "the attending physician, in consultation with his patient." Then at "approximately the end of the first trimester," when by statistical evidence induced abortion became

---

* In legal terminology, such a statute must not only satisfy this "compelling" need of the state; the law must also be "narrowly tailored" to serve the compelling interest. In other words, the statute must satisfy an especially important state need *and* must be carefully written to address that need specifically.

more dangerous to the woman than continuing her pregnancy, the state could use its compelling interest in maternal health—and maternal health *only*—by enacting regulations designed to protect the women themselves: it could require, for example, that all abortions be performed in hospitals.

But the protection of life, which people like Lou DeFeo had always imagined to be one of the most basic functions of any enlightened government, was not to be invoked in abortion laws until the third trimester of pregnancy—and then in so timid a fashion that as DeFeo saw it even those laws could do no real good. "With respect to the State's important and legitimate interest in potential life, the 'compelling' point is at viability," the opinion read:

> This is so because the fetus then presumably has the capability of meaningful life outside the mother's womb. State regulation protective of fetal life after viability thus has both logical and biological justifications. If the State is interested in protecting fetal life after viability, it may go so far as to proscribe abortion during that period, *except when it is necessary to preserve the life or health of the mother.* [emphasis added]

Even in the most pessimistic circles of right-to-life organizations, nothing had prepared people for reasoning quite like this. Requiring every state to allow legal abortion within its borders was bewildering enough, but the way the Supreme Court had done it—that was what struck so powerful a blow. If what Blackmun and his fellow justices had written was meant to be Solomonic, a respectful nod toward all sides, its effect was almost precisely the opposite on those who believed they were battling for the rights of human beings too fragile to defend themselves; from their perspective emergency-room physicians might as well have been ushered into an arena with murderers and told with great cordiality that the Court saw the merit in all their respective positions.

We need not resolve the difficult question of when life begins. What did the justices imagine their job to be? They were the *Supreme Court.* They were *supposed* to resolve difficult questions. If the Supreme Court was going to duck every time "those trained in the respective disciplines of medicine, philosophy, and theology" disagreed with one another, then the justices might as well give up and go home; they had sidestepped, with their seven votes of support for the majority opinion, what right-to-life people saw as the single most important issue before the Court in *Roe* v. *Wade.*

The humanity of the unborn baby—a phrase whose very vocabulary seemed to underscore its internal logic—had been relegated in a single morning to the status of opinion. The sociologist Kristin Luker, whose highly regarded 1984 book *Abortion and the Politics of Motherhood* was based in part on detailed interviews with California activists from both sides, wrote perceptively about the profound sense of disbelief and betrayal her right-to-life interview subjects described:

> Accustomed as they were to thinking that theirs was the majority opinion, the pro-life people we interviewed saw in the Supreme Court decision a way of thinking that seemed bizarre and unreal. Something they believed to be both fundamental and obvious—that the embryo was a human life as valuable as any (and perhaps more valuable than some because it was innocent, fragile, and unable to act on its own behalf)— was now defined as simply one opinion among several. What was more, it was defined as an opinion belonging to the *private* sphere, more like a religious preference than a deeply held social belief, such as belief in the right to free speech. It was as if the Supreme Court had suddenly ruled that a belief in free speech was only one legitimate opinion among many, which could not therefore be given special protection by any state or federal agency.

It offered no solace, in the grim Jefferson City office of the Missouri Catholic Conference, to find saluted in print the abortion opponents' "deep and seemingly absolute convictions" and the state's "important and legitimate interest in potential life." The *law* was what mattered, the written word in the statute books, and the law had just been obliterated by the Supreme Court. There were only two passages in *Roe* v. *Wade* and *Doe* v. *Bolton* that seemed to Lou DeFeo to make any logical or moral sense at all, and those were the dissents: Justices William Rehnquist and Byron White had cast the two no votes, and White in particular used language so barbed that DeFeo repeated it aloud with some satisfaction. "Raw judicial power," White had written, attacking both the substance of the majority opinions and the reasoning his fellow justices had used to shape them:

> The Court for the most part sustains this position: During the period prior to the time the fetus becomes viable, the Constitution of the United States values the convenience, whim, or caprice of the putative mother more than the life or potential life of the fetus. . . .
> With all due respect, I dissent. I find nothing in the language or history of the Constitution to support the Court's judgment. The Court simply fashions and announces a new constitutional right for pregnant mothers and, with scarcely any reason or authority for its action, invests that right with sufficient substance to override most existing state abortion statutes. The upshot is that the people and the legislatures of the 50 States are constitutionally disentitled to weigh the relative importance of

the continued existence and development of the fetus, on the one hand, against a spectrum of possible impacts on the mother, on the other hand. As an exercise of raw judicial power, the Court perhaps has authority to do what it does today; but in my view its judgment is an improvident and extravagant exercise of the power of judicial review that the Constitution extends to this Court.

In the newspapers and the Catholic missives piling up rapidly on the desks of DeFeo and Cox, reaction alone filled columns of newsprint. Bishops denounced the ruling from nearly every Catholic diocese. In St. Louis, Cardinal John Carberry said he was "saddened and appalled by the disregard for human life"; in Jefferson City, Bishop Michael McAuliffe said he was "outraged"; in Philadelphia, Cardinal John J. Krol, president of the National Council of Catholic Bishops, used the same epithet—"unspeakable"—that had been invoked by the Second Vatican Council a decade earlier to describe the crimes of abortion and infanticide. "The Supreme Court's decision today is an unspeakable tragedy for this nation," Cardinal Krol said, in a statement reprinted at length in *The New York Times.* "It is hard to think of any decision in the 200 years of our history which has had more disastrous implications for our stability as a civilized society."

Abortion Ruling Draws Cheers and Dismay Here, the *Globe-Democrat* headlined in St. Louis. The cheers took up some of the newsprint too: there were celebratory remarks from the American Civil Liberties Union and the Planned Parenthood Federation and the Center for Constitutional Rights, where *Roe v. Wade* was publicly hailed as a great victory for the women's liberation movement. Frank Susman, telephoned for reaction by the reporters in Missouri, exulted at the news: "I think it's fantastic," he said. "It will clearly invalidate our present statute, as I understand it. . . . For all practical purposes, the Texas and Missouri laws are identical."

In the general-audience papers the *Roe v. Wade* stories were written with a note of finality: questions remained, individual state statutes must still be examined one by one, but the Supreme Court of the United States had settled the argument. *Resolution* appeared in many of these accounts, both as word and implicit idea. "A few unresolved issues remained after yesterday's long-awaited ruling," declared the January 23 *Washington Post,* offering by way of example one of the smaller matters—veto rights of the father—that the court had skipped over as it put to rest the big ones. In New York, where legislative abortion battles had been busying statehouse reporters for more than four years, the second paragraph of *The New York Times'* principal next-day *Roe* story began like a door swinging shut: "In a historic resolution of a fiercely controversial issue . . ."

But in the *St. Louis Review,* the archdiocesan weekly that was delivered by mail to ninety thousand Catholic households, the coverage was of a battle *begun.* Suddenly everything that preceded the news of January 22 —and there had been many hundreds of articles, in this and every other archdiocesan paper in the country, ranging from state-by-state legislative reports to long moral and political analyses of abortion law—was shoved aside as warm-up. "ABORTION—The Battle Lines Are Drawn," read the headline stripped across the *Review's* February 9 front page, and nothing in that article so much as suggested that the Supreme Court had presented the nation with a "resolution."

> WASHINGTON—The U.S. Supreme Court decision striking down state abortion legislation has nationalized the anti-abortion movement.
>
> Prior to the decision issued late last month, anti-abortion efforts were usually aimed at state legislatures. But with the restrictive statutes supported in state capitals apparently outlawed by the Court decision, the new target has become the Congress.
>
> The aim is to overrule the Supreme Court through enactment of a constitutional amendment which would specifically guarantee the right to life of the fetus. In its decision the Court said that the fetus was not due constitutional protection of life because it is not a person in the constitutional sense.

What the Supreme Court had handed down, this and scores of other post-*Roe* articles urged in their tone and the fervor of their interview quotes, was a massive recruiting call for right-to-life organizations across the United States. In Congress, the staffs of two Republican offices had rushed to work on legislative language that might eventually undo *Roe* v. *Wade* and *Doe* v. *Bolton.* Maryland representative Lawrence J. Hogan, whose brother was the obstetrician who had provided the Willkes with some of their first useful pictures, announced within days of the rulings that he was drafting a constitutional amendment and intended to rally his colleagues to join him. "If I had been alive in Nazi Germany, I like to think I would have had the courage to stand up and oppose the slaughter of Jews," Hogan said. New York senator James Buckley told reporters he was consulting "men of science and law" to begin a parallel campaign in the Senate. Public statements and defiant "resolutions"—the only context in which the *Review* would use that word—were delivered to news services by the National Right to Life Committee, by the National Youth Pro-Life Coalition, by the National Federation of Catholic Physicians' Guilds. "The default of judgment of the Court lies in its own admission of indecision on the very crux of the problem, namely, when does life begin," read the angry statement from the Physicians' Guild. "There is surely irrefutable evidence that life begins long before six months of gestation."

That so many of these dispatches were arriving affixed with the Catholic label had not gone unnoticed, in either the lay or the religious press. The *St. Louis Review* ran an anxious roundup of protest from inside other churches—here was a Lutheran minister from New York Right to Life, a Baptist minister from New Jersey Right to Life, a blast from the Mormons describing abortion as "one of the most revolting and sinful practices of the day." But neither Baptist nor Lutheran church officials had themselves raised public objections to *Roe;* of the commonly known non-Catholic religious denominations, only the Mormons and the Orthodox Church in America joined the Catholic Church in open denunciation of the Supreme Court's rulings.

Even the Orthodox Rabbinical Council of America, the only branch of American Judaism that condemned early abortion as the taking of human life, declared through a spokesman that the Court's authority must be respected even when its decisions appeared flawed. "Non-Catholics Protest Court Ruling," read the *Review,* doggedly, but the problem was impossible to evade: in the press, on the television, and inexorably in the public mind, the first great protest to *Roe* v. *Wade* was being portrayed as a face-off between the Supreme Court and the Catholic Church.

The congressman Lawrence Hogan, working so swiftly in Washington on his constitutional amendment proposal, was identified as Catholic in a wire service story only seven lines long—"Lawrence J. Hogan, Republican of Maryland, a Roman Catholic." It was a lead chain around their ankles. Bella Abzug could address the National Association for Repeal of Abortion Laws and nobody ever hurried back to the newsroom to write: "Rep. Bella Abzug, Democrat of New York, a Jew." The connection was going to have to be severed somehow, or at the very least stretched enough to make people see that the men and women enlisting in right-to-life groups were not simply snapping to attention under orders from their bishop, and one of the first targets was the modest and somewhat ineffectual National Right to Life Committee. In Arizona the internist Carolyn Gerster, who as an articulate and strikingly good-looking Episcopalian was emerging as one of the bright Protestant lights of regional right-to-life activity, was still in a paralysis of post-*Roe* gloom when she was telephoned from Minnesota by Marjorie Mecklenberg, the Methodist who had helped start up the Minnesota organization. The little National Right to Life office in Washington, D.C., was not going to suffice any longer, Gerster recalls Mecklenberg saying briskly: it was too Catholic, it placed multistate meetings on Catholic campuses, its director was a priest.

"She said, 'This is obvious, now that the court decision has occurred —we need a national nonsectarian organization,' " Gerster recalls. "We should have had one *long* before that. We were very slow awakening."

Nonsectarian was the key principle, raised like a national flag as a small collection of men and women from right-to-life organizations around the states began gathering biweekly at the centrally located (and indisputably nonsectarian) O'Hare Airport in Chicago. One or two days at a time, in long, contentious meetings that sometimes left them exhausted, the dozen of them—Gerster, Mecklenberg, and a handful of attorneys and regional leaders from states as far afield as Washington, Texas, and Vermont—argued over exactly what it was that an organization calling itself National Right to Life ought to pursue. That abortion should be returned to the criminal codes was agreed, but how was that to happen? What about the social climate that had helped make abortion acceptable? What about the alarming possibilities implicit in a willingness to abort—euthanasia, for example, or the eugenic elimination of the unfit? Should intrauterine devices and birth control pills, which were believed to work partly because they prevented implantation of already fertilized eggs, be characterized as abortifacients? Should the new organization take a stand on contraception in general, urging it as a preventative to unexpected pregnancy, or condemning it as part of the mentality that had led Americans to believe that sex carried with it no responsibility for the creation of life?

At the height of these meetings a grand disorder of notebooks and hardbound volumes would litter the long table; Gerster brought her embryology texts, the lawyers brought sheafs of legal papers, the Lennart Nilsson article would be opened to the most affecting of the *in utero* pictures. "Some people wanted to get the hungry and homeless in there—Marge, for example, had a much broader scope, human life in every aspect," Gerster recalls. "Some people were opposed to capital punishment, and they wanted to put that into Article One, but people said no, that was a different issue, that we were talking about the taking of innocent human life without due process of law."

Laboriously, many drafts in the making, a Statement of Purpose appeared.

> The purpose of the National Right to Life Committee, Inc., is to engage in educational, charitable, scientific and political activities, projects, or purposes, including specifically, but not limited to, the foregoing:
> 1. To promote respect for the worth and dignity of all human life, including the life of the unborn child from the time of fertilization.
> 2. To promote, encourage, and sponsor such mandatory and statutory measures which will provide protection for human life before and after birth, particularly for the defenseless, the incompetent, the impaired, and the incapacitated.

As a call to arms for a national movement this was not exactly fiery, but its measured tone was deliberate: it was clear, it was calm, it made no mention of God. A person of any faith could find merit in such a statement of purpose. Indeed, as the newly constituted board of the National Right to Life Committee saw it, a person of no faith could be inspired—*should* be inspired, if this person understood as they did both principles of moral decency and the basic mechanics of human biology—by the vision of promoting "respect for the worth and dignity of all human life."

All they needed now was the movement.

The potential was of massive proportions, it seemed to them, and largely untapped: the voting populations of fifty states. Legalized abortion had been offered by state initiative only three times in the decade since that disruptive idea had first been introduced into political discourse, and the initiatives had been voted down twice—both in North Dakota and in Michigan, where public campaigns and effective use of the Willkes' war pictures had ensured that right-to-life activists not be caught offguard the way they had been during the country's first abortion initiative in 1970 in Washington State.*

Legislators too had been made to understand the hazards of allowing legal abortion into their own states' codes; there were many ways to look at NARAL's lists of the new state abortion laws, and one of them was to bear in mind that *more than half* the American state legislatures had so far chosen to leave their criminal abortion laws alone. Public pressure had almost managed to reclaim New York State, the busiest abortion center in the nation: the same legislature that had voted abortion in had turned around two years later, after a ferocious statewide lobbying campaign, and voted abortion out by repealing the 1970 abortion law. Only Governor Nelson Rockefeller's veto had interfered. There were cynics who suggested that a lot of legislators had voted with New York Right to Life precisely *because* the governor had promised to sign a veto so that abortion would remain legal within the state—it was a cheap vote, in other words, satisfying hard-driving lobbyists without actually imperiling legal abortion. But the sentiment for protection of the unborn was out there; National Right to Life was confident about that. The challenge was tapping it: reaching those people, educating them, showing them the war pictures, rousing them to action.

And what were the tools for creating a mass movement where only scattered regional protest had existed before?

They were tangible, some of them, and plain as typewriters: the movement, like nearly any social movement trying to swell both its base and its audience, was going to need typewriters. It was going to need copying machines, and staplers, and human beings to type and copy and staple. It was going to need buildings in which this typing and copying and stapling might be carried out, and it was going to need lecture halls and address lists and a tight communications network and political veterans and an organized chain of command and the quick, reliable mobilization of many volunteers, preferably volunteers whose schedules—home

---

* In November 1970, Washington voters approved a referendum legalizing abortion— for state residents only—through the seventeenth week of pregnancy.

during the workday, with ready access to the telephone—made them available for on-call work. It was going to need ready and regular access to large groups of people, men and women who might already be sympathetic and who could be reached all at once with a brief encouraging speech or a big stack of newsletters to be read and passed on to neighbors.

In many parts of the United States, and particularly in the state of Missouri, only one organization in the early 1970s had at its disposal every one of these tools, all the way down to hand-typed, broad-circulation newsletters that were already distributed once a week.

ST. PETER CHURCH
216 Broadway, Box 174
Jefferson City, Missouri

March 7, 1971
Second Sunday of Lent

DAILY MASSES DURING LENT AT
St. Joseph Cathedral: 6:30 8:00 7:30 P.M. (not on Saturday)
Immaculate Conception: 6:00 8:00 5:15 7:30 P.M.
St. Peter: 6:00 7:00 7:50 12:05

Registration for the First Graders for the School Year 1971–1972 after the Saturday evening Mass March 20 and after all the Masses on Sunday March 21: your cooperation is absolutely important, as this will determine the number of teachers needed for the first grade for the coming year.

We are accepting applications of teachers who may wish to teach in our school for the coming school year: See or call Sr. Cyprian 636–6098 or 636–8922.

FREE CHEST X-RAY for 18 years of age or older: Gerbes Shopping Plaza, March 10 from 2:00 to 8:00: March 11 at Montgomery Ward Plaza from 2:00 to 8:00: March 12 at High and Jefferson from 10:00 to 2:00 and 3:00 to 7:00: March 13 at Jefferson Plaza from 10:00 to 2:00 and 3:00 to 7:00: Russellville on the Square, March 11 from 10:30 to 12:30.

A public Hearing has been Scheduled for 8 P.M. Monday Mar. 22, 1971, in the House of Representatives Lounge, Missouri Capitol Building, Jefferson City, for HB 650 (retyped below) repealing Missouri's abortion law relating to the crime of abortion.

HOUSE BILL NO. 650

To repeal section 559.00 RSMo 1959 relating to the crime of abortion, and to enact in lieu thereof one new section relating to the same subject. . . .

Of *course* Lou DeFeo saw the irony of the situation; a person would have to be pretty dense to miss it. How do you secularize a movement when its single most effective organizing vehicle is the Roman Catholic Church? For four years, ever since his arrival in 1969 at the fledgling Missouri Catholic Conference, DeFeo had fended off jabs from the legal abortion side: How can you take seriously your own argument that this is not a religious crusade when you're up on Church pulpits as you argue? And while DeFeo believed that was not precisely true anyway—more than the pulpits themselves it was the *facilities* of Catholic churches that had proved so useful, the campuses and printing offices and message-relay systems of enormous reach—it was also apparent to him that somebody had to take the lead on this and that he was proud of the Church for stepping up to do it. Was a lecturer with the Willke slides supposed to refuse a free auditorium because the building around it was run by nuns? Was Missouri Citizens Opposed to the Liberalization of Abortion Laws supposed to ignore the fact that a well-placed notice in one Sunday's parish newsletters could alert more than half a million Missourians to an impending public hearing? A single "Reverend and dear Father" letter, signed by Cardinal Carberry and mailed simultaneously to the five hundred parishes in the state of Missouri, could for the price of postage stamps and copy paper sweep as effectively across the state as a lavish advertising campaign:

March 1, 1971

Reverend and dear Father:

A bill has now been introduced into the Missouri legislature to liberalize Missouri's anti-abortion laws. As you know, a number of states have already passed similar liberalized laws. We want our legislators and all Missouri citizens to know the affront to God and the attack on human life that we feel these laws present.

Of special importance in this campaign is the understanding of our own Catholic people and other people of good will who oppose virtually unrestricted abortion. In an attempt to make our position crystal clear, the Bishops of Missouri issued a joint Statement on Abortion in December 1970. Although the statement was printed in the St. Louis Review and fully reported by the secular media, we feel that even wider dissemination of the statement would help our cause. For that reason, we have provided that the statement be reprinted in brochure form for distribution throughout the state.

Your assistance in making copies of this statement available to your parishioners will be deeply appreciated. We hope that every parishioner would be given a copy of the statement. Distribution will be taken care of by the Archdiocesan Council of the Laity, 4140 Lindell, and so your orders should be directed to them.

Please keep this campaign for the protection of human life in your prayers and inspire the people to support it.

> Devotedly yours in Our Lady,
> John Joseph Cardinal Carberry
> Archbishop of St. Louis

A person ought to read that letter carefully before leaping up to shout about church and state, it seemed to DeFeo: "Catholic people *and other people of good will,*" the archbishop had written. That was who they were trying to reach, people of goodwill, and as DeFeo saw it his job was to press the state legislature for causes that Catholics championed but that worked to the benefit of Catholic and non-Catholic alike. Until the thunderbolt of January 1973, abortion had consumed only a modest portion of the Missouri Catholic Conference's efforts; Conference lobbyists worked on prison reform, too, and support for organized labor, and liberalization of the welfare system, and the preservation of the family farm. When DeFeo was hired by the Catholic Conference in 1969 the pressing issue on the legal and legislative agenda was state aid to parochial schools; DeFeo had spent six years with the state attorney general's office, and his credentials for the new job were based in part upon his lawyer's familiarity with the Department of Education. When he thought in later years about his initial interviews, the talks with the bishops and Conference staff about the work that lay before him, DeFeo was not certain the subject of abortion had even come up.

The very notion of a state Catholic Conference—of a Catholic presence placed formally and permanently amid the workings of state government—was in fact relatively new to Missouri. Into the 1960s no single office collectively represented Missouri's four Catholic dioceses; each worked independently, in its own corner of the state, and maintained a separate legal staff. There was no "lobbying," in the conventional sense; nobody worked the state legislature from session to session with a focused eye toward politics and the interests of the Church. And no single office tried to handle paperwork and administrative business (the network of Catholic services, after all, included hospitals, social service agencies, and a bustling school system) for all four dioceses, from powerful what-parish-are-you-from St. Louis to the rural and Southern reaches where Catholics were a tiny minority.

It was not that the Church was wholly aloof from politics, in either Missouri or the rest of the states with a sizable Catholic presence: for more than a century American bishops had been making occasional forays into the political arena, sometimes with resounding success. Certainly there were dioceses around the country in which everyone knew how much clout a particular bishop carried in City Hall or the statehouse, and when Margaret Sanger was campaigning in the 1930s for congressional repeal of birth control restrictions, her most powerful and politically effective opponent was the National Catholic Welfare Conference (NCWC), which had formed during World War I as the country's first bishops' organization and thus the first attempt at a unified voice for the American episcopate. Among the Conference's assignments of responsibility was a department called Social Action, and for a quarter century that department was run by a Washington, D.C., priest and writer, Monsignor John A. Ryan, who worked publicly and energetically in the economic politics of his era: he helped draft state minimum-wage laws, circulated proposals for postwar public works and housing programs, and wrote prolifically in support of New Deal principles.

But the NCWC's support among Catholic bishops was sporadic, and before 1960 only a very few states around the country (no more than three or four, by the reckoning of Catholic historians) had organized formally into statewide conferences that made regular, undisguised efforts to involve the Church in the workings of local government. Direct meddling of this nature had been regarded for many generations as a volatile proposition, complicated by hostility toward Catholic immigrants and accusations that American Catholics were actually dubious patriots who worked at the behest of Rome. It was not until the 1960s that the real change began, prompted by two exceptional events within the first half of the decade: a Boston Irish Catholic was elected president of the United States; and the Second Vatican Council gathered in Rome.

John F. Kennedy was the first Catholic politician ever to reach the American presidency, and his effect on American Catholics was profound. Here, finally, was a national leader who had beaten back the split loyalties arguments and proved, as the political scientist Timothy A. Byrnes has written, that a political leader could be "both Catholic and American, comfortably loyal to his religious roots yet quintessentially American in his politics and lifestyle." And Kennedy's presidency overlapped the onset of the Vatican Council, the three-year convocation of bishops called by Pope John XXIII to bring an ancient Catholic Church into the twentieth century. The mission of Vatican II was massive and complex, with bishops from around the world composing and revising and battling over conciliar documents that addressed every aspect of life in the Church. To Catholic men and women in as far-flung a place as St. Louis or Jefferson City, the Council-approved changes began arriving with unnerving rapidity after 1965: the Mass was to be conducted in English; the priest was to turn around and face the congregation instead of murmuring inaudibly toward the altar; Protestants were no longer to be thought of as doomed rebellious miscreants who had abandoned the one True Church. Matt Backer was so startled the first time he listened to the Gospel of St. John in English that he repeated the words to himself as though he had never really heard them before: *In the beginning was the Word, and Word was with God, and the Word was God.* All Backer's life it had been a church of mystery, of unquestioned decorum, and now amid the sudden turbulence of Vatican II it was supposed to be a church of the people, the windows flung open, the musty air shooed from the room.

"In wonder at their own discoveries and their own might, men are today troubled and perplexed by questions about current trends in the world," read what was to become the most famous of the conciliar documents, released in December of 1965:

> And so the Council, as witness and guide to the faith of the whole people of God, gathered together by Christ, can find no more eloquent expression of its solidarity and respectful affection for the whole human family, to which it belongs, than to enter into dialogue with it about all these different problems.

The title of that conciliar document, the Pastoral Constitution on the Church in the Modern World, included in its nine words the message that helped create the Missouri Catholic Conference. *The Church in the Modern World:* when the bishop of Jefferson City invited the young ex-seminarian Anthony Hiesberger to interview in 1967 for the directorship of a new organization meant to carry all four dioceses directly into the Jefferson City statehouse, Hiesberger was not sure whether it was the job proposal or the invitation itself that surprised him more.

"My upbringing was European traditional, conservative, pre–Vatican II," recalls Hiesberger, who by 1967 had married and gone to work in his family's Jefferson City printing firm. "And to *go to the bishop's office*—it's not like you went down the street to the barbershop. I remember so often our pastor saying, 'When you go to Mass on Sunday, you dress as though you've been invited to the governor's house for dinner.' So when this whole concept was laid out to me—the concept of a much broader agenda, a much larger mandate than I'd ever seen the Church spell out, particularly coming from the bishop himself—that we'd have departments of social concerns, and educational affairs, and some of it was going to be confrontational—I'm thinking, *How is this going to be implemented?* You've got to *implement* it somehow. You can't just have these bishops sitting in a little room, saying, 'How are we going to get involved in the world?' You have to *go to the state capitol dome.* It was terribly exciting. To me, the Church had come alive."

The plan was being duplicated in states all over the country—most of the existing state Catholic conferences were established in the year or two immediately following the close of the Vatican Council. And in Missouri, during the inaugural weeks of the new Catholic Conference, Hiesberger found that his job had very little to do with abortion. The Conference was to serve as the capital city office for the eight hundred thousand Catholic people in the state of Missouri. It was to muddy the Church's feet, deliberately and vigorously, in the roiling waters of the Modern World; its staff would become expert in law, politics, educational policy, and "social concerns," the broad category that might include anything from migrant farmworkers to fair housing. The Conference was to defend the interests of Catholics, particularly when the parochial school system had a chance at state money or aid, and more generally—more excitingly, Hiesberger thought—it was to work for what the Pastoral Constitution had called the "common good," the "sum total of social conditions which allow people, either as groups or individuals, to reach their fulfillment more fully and more easily."

The Pastoral Constitution had gone on to list with some specificity just what some of those social conditions were: the right to be educated, to find work, to be clothed and housed, to build a family, to worship freely, to act according to the dictates of one's conscience. These were rights that applied to *all* people, the Church was now insisting, not simply Catholics. And just as those rights were to be defended, so was the Church in the modern world now commanded to attack offenses to the "human person" —offenses against human dignity, such as slavery or arbitrary imprisonment or degrading working conditions, and offenses, as the Pastoral Constitution phrased it, against "life itself."

Like murder, genocide, euthanasia, and suicide, abortion was in the wording of the Pastoral Constitution an offense against life itself. No distinction of grammar or punctuation separated it from these other offenses; in the thousands of pages of documents issued by the Vatican Council, induced abortion was mentioned only twice, and both those references appeared to take it for granted—just as Matt Backer's Catholic medical journals had taken it for granted—that the word "abortion" was to be uttered in the same breath with acts that *all* civilized people understood to be the most terrible of crimes. Of the three men hired as the early staff of the Missouri Catholic Conference, none of them—not Hiesberger, DeFeo, or Cox—could recall by the time of their adulthood any but the most passing mention of abortion; it was settled business in their youth, and not the sort of subject one brought up for serious debate. Even when Hiesberger thought about his seminary training in St. Louis, it seemed to him that abortion had been dismissed as the kind of evil plausible only for the most abstract moral discussion, to be raised in the same voice one might use to talk about suggesting to a distraught unwed mother that she take her newborn down to the river and drown it.

Then some months into Hiesberger's first year on the job, 1967, as though the American Law Institute and the Second Vatican Council had converged from vastly different directions to collide right in front of him, abortion arrived at the Missouri statehouse. That was the year of Senator Robert Prange's first effort to change the Missouri abortion law; and by 1971, when Prange's modest Model Penal Code health-of-the-mother bill had been superseded by the much broader any-reason-until-viability bill, it was evident to the Catholic Conference staff that part of their responsibility—both to Catholics *and* to the "common good"—included marshaling the Church's ample resources to fight these bills off.

From the office of the Catholic Conference came advisories, bill excerpts, lists of the district address of every member of the House Civil and Criminal Procedures Committee. "IT IS MOST IMPORTANT THAT THE COMMITTEE MEMBERS LISTED BELOW RECEIVE *PERSONAL CONTACT* FROM YOU OR A GROUP IN YOUR AREA WHILE THEY ARE AT HOME THIS WEEKEND," Hiesberger could declare in a single morning's typing to five hundred parishes at once. And they had done a creditable job: for six years, as the statutes changed in one state after another, no legal abortion bill had survived a final vote in the Missouri statehouse.

Surely this was not all the Catholic Conference's doing—Missouri was known both inside and outside its borders as a mulish state, famously reluctant about being hauled anyplace new. It was heavily rural. Its two populous urban regions, Kansas City and St. Louis, tended toward fierce competition both with the farm country and with each other, producing stalemates that worked against political innovation. The state had sizable church affiliation among Protestant groups known generally for their conservatism: Southern Baptists, Assemblies of God, the Missouri Synod of the Lutheran Church. But Protestant organization was not like Catholic organization; even within their denominations no Protestant church fit into a centralized statewide chain of command the way a Catholic parish did. There was not a Protestant church official in Missouri who could at least theoretically reach eight hundred thousand people with the writing of a single letter.

Besides, the Protestants weren't *helping*. Bill Cox used to pack up his slides, after an evening at a Protestant church, and wonder whether some deep residue of anti-Catholicism was holding these people back: they were cordial, they listened to his talk, they put their hands to their mouths when they saw the dismemberment pictures, and then nobody at the Catholic Conference or the Fight for Life Task Force would ever hear from them again. In statehouse testimony they might find the occasional Lutheran or Baptist minister on their side—one Baptist minister from Springfield had concluded his passionate soliloquy against the 1971 liberalization bill by crying at full preacher's throttle, four times in a row, *"Thou shalt not kill!"*—but the leadership, the men who might have flagged the attention of the newspapers or helped even in a haphazard Protestant way to mobilize people by the thousands, had settled in the enemy camp.

The state's Episcopalian bishop had sided publicly with the legal abortion forces. The United Methodist Church had called for dropping abortion entirely from the criminal code. The Southern Baptists in Missouri maintained a regular presence in the statehouse through the lobbying efforts of Hugh Wamble, a tenacious and powerfully articulate university professor, but Wamble seemed to DeFeo and Hiesberger to save all his passion for fighting them on aid to parochial schools. The Southern Baptist Convention leadership had in any case been making overtures toward acceptance of legal abortion—a 1971 national resolution had approved in principle the Model Penal Code categories—and when Southern Baptist newspapers around the country received from Washington, D.C., their national news service's first January 1973 analysis of *Roe* v. *Wade*, the lead sentence read as though the high point of the ruling was its philosophical slap at the Catholic Church: "The U.S. Supreme Court, in a 7–2 decision that overturned a Texas law which denied a woman the right of abortion except to save her life, has advanced the cause of religious liberty, human equality, and justice."

At the Catholic Conference it was a mystery, this abandonment by men who professed to have dedicated their lives to the service of God, and Tony Hiesberger used to lie in bed at night and work it around in his mind, trying to make sense of it. The Vatican Council had exhilarated Hiesberger from the first high-pomp day of its opening ceremonies, and he had seized with particular gratitude the conciliar messages about ecumenism, about people of faith uniting for the betterment of the world. Hiesberger and other Christians had toiled together in the vineyards, literal toil and almost literal vineyards; one of their projects was a nonprofit group to help the farmworkers who migrated in deep poverty from one crop to another, and he knew that "sanctity of life" was a phrase that genuinely inspired the men and women who traveled with him to the agricultural fields.

How could they care so deeply about the sanctity of a human person only *after* the moment of birth? Was the poverty of one child so much more compelling than the death by abortion of another? They could tell him as many times as they wanted that he was parroting the dictates of the Church when he extended "sanctity of life" to the still unborn; Hiesberger didn't believe it. It was like telling him he thought the sky was blue at midday because his Catholic mother had once named the colors for him. Lou DeFeo had read a passage in James Boswell's *Life of Samuel Johnson* about Johnson and Boswell listening to a cleverly structured argument that matter itself exists only in the mind; Boswell says the argument is obviously wrong, but that he cannot think how to refute it, whereupon Johnson kicks the nearest rock as hard as he can. "I refute it *thus*," Johnson declares, and that was how DeFeo felt sometimes, listening again and again to elaborate arguments about how taking an unborn life could be regarded as a moral act. *I refute it thus*, he wanted to say, slapping a Lennart Nilsson picture onto a desktop and walking away; DeFeo was a legislation man and powerful photos were not really his style, but it made him tired listening to people rationalize elaborately about "personal privacy" and "quality of life" when the reality in front of them was as solid as Samuel Johnson's rock.

You couldn't wish reality away, even if you did wear a clerical collar: DeFeo had said something like that to a whole state committee once, when he was testifying on the 1971 bill at a hearing that had gone on until two in the morning and he looked up at the black hands of the big clock above the dais. You can move those hands back if you want, DeFeo had told the exhausted legislators, trying to show what he meant, but that will not change the fact that *it really is two in the morning.* Reality defies our efforts to redefine it, DeFeo said at that hearing, or to twist it around with fancy words. People could substitute "fetus" for "baby," if they insisted on the emotional remove of the clinical terminology, but that did not change what grew inside a woman's womb: it *was* human, it *was* a life, abortion *did* kill it. Just because the Protestant leadership refused to acknowledge that didn't mean their own church members agreed with them.

And if the Catholic Conference could not work with Hugh Wamble or the Episcopalian archbishop or the General Conference of the United Methodist Church—if the big Protestant denominations were going to place in positions of leadership men with selective blindness to the common good—then Lou DeFeo was entirely willing to push on through the statehouse by himself.

SECTION 4. *The General Assembly of the State of Missouri further finds that there is a high degree of danger to maternal health and well-being in abortion techniques commonly referred to as the "Suction Abortion" and the "Dilation and Curettage Abortion," and that these techniques inflict on the unborn life within the woman on whom the abortion is to be performed cruel and inhumane treatment.* Everything went into the first post-*Roe* bills. Every construction a person could think of, every definition and restriction and deliberately worded phrase, every prohibition that was not exactly, technically, a prohibition on abortion itself—all of it was rushed into type, as DeFeo hurried back and forth between the Catholic Conference and the statehouse offices of the legislators he knew would join him, to beat the March 1 deadline for introducing new bills. "If we could have written a *taxing* law that would stop abortion, we would have," Lou DeFeo recalls. "From the beginning, the idea was: You give no gray area on this. You only give what's absolutely decided against you. If the courts want one more inch on this, we're going to force them to write an opinion that *says* so—that says, for example, that fathers don't have rights."

DeFeo and his legislative allies wrote father's consent into an abortion bill. They wrote a hospitals-only requirement into an abortion bill—no abortions to be performed in clinics or doctors' offices. They wrote one bill prohibiting saline abortions, and one bill protecting any person or institution that refused to participate in abortion, and one bill prohibiting saline, dilation and curettage, *and* suction abortions, which would have kept abortions technically legal while eliminating, with the curious exception of hysterotomy, every effective medical method for performing them.

The purpose of this act is to preserve and protect the life and health of pregnant women and the unborn life with which they are pregnant. A week after the deadline—less than a month and a half after the Supreme Court ruling that *The New York Times* had described as "historic resolution"—Louis DeFeo was able to tell the Missouri bishops that seven separate abortion-limiting bills had been introduced in the Missouri General Assembly, five of them written by or with help from the Catholic Conference.

"A number of Representatives were invited to meet in the House Lounge to discuss with us a response to the Supreme Court's action," DeFeo wrote his bishops. "Some 40 to 50 Representatives attended the meeting, many indicating their deep concern and desire to prohibit abortion in Missouri."

Lou DeFeo was a man who chose his vocabulary carefully, and the word "prohibit" was not a slip. It was important to keep the goal in sight. The goal was not regulation; society "regulates" a practice it has formally decided to accept, and if regulation was all they wanted, then they really were composing mechanical details for the Nazi distribution of yellow stars. The goal was *prohibition*, prohibition in the state of Missouri, and in a better world, perhaps, prohibition in the states across the borders as well. The *St. Louis Review* had called it precisely: for six years a nation full of isolated little battalions had been fighting abortion change state by state, with every victory in a place like Missouri made hollow by the legal abortion states a highway or an airplane ride away. In a single day the U.S. Supreme Court had given them all the one thing that could effectively bind them together, a common enemy, and DeFeo advised his bishops that Missouri legislators had already taken the first steps toward a national response—toward a vote that might eventually permit a Jefferson City politician to help make abortion illegal in Manhattan or Los Angeles. A House resolution had been signed by forty Missouri representatives and was aimed, DeFeo wrote, at the U.S. Congress. "If adopted," he wrote, "it would call upon Congress to take appropriate action to amend the United States Constitution to protect the human life of the unborn."

In St. Louis, as Lou DeFeo's memorandum arrived at the Archbishop's office, the chancery was already astir. The *Review* editor was an energetic priest named Edward O'Donnell, a New York–raised Depression baby with a strong handshake and a tough-guy build, and O'Donnell had laid out his

Roe v. *Wade* articles with mounting disbelief: life, liberty, human happiness, what were those words supposed to mean if the country was going to sanction killing before birth?

"It was *clear* to me that we were dealing with human life here," O'Donnell recalls. "The common human experience was that when a woman was pregnant, she was carrying a baby. Words like 'fetus' might have been used out there by medical students, but some of these—'zygote' —were like those little bones of the foot that the poor medical students have to learn. I think I reflected the common wisdom. And I think that's not just wisdom, it's the truth."

At O'Donnell's office the telephone began to ring, insistently, from the hour the news came over the radio. Why people were calling *him* O'Donnell didn't know, except that his name was printed in each week's *Review* and maybe they were shy about demanding to speak directly to the Archbishop. "They'd call and say, 'The Church has got to do something. What are we going to do about this?' And we would say, 'Well, there's this organization called Fight for Life, we urge you to call them.' And people would say, *'That's not good enough.'* I can remember people saying, 'I saw you in your collar in Selma, you're down there with your big collar in defense of all those blacks, where are you in defense of all these babies? *Don't shirk it off.'* "

And they were right: O'Donnell had been in Selma, during the civil rights marches, twice. Once six nuns came with him, in their habits; their pictures had been in the newspaper, and afterward they had gone on the radio to talk about racial discrimination and religious witness. O'Donnell believed the Church had done important civil rights work in St. Louis— the previous Archbishop, Joseph E. Ritter, had defied white opposition by desegregating the Catholic school system long before the Supreme Court prohibited racial segregation in public schools, and in 1963 Ritter had directed his priests to set up a human rights office, a full-time concern, with local committees at each parish and the headquarters in a good window office at the chancery. The symbolism was deliberate and serious; it was the first formal Catholic archdiocesan human rights office in the country, and for ten years one of O'Donnell's multiple job assignments was assisting the office's director as they set up parish committees and tried to respond to matters like job discrimination and fair housing complaints.

Was this the proper province of a religious institution? For a while there were a lot of white bigots who didn't think so, and some of them were Catholic. But Ritter had defied them and O'Donnell admired him for it: the Pastoral Constitution on the Church in the Modern World had urged exactly this upon them, the "solidarity and respectful attention for the whole human family," and as he stacked up the *Roe* v. *Wade* telephone messages, O'Donnell began to wonder whether the archdiocese shouldn't organize as purposefully and ambitiously against abortion as it had against racial discrimination. Occasional announcements in the parish newsletters were useful as far as they went, but what if there was an office in the chancery, a committee in every parish, a roster of steady organizers spread across the breadth of the archdiocese?

O'Donnell talked it through with Joseph Baker, a priest who wrote *Review* editorials and had also been fielding angry telephone calls from parishioners all over St. Louis. Together O'Donnell and Baker examined their resources: the St. Louis archdiocese covered one third of the state of Missouri, 6,000 square miles, 250 parishes, 550,000 Catholics. If they could summon a person or two from every parish they could fill a school auditorium for an organizational meeting, and as they sent out their invitational letters, with Cardinal Carberry's signature suggesting the command performance nature of this event, O'Donnell and Baker made arrangements with the nun in charge of a suburban St. Louis girls' high school.

They were expecting about five hundred people, they said. They were going to need folding chairs—maybe fewer than five hundred, because it would look discouraging to see empty chairs. The diocese would provide soft drinks and cookies for the after-meeting, and they wanted also the use of classrooms so that people could break into separate work groups and plan their first activities.

A small notice appeared, in regular type, in the *St. Louis Review:* at St. Joseph's Academy, on March 14, interested persons were invited to an evening meeting "aimed at ameliorating the effects of the Supreme Court's recent pro-abortion ruling and eventually overturning the Court decision." It was spring already, and warm in St. Louis. The priests had enlisted the help of a St. Louis parishioner named Mary Frances Horgan, a homemaker and experienced Democratic campaign volunteer who had just completed for the diocese a lengthy report on religious education in the local schools, and when Horgan pulled up to St. Joseph's for the March 14 meeting she became aware that she was actually dripping, the warmth was so sticky and still. She looked for a place to put her car. The spaces in front of the school were full, and the spaces a block away were full, and as Horgan kept driving with the windows down and her palms damp, she began to see with a small thrill of understanding that *there was no place to park.* The streets around St. Joseph's were *lined* with cars; there were cars on the sidewalk and cars angled against corners and cars pushed right up on the grass in front of St. Joseph's.

Horgan was a very long way from the school's front door when she finally found a parking place. She walked back quickly, dripping, and saw that she was not going to find a seat in the auditorium either.

The chairs were all taken. The aisles were jammed. Somebody had pulled down the foldaway bleachers to make more seating, but it still was not enough, people were smashed shoulder to shoulder against the walls, they had eaten all the cookies and drunk all the soda, Father O'Donnell was running back and forth trying to hold everything together. Up on the stage was the archbishop, in his "house cassock," the black robes with red buttons, the pastoral cross around his neck. He had not worn the full red robes he would have used for Mass, but neither was he in his community meeting clothes, the good men's suit with a prominent cross. The house cassock was a signal, we are here on Catholic property *giving witness as Catholics*, and when the meeting got underway, Mary Frances Horgan heard the opening words as both prayer and rallying cry: it was Bishop Joseph McNicholas who kicked it off, he was good at this, a vibrant bare-knuckles Spencer Tracy sort of priest who knew how to work a crowd. (McNicholas used to preface Mass during baseball season by promising church newcomers that he would finish in time for the first pitch.) And there were bowed heads now in the St. Joseph's auditorium when McNicholas began to speak, Horgan would remember that afterward, but when he cried *We are going to organize, and we are going to fight,* it was evident at once that nobody in that auditorium mistook it for a church: they cheered.

They cheered exuberantly, it seemed to Horgan, and perhaps with a political audience's premonition of the work that lay before them; her name was announced, as the chair of a new grassroots organization to be run by the Archdiocese of St. Louis, and as she walked toward the stage she heard a woman say quietly, taking in Mary Frances' earnest, rough-featured face, "Too bad she's not photogenic." It was an extraordinary night. Edward O'Donnell had never seen that many Catholics gathered in one place outside a High Mass or a school graduation; this was like a rally for some charismatic political candidate, except that there *was* no charismatic political candidate, there were two long Supreme Court opinions and a couple of bishops on a high school stage. When people were supposed to go home they stayed on anyway, and made a catastrophe of the small-group workshops; the classrooms were too small and the soft drinks had run out and the hallways were crowded, O'Donnell thought ruefully, with many chiefs and no Indians. Lawyers and doctors and parish leaders and political organizing veterans like Mary Frances Horgan—these were not people who were going to be docile about taking orders. It was nearly midnight by the time the school emptied out, and for a while Father O'Donnell and Mary Frances and a few of the others stayed behind in the quiet auditorium, talking, savoring, stacking chairs, and trying to console the St. Joseph's nun who had watched the crush of automobiles ravage the school's front lawn.

"The euphoria just went on from there," O'Donnell recalls. "Because everything we touched, for a long time afterward, turned to gold."

SEVEN

The Target

1973

ON JANUARY 22, 1973, as the first reports of the Supreme Court's abortion decisions were reaching the radio and wire services, a Washington, D.C., clinic director named Harry Levin picked up the telephone and called St. Louis to tell Judy Widdicombe the news.

It was 9:20 A.M. in Missouri, an hour earlier than Washington, and Judy took the call at her desk in the Pregnancy Consultation Service office —they had an office now, a physical headquarters on a public street, with a conference table that was kept clear so that Judy could ask women to lie back for a quick estimation as to how far their pregnancies had progressed. The headquarters looked nothing like a medical office; it was just a small St. Louis house with the furniture moved out and the bedrooms redone into counseling rooms.

Armchairs and a conference table and a set of decorator-colored telephones, that was as much abortion equipment as the law would permit Judy Widdicombe, and for two full years, as she studied the new abortion facilities in Washington and Kansas and California and New York, a private inventory had been multiplying in her head: *this* is how I would outfit it, *this* is how I would staff it, I could *do* this in St. Louis and *I could do it on my own.* A year earlier she had gone so far as to ask her board of directors for a formal vote of support for the eventual creation of a Pregnancy Consultation Service facility where a woman could have an abortion instead of simply talking about it, but the board's yes vote was nothing more than a note in a file drawer; as long as the state abortion law remained intact, all Judy could offer pregnant women was a grim kind of single-purpose travel agency, and in Missouri even that was an invitation to arrest.

Now she listened to Harry Levin, on the telephone from Washington. She made him say it again, to make sure she understood.

Every state? Judy Widdicombe asked.

Every state, Harry Levin said.

"Oh, my God," Judy Widdicombe said.

She called an emergency board meeting, and then she announced a press conference. The St. Louis reporters put Judy in their reaction stories. *A private abortion referral agency, Pregnancy Consultation Service, is making plans to establish its own clinic to provide abortion service, according to its director, Mrs. Judith Widdicombe.* She made up a date for her service's debut, March 8; she had nothing whatever to base it on, no building, no equipment, no budget, no staff. But it sounded so specific and imminent that she thought the reporters might write it into their articles.

"I knew we couldn't do it," she recalls. "But I also knew we needed to hook the press, because they were the best advertisers we had. And I knew the papers outside St. Louis would pick it up. I wanted the women to know that something had happened in Washington that *would* affect their lives in Missouri."

She didn't examine the opinions, as the first typed copies began to circulate around the country, and she was not sure precisely how the Missouri criminal abortion law was now supposed to vanish from the statute books. Judy didn't read statutes and she didn't follow lawsuits; the fortunes of *Roe* v. *Wade* and *Doe* v. *Bolton* had barely caught her attention as the two cases made their way through the federal judiciary, and she kept up with *Rodgers* v. *Danforth*, the Missouri case, only because Frank Susman had asked her to testify for him in court. All Judy knew was that a massive gate had suddenly lifted before her, and she intended to bolt straight through: she was going to build an abortion center, the first legal center in the largest Midwestern city south of Chicago. She had doctors to hire, real estate to examine, instrument catalogues to read. There was probably no one in Missouri who knew as intimately as Judy Widdicombe both the numbers of potential patients and the urgency of their need; if she pushed it, if she started up an abortion service faster than anybody else thought possible, she could keep an extra month or two's worth of women off the airplanes to New York.

But what exactly was that term supposed to mean? *Abortion service.* The only immutables required for an induced abortion were a pregnant woman and an abortionist. You could work an abortion service into a hospital gynecology wing; you could start up a citywide abortion service by trying to convince every gynecologist in town to keep abortion equipment in his office; you could run an efficient and far-reaching abortion service, as Judy knew in particularly graphic detail, from the back bedroom of a rental apartment. There were no textbooks, in January 1973: the concept of elective abortion as a legitimate medical service was still so new that it seemed to Judy they were all making it up as they went along. All she had to work from was what she had already seen, and what Judy Widdicombe had seen so far were direct by-products of pre-1973 state abortion laws—from the Corpsman's operation in Chicago to the bustling clinics of New York.

New York was the nation's real testing ground, certainly. California's abortion business had grown in more gradual and haphazard fashion, as the state's 1967 Model Penal Code law was reinterpreted in stages over the years, but in New York grand-scale legal abortion had arrived all at once, in the summer of 1970, amid enormous publicity, and with the bulk of the practice centered in the relatively small geographical area of New York City.

And because the famous no-questions-asked New York abortion law included no instructions as to how or where abortions were to be performed, the state's medical officials and abortion activists had spent many months in heated argument about what the new abortion practice was going to look like. Was it to follow established medical tradition or break out into something entirely new? What was safe? What was practical? How could a lot of physicians suddenly take on a procedure at which they had almost no experience and keep from creating one massive nationally scrutinized disaster? "Next January, the Legislature will meet again and say, 'See what a mess they've made of the abortion law,' and they'll rescind it," warned Dr. Robert Hall, the obstetrician-gynecologist president of the Association for the Study of Abortion, as he worried in a 1970 *New York Times* interview about the dangerous possibilities of starting up legal abortion the wrong way. "The other 49 states, looking at the New York experience, will see us screwing it up and will stay away from abortion repeal. I'm even naive enough to believe that Supreme Court justices read the papers, and they will wonder why they should legalize abortion repeal for the whole country if this is the way we behave in New York."

At the center of the New York arguments was a single medically and philosophically loaded question: *Where are we going to do this?* For many years the conventional site for a proper medical abortion, in New York and every other state in the country, had been the hospital operating room. If a woman was able to work her way into the medical system—if she was one of the rare patients genuinely endangered by pregnancy, or had somehow managed to convince a hospital committee that her condition required a "therapeutic"—then she was aborted with surgical instruments, often under general anesthesia, in a procedure that required an overnight hospital stay. It was a costly and time-consuming procedure, requiring payments to physician and anesthesiologist and hospital, and it efficiently weeded out the poor; medical studies showed, to no one's surprise, that a private hospital patient was far likelier than her public hospital counterpart to be approved for a therapeutic abortion.

Hospital abortion was expensive, heavily medical, and loaded with disapproving signals about the gravity of the act being undertaken; an abortion patient might find herself attended by a hostile nurse, or bedded in the obstetrics ward next to a woman in labor. And even the more abortion-receptive hospitals had only a limited number of available beds and operating rooms. And instrumental abortion under general anesthetic was not the easiest procedure to carry off. For the physician with sharp curette in hand, performing an instrumental abortion was a little like putting on a blindfold and then scraping out the inside of an orange: too timid a scrape, and the woman might still be pregnant when the operation was finished; too strong a scrape, and the curette could slice up into the pregnancy-softened tissue of the uterine wall.

Then in the mid-1960s, word began to spread—by rumor, by intriguing medical articles from halfway around the world, and finally by firsthand accounts from doctors who grasped at once the implications of what they had just seen—that there was an entirely different way to extract the contents of the uterus. A Los Angeles gynecologist recalls attending a national physicians' meeting in Chicago during these years and listening with astonishment as a Washington, D.C., doctor regaled his colleagues with his story of completing an abortion for a Soviet diplomat's wife: the lady was suffering a miscarriage, the Washington doctor said, and as he began inserting his instruments to remove the last of the pregnancy, he was chided by the Soviet doctor who for protocol reasons had accompanied the patient to the hospital.

"So the Russian doctor was in the operating room," the Los Angeles gynecologist recalls. "And he said, 'How come you don't vacuum her?' None of us had ever heard of this vacuum machine before. I couldn't wait to get my hands on one."

Vacuum aspiration itself was not a wholly new principle in American medical care: suction was already used for tasks like removing fluids during surgery, and gynecologists sometimes pulled tiny amounts of biopsy tissue through a narrow vacuum tube. But the idea of emptying a pregnant uterus with a mechanically driven suction pump did not begin reaching this country's mainstream medical literature until 1967, when influential publications like *Obstetrics and Gynecology* and the British *Lancet* printed the first studies of a decade of suction abortion in Eastern Europe and Asia.* If the international reports were to be believed, a vacuum abortion

---

* Printed vacuum aspiration studies began appearing in foreign medical journals as early as 1958, when a Chinese journal published a report of three hundred abortions performed—with what a translated summary described as "very satisfactory" results —using rubber tubing and a suction pump.

could be completed in *two minutes,* without general anesthetic, and with only a remote chance of accidentally slicing through the wall of the uterus. All the abortionist had to do was insert an open-mouthed tube through the cervix, turn on the suction machine, and guide the tube around the inside of the uterus, as though vacuuming the corners of a room; there was no cutting, no deliberate pressure with a sharpened blade, and far less likelihood of leaving part of the pregnancy behind.

What that meant, if it was true, was extraordinary. It could transform the sterile medical abortion from an expensive operating-room procedure to a ten-minute in-and-out that might—at least in theory—be managed competently by any physician with a vacuum in his office and a working knowledge of basic gynecology. In San Francisco, where sympathetic doctors had made the University of California Medical Center a relatively easy place for German measles–exposed women to receive hospital abortions during the 1960s rubella epidemic, a visiting Israeli physician had so prodded staff gynecologists with his descriptions of vacuum that one of the chief gynecologists enlisted a local engineer named William Murr to try making a pump of his own—to let the medical staff see if this machinery really worked as well as the foreigners seemed to think it did.

Murr did not know very much about abortion; he was a Southern California–born entrepreneur, a Cal Tech dropout who had moved to Berkeley and started a successful business building tonometers, which are used by optometrists to measure pressure in the human eye. Murr did know about vacuum, though, and at the U.C. professor's urging, he traveled to Tokyo to study the desktop-sized pumps that the Japanese were using for abortions. The Japanese metal suction tubes looked awfully narrow and sharp to Murr, which worried him (this equipment was supposed to reduce the chance of perforation, after all), and back in the workshop of his Berkeley tonometer company he began cobbling together his own prototype—a toaster-sized mechanical vacuum pump, improvised from a paint sprayer that was set to operate in reverse, with two glass collection bottles and tubing made of clear plastic.

"I wanted the guy to be able to see what he was doing," Murr recalls. He worked closely with Dr. Edmund Overstreet, a University of California gynecology professor who had become an emphatic critic of criminal abortion laws, and neither the physician nor the engineer had any hesitation at all about the ethics of refining machinery designed to whisk a human pregnancy into a small glass jar. "He was adamant," Murr recalls of Overstreet, who died in 1983. "He said that one way or another, medicine was going to have to give this kind of relief to women who are pregnant and who find it unacceptable—psychologically, socially—that we *have* to give them a safe and easy way out. He regarded it as a fait accompli when it wasn't yet, no question about it."

The machinery worked. There were drawbacks during its initial design stages—Overstreet's colleague Alan Margolis remembers the early cannulas, the open-ended tubes that enter the uterus, as so rigid and wide that he was afraid that if a doctor did perforate the uterine wall, the vacuum would immediately begin sucking up bowel. So Murr kept tinkering with the design, his entrepreneurial instincts now in full swing, and in 1968, as California's year-old Model Penal Code law was settling into place and a few other states were beginning to modify their own criminal abortion laws, the Berkeley Tonometer Company released its first advertising brochure for the "VC-1," available with the trademarked "Vacurette" small plastic cannulas, the conveniently located Teflon suction-control ring, and the Crouse-Hinds explosion-proof plug. "The pump has been selected for absolute minimum maintenance requirements in addition to the capacities mentioned above," read Murr's debut brochure. "It is driven by a rugged, explosion-proof 1/4 hp. motor, and the combination will give years of service without special maintenance or attention."

Murr was eager to show off his handiwork, and he set up a booth the following year at a public health convention in Miami. He might as well have been promoting the public health benefits of guillotines. "Boy, we had sort of a wall of frigid air around us," Murr recalls. "Everybody moved to the other side of the room and walked *fast*. About one public health nurse out of fifty or a hundred would stop by and pick up a leaflet and walk on." But the VC-1 was selling, one at a time; Murr sent one machine to Denver, another to Philadelphia. And Alan Margolis had taken an intense interest in the possibilities of the vacuum machine; he could see at once how profoundly the *idea* of abortion might change if terminations could be rapidly and easily completed inside a doctor's office. Under the new California law it was illegal to perform an abortion outside a hospital, so Margolis set up a Berkeley Tonometer aspirator in his own office— located within the UCSF hospital—and began taking notes on the safety and efficiency of the new mechanical procedures in an office setting.

They went, according to Margolis' memory and written reports, without a hitch. "It was just a hell of a lot easier," Margolis recalls. "And it was obviously a much safer thing to do."

Thus by the summer of 1970, as the state of New York prepared for the July 1 lifting of almost all restrictions on abortion up to twenty-four weeks, a tiny collection of medical equipment companies (at least one East Coast firm had also begun adapting suction devices in response to the international abortion reports) was making machinery that had the potential to help redefine the entire procedure by which pregnancies were brought to an early end. With its Eastern and European competitors pushing hard for the promising New York market, Berkeley Bio-Engineering (Murr had changed the company name to keep up with his more diverse product line) helped set the national standard for a machine that—for most first-trimester abortions, at least—might render the hospital operating room almost completely irrelevant.

This was a deeply unsettling proposition in 1970, and more conservative medical forces dug their heels in to fight it: the machines were still new, medical surveillance was tightest inside hospitals, it was wildly irresponsible to take in thousands of women at once without the attention and emergency backup available in established hospitals. Association for the Study of Abortion president Robert Hall had already faced catcalls from more impatient legal abortion advocates by arguing not only for hospitals-only abortions but also for a state residency requirement, just to keep the whole enterprise manageable during its first trial year.

"To cope with even 50,000 abortions," Hall said in a *New York Times* interview, citing the common estimate of the number of New York women likely to seek abortions the first year after the law changed, "eighty to ninety percent of them will have to be done on an outpatient basis, using suction and minimal anesthesia. There isn't one doctor in the state of New York who has ever done a legal abortion under those three conditions— outpatient, local anesthesia, and suction. To do 50,000 with these strange new techniques will be trouble enough in a hospital."

But if New York imposed no residency requirement, Hall worried, every state in the Northeast was going to begin dumping its pregnant women into the medical facilities of Manhattan. Where were those women going to go? It was not as though the local doctor in Hartford or Cleveland was about to offer up the address of a nice Upper East Side gynecologist with an aspirator in the back room. Hall could see what was coming, and he despaired about it aloud: *clinics.* "Let the floodgates open, let in 500,000, and you will have to have independent clinics to accommodate them," he said. "Then you will have deaths, profiteering, gruesome stories on the front page of the papers."

Hall didn't use the words "abortion mill," but he didn't need to; the image was as vivid as if he had copied out one of the distressing line drawings from the right-to-life leaflets. For a long time it had been standard practice in the state-by-state abortion fights to deliver alarming predictions that loosening state laws would visit upon the region in question an onslaught of "abortion mills," a term that was never explicitly defined but was universally understood to describe brutal, unsanitary institutions in which abortions were cranked out as mechanically as new steel or crushed wheat. It was an image drawn from the illegal days, but what made it upsetting in the public discourse had less to do with legality or hygienic standards than with numbers: an abortion mill was a place that would do a *lot* of abortions, hundreds, perhaps thousands, and worse yet, it would do *nothing else.* An abortion mill would operate without the comfort of euphemism, and in every state in which this specter had been raised, it was instantly understood to be so obviously dreadful—the idea of a single, legal, publicly identified facility that drew crowds of women solely to abort them and send them home—that the standard rebuttal was repeated reassurance to the people: modernizing our statute will not bring abortion mills into the state. Nobody ever stood up in legislative debate to call for *better* abortion mills, *safer* abortion mills, abortion mills for the Chamber of Commerce to brag about.

But the more people argued about it in the state of New York, the surer the outcome appeared to be. Timing, medical intransigence, and the final wording of the law worked together in the summer of 1970 to shape irrevocably the model for abortion service in what was about to become the single most concentrated abortion center in the United States. Because the no-residency-requirement forces prevailed, making abortion legally available to any pregnant woman who could get herself to New York, all the cautious intentions in the world were not going to be able to keep the action inside hospitals. Dr. Bernard Nathanson, a Manhattan ob/gyn with multiple hospital appointments who had joined in the founding of the National Association for Repeal of Abortion Laws, typed out a long 1970 memorandum predicting the chaos that would befall the hospital system if abortion service was not dispersed somehow into the larger medical community:

> It must be clear that the unimaginably large demand for abortion in the very near future will simply swamp the hospitals and pre-empt all hospital beds and operating time. . . . Under this strain, even operating on a twelve-hour day seven-day-a-week basis, hospitals would simply have to close their doors to all but emergency cases in other specialties and categories. In addition to the inability to care for elective medical and surgical patients, the residency teaching programs in the major and minor specialties would cease to exist.

The simplicity of vacuum abortions made their two options simple, Nathanson wrote. Qualified gynecologists could offer abortions in their offices—perhaps a temporary "'cottage-industry' approach to a massive problem," he added, but one that might at least be implemented right away. And although the prospect of doctors' office abortions alarmed health officials, both because of the difficulty of monitoring results and because even simple suction procedures required some form of emergency backup, there was another more fundamental argument in favor of doctors' office abortions. Nathanson did not mention it in his memorandum, but other physicians in a fever of early optimism had already imagined a time when elective vacuum abortions might be seen as ordinary mainstream gynecology—part of the regular array of services any local good ob/gyn or family practitioner might be expected to offer. "There are 22,000 OBG men in this country," one plaintive physician wrote in *Medical Economics*, describing the onslaught of abortion patients referred to his office after the doctor and his partner declared aloud that they were responding to the New York law by including abortion in their gynecological practice. "If only half share the obligation, that will mean two abortions per physician per week."

That the *Medical Economics* author felt compelled to write these words anonymously—his entire article, entitled "Suddenly I'm a Legal Abortionist," was written using pseudonyms for the doctor, his partner, and even the name of his "downstate New York community"—placed the problem in precise focus. Alan Margolis had seen it coming early on, even as he exulted at the practical possibilities of the new vacuum machines: "Colleagues would say, Well, I'll refer patients to you, my hospital board is conservative, I've got plenty to do anyhow," he recalls. "There were a variety of rationalizations behind it. It was quite obvious that they were reluctant to risk being known as abortionists."

A century of criminality had laid down too thick a patina to be stripped away by the sudden rewording of the law: you couldn't *get* abortions into doctors' offices, not enough of them to make a difference, certainly not enough to accommodate the out-of-state women that independent private practitioners would probably be reluctant to treat even under less controversial circumstances. So strong was the stigma—and, just as important, the conviction that others would never let go of the stigma—that a doctor could invite it in force simply by adding a few abortions to his otherwise unremarkable caseload. "Dr. Julienne" of the pseudonymous *Medical Economics* article described episode by episode his own colleagues' icy reaction to his and his partner's decision to offer abortions: "Len is not thin-skinned, nor am I, but both of us have been made uncomfortable by formerly friendly doctors who have passed by empty seats at our table in the coffee shop to squeeze into places at tables considerably more crowded," he wrote. "Colleagues who used to greet me amiably in the corridors now pass by with a curt nod."

For the new abortion patients, then—for the many thousands of pregnant women about to descend upon New York—the only practical option left appeared to be clinics, specialized facilities, medical centers that were going to have to try to recast in the public eye the entire notion of the "abortion mill." In the new era of legal abortion, specialized clinics could prove themselves safer and better equipped than doctors' offices, cheaper and quicker than hospitals; the best abortion clinics might even be staffed by doctors who could take pride in calling themselves experienced abortionists. "The most desirable solution is the construction by private interested groups (with or without government subsidy) of out-patient facilities *totally devoted to abortion*," Bernard Nathanson wrote in his memorandum (emphasis his). "These facilities would naturally be built to strict specifications and periodically rigidly inspected."

It took some time to sway the reluctant medical societies and local boards of health; for some weeks after the July 1 start-up date, New York health officials continued to press for strict guidelines keeping abortions inside the hospitals—or at least inside elaborately equipped clinics with close hospital affiliations. But NARAL and other legal abortion advocates kept agitating for more flexible regulations ("Raped by Medical Bureaucracy," read the picket signs at one of the public hearings on abortion guidelines), and by the time Judy Widdicombe began making the Manhattan rounds in late 1970 and early 1971, there was no question at all about where her New York–bound Missouri women were going to be obtaining their abortions. The list of business names was multiplying rapidly: Eastern Women's Medical Center, Manhattan Women's Center, the Center for Reproductive and Sexual Health, Park East, Park West, Park-Med. An obstetrics and gynecology professor wrote a paper for a New York medical conference in November 1971 entitled "Free Standing Abortion Clinics: A New Phenomenon," and the doctor marveled at their scope:

> Of the some 26-odd clinics that exist in metropolitan New York, the range extends from a plush, converted doctor's office on Fifth Avenue —replete with African sculpture and copies of *Vogue* magazine—to a 20,000-square-foot complex on the East Side, which is now involved in the termination of pregnancy and is beginning to become involved in such things as ligations of the vasa and outpatient tubal ligation by laparoscopy. . . . It makes little difference to argue whether we should give care to patients who come from other parts of the country or whether we can; the fact of the matter is that we did and we are doing so, and the clinics have made this possible. Patients have always flocked to areas where superior kinds of medical care were available.

Health officials had settled on their formal standards for what was acceptable: within the five boroughs of New York, for example, the city health code now required abortions to be performed "where there is qualified supervision in obstetrics or surgery, and where equipment, staff and facilities are provided to handle hemorrhage, shock, cardiac arrest and other emergencies." That satisfied NARAL and cleared the way both for reasonably well-equipped clinics and for the small private hospitals that were enthusiastically courting the high-volume new abortion business. The code did prohibit doctors' office abortions within the city boundaries ("no restraint on the city physician who doesn't mind a little commuting," *Medical Economics* pointed out), but Judy had no interest in some lone doctor's abortion practice anyway. She needed volume, a place that could accommodate a whole carful of out-of-state women at a time, and she needed fees so low that women could afford them after already laying out $300 for the airfare to New York. The clinic system made sense, for a Midwestern abortion conspirator with no particular reverence for either doctors or hospitals: they were creating something new, all the impatient men and women of the legal abortion movement, and it seemed to Judy that they might as well create a new kind of medical service to accommodate it—even if the very service they created continued to exile elective abortion from conventional gynecological care. "*We* needed to write the textbook," she recalls. "*Then* we could think about mainstreaming it."

So Judy knew exactly what she wanted, the morning the call came from Washington, D.C.: she wanted a clinic, a Free Standing Abortion Clinic, and she wanted to run it herself. The lawyer Frank Susman said she couldn't do it and some of the Pregnancy Consultation Service board members agreed with him, but Judy didn't care, she knew she wasn't a doctor or a businesswoman or some high-salaried hospital administrator, she was just a St. Louis registered nurse with a big voice and a frantic feeling about a particular thing that happened only to women, and *she was going to do this*. She wrote down a name: Reproductive Health Services. She took $500 out of a savings account and bought stock in Berkeley Bio-Engineering. She got on the telephone and began calling the managers of medical buildings, asking questions, comparison shopping, looking for three thousand square feet with satisfactory plumbing and room to expand.

SHE WANTED it near a hospital, on a major bus line, centrally located, with an entrance big enough to allow an emergency stretcher through. She wanted it in a multistoried building, preferably up on a higher floor, and surrounded by other medical offices so that a woman could walk in and out without being pegged as an abortion patient. Nobody else liked the name "Reproductive Health Services"; they thought it would be confused with the local Masters and Johnson enterprise called the Reproductive Biology Research Foundation, but Judy was adamant. She wanted the name to have breadth beyond abortion, to suggest the larger picture. She was adamant about everything. She knew how she wanted the furniture to look, she knew she wanted running water in every procedure room, she knew she wanted volunteer counselors working side by side with paid staff. In later years people like Frank Susman would shake their heads in exasperated admiration, remembering Judy in full steamroller mode; the complete *Roe* and *Doe* opinions had barely made it into the law libraries when Judy was hunching over her office desk with catalogues full of pictures and fine-print technical descriptions of surgical instruments and examining tables and sterilizing machines. The supply salesman working with her was sixty years old and had been handling medical equipment since 1929, but he had never encountered anyone quite like Judy; he used to say to people that she could stand up to the president of the United States and tell him to go fly a kite.

Judy worked from models, the clinics she had watched in action, picking and choosing the parts she liked. New York's Women's Services didn't charge very much for its abortions, she liked that, but the waiting room and halls were so crowded that she always felt exhausted just walking in the front door. Preterm, which was Harry Levin's excellent clinic in Washington, D.C., ran its business as a nonprofit and printed up useful brochures about the importance of counseling; Judy was adamant about that, too, that her clinic should run as a nonprofit community service and not as some investor's guaranteed twenty percent return. But the Preterm administrators had tried so hard to keep the clinic from looking sterile and forbidding that Judy thought some of the signals had gotten mixed: the recovery area looked like a living room, comfortable couches and overstuffed chairs, and what did that suggest to a woman just emerging from her abortion? These were medical patients, not houseguests. It was a clinic's job to remind each woman that this was serious business, deciding to undergo an elective abortion; it was minor surgery but it was surgery nonetheless, there were physical complications sometimes, and a woman must understand how important it was to watch herself afterward if she bled too long or her temperature began to climb.

Making it look warm and medical at the same time, that was the trick. And Judy didn't want the walls lined with Women's Liberation posters, either. She was interested in what the feminists had to say; she liked some of the provocative articles she saw in the defiant new glossy magazine called *Ms.*, and she had read certain passages in *The Feminine Mystique* so many times that she kept her dog-eared copy in the kitchen, with her cookbooks. Judy had understood exactly what Betty Friedan meant when she wrote about The Problem That Has No Name; it seemed to Judy that some years back she had smacked right into the Problem, and she had been vastly relieved to learn that she was not the only one who had it, that she was not crazy, that a woman could love her husband and her children and also want to leave them every day for the other life that was her volunteer work or her academic studies or her job. The Widdicombes had always needed Judy's nursing income, because Art had never made much money with his newspaper delivery routes and his occasional business schemes, but if they had been wealthy Judy would have found some other reason to put on a uniform and go to work. She *liked* going to work. She loved nursing, the urgency and seriousness of the work, the friendship with other nurses, the reassuring touch on a patient's arm. At work she always felt that she was stretching to her full height, which was substantial, five foot nine, with the wide Judy shoulders and the now-familiar impatience with authority in general and male authority in particular. Judy liked to suppose that doctors rolled their eyes behind her back and said to each other, *Well, I just don't think I'm going to try to intimidate this broad,* and when she read some of the modern feminist writing she was grateful for what seemed to her the central premise, that this might be an admirable thing for women to imagine, that women had as much right as men to power and opportunity in the world outside the home.

It was when they slid into revolutionist broadside that they lost her. Now and then someone would pass Judy a sampling of the more radical women's papers, with their uneven typeface and their fierce political cartoons and the title pages that suggested the vigor and inventiveness with which the Problem was being named: *Off Our Backs; It Ain't Me, Babe; Goodbye to All That*. Some of these publications were originating in the Midwest (*Ain't I a Woman*, with reports from the Red Women's Detachment and predictions of the free nationwide medical care that would come with the Revolution, was published by a collective in Iowa City), but the most active centers of rebellion appeared to be in California and the Northeast, and when Judy looked at the radical journals she felt like some prim Missouri nurse ducking fusillades from both coasts. Everything seemed to be making these women angry, not just housewifely frustration or unfair treatment in the workplace, and it mystified Judy to imagine that she was meant to feel oppressed or defensive because she drove a station wagon and liked fixing up her living room and paid off the boys' Little League dues by working the concession booth on Saturday afternoons. In the pages of these journals legal abortion was perceived not only as a woman's right but as emblematic of a grander battle against the male world in general and the male medical world in particular:

> A group of sisters did a witch action (witch doctors performing last rites over the expired Dr. AMA),

read a 1970 account, published in *Ain't I a Woman*, of a women's Chicago protest against the American Medical Association.

> . . . the rest of us milled through the audience and out in the halls of the Palmer House Hotel handing out leaflets and rapping with people. At one point a particularly arrogant, obese meatball representing the Society for the Protection of the Unborn started giving an anti-abortion talk ("then the doctor pulls out a tiny arm, perhaps still wiggling, torn loose at the elbow") which tried to equate (Dig This!) Women's Liberationists with the Marquis de Sade. From where we were scattered in the audience, we couldn't get him off the stage, so about ten of us finally stalked up to him, surrounded him at the microphone, and just stood there, with our arms folded across our chests, looking and feeling very menacing. Someone started pushing him, and a sister took the mike, and with loud support from most of the audience, gave our position, stressing the idea of voluntary maternity as a basic right.

Here was a starting principle they could all agree upon, surely: voluntary maternity as a basic right. If she had been pressed on it Judy might have been amused by the witch doctors and the menacing women with their arms across their chests, but tactically, philosophically, they were marching around on another *planet*. From the decorous remove of the St. Louis suburbs it was an effort simply to keep current with the vocabulary. There were complicated subdivisions to the women's movement now, the Women's Rights groups and the Women's Liberation groups; the Rights groups wanted access to the things men got, and the Liberation groups wanted to smash the system and start all over; and "women's health," which Judy was under the impression she had studied rather extensively as a registered nurse with a specialty in obstetrics and gynecology, was now being capitalized into new and ambitious meaning so that it was a movement too, with milestones and manifestos and publications of its own. "Learning about our womanhood from the inside out has allowed us to cross over some of the socially created barriers of race, color, income and class, and to feel a sense of identity with all women in the experience of being female," the women of the Boston Women's Health Collective wrote in the introduction to *Our Bodies, Ourselves*, which grew from a 1969 set of mimeographed teaching materials to a large-format paperback that was to sell three million copies:

> We have been asked why this is exclusively a book about women, why we have restricted our course to women. Our answer is that we are women and, as women, do not consider ourselves experts on men (as men through the centuries have presumed to be experts on us). . . . For some of us it was the first time we had looked critically, and with strength, at the existing institutions serving us. The experience of learning just how little control we had over our lives and bodies, the coming together out of isolation to learn from each other in order to define what we needed, and the experience of supporting one another in demanding the changes that grew out of our developing critique—all were crucial and formative political experiences for us. We have felt our potential power as a force for political and social change.

"Power" was a recurrent theme in the writings of the new women's health activists; so was "control." The American health-care system, with its corporate bureaucracies and its positions of authority filled principally by men, had provided an ideal landing spot for women interested both in feminist ideas and in the revolutionary societal transformation envisioned by the New Left: if one was seeking out evidence of the ways modern society could be said to demean women, especially poor women, it was hard to find a more vivid setting than a hospital or doctor's office. "We're drugged, strapped, cut, ignored, enemaed, probed, shaved—all in the name of 'superior care,' " declared a Los Angeles activist named Carol Downer in a 1972 speech to the American Psychological Association. "How can we rescue ourselves from the dilemma that male supremacy has landed us in? The solution is simple. We women must take women's medicine back into our own hands."

The chapter headings in *Our Bodies, Ourselves* formed a kind of index to some of the early concerns of the women's health activists: "Sexuality," "Venereal Disease," "Medical Institutions," "Prepared Childbirth." But for the women performing the physical work of the movement—writing the literature, working the telephones, attending the conferences and discussion groups—no other issue had the immediate and visceral drawing power of abortion, which after all was not simply mishandled by a callous and male-dominated medical system, but was explicitly *illegal* in most states. Carol Downer had become one of the central figures in West Coast feminist women's health groups by the time she gave that 1972 speech, and her own initiation was instructive: four years earlier, when she was a Los Angeles homemaker raising six children, Downer had attended a NOW chapter meeting chiefly to satisfy her curiosity about the women's rights activists whose exploits were beginning to show up in news reports. She picked abortion, when she was offered her choice of committees to join, and she picked it because she understood instantly how important it was: Downer had undergone an illegal abortion herself, when she had four children already and was divorcing her first husband, and the second-floor walk-up doctor, or whatever he was, had stuffed into her a gauze that dried inside her cervix to such a sharp hard edge that when she began to ease it out the next day, completing the abortion in her bathroom, she felt as though she were pulling a razor blade through her vagina. "I couldn't get more than an inch before I'd be in a cold sweat and ready to faint," Downer recalls. "It took me *hours* to pull this gauze out of me. But I did."

One of Carol Downer's chief companions, on her Los Angeles NOW abortion committee, was the savagely articulate California abortion crusader Lana Phelan, whose public account of her own illegal abortion had been transfixing lecture audiences since the mid-1960s. Phelan and her Northern California colleague Pat Maginnis co-wrote a small underground classic called *The Abortion Handbook*, which managed within its two hundred pages both to assail the American medical and legal system and to provide angry, entertaining, deeply cynical advice to women seeking abortion themselves. "Abortion laws are woman-control laws, or chattel laws, if you prefer," they declared in the introduction, and the chapters proceeded into the practical: "How to Reach Your Backyard Abortionist"; "Faking the Hemorrhage"; "Instant Psychoses for the Pregnant Woman." Here was the *Handbook* on some suitable behavior for the woman trying for one of the Model Penal Code's "preservation of mental health" abortions:

Now you are in the psychiatrist's office. Rule No. 1 is do not exhibit one glimmer of good health or mental stability. These two attributes on your part can get you scratched right off the Doctor's Hospital Abortion Committee Quota List. Isn't that an impressive title? They mean it to be. It can well sentence you to death if you don't respect it properly.

Tell the psychiatrist you want your pregnancy terminated, aborted, whatever euphemism you choose. He will proceed to reiterate the statements of your ob-gyn, i.e., "You'll just love it when it's born." Women call this "the big lie." He knows better, you know better, but don't argue the point right now. If you feel sufficiently nauseated, be sick on his nice rug. Tell him you often get sick like this; you did the other day when you flushed the new kittens down the toilet. This is a good opener, and will shake him up regarding your maternal qualifications.

Carol Downer apprenticed herself to Lana Phelan, listening to Phelan's talks and studying her arguments, and through the NOW committee Downer was introduced to the West Los Angeles clinic of a charismatic young abortionist named Harvey Karman. Karman was in the process of making a controversial name for himself among the nation's abortion pioneers: he performed abortions despite the fact that he was not a medical doctor (Karman identified himself as a psychologist), and in subsequent years was to come under criticism for promoting what was later judged to be an unsuccessful and apparently hazardous second-trimester abortion method.* But during the initial period of state abortion law change Karman had designed some medical devices whose merits were evident to many physicians trying to learn and improve upon existing technique. Karman made special cannulas, the open-ended tubes that are inserted into the uterus during abortion, using a plastic that was flexible rather than rigid; in the Karman cannula (the design is still widely used, and the name has stuck to this day), the openings were made not at the tip of the tube but just at either side of the tip, so that the tube end was blunt and less likely to perforate the uterine wall. The smallest of the cannulas were narrow enough to enter the cervical opening without the use of dilators, the graduated-width metal tubes that were usually inserted into the cervix to open it slowly and amid considerable discomfort. With his cannulas, Karman used an aspiration device so basic that it required no electricity: he had attached the tubing to an oversized syringe that created a vacuum

---

* During the early 1970s Karman advocated a second-trimester technique he referred to as the "supercoil," which involved inserting a plastic coil into the uterus, awaiting spontaneous miscarriage, and completing the abortion with suction. The procedure was supposed to be simpler than the standard second-trimester methods of the time; but feminist newspapers denounced the "supercoil" technique as painful and dangerous, and health professionals warned in medical journals of the perils of promoting "supercoils" without further medical studies.

when its plunger was withdrawn, so that an early pregnancy could essentially be sucked out quickly by hand.

With no concessions whatever to the 1967 California abortion statute —no medical degree, no hospital facility, no pretense of screening women for their pregnancy-related physical or mental health problems—Karman was apparently breaking the law with every one of his abortions. But his devices were said to work reasonably well, for early pregnancy, and because they were simple and cheap, they had a resonant impact on a movement with such an abiding interest in matters of power and control. Just as electrical-pump vacuum aspirators could bring abortion out of operating rooms and into clinics or doctors' offices, so hand-held suction aspirators could bring abortion out of the conventional medical system and into the community of women, lay women, women who might make the leap from counseling and referral to *performing the abortions themselves*. Carol Downer remembers as epiphany the first time she saw, in a procedure room at Karman's clinic, the exposed cervical opening of another woman's uterus: "I was utterly enchanted," she recalls. "I think I expected something much more ornate, more complicated, more bloody, not this very fresh, pink little donut thing. . . . And I realized how absurd it was that we had been blocked—that the solution was in our own hands: look, we can control our own bodies. Why are we letting these people put coat hangers in there? Why are we not just *doing* it?"

In April 1971, with this small revelation and Karman's abortion devices much on her mind, Carol Downer helped organize a meeting in a Los Angeles house that had been converted to a bookstore resolutely named Everywoman. Among women with feminist sympathies, one of the exhilarating symbols of the movement was to be the vaginal speculum, that fearsome-looking medical instrument that spreads apart the vaginal walls and locks into place to bring the cervix into view, and it was almost certainly during this Everywoman bookstore meeting that the speculum began its odd ascendancy as a political icon. There was a preliminary discussion, some of the women present recall, about abortion and reproductive control and the significance of Karman's devices. There was an adjournment then, to a back room with a desk big enough to lie on. "And I took off my pants and showed my cervix to people," Downer recalls. "And it confirmed for me that *how* you see your cervix is very important as to what the impact of the experience is. If you see it at your doctor's office, it's just a curiosity, it's nice, it's interesting, but it's not *empowering*. If you see it on your own, or with a group of women, it's very empowering."

A literal and metaphorical barrier had been pushed away by Carol Downer's speculum, and over the next two years she and some of her Los Angeles colleagues crisscrossed the country to spread the gospel of what were now being referred to as "self-help" groups: gatherings of women, with no gynecologist and certainly no man present, who could learn to use a speculum, lamp, and hand mirror to examine themselves and each other for signs of infection or pregnancy. Part of the appeal lay simply in the rebellion from conventional modern gynecology—there was a national flurry of fund-raising and feminist outcry when Downer was arrested in 1972 at her Los Angeles Feminist Health Center for "practicing medicine without a license," an umbrella charge whose particulars in this case included accusations of unlawful diaphragm-fitting and yogurt-applying. (The yogurt was being offered as a remedy for vaginal yeast infections, but when an arresting officer began to confiscate a refrigerated carton as evidence, one of the women at the center observed that he was about to impound her lunch.) "What man would have to spend $20,000 and two months in court for looking at the penis of his brother?" a feminist writer wondered after Downer's acquittal by a jury, sounding the theme that the self-help movement threatened the established order simply by existing, noisily, with its insistent message that women could look after each other without depending on professional men.

But there was more to the self-examination groups than revelation and the occasional carton of yogurt. Carol Downer and her NOW women's health ally Lorraine Rothman, a Southern California schoolteacher with four children of her own, had come up with a vacuum device meant to improve upon the principles of Karman's abortion syringe. Lorraine Rothman had collected the separate pieces from local shops—a Mason jar from the grocery store, plastic tubing from the tropical fish store, a one-way air valve from an industrial supply house—and had put them together at her own kitchen table. After some experimentation, she determined that what she had fashioned was not only safer than Karman's vacuum syringe, because she had built in an extra valve and tube to reduce the chance that air might be accidentally introduced into the uterus, but also so easily constructed that just about anybody could build one. That was precisely the point, and when Rothman applied in 1971 for a U.S. patent (she was awarded one three years later), her own description of the Rothman Method for Withdrawing Menstrual Fluid stressed its simplicity:

A method and apparatus whereby substantially all of the menstrual fluid incident to a normal monthly "period" may be removed in a small fraction of an hour. A simple plastic syringe is employed in combination with a valve to create a suction pump incapable of injecting air into the uterus. . . . The menstrual extraction procedure is performed at the time when the normal monthly period starts, or is estimated to start.

The last five words were critical. The Del-Em, as Rothman named her kitchen table apparatus, could pull into its Mason jar the entire contents of a woman's uterus, including a pregnancy, if the embryo had been caught early enough to pass through the tubing. When they spoke about it publicly, two women without medical licenses traveling from one state jurisdiction to the next, Downer and Rothman always used the words "menstrual extraction" to describe their procedure, and the vocabulary was cast in grand and deliberately careful terms: sisters, we *can* control our biological processes, we can rid ourselves of a menstrual period if we so choose. But plainly it was an abortion device they had fashioned, and more to the point, it was an abortion device intended from its inception to meet the philosophical demands of the women's health movement: women were to use this on each other, Rothman and Downer instructed, in group settings, perhaps a women's center or a comfortable living room, the more experienced women passing on details of technique and sterility. In the spirit of the Del-Em, a woman who had missed a menstrual period might attend to her own dilemma without ever setting foot inside the mainstream medical system—without even undergoing a clinical test that would force her to learn whether she was pregnant.

"We didn't ever think in terms of giving it to the medical profession," Lorraine Rothman recalls. "Our idea was that it belonged to *us*. It belonged to women. So women could take care of the whole business."

No written documentation was ever compiled as to how many American women actually used either the Del-Em or its various homemade adaptations—word traveled, among women attending the self-examination groups around the country, of suction kits fit together from laboratory hand pumps, converted bicycle pumps, vacuum cleaners, and milking machines. Downer and Rothman themselves toured twenty-three American cities, prepared at each stop with a film, some Del-Ems, and a demonstration routine that involved pumping water out of a glass; and Downer recalls that in many of the cities women from the audience would appeal to them urgently in private after the presentations were done. "They would say, I'm late on my period, would you help me out?" Downer recalls. "These were in states where abortion was illegal, and there was quite a bit of demand."

Usually they obliged, and around the country arguments rippled around them. Even licensed doctors who approved of legal abortion were appalled by menstrual extraction: lay women introducing cannulas into the uterus were surely going to find themselves confronted with infection, or uterine perforation, or abortions in which part of the pregnancy had been left behind to set off infection later. Writers in the feminist press were not necessarily enthusiastic about it either. Ellen Frankfort, whose health column in *The Village Voice* reached a large audience with its thoughtful discussions of the complex new overlapping of medicine and feminist politics, worried in print both about safety and the general wisdom of proposing that women make wide-scale use of a device that had never undergone formal testing—particularly since health writers were raising alarms even then about inadequate examination of the risks of birth control pills and IUDs. "I would still hesitate to extract my own period and I am not convinced of its safety," Frankfort wrote in her 1972 book *Vaginal Politics*, but she added that she found illuminating the outraged tone of the letters dozens of doctors (most of them presumably from New York, where abortion was already legal) had written her

about the number of women who would die if this thing caught on, and how terribly dangerous and irresponsible it was for me to quote nonprofessional women in something that would be read by many women. . . . The doctors were more upset at the independence that period extraction in particular and self-help in general gave women rather than at the dangers of either. Few argued or even considered that the device could be made safer; none remembered that the first camera or radio or airplane was not perfect and that only in time could a new invention be judged. Had women called in a board of doctors to set up a clinic for period extraction in which the doctors would be the ones to do it, I am sure they would have responded differently. But they saw, quite correctly, that period extraction was developed precisely to give women autonomy over their own bodies and that it was inseparable from the concept of self-help.

No record of a menstrual extraction death ever surfaced in the public debate, and the sociologist Sheryl Burt Ruzek, who examined some of Downer and Rothman's data for her 1978 book *The Women's Health Movement*, recalls that the menstrual extraction groups seemed to her to have monitored carefully for complications that would require either a repeat aspiration or referral to a doctor. There was a brief but intense in-house debate over the question of women using menstrual extraction kits on *themselves*, a prospect that was arresting in its keep-one-in-every-medicine-cabinet implications, but was dismissed by the godmothers of self-examination as both unsafe and exceedingly awkward in practice: "You'd have to be a contortionist," Rothman says. In any case this was the wrong image entirely, some woman trying all alone to angle a mirror adeptly enough to steer a cannula into her own cervix. The image was supposed to be group defiance, and the solidarity of women, and—because so many of these women also spoke in the vocabulary of the New Left, with its calls to class consciousness and its general hostility toward the professional elite—the "demystification" of a medical procedure that even after state-by-state legalization still seemed to place all the power in the hands of white-coated men.

The white coats rankled nearly as much as the fact that they were men. In Illinois, where the networks of the Chicago Women's Liberation

Union had helped build the abortion referral service whose volunteers all assumed the pseudonym "Jane," word spread by 1972 that the Jane women had learned not only the basics of suction abortion but also the full technique for instrumental dilation and curettage. The Jane women traveled from one "safehouse" apartment to another, their D&C instruments packed into suitcases, and such was their determination to upend the traditional that they avoided using the word "patients" to describe the women they aborted. A Jane abortion (more than eleven thousand had been completed by 1973, according to what documentation the Jane women kept) was performed on a bed, not an examining table, and part of the presentation was the confounding of roles: a woman might switch in a matter of moments from head-stroker and hand-holder to abortionist with sharp curette. "From their first contact with the service, women were told, 'This makes you complicitous. We don't do this *to* you, but *with* you,' " one Jane wrote nearly two decades later, still identifying herself only by the pseudonym:

> We believed that information was power, so we shared everything we knew and hid nothing. Women were expected to participate and encouraged to help one another. Any woman could do what we were doing, and, in fact, she did. . . . By taking the abortion tools—curettes, forceps, dilators—in our own hands, we had effectively demystified medical practice. No longer would we see doctors as gods but rather as skilled practitioners, just like us.

For sheer audacity Judy Widdicombe liked the image of the Jane collective; she had never met any of them personally, but she had heard about them, and for two years she had been hearing also about the California women who traveled from city to city with their vaginal speculums and their jar-and-tube extraction kits. They were subversive, all these women, and there was a certain brilliance to their subversion. Nobody had to spell out for Judy the nature of power in a traditional medical setting: as a registered nurse she had trained in an era when nurses were instructed to leap to their feet every time a doctor walked into the lounge; and all her working life she had been first-named by doctors and ordered around by doctors and generally treated with that cordial, dense, dismissive air that doctors tended to assume in the presence of persons less glorious than themselves. On obstetrical rounds, in the St. Louis hospitals where Judy worked, the doctors walked briskly along while the nurses pushed the supply carts behind them, like train station porters. The nurses were women and the doctors were men, and legalizing abortion had done nothing about *that*; all Judy had to do was tour the New York clinics to see how swiftly and adeptly men with medical degrees had stepped in to take over the role once held by criminal abortionists. Men owned the clinics, men worked the aspirators, and men took in most of the cash that could pile up with dizzying speed in the busiest clinics. Some years hence an unsavory description of this process was to appear in the memoir of Bernard Nathanson, the National Association for Repeal of Abortion Laws co-founder, who spent nineteen months in 1971 and 1972 as medical director of the high-volume clinic called Women's Services:

> The clinic [at the time of Nathanson's arrival in 1971] was paying physicians $75 per abortion. While I was, and still am, a firm believer in the capitalist dogma of the free marketplace, that figure seemed excessive, even to me. Doctors were earning more than $1000 on an eight-hour shift, with two abortions an hour and, in some cases, three. Some worked two shifts a day and doubled their income. Not only did $2,000 a day reinforce the disreputable picture the public already had of abortion, but the per-abortion pay system encouraged abuses. Physicians would fight for the paying cases, find reasons not to do the patients whom the clergy had sent with a request for reduced rates, and disappear altogether when asked to do the free cases.
>
> The pay system inspired incredible situations. Bill Walden, one of the skilled board-certified obstetrician-gynecologists I kept from [previous director Hale Harvey's] staff, later told me that when he first started, the practice was that a "senior" doctor with more experience at the clinic would have to confirm the newcomer's estimate on how far along the pregnancy was. One day he estimated a woman at sixteen weeks, and called in a "senior" with a solid reputation outside the clinic who told him, "She's only ten weeks. You can do her."
>
> Walden started to work and soon was in the middle of a treacherous sixteen-week abortion that took him an hour and a half, with blood, bone and fetal parts all over the room. Meanwhile, the "senior" doctor was running through three women and earning three times the pay while he was tied up. Walden told me that the old hands pulled this on a lot of the new boys, to tangle them in impossible cases and reduce the competition for fees. On the side, of course, there was the trifling matter that they were putting the women patients in unnecessary danger.
>
> It was also brought to my attention subsequently that the doctors would keep double books. Harvey had put them on an honor system. They would keep a personal list of women who had been done and drop it into a hopper, which [the clinic administrator] would use to figure out the paychecks. The doctors did not defraud the clinic by padding the lists, but they would often keep a phony list to show to their colleagues, so it would look like they had done fewer abortions each day than they had actually performed. When a patient would arrive, a doctor would say, "I've only done four today; why don't you let me take this one?" The honest men got the short end.

Nathanson wrote that passage in 1978, after a highly public change of heart—he had resigned from the board of NARAL and was offering his controversial book-length memoir as an argument against legalized abortion.* But his insider's details (Bill Walden confirms the story as generally accurate) would not have struck Judy as shocking or difficult to believe, if she had read them in the early months of 1973; she had watched enough clinics in action to understand the unpleasant paradox before them. Because properly managed early abortion was simple, and because it could be done quickly, and because so many women wanted it, the procedure itself was *boring*. Legal freestanding clinics, with their fine modern equipment and their steady stream of patients requiring one service alone, had turned abortion practice into a numbing legs-up assembly line: insert the speculum, inject some anesthetic, dilate the cervix, insert the cannula, turn on the switch, vacuum out the uterus, pull out the cannula, check for residual tissue, proceed to the next room. There were physicians who did this out of what appeared to be motives of genuine moral conviction— Judy had met a few of them in New York and California—but even for the heroic medical personality it was not the most stimulating line of work. Its only real excitement arose from the sudden crisis, the perforation, the bad reaction to anesthetic, the unexpected discovery of fibroids or polyps or some congenital anomaly of the uterus. Its principal attraction, once all the high-ground rhetoric had played itself out, was money. If you wanted to bring in the doctors, you had to pay them, and pay them well enough to keep them in the building when their interest in the work was flagging and their colleagues outside were snubbing them as Abortionists.

But *you didn't have to give them the reins.* Judy Widdicombe had her own ideas about power, and she was not interested in exercising these ideas in an apartment bedroom or the back of a feminist bookstore. The clinic she intended to build would be run by women and staffed by women: women would handle the books, women would manage the meetings, a woman would occupy the director's office. Women would conduct the lengthy conversations with patients, counseling them, explaining the procedure, staying close by the examining table while the abortion was underway. Men would be hired, because Judy's clinic would promise abortions by licensed gynecologists and in St. Louis most licensed gynecologists were men, but women would hire them—*Judy* would hire them, to be precise about it, and if their performance was unsatisfactory, Judy would fire them.

The doctor as employee, as skilled technician, as cog in a good machine, neither more nor less valuable than the women working alongside him: the personnel arrangement at Reproductive Health Services might in

---

* The payment-per-abortion system at the clinic changed, shortly after Nathanson's arrival, to a by-the-hour arrangement that still paid physicians handsomely for their work.

its own modest way prove nearly as innovative a challenge to the established order as gatherings of women with self-help manuals and Mason jars. Certainly the doctors would be paid, and paid respectably enough to keep qualified gynecologists on the start-up staff. But they would understand at the outset that they were working for a nonprofit, for a facility that intended to invest its earnings in clinic improvements and subsidies for poor women's abortions. It was a remarkable sort of resume that Judy had in mind, and by the end of January 1973, less than a week after the decisions in *Roe* v. *Wade* and *Doe* v. *Bolton*, she had her first active recruit.

His name was Michael Freiman. He was forty-four years old that winter, an obstetrician-gynecologist with three young children and a vibrant wife with the unusual first name of Sarijane ("My mother being dramatic," Sarijane Freiman would say cheerfully, when people asked). Judy had added Michael Freiman to her roster of Clergy Service doctors nearly three years earlier, after she listened to him deliver a Jewish Hospital speech entitled "Be Fruitful and Multiply—*But*." In his speech Freiman talked about criminal abortion laws, cautiously, but with obvious disapproval, invoking an Abraham Lincoln quote about the need to look for new solutions to old problems. Judy telephoned him after that, to tell him about the Service, and over the next months they befriended each other on the phone, Freiman sending occasional patients through Judy to the New York clinics and then examining them, always with some astonishment at how clean the results appeared to be, when they came back. He called her again two days after *Roe*. "Well?" Freiman demanded. "What are you going to do?"

Michael Freiman had carried with him since medical school a deep fury about criminal abortion law, and the story he would tell—the story Sarijane would tell also, because she was home the night it happened and she remembered the terrible expression on her young husband's face— was about the college student who arrived in the emergency room with something hanging from her vagina. The chief complaint was nausea, but when Freiman was called in he looked at the young woman on the examining table and felt suddenly sick himself. "That's a loop of bowel," Freiman said. He was told the young woman had used a coat hanger. When he opened her up and pulled the bowel back in, he saw that she was not even pregnant, and somehow this made him the angriest, so that when he told Sarijane about it afterward and she wondered why the young woman had not at least gone for a pregnancy test, Freiman grabbed his wife's arm with startling ferocity and told her to imagine, just for *one minute*, how frightened a person must be to do something like that to herself. That was in 1961, when Freiman was in his obstetrics and gynecology training, and other abortion wrecks had come through before and since; there was one woman who lived long enough to tell hospital people that the abortionist had injected soap suds into her uterus, and what Freiman remembered specifically about this patient was the cause of death, intravascular coagulation, which in plainer English meant all her blood components had clotted at once.

He learned, as a resident, to do a tidy dilation and curettage. In the hospital gynecology service they never referred to the emptying of a uterus as anything else; the patient records read "D&C," with no additional information that might invite troublesome inquiry, and when a senior physician on certain occasions asked Freiman to step in as operating surgeon for a patient they both knew was pregnant—left instructions for Freiman to step in, actually, the senior physician himself having conveniently been called away the moment the procedure was scheduled—Freiman did as he was told, attending to the lady's difficulty and writing into the chart the terminology he had been instructed to use. And what disturbed him in retrospect was not that he had done the procedures, D&C's, abortions, whatever one chose to call them, but that he had done them only on the wealthy. The poor pulled their innards out or got soap shot up inside them. In his private practice, during the years before the Clergy Service, Freiman used to keep a list of the reputable illegals, Mrs. Vinyard and the best names from Cuba and later the tolerant doctors of Sweden or Japan; that was as much guerrilla medicine as he was willing to take on, but for many months before *Roe* Michael Freiman had believed that he would work abortion into his practice the day the law permitted it.

Would he do them in his own office, with the obstetrics patients and the menopausal ladies in the examining rooms next door? He couldn't see it. Even after he knew that he would perform abortions someday, Freiman had in his mind an image that would not recede, the abortionist with a cigar in his mouth and ashes falling on his big sweaty stomach and soiled things piled up nearby in the sink. There was a purity to the idea of a clinic, sterile white rooms and trained staff people bustling about, and the day after *Roe*, Freiman had called the local Planned Parenthood chapter to see how quickly they planned to begin offering abortions. The Planned Parenthood people cleared their throats and said, Well, goodness, very complex, the board just isn't going to take this up for a while, so Freiman hung up in irritation and called Judy Widdicombe.

They met at the Freimans' house, Art and Judy and Michael and Sarijane. This was the first time Sarijane had ever seen Judy Widdicombe, a person to go with the honed alto voice on the telephone, and she was taken aback; she had expected someone more operatic-looking, more theatrical. Judy still had a big blond softness when you first looked at her, but she had her own way of invisibly clearing a space before her when she came into a room, and Sarijane, who was a homemaker still (a late bloomer, she would say some years later, once the children were older and she had gone to work), was not sure she had ever encountered such self-confidence in a woman who was willing to sit there in her living room and talk to her. Judy appeared to be walking around with a full set of blueprints in her head. She had no intention of negotiating this or that aspect of Reproductive Health Services: it *would* be a nonprofit, she *would* be the director, and she was as pleasant and warm as could be as she reeled in Michael Freiman. By the end of the evening Freiman was staff to a clinic that did not yet exist, teaching physician for a procedure he did not really know how to do, and chief medical adviser in the selection and procurement of a lot of mechanical and instrumental devices he had never used: Karman cannulas and motor-driven cervical dilators and electric vacuum machines.

Afterward the Freimans spent some time wondering which worry should alarm them most, Michael's obstetrical practice or their children in public school or Sarijane's parents, who were mildly religious Jews in their sixties and might be distraught at seeing their son-in-law's name appear in the morning newspaper alongside somebody's incendiary quote about baby murder. The Freimans decided to bring Sarijane's parents to the house, to tell them delicately, and as Sarijane was trying to feel her way into her explanation, to talk about why this meant so much to Michael, her sixty-five-year-old mother raised her hand and told Sarijane to stop. "I'm so proud of him," Sarijane's mother said, and then, matter-of-factly: "You know, years ago, I had an abortion." She said it was between the oldest boy and Sarijane, that they were young and poor, that someone had shown her where to go, that this was what women did in those days, and Sarijane stared at her mother, to whom she had been about to *condescend*, to offer some patronizing discourse about desperate women turning to criminal abortionists. Later when she remembered that evening it was her mother's calmness that stunned her still, the quiet way she made it plain to Michael and Sarijane that the details were grim and none of their business. "I said, 'I never knew this, I can't believe this,' " recalls Sarijane Freiman. "She said it was not nearly as uncommon in those days as I probably thought it was."

Publicly, under the terms of the handshake agreement that now linked Michael Freiman and Judy Widdicombe, Freiman was bypassing the lead medical job at Reproductive Health Services. "Medical Director" was the standard title for the commanding physician at an abortion clinic, and at Judy's clinic the medical directorship was to be assumed by the Washington University obstetrics professor George Wulff, who had already been volunteering his time as one of the Clergy Service examining physicians.

Wulff was a perfect shield, as far as Freiman was concerned: sixty-four years old, tall, silver-haired, patrician, WASP. He had his own collection of illegal abortion stories—Wulff was particularly good at talking about the pre-penicillin years before World War II, when the infected abortion patients at Kansas City General were kept in a ward that was regularly cleared of bedbugs by a man who lifted mattresses from the wire springs and flamed them with a blowtorch—and there was an unflappable, country club presence about him that made him appear impervious to what was likely to come. Wulff had no young children to be taunted at school; he had retired from private practice; nobody was going to mail him (as they would to Freiman, later, in envelopes with no return address) carefully folded pieces of paper containing anti-Semitic epithets and hand-drawn pictures of people with hooked noses. And Wulff plainly had no intention of vying with Freiman for the hands-on challenges of abortion clinic medicine. Wulff was vague about mechanical detail and seemed content to watch from a distance while Freiman pushed ahead, the triggerman hard at work, and the truth was that Freiman had always loved this part of medicine anyway, the exploratory part, learning the machinery, mastering the technique.

He flew to New York for his tutorial. By the second half of 1972 Judy had taken to sending nearly all her early abortion patients to one clinic in Westchester County, out in the suburbs, where the doctors were efficient and the atmosphere reasonably upbeat and serene. For five straight days Michael Freiman haunted that clinic, watching and asking questions and taking photographs of everything he saw—the sterilizers, the aspirators, the different sizes of forceps. He angled the camera to put the brand names in the pictures, so that Judy would know what to order. He stood in the procedure rooms while the New York doctors worked, and when one of them turned to him and said, Here, you do it this time, Freiman looked at him quickly and then picked up a cannula, trying not to feel frightened. The doctor studied Freiman's hands, the way they balanced the instruments at once gently and easily, the quick light way he guided the cannula in, and nodded over Freiman's head at the nurse. "He's got the touch," the doctor said.

He *did.* It took Freiman aback. In obstetrics he had never been forced before to size early pregnancies precisely, but he saw how to do that, too, and how much it mattered to get it right. In the evening the doctors entertained him in Manhattan and Freiman sat in their grand apartments drinking champagne and eating caviar, which they had piled upon a plate; Freiman had never seen that much caviar in one place before. He had an uneasy feeling about this, sitting there overlooking Riverside Drive with the shimmering pile of abortion money caviar, especially after one of the clinic investors, who had also invested money in the pornographic movie *Deep Throat,* took him aside and suggested to Freiman that perhaps he was a shmuck for not opening his own clinic. Freiman said he appreciated the man's viewpoint but that money in this instance was not actually his primary goal. At the end of the week he came home to St. Louis, understanding that he probably knew more about induced suction abortion than any other physician in the state of Missouri, and gave Judy the pictures so that she could spread them before her, equipment catalogues opened for ready comparison, and get to work.

THEY *would* need money, but Judy had fixed that, too. By herself, because the Pregnancy Consultation Service board was not willing to back her as director, she went to Joe Sunnen, again, for help.

For more than half a decade now, a small contingent of financial angels had been offering up most of the seed money for abortion activists and researchers around the country: the East Coast had its private foundations and individual donors (many of them, like the Lalor Foundation and the General Motors heir Stewart Mott,* veterans of the population control movement); the West Coast had a steady funding source in the Santa Barbara–based Hopkins Charitable Fund, which provided grants to NARAL and the Association for the Study of Abortion. And from St. Louis, the industrialist Joseph Sunnen had been doling out national abortion support money ever since the Model Penal Code laws first turned up on state legislators' desks. The Clergy Service received money from Joe Sunnen; NARAL received money from Joe Sunnen; the Sunnen Foundation gave money to clinics and referral centers and teaching symposia on abortion techniques. When the nine San Francisco obstetricians faced state medical board charges for performing rubella abortions in 1966, it was Sunnen who had offered to pay the defense fees if they would pursue a full-blown legal confrontation; even after his offer was declined, the rumor traveled in abortion circles that Joe Sunnen was ready to pay a million dollars to any physician willing to sacrifice his career in a public challenge of criminal abortion laws.

In sheer dollar value that rumor was almost certainly an exaggeration, but there was no question about either Sunnen's fervor or the reach of his money. By the time *Roe* v. *Wade* was handed down, Sunnen was seventy-five years old, a white-haired gentleman who had added to his St. Louis factories a subsidiary business unrelated to the machinery that had made him a wealthy man. The Sunnen Products Company built devices useful for automobile repair, Adjustable Valve Lifters and Precision Gauges and Rod Reconditioning Honing Units. But ten miles from the main plant was the international headquarters for Joseph Sunnen's newer enterprise, the Emko contraceptive foam company. Around the Sunnen offices the Emko story was always told as the philanthropic inspiration of a self-made man: Joseph Sunnen and his wife had been invited to Puerto Rico in 1957, the story went, and amid the squalor of the shanty towns they contemplated what was described in an admiring Sunnen biography as "the miserable and frightening consequences of uncontrolled population growth":

Families of ten or more living in one-room shacks, sparse or no bathroom facilities, hunger and apathy were evident on every hand. . . . A Puerto

---

* Stewart Mott, then a young General Motors heir (his father had been a GM founder and major stockholder) who went on to become one of the chief personal donors to NARAL and Planned Parenthood, recalls that it was Joe Sunnen who first impressed Mott with the urgency of the cause; Mott was working in St. Louis in 1963, and over a lunch with Sunnen listened to the older man expound with such enthusiasm that Mott went back to Michigan and helped found a Planned Parenthood affiliate in Flint.

Rican lady physician and government official who was accompanying the couple on the tour, complained of the lack of existing contraceptive methods. Then she hurled a challenge: "Mr. Sunnen, you are an inventor. Why don't you invent something that will help our poor people?"

Joe Sunnen went home to St. Louis, according to the company account, and convinced local pharmacy researchers to help him invent something. The result of their efforts was a shaving-cream sort of spermicide that could be foamed directly into the vagina, and so consumed was Sunnen by the wonders of his foam that he took to carrying sample cans of it in his pockets so that he could pull them out and press them into the hands of women he had met only moments before. Sunnen startled flight attendants, tour guides, women who happened to be standing beside him at stoplights; in his later years, his secretary would watch him load up on samples until his pockets were bulging and then come back at the end of the day satisfactorily relieved of all his giveaway cans. He had begun to see population control as an international imperative, and the name "Emko" was chosen not because it meant anything but because its syllables (which turned out, to the vast amusement of one potential market country, to translate to "Happy Girl" in Japanese) were readily pronounceable in many languages. For a time Sunnen's Emko company distributed a newsletter along with its products, the columns lively with barbs against those who stood in the way of world population control:

In-case-you-didn't-know-it department: Lemmings *have* no population problems. What are lemmings? Why, those petite but prolific arctic rodents, of course. When threatened by crowds or starvation, thousands accommodate their brethren by marching to the sea and drowning themselves.

The links between population control and legalized abortion were evidently obvious to Joseph Sunnen as early as the mid-1960s, when he made his defense funding offer to the San Francisco doctors, and by the close of the decade the Sunnen name was often near the top in tally sheets of the foundation and individual donor money that was pushing the abortion movement along. Judy Widdicombe had encountered her first Sunnen money in the early months of the Clergy Service, when Sunnen's foundation agreed to help pay for telephone bills and office supplies; Sunnen's support kept coming as the counseling service grew, and finally Sunnen put Judy on the Emko payroll and let her set up the Pregnancy Consultation Service in a company-owned house near the Emko plant. Judy was supposed to spend half her time on Emko product development, but the Service was taking over her life and Joe Sunnen did not seem to mind; his own son Robert, who was poised to inherit the company, had developed such an intense interest in the counseling service that he sat through the volunteers' training program and began reporting to Judy's office to counsel pregnant women himself.

Now Judy needed more from Sunnen—considerably more—and she got it: fifty thousand dollars, granted as a Reproductive Health Services loan (which Sunnen would decline to collect on, in later years, thus turning it into a gift). Here was Judy's start-up budget, and she and Freiman together walked the streets of the St. Louis neighborhood called the West End, where Barnes and Jewish hospitals formed a massive medical complex that fanned into many private office buildings. They drove to and from the Barnes emergency room, marking distances, timing drives, imagining the practicalities of hurrying a perforation case or a cardiac arrest through the two-lane streets with the restaurants and bars lined up on either side, and finally Judy signed a lease for half the second floor of a twelve-story office building exactly four-tenths of a mile from the emergency-room door. The central religious and administrative buildings of the Catholic Archdiocese were so close by that some wondered at Judy's judgment, the cathedral was practically going to throw its great St. Louis shadow across her abortion clinic's front door, but Judy brushed them off; some of those Pregnancy Consultation Service board members didn't seem to think she could do anything right, run a clinic or name a clinic or even settle on the proper block of St. Louis in which to house the clinic they doubted she could run or name.

Watch me, Judy had begun to say to herself, gathering momentum now, purchasing, ordering by telephone, pacing off the length and breadth of future procedure rooms. *Watch me.* The medical supply salesman showed her file cabinets, supply carts, autoclave sterilizers, reclining chairs for the recovery room. Michael Freiman checked off the equipment list for each room's outfit trays: speculum, gauze, cotton-tipped applicators, sponge forceps, curettes. A registered nurse named Vivian Rosenberg helped bring in nursing candidates; Vivian had known Judy Widdicombe since the early months of the Clergy Service, when Vivian had signed on as one of the early volunteers, and the two of them had a relationship of powerful, strong-tempered loyalty despite their Mutt-and-Jeff disparity of size. Vivian was five feet tall and had recently divorced her husband, which had temporarily made her thinner and frailer than usual, and when she and Judy fought, they stood glaring at each other with the top of Vivian's head barely reaching Judy's chin. Long afterward Vivian would say working alongside Judy was the best job she had ever had, even when she was doing it at the Clergy Service for free; she liked the way Judy nurtured and bullied at the same time, and even before *Roe* v. *Wade*, Vivian had flown to New York to study the technique of the suburban abortion doctors, because Judy asked her to.

The first time she watched one of the New York doctors switch on an aspirator and begin vacuuming blood through the plastic tubing, Vivian thought for a moment that she was going to pass out, but surgery always made her queasy when she hadn't attended in a while. What was inside the tubing did not much affect her except to make her feel sad for the grimacing half-draped woman before them; Vivian had three children of her own, but when she was in her twenties she had helped another young nurse through an illegal St. Louis vagina-packing abortion, offering her a bed on the sofa for a couple of days afterward, watching her cramp and writhe and turn white and spike a fever until the young woman finally put on her uniform and went back to work, still cramping, so that the next day she aborted an intact fetus in the hospital bathroom. The abortion cost $350 and involved catheters and gauze stuffed into the cervix and *castor oil*, a registered nurse drinking castor oil to get rid of a baby from an older man who wouldn't marry her, and when Vivian Rosenberg was with the Clergy Service she was very damn effective at describing for potential new counselors the emotional state of the women they might be asked to help.

She had no problem finding nurses for Reproductive Health Services. They recruited by word of mouth; nobody was foolish enough to take out a Now Hiring Abortion Nurses advertisement with a telephone number printed right next to it. Judy was trying to manipulate as precisely as she could just what the public knew about Reproductive Health Services anyway, and she had taken to offering reporters independent tidbits of information, one newspaper at a time. There were two daily newspapers in town then, the *Globe-Democrat* and the *Post-Dispatch*, and although their editorial pages clashed on nearly everything, including abortion (the *Post-Dispatch* approved of *Roe* and the *Globe-Democrat* emphatically did not), Judy had been quoted without rancor in both papers and knew that each of their feature sections had already carried long and thoughtful abortion articles. She had come up with a plan of minor deception, useful both for the cultivating of certain reporters and the temporary postponing of possible opening-day unpleasantness: as Reproductive Health Services readied itself for business, the first staff hired and the delivery trucks pulling up alongside 100 North Euclid Avenue, Judy lied to most of the press about the start-up date.

It was a very small lie, certainly, merely a modest rearrangement of the calendar. Already it was May, and Judy was chafing with every delay; the clinic might have opened a full month earlier if Michael Freiman had not insisted on waiting for the federal court to make its formal pronouncement on the post-*Roe* status of Missouri's abortion law. The ruling came in early May—a three-judge panel, as expected, found the Missouri statute unconstitutional under the new guidelines of *Roe* v. *Wade* and *Doe* v. *Bolton*—and Judy answered the ensuing press calls by declaring that on the morning of Thursday, May 24, the medical staff of Reproductive Health Services would accept the clinic's first patients.

But she opened on Wednesday instead. At the last possible moment Judy called one *Globe-Democrat* reporter to confide that she had pushed up the opening, which meant that at 9:00 A.M. on the start-up date for the first legal abortion clinic in the state of Missouri, only the presence of a lone newspaperwoman suggested that anything out of the ordinary was about to take place inside the big brick medical building at the corner of North Euclid and West Pine.

There were thunderstorms that morning, the rainy close of spring in St. Louis, and once her small medical and counseling crew had taken off raincoats, Judy insisted they put on laboratory jackets, so that all of them stood nervously in their white coats, straightening lab trays, brushing imaginary dust off countertops, and waiting for the first patient to be led into a procedure room. Afterward nobody would remember the name or the personal circumstances of Patient Number One; she was ten weeks pregnant and of suitable age, young enough to be in excellent physical health but old enough to raise no eyebrows about abortions on minors, and what Vivian Rosenberg and Michael Freiman noticed in particular was that Vivian, who had agreed to assist during the day's first procedure, was having a difficult time keeping her hands from shaking as she handed Freiman the instruments. Vivian kept calling Michael "Doctor," which unnerved him because until then she had always addressed him by his first name. Michael kept trying to loosen her up, tell her everything was going to be all right, but the truth was that his own wife was spending the day at Art Widdicombe's house with the lawyer's telephone number and what they hoped was enough cash to bail Michael and Judy out of jail.

The abortion went perfectly. Nobody tried to arrest them. They did seven patients that day; Michael Freiman said afterward that it must be like soloing an airplane for the first time, you keep expecting it to waver and crash but it doesn't. Judy went in and out of her office, sitting down, standing up, walking the hallways, studying the waiting-room furniture, stepping into procedure rooms to watch Freiman work; she loved how everything looked but she saw at once that the complex was too small, that she had been right to move into a building that had more office space available across the hall, that it was going to take them no time at all to spread out across the entire second floor.

The next day she came in early, before anyone else had arrived. She put on her laboratory coat. She looked out the window, and busied herself at her desk, and waited

Nobody knew what to expect. There were rumors that the cathedral had augmented its May 24 morning Mass to direct prayer specifically toward the inauguration of Reproductive Health Services. There were rumors that something was supposed to happen after that, something that would start at the cathedral and make its way up the street toward 100 North Euclid Avenue, and it was not until the middle of the morning, when Judy was standing in a procedure room near a window that had been opened to let in the warming spring air, that she first heard the noise. In the beginning it was faint and faraway and she supposed she must know what it was, but she stuck to what she was doing, arranging instruments and checking sterilizers and keeping herself occupied as the noise swelled slowly, growing richer and more distinct, until from inside her clinic Judy heard someone shout, in a voice that carried through the halls: "They're coming."

She went back to the window then, and looked down onto the street. They were coming up Euclid. *Storm the walls* was the phrase that would occur to Vivian Rosenberg many years later, remembering how they looked, hundreds of people pushing steadily up the two-lane street as though they intended to fling up grappling hooks and begin scaling the brick walls before them; she remembered also the stiffness of Judy Widdicombe's back, and the way Judy stood for a long time at her window and watched the people keep coming, spilling off the sidewalk and into the traffic lanes, massing around the corners and the parking lot two stories below. She had given them a target and they were marching upon it together, heads up, shoulder to shoulder. They were singing as they came.

PART TWO

A House Afire

EIGHT

The Education of the Pharisee

1978

SAMUEL LEE came to the city of St. Louis in a dented red Volkswagen bug with his worldly possessions piled up on the back seat. He arrived on a snowless afternoon in January 1978 and drove west on the interstate until he found the turnoff for South St. Louis, where there was a small vacant apartment waiting for him across the street from a Sidney Street grocery store. In the Volkswagen he chugged slowly through South St. Louis, past the brick Italian bakeries and the narrow metal awnings and the grim little winter-bleached front-yard lawns; he had unkempt black hair and a beard he trimmed only occasionally and he wore glasses with plain black frames, the kind from the twenty-nine-dollar rack at the discount optician, so that people who saw him from a distance would say to themselves afterward that except for the glasses he looked like an Old Testament prophet or a Russian monk, with a torso so long and slender that when he lifted his shirt the ribs showed under his skin.

Inside his automobile, which during the twenty-first year of Sam Lee's life was the only item he owned that had cost more than thirty dollars to buy, an unsteadily balanced assortment of cardboard supermarket boxes held his clothing, the pots and pans his mother had pressed upon him when he told her he intended to live alone, and approximately a hundred books of philosophy and theology: *Sacraments of Initiation, The Trinity and the Religious Experience of Man, The Moral Teaching of the New Testament.* Many of these books bore on their inside covers the imprint of Mount St. Francis, the Southern Indiana monastery where Sam had worked off and on since graduating from high school; the monastery had sold off part of its library collection, and for some days before the sale Sam had hauled books armload by armload over to the display tables, setting aside for his own purchase the titles that interested him. At the book sale he had found what looked to him to be G. K. Chesterton's entire Father Brown series, the early 1900 English short-story collections about a Catholic detective-priest, and Sam liked to think of the murder mysteries tucked in like a kind of lowbrow leavening amid the sober texts he was bringing with him to St. Louis.

He had come, after three years of trying to understand his obligations both to God and to his fellow man, to study for the priesthood. St. Louis University had accepted his application as a pre-seminary university student, and from his newly rented bare two-room apartment in South St. Louis, with the fluorescent lights of Mike's Market shining coldly outside his front window, Sam intended to complete the basic philosophy and theology courses that would allow him to enter the seminary and emerge as an ordained Catholic priest. He had a hazy notion of the priestly life before him and was surer about what he didn't want than what he did: he didn't want to teach, even though his favorite priest had directed the seminary for high school–age boys at Mount St. Francis; and he didn't want to run a parish and be isolated in some lone neighborhood church, overseeing the Bingo games and surviving diocesan politics. He was drawn to the community and piety of monastery priests, but he didn't want to live wholly cloistered from the world, either, and in the months before enrolling at St. Louis University, Sam had shopped around among some of the smaller religious orders whose members made it their mission to go out and live the Gospel among the people. He liked the Divine Word Missionaries, but not enough to think seriously about joining them; the Little Brothers of Jesus impressed him because up in Detroit, where he'd gone to talk to them, they set themselves so tenaciously into the world that one of them had taken an assembly-line job in an automobile plant.

But the Little Brothers of Jesus told Sam he was too young to join them, that he had to finish college and be seasoned for a while by life in the outside world. It was the Franciscans who still seemed to Sam the most plausible match: the priests he had known at Mount St. Francis were Franciscans, and even though Sam had never lived among them as a professed brother, he had watched them carefully and listened to them describe the Franciscan ethic of strong community and work among the poor. In St. Louis the Franciscans ran a residential and spiritual center, a big brick building three blocks from the St. Louis University campus, and Sam sometimes thought that if he had been *sure* about his vocation—if the path before him had been illuminated in some wondrous fashion, so that he could present himself without misgivings to the friars who selected candidates for entry into the Franciscan order—then he might have been invited to live there, with a furnished dormitory room and meals served daily in the dining hall.

It was a lively place, this Franciscan house, its walls decorated with pictures of St. Francis and its bulletin board crowded with clean-up-the-vacant-lots petitions to City Hall or flyers for the weapons-to-peace conversion project. Neighbors stopped by to read the pinned-up notices and talk; homeless people appeared at the door for free bag lunches; local schoolchildren came in to do their homework in a room the Franciscans had set up as an impromptu study hall. Sam was friendly with young seminarians who had moved to the house from their years at Mount St. Francis, and he had decided finally to find his own lodgings only because he thought it somehow dishonest to apply for residence among the Franciscans—asking them to invest in him, it seemed to Sam, before he was prepared to settle fully among them.

So a Franciscan friar had introduced Sam Lee to Mike, the grocer, who was Catholic like nearly everybody else in South St. Louis, and knew of a vacant apartment across the street. On slow business days Mike cooked in the back of his store, rich-smelling food like chili or cabbage rolls with ground meat, and when Sam pulled up with his sleeping bag and his boxes of books, the grocer made sure there was something good to eat for the thin young man who looked as though his profile ought to be up in stained glass at St. Anthony's. Unceremoniously Sam carried his belongings upstairs and piled them on the floor; that was how he intended to leave things, more or less, with his sleeping bag against the wall and his books lined up with their titles showing. He spent a day getting his bearings, driving back and forth to the campus, registering for class, and then he stopped by the Franciscan house to look in on his friends.

A meeting was in progress the afternoon Sam arrived.

A long time afterward the seminarian who was conducting the meeting, a young Minnesota man named Vince Petersen, would remember looking up from his chair in the Franciscan house conference room and saying almost audibly to himself, Who is this? The shabby clothes, the ragged beard, the gaunt and serious face. Surely not a university student, Petersen said to himself, and turned back to the men and women at his meeting. But some time later Sam was still there, smoking a cigarette, listening to every word. He had moved into the conference room and seated himself quietly on a folding chair in the back. He didn't say anything or raise a hand to interrupt, but Petersen could see that he was intensely absorbed and that he had a way of appearing at once entirely focused and entirely relaxed—something in his eyes, or the set of his shoulders. At the meeting they had to talk about tactical strategy as well as philosophy, and Petersen wondered briefly whether he ought to ask the newcomer to leave, but he decided to let it go, trust in God maybe, or at least in benign luck.

Thus Sam Lee stayed, and listened to the second of three meetings called to plan an organized act of civil disobedience in the doorway of a St. Louis abortion facility, and over the span of a single afternoon the rest of his life changed course.

SAMUEL HENRY LEE was the fourth of sixth children, and after he was born his skin was so jaundiced that the doctors transfused all his blood twice and told his mother that her baby was as likely to die as to live. When she received this news Sam's mother offered him to God. She was Lithuanian, but she avoided speaking or reading Lithuanian because Sam's American father had urged her to improve her English, and in English she articulated a direct appeal: Whatever You wish to do with him, Lord, I offer him to You. In later years the inherent tension of the transaction would be heightened somewhat by family mythology, so that Sam and his brothers and sisters were under the impression that there had been four or five or perhaps a dozen transfusions, and in any case the outcome was evident before them: God had given the baby back.

The older children remembered praying for Sam during the days after he was born. They lived then in Ossining, New York, but when Sam was still small they moved to New Jersey; Sam's father was an engineer and every few years he would change employers, airline industry to defense contractor to naval yard. The children were raised to understand these changes in a particular way. Their mother Sophie was Catholic, unwavering and profoundly rooted in the faith in which she had been raised, and their father Robert (who had met Sophie in Germany, after the war, in an American mess hall, when he was still attractive in his officer's uniform and they found a friend to translate for them because neither could speak the other's native language) was an agnostic who would not embrace God but had instead embraced morality as though it were a kind of religion in itself. His own parents had been strict and humorless Methodists, and when Robert Lee examined the Catholicism of his new European wife, he appeared to take great satisfaction in the logic and precision of Catholic morality: he read Catholic ethics and moral instruction as deliberately as if he were studying engineering texts, and when he talked to his children in a fatherly way it was often to explain to them what was morally acceptable, within the structure of this Catholic teaching, and what was not. "Black and white," the brothers and sisters all said, remembering their father's unblurred delineation of right and wrong. He had left this job because his superiors instructed him to lie about something, that job because he could see people cutting corners. In the midst of jobs he would grouse aloud about the moral and ethical failings of the employees around him.

"He seemed to think there was some systematic way of getting the answer to a difficult problem," recalls Sam's older brother, Mark. "He was looking for manuals, almost. A blueprint, a manual, a cookbook. 'What do you do in *this* situation?' He wanted to be absolutely certain he was doing the right thing."

For the children Robert Lee was a difficult man to be around, sharp and critical and quick to let loose with an acid outburst or a sudden angry whap to the head. They were too young to wonder at the walking contradiction, the lectures on moral behavior from a man who could scare away his own sons and daughters, and each of the children in turn learned to sidestep him when possible and bask in the comfort of their affectionate multilingual mother and her surpassing trust in God. On Sundays Sophie took the children to Mass, all of them squeezed into the car together, and because Robert always stayed home to tinker in his workshop, the children remembered these Sunday trips not as duties to be dispensed with but as cheerful time alone with their mother. When he was young Sam was a quiet figure among them, skinny and a little nervous but startlingly bright; on the sports field he was always picked last for teams, *oh well okay I guess Sam can play too*, but at school he was a math whiz and whenever the Scouts or the parish volunteers were raising funds, it was Sam who worked the door-to-door sales with such instinct and agility that his oldest brother, Richard, would call Sam the smartest of them all. Richard and Mark said Sam was going to be a businessman, and he was going to be awfully good at business, too. He was going to make a lot of money.

But when he was fifteen years old Sam had an experience—not exactly a conversion, because he was Roman Catholic before and Roman Catholic afterward, but something that seemed to rearrange him inside and begin nudging him toward the priesthood. The Lees were living in southern Indiana then, not far from the Franciscan friary that stood at the center of the splendid orchards and rolling hills of Mount St. Francis, and a girl Sam liked told him that a Mount St. Francis brother was inviting young people to prayer meetings at the friary. As it happened these meetings were part of a national Catholic movement, gently revolutionary in its own cautious way, in which young men and women would sit together in some comfortable place, away from the formal arrangements of altar and pew and printed ritual prayer, and express in their personal words and songs the presence of God—"charismatic" was the word they used, from the Greek for gifts, the gifts of the Holy Spirit. It was more like Pentecostal behavior than the decorum of a conventional modern Catholic church; hardly anyone leaped up and shouted in holy ecstasy during a Catholic charismatic meeting, but the prayer was improvised, inelegant, and direct from the heart, and speaking in tongues was not unheard of. At Mount St. Francis the priest who conducted the charismatic groups was a Franciscan named Christian Moore, and many years afterward Father Christian would also remember young Sam, who did seem to have this effect of memorable first impression on people around him.

"We were having prayer group one night," Father Christian recalls. "And all of a sudden, here comes this guy and starts hugging everybody, saying, 'God bless you, God bless you.' " A young man named Dan Foley, who was assisting Father Christian at the meeting, found Sam grasping his hand with such fervent enthusiasm that Foley wondered whether Sam intended to let go, "pumping my hand, literally shaking my hand for five minutes, and he kept talking about how *wonderful* this was," Foley recalls. Foley himself had experienced a religious conversion the previous year, from the Reform Judaism of his upbringing to a radical Christianity that took literally the concept of living among the poor; Foley had given away his entire inheritance, a quarter of a million dollars, contacting charity after charity until all of the money was gone. He lived in an austere room at Mount St. Francis, and the things he said made a deep impression on Sam, who was seven years younger and felt himself made stronger and different by whatever it was that had settled upon him at these charismatic meetings.

In later years Sam was good at describing things that had happened to him, but when he talked about the charismatic meetings, it was hard for him to locate the right words. There was no sudden vision or audible Voice of God, but as he listened to the spontaneous prayer, and found himself also speaking directly and without the mechanical repetition of memorized verses from the missal, he was suffused with a feeling of deep pleasure, a grounding, a clarity of purpose. His sister Roberta, who was three years older, saw the change in him right away. Before he went to Mount St. Francis Sam had been wary around their father, as all of them were; there was an uneasy fragility about Sam that had always made Roberta feel protective, but this vanished over a matter of weeks and she saw that her younger brother could look right at their father without expecting to be shouted at for doing something wrong. "The fear was all gone," she recalls. "He would just look at him with deep love in his eyes, and he would say, 'Yes, Dad. Yes, Dad.' And the nervousness and tension were not there anymore. It was incredible."

Sam went back to high school with his newly deepened faith displayed so proudly that he might as well have been wearing a signboard. Even Foley was surprised to see a fifteen-year-old boy strolling without embarrassment into school with a button on his denim jacket lapel that read, "A Smile a Day Keeps the Devil Away," but Sam was deeply happy and oblivious to mockery. His grades stayed high. He knew how to tell a good joke. Later he knew his friends sometimes thought of him as a kind of Pharisee, one of the Old Testament Jews who were quick to criticize others for inadequate observance of religious law. The students had a song they used to sing around him, grinning, to the tune of an old Catholic hymn:

Sambo was a Pha-ri-see
Holier than thou or thee

Father Christian kept a watchful eye on Sam and wished he would slow down a little, so starry-eyed was the boy with his powerful love for God. He hoped that if Sam went to the seminary he would hold off for a while, maybe date some girls first, attach himself more firmly to the ground. But he did believe Sam had been somehow taken up by God in an unusual and genuine way, and the priest would study Sam as their prayer group made its regular Sunday fellowship visits to nursing homes. Sam *cared* for these people, Father Christian saw; he remembered their names, he befriended the elderly blind lady, he followed the visits with telephone calls to say hello. The priest began to trust him, seeing that Sam could be relied on to help with practical tasks, that he wasn't just showing off for the friars.

Finally Sam moved onto the Mount St. Francis property, after his graduation from high school. The friars gave him food and a dormitory room and paid him forty dollars a week to cut grass or paint walls or rebuild the wooden boat dock. It was 1975 and the Church was still reeling from the Second Vatican Council, but when his chores were done Sam would walk the acreage of Mount St. Francis, past the vegetable garden and the water tower and the small still lake, and he understood that he was trying to find a way to live.

He hitchhiked away for months at a time, as though he meant to study different parts of the human spectrum in limited doses: he went to a New York State homestead, where Dan Foley was staying with a severe back-to-the-land Christian homesteading family that disapproved of electricity, telephones, white sugar, and women in pants; he went to New York City and took arduous day jobs to make enough money to eat. In New York Sam lived in a grimy one-room walk-up with a shower that didn't work and street people stretched across the hallway. He thought at first that he would try to make his way to Europe, but after a while Sam came to understand how much he was going to learn about the world without ever leaving Manhattan. He was only 120 pounds, but he lifted rubber from reeking factory ovens and stacked heavy boxes full of plumbing parts and pushed dollies of bolted cloth around the garment district. Because Sam was so thin the dollies would bump up on the curb and fall back on him, knocking him out toward the traffic, so he learned to jam himself against them and push them again, and when he studied the beaten-down men and women around him he began to feel that all of these aspects of life were things a person could *choose* to avoid, if he had the education and the means to do so. But this would not be a moral choice.

Moral choice consumed him. The city was so full of possibilities and Sam wrestled with them, astonished by his own innocence. A beggar tried to get fifty cents off him in the subway by pleading that he needed cash to buy a token to go visit his sick grandmother, and Sam, thinking, *Ha, I know a line when I hear it*, insisted on going with the man to buy the token for him. Sam paid for two instead of one, handed them to the man with a nice feeling of informed benevolence, and walked just far enough away to watch the man turn around and convince the token seller to take them back for cash. That was supposed to be against the rules, but the beggar did it anyway, and practically under Sam's nose, too. So now Sam had a problem, his father's son with the grace of God resonating inside him: what was he going to do when people approached him for money? What was the moral act? Sam wasn't looking for a philosophy here, some conscience-assuaging way to think about giving a quarter or not giving a quarter; he needed to know what to *do*, how to *act*, how to make sure the needy human being before him actually received, at that moment, a tangible bit of help.

He began to engage street people in conversation when they approached him for money, and then walk with them so that he could buy the food they needed and stay around long enough to make sure it got eaten. He was nineteen years old in New York City and living on so little money himself that he had to go to the Catholic Worker house for his showers, but he would pay for the homeless person's coffee or sandwich and then sit with him for a while, share the food, try to elevate the condescension of charity into a modest effort at momentary dignity or human connection. It seemed to Sam that this had something to do with the priesthood, that if he did go to the seminary he would want to emerge to live simply and in a manner that never took him entirely away from a daily ministry among the poor, and when he came home from New York he began to organize his studies so that a proper seminary would eventually let him in.

He read theology, religious history, biography. He took basic courses at a local branch of Indiana University. He read books about the lives of the saints, about Mohandas Gandhi, about Dr. Martin Luther King, Jr., about the nineteenth-century Frenchman Charles-Eugene Foucauld, who had abandoned his nobleman's birthright to become a Trappist monk and spent his last years alone in a hillside hermitage in the Algerian Sahara. Sam was fascinated by these stories; he didn't think he was a holy man and he didn't want to live in a hermitage, but he wanted to understand spiritual commitment, and in particular the kind of commitment that makes tangible, dramatic changes in individual human lives. Francis of Assisi, the saint from whom the Franciscan order took its name, had left behind a comfortable bourgeois patrimony to preach and live in community with the lepers and the marginalized of the world. Something was at work in these men and had led them to root their daily existence deep within their moral beliefs, to follow the teachings of Jesus in a way that was literal, all-consuming, and direct.

And this was what Sam heard, at the St. Louis house of the Franciscan brothers, on the January evening in 1978 when Vince Petersen watched him from the other side of the room, and why Sam stayed to listen until the meeting had come to a close. They were talking about civil disobedi-ence. Gandhian principles, nonviolent resistance, fasting and prayer, a peaceful presence in the face of an unjust law—Sam had read about these ideas and admired the people who put them into practice, but he had never sat through a meeting at which men and women appeared to be working them out in advance. The conversation was as practical as a theater re-hearsal, but every practicality arose from questions of moral and tactical direction. Who would sit down at the building's entrance? Who would link arms together? Who would stand to one side and be designated to talk to women as they arrived for their abortions? Would they go limp during arrest? How was the bond to be posted, and what were the preparations for the legal defense?

It exhilarated him. Afterward a more mystical person might have remarked at the precision of the timing: here was this meeting and here by happenstance was Sam, who some years ago had been handed back by God, perhaps for some purpose that was only now making itself clear, but Sam was all tangled up with more pressing questions and went away to think. Civil disobedience in the cause of abortion protest—this was new to Sam, a novel idea, and he had to decide what he believed about it. What *was* the moral nature of an induced abortion? How grave a wrong was it to use a machine or instruments to end a pregnancy? What was the obligation of the bystander who wished to do the ethically acceptable thing?

Sam walked to the St. Louis University library, asked for directions to the card catalogue, and began pulling volumes off the shelves.

He found abortion books, abortion magazine articles, bound collec-tions of abortion reprints. He found legal studies, medical studies, histori-cal studies, theological studies, and arguments over philosophy and scriptural interpretation and the use of statistics. For a week he haunted the campus library, pacing back and forth between the fluorescent light of the stacks and the long wooden table where the volumes were piling up one after another; Sam had never been methodical in his study habits and tended to inhale subjects all at once, without taking notes, reading page after page, so that now he worked rapidly through essays like "A Protestant Ethical Approach," and "Civil Law and Christian Morality," and "The Indications for Induced Abortion," and "The Legal Case for the Unborn Child."

He studied points of view, set them out against each other, tried to work them around in his mind. Because St. Louis University was run by Jesuits, there was quite a lot of Catholic abortion writing in the library stacks, but Sam was interested in all of it and paid close attention to the philosophical arguments defending abortion, since most of these were arguments he had never really considered before. In truth Sam had never really considered *any* abortion argument before, not with the kind of seriousness he gave to more immediate matters like whether or not to hand money to street people; in January 1973, when the Supreme Court released the *Roe* and *Doe* opinions, Sam had been a sophomore in high school and only dimly aware that some momentous news was perceived to have emanated from Washington. Their mother had grieved aloud at home, so that later Sam's younger sister Emily would remember the *Roe* announcement chiefly for their mother's sorrowful countenance as she stood in the kitchen and told her children what the Supreme Court had said, and once after that Sophie had taken Sam along with her to a right-to-life meeting in the back room of a local bank. There were some Willke war pictures at the meeting and of course everyone was horrified. But Sam was deep in his Mount St. Francis prayer group by that time, and even within the group, which included some Protestants, the students had argued about the morality of what the Court had done. For himself Sam had put it aside; there didn't seem any question that abortion was wrong, but he had no particular reason to mull over what *kind* of wrong, and whether two Supreme Court rulings mattered enough to worry about.

Now he saw before him, spread out in the scholarly disarray of his library abortion shelf collections, the chronicles of a multistate battle that had already been underway for more than half a decade by the time the Supreme Court formally sanctioned the destruction of unborn human life. That was how the *Roe* and *Doe* opinions were characterized in some of the volumes Sam pulled out to read; it was a disturbing way to summarize 150 pages of line-by-line judicial reasoning, but it was meant to be disturbing. "No amount of statistical calculation, moral protestation or subtle legal argument can change the fact that an abortion destroys a human life," read a U.S. Catholic Conference paper prepared for testimony before a congressional subcommittee in 1976. "What moral principle can equitably and justly balance the potential accomplishment of social good with the direct and deliberate destruction of one million or more unborn human beings each year? *There is no such principle*" (emphasis added).

The more he read—the more books he took down and examined and closed to put back up on the shelves—the more it seemed to Sam that this was true: there was no such principle. No one was able to propose a more logical time for the moment of an individual person's beginnings than the joining of egg and sperm; no one had proposed it, Sam decided as he read, because it was not possible to do so. The first beating of the heart, the first brain waves, the first perceptible movement in the womb, the emergence from the birth canal, "viability"—each of these suggested cutoff points seemed specious to Sam, an effort to put some pseudoscientific ballast into what was really being proposed, which was this: a human life begins when the pregnant woman and her doctor decide it does.

Well, this was just wrong. Some things were right and some things were wrong, Sam in his childhood had learned no lesson more deeply than that, and here was a wrong so vast and so immediately real that because of it human infants—for Sam now believed it made no sense to think of embryonic life as anything else—were being, to use the clinicians' own remarkably chilly language, "terminated." *Terminated*. In later years Sam would talk about something snapping inside him, his moral duty suddenly unshakably plain, and when he had to explain it to people who didn't understand, he would pull out a lesson that no one could miss: a man is standing beside me, beating his child with a club, and the child is crying out in pain. What am I supposed to do?

Of all the writings Sam Lee read, all the arguments and exhortations and accumulations of data, one essay stopped him. Among abortion activists this was quite a famous essay, printed originally in 1971 in the journal *Philosophy and Public Affairs*, and it had already pushed more than a few committed right-to-life volunteers to think carefully about the extended logic of their own positions. The author was a Massachusetts Institute of Technology philosophy professor named Judith Jarvis Thompson, and although her essay was entitled "A Defense of Abortion," it was often referred to as the Violinist Argument because of the provocative fable it used as its starting point.

Let us assume, Thompson had written, that every person has a right to life and that life begins at conception. And now let us assume further that you, the reader, have been kidnapped by the Society of Music Lovers and have awakened to find yourself in a bed with an unconscious violinist, a famous musician of indisputable value to the world, plugged into your kidneys. The Music Lovers explain that you need only lend the violinist your kidneys for nine months, but that without this temporary hookup the violinist will die.

"Is it morally incumbent on you to accede to this situation?" Thompson wrote.

No doubt it would be very nice if you did, a great kindness. But do you *have* to accede to it? What if it were not nine months, but nine years? Or longer still? What if the director of the hospital says, "Tough luck, I agree, but now you've got to stay in bed, with the violinist plugged into you, for the rest of your life. Because remember this. All persons have a right to life, and violinists are persons. Granted you have a right to decide what happens in and to your body, but a person's right to life outweighs your right to decide what happens in and to your body. So you cannot ever be unplugged from him." I imagine you would regard this as outrageous, which suggests something that really is wrong with this plausible-sounding argument I mentioned a moment ago.

It was manifestly true, Thompson wrote, that if you acted to save your own life, by reaching around and unplugging the violinist, then you were not committing murder. And it was equally true, she argued, that neither the violinist nor the Society of Music Lovers could demand the use of your kidneys for a shorter time: "I am arguing only that having a right to life does not guarantee having either a right to be given the use of or a right to be allowed continued use of another person's body, even if one needs it for life itself."

Sam was stumped by this argument. Instinctively he felt there was something wrong with it, but he couldn't see what it was, and it troubled him so much that he borrowed from the library the collection that contained the essay and took it with him to his Introductory Philosophy class, so that he could show it to the teacher. The teacher was a young graduate student named Mary Ducey, herself already a low-key veteran of regional right-to-life work—in the summer, during their undergraduate years, she and her sister had joined the picketers at Reproductive Health Services, where Mary liked to carry a sign she had made that said UNBORN WOMEN HAVE RIGHTS TOO—and many years later Mary Ducey remembered glancing up at the end of class to see Sam making his way toward her around the desks with a hardbound book in his hand.

"He said he had successfully refuted all the arguments in there except one," Ducey recalls. "But he was having trouble with this one."

Mary Ducey already liked Sam Lee, who was clearly one of the brightest students in her class, and she told him to give her a day to read Thompson's essay, which she had heard about but never seen in its original form. When she read it, she saw the right-to-life hole in the argument: the Society of Music Lovers has conducted a *kidnap*, nabbing this hapless person's kidneys by force. Most women who have sex have not been forced into the act. You could overlay the analogy on forcible rape, perhaps, but not on consensual sex, and thus not on the vast majority of abortions being performed at places like Reproductive Health Services. Thompson had anticipated this rebuttal in her essay and had come up with yet another analogy, suggesting that a woman who voluntarily has sex and gets pregnant is like a person who opens a window for fresh air and discovers that a burglar has climbed in; surely the opening of the window, Thompson had written, does not give the burglar the right to make use of the house.

It was a flimsy analogy, Ducey decided; how could embryonic human life be imagined as a burglar? Procreation was a *part* of sex, whether people wished to characterize it that way or not. If you wanted to get carried away in analogy it made more sense to suggest that getting pregnant after voluntary sex was like opening a first-floor window in a neighborhood that was constantly teeming with curious and intrusive children. Did that give a person the right to kill every child who climbed in? There was only so far a philosopher could go with these abstractions, and when Sam came back to class the next day, Ducey told him what she thought. He listened carefully and without argument. "I remember him looking up," Ducey recalls, "and it was like a light bulb coming on."

Sam thanked his teacher for her time, returned the book to the library, and went back to the Franciscan house to volunteer for arrest.

• • •

FOR FIVE YEARS after January 1973, from the first stunned reading of *Roe* v. *Wade* and *Doe* v. *Bolton,* two powerful convictions had driven the men and women in American right-to-life organizations—both regional and national, both Church-affiliated and secular. The first conviction was this:

> *As soon as the American people understand the terrible thing that has happened, they will demand together that it be reversed—legally, politically, and morally.*

And the second conviction was this:

The longer it takes, the harder it will be.

By 1978, the paradox was evident to a lot of people and made them at once passionate in their commitment and desperate about the stakes. Every march and meeting and newsletter, every hand-lettered signboard and four-color brochure, every speech and photo show and right-to-life convention schedule was now assembled with these dual suppositions simultaneously at work. For years right-to-life leaders had argued privately and publicly that the slow slide toward state-by-state abortion law change could be stopped at once if people could simply be made to *see*—if newspapers would print the abortion pictures, if radio stations would play the recordings of first-trimester fetal heartbeats, if ordinary Americans could be confronted in their own hometowns with the prospect of licensed medical facilities being used with government approval to destroy unborn human life.

How could this possibly *not* be so? "It was sort of a Children's Crusade," recalls Carolyn Gerster, the Arizona internist who attended the 1973 organizational meetings of the newly reconstituted National Right to Life Committee, remembering the straightforward confidence with which she and some of her colleagues had once assessed their own future. In the year before *Roe,* Gerster remembers, she and other right-to-life activists had been resolutely calculating the number of months it would take them to reclaim the states that had changed their abortion statutes. The goal they had set for themselves was eighteen—eighteen months, they believed, to reverse all the changes, to undo New York and take back California and repair by referendum or legislative vote every weakened new statute that had come to mean an unborn child was a person in one state but not in another.

Certainly they had anticipated some resistance: Washington State was probably going to be a tough sell, for example, with its history of legal abortion by initiative and a statewide religious presence too weak to be of much organizational use. But the Washington strategy would build, it was imagined, upon the same basic principle that right-to-life activists articulated in every state: *Explain* it to people. Explain it to them courteously if you can, more noisily if you need to, rudely and at full volume if you must; bring out the war pictures, bring out the pamphlets, fan out beyond the slide show auditoriums to reach directly to the still misled. By the end of 1972, any regional right-to-life group could send away to Hiltz Publishing, the busy Cincinnati company that was turning out Jack and Barbara Willke's work, and receive by return mail whole boxloads of materials that were lightweight and eye-catching and suitable for pressing into the hands of a passing legislator or shopping-center customer. There was "Life or Death," a full-color flyer that Hiltz would print up in bulk orders for five cents apiece; the cover showed one of the Willkes' bloodred "candy apple babies," still intact from an apparent saline abortion and curled up against its own umbilical cord, and inside, also in color that showed up well on the glossy paper, the Willkes had laid out some of the magnified suctioned-body-parts pictures alongside the skillfully illuminated enlargement they always referred to as Tiny Feet.

The text of "Life or Death" was written in language that was clearly intended to sound explanatory and matter-of-fact. "Suction abortion is performed between six and 12 weeks," read the caption under one of the photographs:

> This method involves the insertion of a tube through the cervix (mouth of the womb). Connected to a powerful suction, this tears apart the body of the developing baby and sucks it out. Then, either with this tube or with a curette (a loop shaped steel knife), the abortionist cuts and slices the deeply rooted placenta, the afterbirth, from the inner walls of the womb.

On the back was additional text, Q and A style, in the manner of a professor satisfying a curious but respectful student.

> Q. Abortion is only a religious question, isn't it?
> A. No. Theology certainly concerns itself with respect for human life. It must turn to science, however, to tell it when life begins. The question of abortion is a basic human question that concerns the entire civilized society in which we live. It is not just a Catholic, or Protestant, or Jewish issue. It is a civil rights question, a human rights question, a question of who lives and who can be killed.

Q. A civil rights question? How so?

A. 1) The first question to be asked is: What is this inside of her womb? Is it a human life? The answer is found in natural science, medicine and biology. At the first cell stage, fertilization, this being is alive, not dead. Human? yes, not another species. Sexed? yes, male or female from fertilization. Complete? yes, nothing has been added to the single cell, whom each of us once was, nothing except nutrition and oxygen.

Science has long shown conclusively that this is a human life from the beginning.

2) The second question is: Should there be equal protection by law for all living humans, or should the law discriminate, fatally, against an entire class of living humans as with abortion, which discriminates on the basis of age (too young) and place-of-residence (living in the womb).

So, abortion is a violation of human rights, of civil rights.

Some regional organizations, during what the veterans would come to remember as the initiation years of the very early 1970s, had worked up their own materials. The Right to Life League of Southern California handed out black-and-white pamphlets featuring the Missouri doctor William Drake's bucket shot and two pages of description of dilation and curettage and hysterotomy under the headline "A DOCTOR TELLS—HOW A CHILD IS ABORTED!" In Northern California, a committee called Concerned for Life led off its modest little mimeographed and stapled flyer with a blurry reprint of one of the *Life* magazine fetal pictures and a typed outline, the uneven typeface suggesting some lone person with the opened magazine and a manual Remington, entitled "THE DEVELOPING BABY":

18–24 days   Has a heartbeat.

30 days   Has a brain, eyes, ears, mouth, kidneys, liver, umbilical cord & heart pumping his own blood.

43 days   Electrical brain waves.

6 weeks   Skeleton complete, in cartilage, buds of milk teeth appear and he or she makes first movements of the limbs ("quickening"), though mother doesn't feel this for at least another 10 weeks.

9 weeks   Will grasp an object placed in his palm and can make a fist.

11–12 weeks   Breathes fluid, sucks thumb.

Implicit in the text of every one of these handouts was the logic that inspired them in the first place. If human life began at conception (as obviously it did), and if it was nonsense to carry on about "sort-of" life or "semihuman" life (as obviously it was), and if most Americans were good

people uninterested either in promoting a radical social agenda or in making money off abortion centers (as surely they must be), then for the first
generation of right-to-life volunteers the conclusion was inescapable: a
person who was not distressed by the idea of legally sanctioned abortionists probably just didn't *know*. Clearly there was something this person
had missed. Perhaps the humanity of the fetus had not been properly
explained, or the details of abortion itself were still vague; in any case
it was the responsibility of right-to-life activists to supply the missing
information, to be the bearer of unpleasant truth.

On January 22, 1973, the magnitude of this task had swelled with
such spectacular swiftness that for many years afterward women and men
with one right-to-life group or another would remember feeling that they
ought to turn to strangers on the street and cry, Can you *believe* it? Can
you believe what the Supreme Court has done? In St. Louis a homemaker
named Mary Ann Johanek, who had presented lectures in Chicago for
Illinois Right to Life before the family moved to Missouri, had picked up
the youngest of her six children from school the day of the *Roe* v. *Wade*
announcement with the car radio still squawking and her face so drained
of color that her boy asked her quickly what was wrong. "This is a very
sad day for our country," Johanek had said carefully, not wanting to weep
in front of her son, and she began to recover from her shock only as she
repeated to herself the reassurance that had kept her from discouragement
in Chicago: they don't understand.

"We'd always come back to *we have to educate*," recalls Johanek, who
helped start up Missouri Citizens for Life (and went on in later years,
pushed partly by the fervor of the abortion debates, to become a practicing
attorney). "We'd say, Why isn't everybody as upset as we are? Maybe
they're not as upset as we are because they don't know. So that was our
big thrust. People have to know."

And now it was not simply abortion that people had to know; they
were going to have to understand a Supreme Court ruling, too, and what
the ruling meant in their own state, and why the reasoning was shameful,
and how the justices had simply declined to take on the most important
question before them, which was the determination of what is and is not
unborn human life. Within right-to-life organizations there was a certain
perverse relief, although no one in the trenches would phrase it quite that
way, in what they read as the sheer outrageousness of the *Roe* and *Doe*
opinions: they were so bad that they could almost be said to simplify the
work ahead. All that was required was to teach the public what the Court
had done and then channel the anger to pass a constitutional amendment
protecting human life from conception on, and then everybody in the
state-by-state organizations could pack up and go home.

Among all but the political sophisticates of the movement, it was
difficult at first to imagine that this process would take very long—a few
years, possibly, an electoral cycle or two, certainly no longer than the close

of the decade. In Kansas City, the right-to-life organizer Mary Kay Culp, who had been pulled into the work herself by a "Life or Death" pamphlet handed to her in a shopping center, listened in disbelief three years after *Roe* to a talk by Carolyn Gerster, who had come to Kansas City bearing startling news: it was possible, Gerster said, that the battle might go on past 1980. The right-to-life people were all listening to Gerster in a steak restaurant called the Golden Ox—Mary Kay Culp would remember this later in considerable detail—and Culp scoffed and said to herself, simultaneously, *That's ridiculous* and *Oh, my God.*

"You have to understand," Culp recalls. "We thought this was going to be a couple of years, and we'd make it right, and it would be over."

But Carolyn Gerster was part of the national leadership by that time, about to commence a two-year tenure as president of the National Right to Life Committee, and the organizers who worked Washington, D.C., and traveled from state to state had come to believe that in a certain dreadful way they were racing against time. Gerster had phrased it to herself early on, amid the first swells of righteous post-*Roe* enthusiasm, as she looked around the meeting rooms at her comrades in the Children's Crusade: "Even then there was a little voice in the back of my mind," Gerster recalls. "And it said: 'But people will get used to this.' "

Here was the catch, the fear, the first deep crack in the optimism of even the most stalwart. *People will get used to this.* If the Supreme Court refused to condemn abortion, and every state was forced to accommodate abortion, and abortion centers across the United States were permitted to advertise their services as government-sanctioned medical care, then sooner or later principled objections were going to weaken amid the steady lure of convenience. Rationalizations would multiply. Eyes would be averted. Young women would come of age assuming that society and the American judicial system stood ready to offer them this "choice." Definitions would appear out of thin air, twisting things this way and that, until the act of ending a human life was called something else entirely, something designed to soothe and sanitize. So pervasive might this become, this defining and rationalizing and averting of eyes, that some of the hardest-working activists began in spite of themselves to imagine a time in the not too distant future when Americans *would* understand, when they would see abortion as the killing of a human being on its way toward birth, and *they would approve of it anyway;* they would shrug, collectively, and shake it off as an unpleasant necessity of modern life.

It is almost impossible to overstate the despair these images set off in dedicated right-to-life circles. From their vantage it was as though some mad cabal of American professionals had presented various pragmatic arguments for legalizing the selective elimination of unwanted children under the age of three (it's a painful choice, but really it must remain the mother's personal decision, and here is the machinery with which the children can be most efficiently dispatched), and then gradually, unbelievably, persuaded more and more of the populace to settle back and *agree* with them. In the years leading up to the fifth anniversary of *Roe v. Wade,* the urgency of this despair was palpable in nearly every public display of abortion protest: in the annual March for Life in Washington, when demonstrators converged by the tens of thousands on the boulevard leading up toward the Capitol; and in the regional rallies around the states, when men and women gathered with armbands and black cardboard coffins and signs reading *Innocents Slain by Abortion* or *Nazis Killed Jews, Abortionists Kill Babies, What's the Difference?* In St. Louis the fired-up organizers of Missouri Citizens for Life had staged their first big rally for October 1973, ten months after *Roe,* and when they saw the aerial photographs afterward they smoothed them into keepsake books and put them up framed on their living-room walls so that the inspiration would stay fresh and keen: thirty thousand people, that was the estimate in the newspapers, and a Catholic choir and two Baptist choirs and the archbishop and the state attorney general, all massed before the old St. Louis courthouse in the solidarity of their anger at the U.S. Supreme Court.

The courthouse site had been selected deliberately, at that St. Louis rally, and the Missouri Citizens for Life organizers were proud of themselves for guessing in advance how powerfully well received its symbolism would prove to be. *Dred Scott v. Sanford,* the slaves-as-private-property case that resulted in one of the infamous Supreme Court rulings of the nineteenth century, had commenced its legal journey at the St. Louis courthouse; Dred Scott was a Missouri slave who tried for more than a decade to sue for his freedom on the grounds that he and his former master had lived for a time in states where slavery was illegal. The Supreme Court's 1857 *Dred Scott* ruling was noted in the history books for its declaration that the Congress could not bar slavery in any territory under its jurisdiction, and right-to-life organizations liked to remind people that the Court had based much of its ruling on a finding that more enlightened generations now viewed with contempt: the Negro slave, Chief Justice Roger B. Taney had written, was not an American citizen at all, but was in fact the slaveowner's property and therefore entitled under the Constitution to "no rights which the white man was bound to respect."

The connection, lest it prove too subtle for the average passerby, was condensed into billboard format at the St. Louis courthouse rally. *Black Non-person; Unborn Non-person,* one of Jack Willke's books was to put it some years later, repeating in paperback the *Dred Scott–Roe* analogy displayed for the crowd in St. Louis. The reasoning, under this analogy, was equally damnable in both cases: *Slavery Is Legal; Abortion Is Legal. Choose to Buy-Sell-Kill; Choose to Keep or Kill. Abolitionists Should Not Impose Morality on Slaveholder; Pro-lifers Should Not Impose Morality on Mother.*

The capitalized phrases were attention-grabbers, not the stuff of law journal research, but as attention-grabbers they served their purpose. By the mid-1970s the right-to-life movement was still overwhelmingly white and Catholic (the president of the National Right to Life Committee was an eloquent black physician named Mildred Jefferson, and even she understood herself to be an exception), but some of the most enduring in-house imagery was of early American abolitionists—forward-thinking men and women, the argument suggested, who had defied politically popular prejudice to articulate and crusade for the rights of a dehumanized minority. "Was it surprising," Willke wrote in his slim treatise, *Abortion and Slavery,*

> that, being property, slaves were treated as "livestock," their children called "increase," their mothers "breeders," the men "drivers"? That they were advertised and sold with hogs and sometimes treated far worse?
>
> Dehumanization was necessary in order to keep others from sympathizing with them. It is always easier to exploit, injure or kill if those in power feel superior and those beneath are called non-human names. The game of semantics is not new to us. Did your mother, who was pregnant and wanted to tell you children of the coming happy event, ever place her hands gently upon her abdomen and say to you, "I have a feto-placental unit in my uterus"? Why the insistence of the pro-abortionists on dehumanizing an unborn child by calling him a "fetus," and "embryo"? Why does the Planned Parenthood counselor speak only of "pregnancy tissue"? Why use the words "terminate pregnancy" which speaks only to the mother's condition and ignores the baby she carries? All of us terminated our mother's pregnancy, probably around nine months, but none of us are dead.

The heroism of the abolitionists, the analogy asserted, lay in their ability to see human beings where others around them saw something else—something different, expendable, not quite human. And when the abortion clinic sit-ins began, two years after *Roe*, some of the philosophical groundwork—there was quite a bit of philosophical groundwork, in leaflets and position papers and long typewritten proposals for legal defense—drew from the abolitionists and the modern civil rights movement in the South. "A sit-in involves our implicit assertion that our capacity to endure suffering is greater than the abortionist's capacity to destroy," read a widely distributed pamphlet printed in early 1978. "With Rev. Martin Luther King, Jr., we must come to believe that 'unearned suffering is somehow redemptive.' "

It was a natural escalation, the clinic sit-in, and as public protest technique it caught the imagination of the young and university-oriented wing of the right-to-life movement. Since 1973 the National Right to Life Committee had been the movement's heavyweight, with a big national board and regional affiliates all over the country, but there were smaller independent organizations, too, some of them aimed directly at the kind of bright and ardent college student who in an earlier generation might have been intensely interested in campus rebellion or demonstrations against the Vietnam War. "A simple statement, made honestly and courageously, can have immediate and revolutionary repercussions," read that 1978 pamphlet, which had been written by young activists whose mailing address was a residence hall at Harvard University. "The core of nonviolent direct action is very simple. You say, not just in word but in act: 'These children are my brothers and sisters and I am with them. I link my fate with their fate.' "

According to the handful of young men and women who came to be regarded as guardians of this corner of right-to-life history, the first real sit-in—a prearranged trespass at the entrance to an abortion clinic, with arrest as a deliberate part of the plan—had taken place in the Washington, D.C., suburb of Rockville, Maryland, in the summer of 1975. The six women who had participated were arrested, as they had wished, but their illegal trespass cases were dispensed with quickly and without much publicity (each was sentenced to six months of unsupervised probation), and it was a good two years before the idea began to spread west and north. By 1977 word was traveling in right-to-life circles of sit-ins in Kentucky and Massachusetts and Connecticut, a dozen reported around the country and perhaps more that had escaped anything beyond local notice; and in St. Louis, the first chief orchestrator of sit-in protest was the young Franciscan Vince Petersen, who had begun studying the intriguing bulletins that were arriving in the Franciscan house mail.

Here was the monthly *National Right to Life News*, for example, in February 1977, its pages lively with updates on legal and political developments a long way from eastern Missouri. The Supreme Court was preparing to rule on the issue of states' right to refuse public funding for abortions. States as far afield as Nevada, Louisiana, and Maine, twenty-four in all, had adopted formal declarations of support for a Human Life Amendment to the Constitution. The fourth annual March for Life had gathered many thousands of people in Washington even though the weather was bitterly cold and windy, and as evidence the paper ran a big front-page picture of a massive crowd of people smiling out from inside thick coats and hats. "When it was over, the crowd of chilled pro-lifers didn't want to go home," the *News* reported. "They asked for one more song."

And on page three, under the headline "Protestors' Trial in Cleveland Is Cancelled," there was this:

CLEVELAND, OH—The trial of the eight women arrested for trespassing at an abortion clinic here has ended with an acquittal. The defense lawyers submitted a motion for acquittal when there was insufficient evidence to prove adequate notice of trespassing. The move was a disappointment to the protestors.

The eight women, members of People Expressing Concern for Everyone (PEACE), were arrested when they blocked the hallways of the clinic and were charged with criminal trespassing.

Under normal circumstances the trial would have been held in front of a judge only, but the lawyers, members of Lawyers for Life, requested a trial by jury, assured that the jury would have acquitted the women. The protestors hoped to use the trial to establish the personhood of the unborn. . . .

The defendants, who had stated that they were willing to keep appealing a conviction all the way to the Supreme Court, said they will keep on defending the rights of the unborn.

"That's why we did it," said defendant Mary O'Malley, "to protect human beings and to restore their legal rights."

In 1977 Vince Petersen was twenty-six years old and completing his first year with the Franciscan order, although he had considerable work to do before taking vows as a friar. He thought of himself less as a priest than as an artist; Petersen was taking studio classes at St. Louis University and planning out a wall mural on an abandoned brick building near the Franciscan house, where he intended to paint St. Francis as a black person in the traditional hooded habit, blessing all of creation around him.

Petersen was excited about this project because the neighborhood was poor and mostly black, profoundly different from the Minneapolis surroundings in which he had been raised; he was a slight young man, not quite five feet seven inches tall, with a brown mustache and a soft-spoken Minnesota-bred gentility that sometimes made Petersen impatient with himself. In later years he would grouse about the stifling climate of "niceness" in which he had been raised, the instruction that being nice occupied the primary spot in some grand moral scale of values (smiling and waving at a person, Petersen would explain by way of example, while mentally giving him the finger for cutting you off in traffic). This was very Minnesotan behavior, and Petersen sometimes felt that he had been retreating into it off and on all his life; he had attended junior college outside Minneapolis and joined the campus antiwar and environmental activists, marching, organizing strikes, and generally railing against the establishment until he exhausted himself and took cover again, this time more definitively, in niceness.

"My way of hiding was to go into religious life," Petersen recalls. "And I thought, I will be safe here. I don't have to get *involved* in stuff."

But a few of his fellow Franciscans eventually made him see that he was wrong to believe this. "I had certain friars in my community who kicked me in the butt and told me that the seminary was no place to run away from the world, but that it was a place where I really had to encounter reality," Petersen recalls. "And that social activism has everything to do with—well, with a life of faith. If you see something happening that is evil, in a life of faith, you're called to speak up against it."

At the Franciscan house community programs were already under way: the vacant-lot cleanups, the weapons-to-peace conversion project. An occasional Missouri Citizens for Life flyer found its way to the Franciscan bulletin board, too, but beyond that the Franciscans had organized no special presence in the right-to-life campaigns until Vince Petersen began growing uneasy about his own passivity in the face of the abortion issue. As a student in Minneapolis he had worked for a while with a University of Minnesota group called Save Our Unwanted Life (SOUL), and the SOUL philosophy seemed to Petersen so consistent and obvious that he had a difficult time understanding why the students who joined were considered eccentrics by the rest of the antiwar left. It had always struck him as paradoxical that the same people could both deplore the Vietnam War and demand that abortion be legalized. Wartime killing, capital punishment, the motorized machines built for the sole purpose of sucking out human life before birth—all of these seemed to Petersen to form an inseparable chain of violence and disregard for life, which was God's greatest gift to man, and he was gratified to see that the new sit-in demonstrators at abortion clinics appeared to be lay Catholics who believed as he did.

What was it that he had in mind? Petersen wasn't entirely sure at first; he had read Saul Alinsky, the old Chicago prophet of radical organizing ("revolution by the Have-Nots has a way of inducing a moral revelation among the Haves"); he knew as much as the next literate ex-demonstrator about the tactics of Dr. Martin Luther King, Jr.; and he had some passing familiarity with *The Politics of Nonviolent Action,* an imposing three-volume study by a Massachusetts political science professor named Gene Sharp. Sharp's books were being read carefully by some of the sit-in organizers on the East Coast; their sweep was extravagant, detailing nonviolent protest over the centuries, and several hundred pages were given over to specific examples of practical approaches in action. The chapter on Nonviolent Intervention (as distinct from the chapters on Political Noncooperation or Economic Noncooperation) explained forty-two different techniques of protest, each already accomplished in India, East Germany, Mississippi, fourth-century Italy, or somewhere else in the great span of world history, beginning with "Self-exposure to the elements" and running through the Sit-in, the Stand-in, the Wade-in, the Mill-in, the Pray-in, Guerrilla Theater, Nonviolent Land Seizure, Selective Patronage, and Seeking Imprisonment.

"Sometimes civil disobedience is seen to be called for," Sharp wrote in his Part Two, because of a belief that a certain law is illegitimate, the body or person which instituted the law having had no authority to do so. . . . One of the most drastic forms of political noncooperation, civil disobedience is an expression of the doctrine that there are times when men have a moral reponsibility to disobey "man-made" laws in obedience to "higher" laws.

Such a time was upon them now; there was no question about that. But Petersen wanted to organize something that seemed to him more philosophically Franciscan than a trained squad of Arrest Me volunteers. "We are engaged in a struggle to change people's hearts," read the Harvard sit-in pamphlet, which reached St. Louis as Vince Petersen invited to the Franciscan house the first participants in the plans for a sit-in to mark the five-year anniversary of *Roe* v. *Wade*.

And it was this pamphlet's model, more or less, that Petersen hoped to follow: explain your presence, to yourself and those who come in contact with you, as a defense of living persons who can be visualized *through* your presence. Some people might be too misguided, too willfully blind, or too distraught themselves to see with their imaginations the growing child inside a pregnant woman's body; but nobody could miss a twenty-five-year-old protestor with his back against the clinic's front door. " 'I am with him,' " the pamphlet advised the protestor to declare, wordlessly, by positioning himself in the proper place. " 'Before you kill him, you must deal with me.' "

Probably they would be arrested. It would be good if they were arrested. But getting arrested was not the first goal; the first goal was placing a grownup body between the intended victim and the abortion equipment. Some of the written materials referred to vacuum aspirators as "killing machines" and the rooms in which they were placed as "chambers," but among groups like the one Petersen now assembled, which was composed principally of Franciscans and students from St. Louis University Pro-Life, there was sober conversation about the philosophy of Christian nonviolence and the appropriate way to think about the people who physically worked these machines. Were the abortionists and their assistants intrinsically evil? Must the protestor find it in his heart—or her heart, for Petersen had as many women at his organizing meetings as men —to love the abortionist and hope for his conversion too? What about the police themselves? Should they be condemned as collaborators? Was it fair to police officers, since after all they were men in service to St. Louis and had families who needed them, to go limp upon arrest and turn the removal-to-paddy-wagon process into a back-straining job of hauling 150 pounds of full-grown noncooperating adult?

By the third of Petersen's meetings Sam Lee was speaking up in these discussions, his voice quiet but emphatic, his longish hair loose around his face. "He was kicking me in the butt," Petersen recalls, "like, *let's get moving,* and *we need to act on this, this really needs to take place."* They zeroed in on a clinic, thinking pragmatically, with an eye to their own inexperience. Reproductive Health Services was the busiest abortion center in the state and had been picketed with dedicated regularity nearly every weekend of the last five years, but because the clinic itself was on the second floor of a large building, the picketers had to stand down on the sidewalk and crane their necks up toward the windows that they imagined looked out from abortion rooms. If people were going to try physically blocking a clinic's doorway, they needed a less fortresslike place to begin. Someone suggested the Regency Park Gynecological Center, out in the western suburbs of St. Louis, where a Peruvian doctor named Bolivar Escobedo ran an abortion clinic in a modestly sized one-story building that Petersen could see might prove an ideal site for a trial-run sit-in: the parking lot was wide and unguarded, providing easy access, and to reach the front and side doors of Escobedo's offices, a person simply had to walk into the building and past an indoor atrium full of potted plants.

At the Franciscan house they rehearsed their moves, imagining where the atrium would be, assigning individual people to separate doors, play-acting the roles of protestor and clinic employee and abortion patient. They reminded each other to regard everyone as victim, to remember that the Supreme Court decision was damaging doctors and pregnant women and police officers, too. They stressed the importance of prayer throughout the entire sit-in: prayer for the women, prayer for the abortionist, prayer for the unborn. They considered the example of the sit-in protestors who had preceded them in other parts of the country and decided that they *would* go limp during arrest, since going limp appeared to follow most precisely the tradition of nonviolent intervention, and someone thought to advise the women not to wear open-toed shoes, even inside the heated clinic, so that their feet would be protected in case the police dragged them over concrete.

On January 22, 1978, at 7:00 A.M., the first carful of protestors arrived at the parking lot outside the Regency Park Gynecological Center. Sam Lee was nervous and excited and wondered whether he had been right to decide against telling his parents what he was about to do, but when it was over at midafternoon the most memorable part was how easy it had been, how essentially uneventful, with two demonstrators sitting at each doorway and the protestors outside singing "We Shall Overcome" and Sam hurrying back and forth between them, bearing messages and sandwiches and cups of hot coffee. The doctor had called the police fairly quickly and the police had come out and stood around looking at them, nonplussed, but then the doctor had decided to shut down his office for the rest of the day and go home. So no one was arrested and they had *closed an abortion clinic,* only a small clinic and only for one afternoon, but they had done it and in their exhilaration they began planning right away to do it again. When they came back to Escobedo's office six weeks later they were eager, energetic, ready to pick up the pace a little bit.

This time they earned a picture in the newspaper. The picture showed two police officers standing, their backs to the camera, and below them on the hallway floor a young woman who was looking not at the officers but down in the general direction of her own crossed legs, as though she were midprayer even with the policemen up there with their hands on their hips and their exasperated posture indicating *All right honey come on now time's up.* The young woman was identified as "Helen Ann Egan, a student at St. Louis University," and it was Helen Ann who saw the open door to the waiting room and decided just to take the sit-in one step farther by going on inside and talking to the women who were in there waiting for their abortions. She was arrested immediately. That was when they knew it was not going to be like the first sit-in. The police were not overly rude or rough, but they hauled them out one at a time, two policemen to a protestor, the protestors trying to make their bodies go completely limp the way they had practiced, which as it turned out was not so easy once they were outside being pulled facedown through the cold. In the days before they had talked quite a bit about the symbolism of sit-in, the principle that the protestor must absorb within himself the vulnerability of the unborn child, and it seemed to Sam that for a long time afterward he was going to remember the feel of his wrists being pulled together behind his back and the small sharp click of the handcuffs around them. He understood that something new was happening to him, that he had made a decision he was not really going to be able to undo; and when the police got him into the back of the paddy wagon he stopped being limp and sat up, Helen Ann and the others pressed around to either side of him, so that they could watch each other's faces as the paddy wagon pulled out of the clinic parking lot and carried them to jail.

THE NEWS STORY ran on the tenth page of the afternoon paper: *"6 Booked After Anti-Abortion Protest."* The names of the six were listed, with "student" the only identified job description, and within the ranks of Missouri Citizens for Life there was a complicated flutter as middle-class men and women, most of them homemakers or retirees or conventionally employed jobholders, tried to settle on an appropriate reaction to the picture of Helen Ann Egan being glared at by the police. Was this what the movement needed—lawbreakers? Men with beards and women with long straight hair and photographs reminiscent of pacifist demonstrations at the Pentagon—these were disquieting images, for an organization whose most confrontational public gesture so far had been its rally at the St. Louis courthouse, and from inside MCL no one rushed to commend the sit-in demonstrators or put out the call for additional comrades in the clinic doorways.

Loretto Wagner, whose formal title at MCL was chairman of public relations, felt rattled by the first sit-ins and went out to watch them in person. The right-to-life rumor network had rapidly spread word of the plans for civil disobedience, even though the time and place were supposed to be somewhat clandestine, and when Loretto saw Sam Lee and the other young people being taken away in handcuffs, she wondered how much time they might spend in jail. It occurred to her to wonder also whether the six young people arrested at Regency Park were the only right-to-life volunteers in Missouri who were genuinely behaving as though human beings were being killed; and in the spring of 1978 Loretto Wagner found herself in some heated exchanges with MCL colleagues who fretted at the idea of right-to-life protestors deliberately breaking the law.

"I remember saying to people: What if this was a two-year-old?" Wagner recalls. "Would you stand back and say that you couldn't do anything about it until a law got passed? If this was a two-year-old, and they were about to murder it right before your eyes? *No.* You'd pick it up and run right down the street with it."

Loretto Wagner was a woman of imposing size and patience and wielded considerable influence in right-to-life circles around the eastern half of the state. She always dated her active work from the week after *Roe* v. *Wade,* when her parish church had invited a guest pastor to speak and the pastor had stared at them all from the pulpit, his voice tremulous with emotion, and cried, Where are all those antiwar protestors *now?* The priest had declared that a new war was being unleashed upon them, a war against the unborn, and Loretto was disturbed enough by his warning that she went to talk to him afterward, picked up a "Life or Death" flyer, and for the first time in her life looked carefully and up close at the pictures. For some time she had stood there with the glossy pamphlet in her hand and all the air knocked out of her, and in later years when the work began to pile up in a deadening kind of way, so that sometimes it felt as though it was all about voting records or legislative strategies or rally permits from City Hall, Loretto would pull out the war pictures and sit with them for a while to make sure she remembered what they were fighting about. "I have to wake up all those little nerve endings," Loretto would say, and it would relieve her to find that the pictures still made her cry. On the picket lines she was never very good at making her own signs (Loretto was more of a full-paragraph person, although she did design a good flyer once, with a snapshot of a friend's baby and a line that read, *Kill her now, it's called murder; kill her last week, it's called abortion*), but she would often pick up one of the picket signs that showed a war picture or the bucket shot, knowing people didn't like to see that out in public but that it was the truth, and showing it was part of her job.

What *was* her job, exactly? The young sit-in people made her think about it. For five years Loretto Wagner had been typing and telephoning and folding press releases into envelopes and working the long-into-the-evening hours of the dependable right-to-life volunteer. She had six children and a husband to look after, so that it was only during school or after bedtimes that she could haul the big electric typewriter over to the dining table and begin messing things up; her file folders bulged with lists and clippings and sample legislation, and she would spread the sheets of paper around her, organizing, the telephone crooked into one shoulder. She was quite a heavy woman, with a pleasant bespectacled face and a wide carriage that she tended to tuck away beneath large simple dresses, and in her right-to-life work the combination of her shape and her tenacity had given her the aspect of a slowly advancing tank: Loretto never shouted at reporters or at legislators, but they learned to brace by instinct when they saw her coming because she was tireless and driven and could outtalk nearly anybody in a courteous Midwestern voice that just kept rolling forward with all its anger and sorrow subsumed into the urgency of the mission at hand.

Loretto Wagner was mainstream, the year the radicals began to get arrested in St. Louis, and as mainstream she was managing extremely well. In the tripod that held up Missouri Citizens for Life, Loretto was one strong leg; the others were also women, one from the West and one from the East, and each carried in her own way as formidable a reputation as Loretto's. In Kansas City, the chief MCL organizer was Kathy Edwards, a sharp young woman with fashionably cropped blond hair and a trim, businesslike presence that she liked to think was a special rebuke to the magazine and television people who seemed to think *Roe* v. *Wade* only bothered frumpy-looking protestors in sagging hose. Kathy had seen pictures suggesting this, side-by-side photographs in a magazine apparently keen to illustrate reaction to *Roe*, and of course the right-to-life woman was fat and disheveled while the pro-abortion woman was attractive and slender and carried a professionally printed sign. It made Kathy furious to imagine that somebody thought readers could be manipulated this way, and after she organized a right-to-life chapter in her suburb (commencing, in sequence, with a classified newspaper advertisement, six women around Kathy's kitchen table, and an eleven-dollar treasury in a margarine tub), she found that she was happiest when she was defying stereotype: striding into debate halls where she could mix it up in her articulate way with people who wanted an argument and were a little unsettled, perhaps, to see Kathy looking like that woman on the opposite side of the magazine page.

The third leg in the tripod was Ann O'Donnell, a wildly capable Irish-Greek registered nurse with a face full of freckles and a temperament that left a lot of stories in its wake. In her St. Louis Catholic high school O'Donnell had been a star athlete, swimming and basketball and field hockey; a male lawyer who worked with O'Donnell at MCL remembered the evening many years back when they had all been teenagers together and Ann had showed up at a raucous party at someone's house. The boys had ended up in the basement, lifting weights and horsing around in a muscular sort of way, and then Ann appeared among them and began pressing barbells too—*heavier* barbells, the lawyer remembered, besting the boys. She was coy about her strength and sometimes used it to amuse herself; one night at some overly long right-to-life dinner she and her husband found themselves seated beside a man who was bragging that he could beat anybody at arm wrestling. "I bet you can't beat my wife," Eddie O'Donnell said. Ann reached over at once, planted her elbow, seized the man's hand in classic arm-wrestling position, and in one swift push thumped it onto the tablecloth. The man began to sputter, protesting that he wasn't ready, and Ann smiled sweetly at him before she turned away. "I only do it once," she said.

Around St. Louis Ann and Loretto worked off each other with an extraordinary efficiency, one fiery, the other dogged. Ann made television appearances and took airplanes across the country for National Right to Life Committee meetings; Loretto worked the telephones and wrote press releases and rounded up the buses and carpools for the Jefferson City lobby days or the annual March for Life in Washington. Loretto worried sometimes about her family, her children so accustomed by now to the sight of her writing rapidly into notepads with a telephone receiver pressed to one ear; it seemed to Loretto that she had once spent her afternoons in more traditional housewifely fashion, baking Toll House cookies and cake recipes she tore from the pages of magazines, but that this part of her life was gradually vanishing beneath the great stacks of paper on her dining table. There were afternoons when her children invited friends over for dinner and they would walk into the house and cry *Mom, get your stuff out of here,* and Loretto would unplug the typewriter and sweep everything into boxes and drag the boxes to her bedroom. I am giving them something more valuable than Toll House cookies, she would lecture herself on these frantic afternoons, as the mailing lists and flyers piled up on her bedroom floor and the back corner of the kitchen: I am giving them, if they will look around and listen, a social conscience.

Loretto Wagner had not held a paying job since she worked as a secretary before her first baby was born. As a child she had grown up eager to commence what she understood would be her principal and most fulfilling work, the raising of a family; she sometimes thought she had a particularly keen feeling about this because her own parents had divorced when Loretto was very young, and her mother had been obliged to work full time as a bookkeeper to support Loretto and her sister, the only two children still living at home. In Loretto's St. Louis Catholic schools of the late 1940s, a divorced working mother was regarded as exotic and alarming; Loretto had not liked this aspect of her childhood much, either the embarrassment or the lonely late afternoons waiting for her mother to come home from work, and when she listened now to the elaborate arguments being constructed in defense of legalized abortion, the one she found most perplexing was the proposition that women required abortion clinics so that they could "liberate" themselves into the work world and in this way make themselves somehow the true "equals" of men.

Loretto tried to be open-minded about these arguments; she knew better than a lot of people how crucial salaried work was for women who needed the income to buy food and pay the rent. But that was not what the angry women on television and in the magazines appeared to be interested in. Loretto kept hearing them talk about their *longings,* their *frustrations,* their need to be *fulfilled,* and these complaints made Loretto feel at once demeaned and a little frightened, as though an entire generation of American women was learning to regard motherhood as a tiresome obstacle in the way of this personal fulfillment. "Biology is not destiny," that was another line Loretto had heard more than once, and every time she heard it she wanted to retort: Excuse me, could we look in the mirror for a moment? You didn't have to be a religious person to see that there was some overarching plan in the design of human beings: women bore children, men did not, and Loretto continued to believe that the bearing and raising of children was genuinely the most important work a woman could undertake. Of course it was not the *only* work a woman could do; women were lawyers and accountants and business owners and police officers, and that was certainly their prerogative, Loretto supposed she wished them well. But four of Loretto's six children were daughters, and she knew what she wanted most deeply for all four of those girls. She hoped they would go to college, study hard, challenge their minds, marry happily—preferably nice Catholic boys, but Loretto liked to think she would be flexible about that—and stay home when their children began to arrive.

Was this so antiquated, believing that women cheated both themselves and society when they tried to make motherhood "optional"? Loretto was not so naive as to assume that pregancy was always wonderful, or that a woman could always be made to see how much she would love the baby when it came. But it seemed to her that over the last decade an awful coarsening had taken place, so that women were being encouraged to say to themselves, upon learning of a pregnancy they had not expected: Let me see, do I want this baby or not? Will this baby interfere with my Plans? Perhaps I should kill this baby, perhaps I should keep it; I have my career to think of, after all, or my education, or my fulfillment. No wonder the abortion proponents went ballistic when a right-to-life debater pulled out war pictures in front of an audience; the entire construct of the pro-abortion argument fell apart the instant they were forced to look at what it was they were defending. Loretto had heard about this happening in auditoriums and on television programs, and she marveled at the tenacity of the skilled pro-abortion debaters who managed to slog on anyway: we *need* to be able to cut fetuses up into small pieces, you see, so that women can achieve this equality they so deeply desire.

People will get used to this. Loretto, too, often found it bewildering that the nation had not yet come to its senses, that she had already invested five years in a movement she had once imagined would complete its work within two or three sessions of Congress. Her on-the-job training had accumulated steadily from one crisis and project to the next; by 1978 Loretto had acquired an array of organizational and political skills that would have filled out a formidable resume. She could type out a press release in lively newspaper fashion, putting the best quotes up top and ending each page with the newspapery "more." With a single set of phone calls she could activate an extensive telephone tree to spread news quickly or summon an emergency meeting or mobilize volunteers. She had grasped early on the strategic use to which the symbolic gesture might be put: for a while after *Roe*, the West County Chapter of MCL, which Loretto had helped organize, swept from one small municipality to the next with proposals that the local governing board, the aldermen or council or whatever, pass a resolution condemning the Supreme Court—declaring that *Roe* was wrongly decided and calling for a Human Life Amendment to the Constitution. Since these resolutions had terrific moral fervor and no legal weight at all, they were remarkably easy to pass, and every local campaign produced more headlines and more volunteers. Every volunteer was a potential body on a bus to the capitol, and by the time the resolutions campaign petered out, thirty-two separate municipalities had formally expressed deep chagrin at *Roe* v. *Wade*, which meant the buses to the capitol were full, when they needed to be, and lively with the high spirits of enthusiastic new recruits.

"Lobby days" was what they called their seasonal assaults on the capitol, and Loretto was one of the MCL stalwarts who knew how to make a lobby day work. Jefferson City was almost directly in the middle of the state, so the buses and carpools would arrive from both directions and pull directly into the capitol parking lot, where they would disgorge several hundred earnest-looking women and men who might as well have had CONSTITUENT spray-painted across their foreheads. As legislative chairman of MCL, Kathy Edwards supervised these lobby days, having learned the nuances of the capitol calendar well enough to know precisely when a certain bill might be hitting a critical step in its passage through committee, and there were moments when Kathy thought, as she watched the volunteers start up the great white steps of the building: Oh boy, they *do* look frumpy. There was a lot of polyester. One lady marched into the capitol wearing rubber boots. Kathy looked at the boots and thought, Well, the hell with it, they're her boots, and a legislator who lays eyes on this lady is not going to mistake her for some paid lobbyist in a suit.

That was the genius of it, of course. That was their strength. When the MCL people worked the capitol they had nothing, in the conventional armament of lobbyists' persuasions; no money, no liquor, no promises of generous corporate donations or comely women in nearby hotel rooms. All they had was constituents. If some of the constituents happened to look as though they had just finished up with the supermarket shopping or a Cub Scout meeting, so much the better; in the daily bustle of capitol business the lobby day volunteers were instantly identifiable from the other end of the hallway, clustered in groups as they moved from one legislator's office to another. Sometimes they brought pies or cookies along as gifts, friendly reminders, and left them on the secretaries' desks. Both MCL and the Missouri Catholic Conference would print up hundreds of brisk instructive flyers, no more than a page or two long, so that novices to the capitol might carry in their pockets folded pieces of paper with the latest updates and the office numbers of the key legislators and large arrows overlaid with urgent directives like ACTION REQUESTED or PRO-LIFE ALERT: GO.

The groundwork was exhaustive, sometimes tedious, sometimes orchestrated to make a legislator buckle under the sheer pressure of hundreds and hundreds of tiny repetitive taps on the shoulder. Loretto liked to tell the story of the bill they saved when an unfriendly committee chairman was sitting on it; MCL wanted it out for a vote, and the committee chairman didn't want it given the chance, and finally the MCL leaders lit up the telephone trees to pass the word that for three days straight every right-to-life volunteer in the state should call or telegram the Speaker of the House and urge him to use his political leverage to force this bill out of committee. Each call or telegram was to start off in exactly the same way: "Mr. Speaker: You are the key." No matter what was said afterward, every message was to begin with those words, so that when the Speaker looked at his mailgrams and picked up his big pile of phone messages, the same phrase kept repeating itself over and over, *You are the key,* until he must have understood that there was a phalanx out there and that Loretto Wagner, sitting placidly in the wooden chair just outside his office, was summoning it straight toward him.

Quietly, without any formal acknowledgment at all as to what was really going on, the bill was let out of committee for a vote. Loretto heard third-hand that the Speaker, a Democrat named Kenny Rothman who almost always voted with MCL anyway, had turned to an intern in his office and growled *Get these goddamn pro-lifers off my back.* But naturally he never said any such thing to Loretto, and when she and Ann O'Donnell watched him stroll into and out of the committee room, stopping long enough to bend down and whisper something in the ear of the chairman who was sitting on their bill, the Speaker nodded pleasantly at them as he departed. "Hello, ladies," he said. It was a few minutes later that their bill magically surfaced for its hearing, and this was really an extremely satisfactory little episode except that what it was about, ultimately, was politics. It was about stapled pieces of paper with words printed on them. The state of Missouri alone was recording approximately twenty thousand abortions per year in the late 1970s, which in crude arithmetic meant that during the week it might take to shove along a single bill—a decent bill, certainly, they were all decent bills, with their public funding cutoffs or their consent requirements or whatever it was that the movement's national legislative people were promoting that season to slow abortion within the constraints of *Roe* v. *Wade*—385 human lives had been "terminated" in offices like the one where Sam Lee and his university companions had sat down together in the doorway.

Loretto Wagner was famous for resigning, on lousy days or in the midst of meetings when the political nitpicking was suddenly too much for her; she would write a note, *I'm resigning everything*, and pass it to the person running the meeting and walk out. Usually she never even reached the point of unplugging the typewriter on her dining table before she changed her mind and went back to work. But when Sam and the other sit-in demonstrators were arrested, when they actually made an abortionist close his business for a few hours, Loretto saw that something vigorous and new was upon them and that the students were right to call it "direct action." Civil disobedience *was* direct, it had a power and an immediacy that no ordinary human being could possibly extract from a Jefferson City committee hearing, and sooner or later the arrests were going to lead to courtroom trials—public prosecutions of men and women who had chosen, like the lunch-counter sit-in demonstrators of the civil rights movement, to defy the law for a greater moral good.

The protestors' only real liability, it seemed to Loretto, was their youth. In the world she knew, nobody paid very serious attention to a college student being hauled away in handcuffs. If the protestors wanted to attract real interest, they needed grownups with them at the clinic doorways—people with families and mortgages and grownup reputations that might be genuinely placed at risk by the prospect of criminal arrest.

So Loretto Wagner, who in the spring of 1978 was forty-four years old, married to an aeronautical engineer, and the mother of six St. Louis schoolchildren, telephoned Sam Lee and his Franciscan house friends to tell them that she was going to get herself arrested too. She would follow Vince Petersen's lead, Loretto declared; she would sit in a doorway, hold the hands of her fellow blockaders, sing the protest songs, confront the police. She had only one reservation, being as large as she was, and she liked to think it was both pragmatic and humane: Loretto had no intention of going limp and breaking some police officer's back. When they came to arrest her, she planned to walk to the paddy wagon herself.

THE NATIONAL RIGHT TO LIFE COMMITTEE held its annual convention in St. Louis that summer and argued back and forth, heatedly and publicly, about sit-ins. It was a big, bustling meeting, with the standard workshop items listed in the two-day program, Grassroots Political Action and Working with the Media and How to Pass a Call for a Constitutional Convention, but the television cameras pushed into the seminar room in which the lead speaker was John Cavanaugh-O'Keefe, a twenty-eight-year-old Harvard graduate who was one of the philosophical guides of the sit-in movement.

It was Cavanaugh-O'Keefe who had written the text of the much-read sit-in pamphlet from Harvard, and he made a compelling presence there at the front of the meeting room in the convention hotel: he had keen dark eyes and looked like an Irish poet and managed, even before opening his mouth, to throw into rich confusion the popular right-to-life stereotype that so galled Kathy Edwards.

He had studied feminist theory at Harvard. He had hyphenated his last name to merge it with his wife's. He was a Vietnam War conscientious objector, recognized by the draft board; Cavanaugh-O'Keefe had fulfilled his service obligations by working as an aide in psychiatric hospitals, and after the war he had joined in demonstrations against construction of the Trident nuclear submarine. "An unjust law does not bind the conscience of any person"—this was John Cavanaugh-O'Keefe's phrasing, in the text of his pamphlet. "We do not believe that saving lives is illegal, but even if it were we would still try to do it."

Sam Lee drove his Volkswagen down to the St. Louis convention hotel and was absorbed by everything, the handouts and the posters and the crowded hallways full of people wearing Circle of Life bracelets or small silver lapel pins in the shape of the Tiny Feet from Jack Willke's pictures. During the convention, a *St. Louis Post-Dispatch* reporter wrote a thoughtful article about the consternation the sit-in demonstrators were stirring up, with leaders like Carolyn Gerster declaring that the National Right to Life Committee would never condone illegal activity, and Sam pushed his way into the crowded Nonviolent Direct Action seminar and listened intently to John Cavanaugh-O'Keefe. The audience members were prodding Cavanaugh-O'Keefe and the other speakers, testing their principles, debating: Where was the dividing line for nonviolent acts? Was it "violent" to destroy property without injuring people? Why stop at nonviolence when the cause was so important?

Everyone in the room knew that several clinics around the country had already been destroyed or damaged by arson; the previous February, in an incident that received considerable publicity, someone posing as a deliveryman had walked into a clinic in Cleveland and flung a bag of flammable liquid into the receptionist's face before setting fire to the clinic's interior. The clinic was full of patients who had to run from the fire, and the receptionist was temporarily blinded. Every right-to-life organization in the area denounced the attack, and the police arson investigator said the fake deliveryman appeared to have been "some guy who just went off his rocker," but it was a bad blow for the movement, and now at the St. Louis convention the sit-in radicals had to work to defend their positions even amid sympathetic colleagues. The *Post-Dispatch* article described the troubled convention exchange between a worried Texas philosophy student named John Barger and the sit-in advocates who were trying to reassure Barger by listing the many reasons they were determined to keep their tactics nonviolent:

Violence would alienate the broad middle class where pro-life has its basic support. All the political gains could be lost. Fair-minded folk by the millions would be turned off. And so on.

The basic answer was that violence does not "work" in the short or long run. It is counterproductive. A pro-lifer cannot ultimately save the baby without converting the woman about to have an abortion because the fetus is totally dependent on the mother, said one sitter, suggesting that an isolated, violent act to save a two-year-old might be justified because that person is a separate entity.

"They're insured. They'd just get another and bigger abortion clinic," another told Barger after the seminar.

Barger told a reporter that he remained intellectually unconvinced. . . . There was inadequate awareness in the right to life camp "that we are the radicals of our time," he said, and it would be very easy for hardcore pro-lifers "to talk themselves into a whole lot of trouble."

Sam heard these reservations expressed, as he sat inside the seminar room and walked through the hallways of the convention hotel, but he was not worried about the temptation to slide toward violence himself— to talk himself into a whole lot of trouble, as the Texas philosophy student had put it. Sam had never believed himself to be an absolute pacifist; he had always assumed that if the occasion ever arose, he would strike out to defend himself and his family from outside attack. He knew there was a perilous elasticity to the concept of "self-defense," too: if I can take up arms to defend my family when the enemy soldiers are at the door, then probably I can take up arms to defend my community when the enemy soldiers are at the outskirts of town. And if there was sound moral justification for violent defense of one's community, then perhaps it was also moral—so the argument could go, if a person was determined to push it there—to take up arms, or stinkbombs or torches or bags of gasoline, in defense of the community's threatened unborn children, who after all had no one else to protect them from death.

There was a certain steely logic to this reasoning, Sam could see that, but he didn't buy it and it appeared to him that none of the 1978 sit-in leaders did either. (Many years later John Cavanaugh-O'Keefe still remembered the evident disappointment of the television people who had crowded into his convention seminar with their camera lights ablaze: "In the workshop it became clear that we were not going to advocate violence," Cavanaugh-O'Keefe recalls, "and at that point all the lights went out.") Violence was pointless for the kind of protest they had in mind. The mission was to portray *abortion* as a violent act. The abortionists committed violence and the protestors confronted them peacefully, that was the message that needed to be conveyed, and for Sam it was precisely this clarity of principle that had drawn him with such force to the idea of clinic sit-ins. He was a Christian on his way to the priesthood, and what inspired him now was the strength of Christian love, a force that had been perverted for violence before and surely would be again; nonetheless, Sam could see that the fear of this perversion must not stop him from following the dictates of his own conscience. Was it wise or honorable to remain passive in the face of a great wrong simply because some crackpot might twist your argument into a rationalization for harming others?

Still Sam was restless. Vince Petersen's group was lagging about organizing its next sit-in. As he listened to Cavanaugh-O'Keefe and his East Coast colleagues, Sam found that he was frustrated by the knots in their own reasoning: if they were blocking doorways to stop killings, if every sit-in protestor was the moral equivalent of the German who speaks up and puts his body between the ovens and the Jews, then why were they not in clinic doorways *every single day?* What other responsibilities could possibly be more pressing? Why hadn't the St. Louis protestors walked out of the county jail, the moment some sympathetic right-to-life people had posted their bond, and driven straight back to the Regency Park Gynecological Center to block Escobedo's office again?

Sam needed to do something. He had an obligation to the leadership and the defense attorneys, he knew that, and he could see that a one-man sit-in was not going to accomplish anything at all. But the spring semester was over and his own energy was eating at him, making him uncomfortable. Finally he decided to put a change of clothes in the back of his car and drive to California, where Berkeley Bio-Engineering made its vacuum aspirators, so that he could persuade the factory workers to lay down their equipment and stop making the machines. It was a notion of such goofy idealism that when Sam talked about it in later years he would smile in embarrassment, but he did it anyway, accompanied by a man and woman from the East Coast sit-ins. On their drive west they slept in campgrounds and ate fruit and cheese from grocery stores, and when they reached San Francisco they laid out their sleeping bags on the apartment floor of a woman named Sunshine, who lived near the Zen Center, was married to a Buddhist taxi driver, and knew some of the sit-in protestors from the Trident missile demonstrations. Every morning Sam and his friends took the rapid-transit train across the bay to San Leandro, where they stood outside the doors of the Berkeley Bio-Engineering plant and tried to hand the employees nicely printed pamphlets that explained the terrible purpose to which their products were put, but the employees were not as alarmed as Sam had hoped and would say things like, Oh, I work in another division of the plant, or Well if I don't do it somebody else will.

After a week they gave up and went home. The director of the company did meet them for a while, inviting them into his office and asking them please to stop disturbing his employees, but it was obvious to Sam that the director had no intention of changing his business. On their way back east Sam and his friends told each other that they must have made an impression on those California factory workers, that perhaps they had planted something that would take root in the months to come, and anyway Sam had already begun working up another idea for the later weeks of summer. He was thinking, he said, about Reproductive Health Services —about the fortress the St. Louis sit-in protestors had so far been too timid to approach.

Sam was having some ethical difficulty with this. He had a plan in mind, and he was trying to see how it might be accomplished without looking anyone in the face and telling a flat-out lie. He supposed he could invent for himself some useful rationalizations about the distinction between active lying and passive deception, but he found with regret that he was going to have to settle for something less than complete honesty because what he was proposing was a kind of espionage, after all, involving switched urine and minor theatrics and a small tape recorder hidden inside a handbag. Sam was going to get a detailed firsthand look at Reproductive Health Services, and he intended to do it the only way he believed he could. He was going to fake his way inside.

NINE

A Lesser of Evils

1978

SYLVIA HAMPTON was the first person to notice him. She had put on her white laboratory coat and he was wearing something threadbare, a wrinkled shirt or a sweater that wanted patching, so that standing beside him made her feel that he was fragile and needed special tending to. She had walked into the clinic waiting room and made her usual introductory remarks, Good morning, and Those of you who have accompanied a patient today may wish to join us in a few minutes for an information session, and when she waited the requisite few minutes and returned to the waiting room, she saw that the bearded young man was getting up with the others and preparing to follow her in.

"This way, please," Sylvia Hampton said.

In the summer of 1978 Sylvia Hampton was one year into her tenure at Reproductive Health Services and liked it more than any other place she had ever worked. Judy Widdicombe had hired her away from the local Planned Parenthood office, and from the instant of the job offer Sylvia saw with some amusement that Judy had developed a memorable management style; they were standing beside each other in a women's bathroom, adjusting their lipstick in the midst of some public event in St. Louis, and Sylvia, who was forty then and ran Planned Parenthood's telephone hotline service, said her work was not going as smoothly as she would like. "Come work for me," Judy said at once, right there in the women's room, meaning it. And Sylvia was uneasy at first about taking the job, because she knew Judy well enough both to admire her and to hesitate at the prospect of Judy as a full-time boss. Sylvia knew one woman who had quit Reproductive Health Services in a rage at Judy Widdicombe: overbearing, obnoxious, insufferable, this woman had complained to Sylvia, describing Judy as a presence at her own clinic; and there had been a second woman, too, another former employee, who said carefully to Sylvia before she accepted the new job: "I've got just one piece of advice for you. If you get in trouble with Judy, don't cry. She'll lose respect for you."

So Sylvia was braced, the week she began at Reproductive Health Services, and watched Judy closely to see what it was going to be like. The other employees showed her around. Here was the waiting room, here was the counseling area, here were the medical rooms with the aspirators and the quilted pads on the table stirrups and the pictures mounted up on the ceiling, so the women on their backs would have something to look at. Each patient was to have a counselor with her, holding her hand or talking quietly to her or whatever the patient seemed to want most, all the way through the procedure. Each patient was to be sitting up when the physician first walked into the procedure room, so that patient and doctor could look each other levelly in the eye while the doctor introduced himself. Each doctor was supposed to introduce himself sincerely, not like the Almighty descending from the heavens or some charity physician with an expression on his face like *here's another one too dumb to use birth control right*. In the midst of her orientation Sylvia stopped for a moment and realized that she was standing outside a closed door listening to the unmistakable, richly audible sound of Judy Widdicombe haranguing someone for unacceptable work; and the longer she listened, the clearer it became that the quiet person at the receiving end of this critique was in fact a doctor, a male physician, and that Judy was berating this doctor for the insensitivity of some remark he had made or some boorish way he had behaved around patients at the clinic.

"And when he walked out of the office, he just had this hangdog look to him," Sylvia Hampton recalls. "And this was a woman who was a *nurse*. Of course she was almost a head taller than he was. That helped. But I looked over my shoulder, and I saw this happening, and I thought, *yes*."

It was a wonderful job. In later years Sylvia would say that if Judy had been a man she would probably have been elected governor of Missouri, and the state would have been a damn sight better off, too; what upset people about Judy looked to Sylvia like qualities that are supposed to be admirable in a man. Judy was ambitious, restless, unapologetic, demanding, and blunt. If she disliked you or found you incompetent, Sylvia could see that Judy was capable of swatting you aside swiftly and without regret, but she and Judy latched onto each other immediately and Sylvia was placed in charge of the S.O. groups, which was a position requiring both diplomatic skill and a strong backbone. The S.O.'s were

Significant Others, the boyfriends or mothers or most-trusted-aunts who sometimes accompanied patients into the clinic, and once they had lowered themselves into armchairs in the sober quiet of the waiting room, it was Sylvia's job to invite the S.O.'s inside the clinic so that they could sit around a conference table and talk. Sometimes they ignored her, or shook their heads and went back to turning magazine pages or staring impassively at the walls; but more often the S.O.'s would look up in surprise and then rise, tentatively, when Sylvia came back to collect them. That was when the delicate part began. Over the months she had learned something about the barrage of emotions she was likely to encounter in any given hour with a roomful of S.O's, and she was supposed to sit there at the table in her laboratory coat and work somehow with all of it, the fear and guilt and hostility and grief, so that when the abortions were over the women would emerge to companions who knew how to be a comfort to them. "The amazing thing to me was how often someone would come into one of these groups and have like this steely thing about them, like *I don't like you, I don't like this whole thing, I'm just going to have to put up with it until I get out of this sleazy abortion mill*, they would have this whole *attitude*," Hampton recalls. "And then they would leave with a very different attitude, they would pump your hand as they went out."

She worked up answers, palliatives, lines of reasoning. A low-key question might start them off: Did you ever think you would be spending the day in an abortion clinic? Usually the questions would begin to come rapid-fire after that, and Sylvia would try to take them on one by one, explaining the procedure, how it worked, what to watch for afterward, how to check for fever, how much bleeding was normal, how the disruption of hormones had been observed on occasion to add a short-lived form of chemical depression to whatever sadness the woman might be feeling already. "Sometimes they would say, 'Have you ever seen the abortion afterward?' " Sylvia Hampton recalls,

and I would say, "Yes, I have." Then they would say, "Well, what does it look like?" And I would say, "Well, it depends on the stage of the pregnancy." "Does it have little feet and a heartbeat?" And I would say, "Yes, at the early stages it does. But you have to have a magnifying glass to see it. And that's beside the point. The point is that this is a developing embryo that is going to become a fetus that is going to become a baby that is going to become a child, a teenager, an adult. Is this what this woman wants? Is this what this woman is ready for? Is this how she wants this outcome? And *why did she make the decision?*" I would kind of put it back on them: Yeah, it *is* a developing human being, but why isn't she carrying it to term? And then they would start to talk about that.

Sylvia liked watching them talk to each other; often these were men and women who had told no one, who had kept their big raucous feelings jammed up inside themselves. Once she sat at the table with a group that included a middle-aged woman and a teenaged boy, and as the session progressed Sylvia saw that something powerful was happening between the woman and the boy even though they had been strangers to each other moments before. When they finally spoke the woman said that her daughter was having an abortion and that she, the mother, was so angry at her daughter's boyfriend that she had refused to see him or let him talk to anyone in the family. The teenaged boy said that this was exactly what his girlfriend's parents had done to him, and that when they found out about the pregnancy, his own parents had shut him out too, so that nobody would talk to him about the worst thing that had ever happened to him. When the boy saw that the middle-aged woman was listening, that he was actually talking to someone who looked a little like his mother and his girlfriend's mother, he began to cry. By the end of the group everyone at the table was in tears and Sylvia had to walk up and down the clinic halls afterward to collect herself. It relieved her to feel that she had reached somebody, that a person had left the S.O. group knowing more about what to do and also seeing some of the bigger picture, and what alerted her first to the bearded young man in her morning S.O. group that day was a curious sense that she was reaching him in a way she did not quite understand.

She liked him, she could see that right away, and she felt that they had connected, that they were speaking directly to each other even with others around them at the conference table. She tried to answer his questions as precisely as she could. He wanted details on the abortion procedure, the medical aspects of it. He wondered how dangerous it was, how each part of the procedure was accomplished, what the complications might be, what Sylvia could tell him about fetal development. It seemed to her that he was struggling terribly with this and that his girlfriend's abortion was causing him very deep unhappiness; but there appeared to be no hostility in him at all, and Sylvia supposed it was useless to pay any attention to the gut feeling that there was something different about him.

He *was* awfully thin, and he smelled like he needed a bath.

Forty-five minutes into the S.O. group Sylvia stood, as was her custom, and announced that she would return after checking briefly on the patients. She liked to do this to give the S.O.'s a few minutes to talk among themselves, and when she stepped into the medical wing of the clinic she found the nurses abuzz with the story of what had just happened in Dr. Schwartz's procedure room. A young woman had reported for her abortion, they said—a woman who had already received her pregnancy test results and completed an individual session with a counselor—and Schwartz had gone in to take care of the termination. But Schwartz was a veteran who had been checking over pre-abortion patients since his volunteer days with the Clergy Consultation Service, and as he examined the young woman he had reared back in surprise.

"Old smart Schwartz," Hampton recalls. "Nothing ever got past him. He really let her have it with both barrels."

The woman was not pregnant. She had handed her first tester a vial of urine that indicated pregnancy, but the pelvic examination showed that the urine must have belonged to someone else—hand-carried to the clinic, apparently, solely to bring off the deception. And she had lied about this supposed pregnancy being her first; Schwartz could tell from the examination that she had already given birth at least once. And she needed to go see her own physician immediately because she had growths in her uterus that might be dangerous and required further examination. So Schwartz and the staff had sent the young woman packing, and it was not until someone looked out the clinic windows toward North Euclid Avenue that Sylvia understood what had happened.

The bearded young man and the woman who was not pregnant were standing together on the sidewalk, conferring. "I don't think the full impact of the scam hit us until they were away from the clinic," Hampton recalls. "At first we thought they were weird. But then we thought, Oh my God, this was a scheme to find out if we were honest. And we thought, Well, we sure gave them a dose of our honesty. Too bad they couldn't do the same for us."

A few days later, in her office at Reproductive Health Services, Judy Widdicombe received her first telephone call from Samuel Lee.

AT THE CLINIC some of the staff members thought Sylvia was being foolish, and they would chide her about it: he won't listen, this is nonsense, why are you wasting your time? But she shook them off. She was moved by Sam Lee. Everything she watched this young man do struck her as so serious and purposeful that she felt obliged to listen to him, to acknowledge his gravity. When Sam called Judy Widdicombe and confessed that he had been spying, that the young woman with the faked pregnancy was a right-to-life friend of his and that he was interested in meeting with someone from Reproductive Health Services so that they could talk about what he had seen at the clinic, it was Sylvia Hampton who volunteered for the meetings—who met with Sam over and over, more than a dozen times in all, during the autumn and winter of 1978.

They met in groups of four: Sylvia, Sam, a right-to-life volunteer named Linda Hatch, and Maureen McCarthy, who ran the local National Abortion Rights Action League affiliate. They would sit in someone's living room, or a cheap lunch restaurant, or even an office at Reproductive Health Services; Sylvia's colleagues might shake their heads, watching the four of them walk into a clinic office together, but the conversations were determined and friendly, even if they did seem to have an exhausting circular quality to them. Sam and Linda would try to convince Sylvia and Maureen. Maureen and Sylvia would try to convince Linda and Sam. Sometimes Sylvia would bring out her full arsenal of logical argument and maneuver furiously this way and that and she would just think that she *had* him—they were walking down the street once, on their way back from lunch and deep into agitated discussion about evolving law and medical safety, and Sylvia cried triumphantly, "Well, if you believe that, Sam, then you believe abortion should be legal!"

But he would not budge. He was gentle and passionate and would not budge. "He was enough to bring you to tears," Sylvia Hampton recalls. Winter came on and at the Reproductive Health Services offices someone came to work one day with the Circle of Life newsletter, typed and stapled by students at St. Louis University, its last two and a half pages composed by one Samuel H. Lee.

"This summer I approached by telephone the director of an abortion clinic in St. Louis and asked to meet with her," the article began.

> I explained to her who I was, and my involvement in sit-ins at abortion clinics, and my desire for rational, non-threatening dialogue with her as opposed to emotional debate. To my delight (and surprise) she agreed to meet with me for lunch the next day. . . . I discovered that she was not a two-headed ogre with a bloody curette in one hand and a wad of dirty twenties and fifties in the other. Instead she is basically a good but misguided person who believes that abortion is a lesser of evils if a woman feels she is subjected to an unwanted pregnancy. And she discovered that I was not a right-wing, reactionary, terrorist, sexist anti-abortionist, but instead (I hope), a basically concerned, caring individual who believes abortion is killing and not justified. It is interesting to note how far the false stereotypes and fears have descended. When the director of the clinic agreed to meet with me she had visions of another Cleveland incident with me dousing her and her clinic with gasoline and lighting a match or else pulling a gun during lunch and shooting her. When she told her clinic personnel that she was meeting me, some begged her not to. She did not sleep the night before we met. Fortunately, no violence took place, and her fears were relieved.
>
> Secondly, I have developed a better perspective of why pro-choicers believe what they do. Some have experienced illegal abortions, either themselves, or the traumatic, sometimes tragic back alley butchery they have seen performed on other women. Therefore if women are going to have abortions, whether it is the killing of an unborn child or not, it is better for the abortions to be safe and legal. Some have been in the forefront of the women's liberation movement and see abortion as a means for women to free themselves from the sexual subjugation by

men which has taken place for centuries. Their fears are that if legal abortions are denied women, then all strides women have made for justice and equality will be lost also. A few are pro-choice partly because they are against pro-lifers—seeing the pro-life movement as a religious and/or patriarchal institution bent on domination by the right wing. If abortion is made illegal, they believe, then all civil rights will be lost also. Those who believe in abortion primarily for these and other reasons, that is for selfish and self-serving motives, I would place in the pro-abortion camp. I have not yet met a true pro-abortionist, i.e. one who believes in abortion as a coercive means or a profit making venture. These people do exist, but I suspect conversion of hearts in these will be more difficult.

Third, my dialogue with pro-choicers has helped to further the truth in me. Pro-choice people are by nature critical of pro-lifers, and are quick to point out our inconsistencies. I have been challenged by these people regarding whether or not I smoke, drive with a seat belt, and about my concern for unwed mothers, about my sexism, and even about whether I pay taxes (that support war and abortion). What develops from these challenges is that I am constantly analyzing my "pro-lifeness" to see if I am consistent or not. I am a better (and humbler) pro-lifer for having truths pointed out to me by pro-choice people.

Finally, I find myself no longer dealing in the abstract on abortion, but in the concrete, in the flesh and blood so to speak. I have talked with people who perform abortions or support the killing; I have eaten meals, joked, laughed with them. I have seen them in clinics where they work, their counseling, their vacuum machines of death. And I have agonized over the question of how these people, who appear basically good, caring, intelligent people can be involved in such a violent, abhorrent activity as abortion. I have no answer yet. But I intend to continue dialoguing, continue sitting-in, continue loving, with the hope that the love that is in their hearts will be transformed into a compassion for the unborn.

Sylvia Hampton's heart constricted when she read Sam's essay; how could you not like such a person? What he saw at Reproductive Health Services had startled Sam, he had declared that straightaway both to Sylvia and to Judy, and Sylvia found that she was worrying about him in a motherly way as she watched Sam struggle to reconcile his convictions and his honesty. "What he wanted to prove was the basic evil of abortion *and* anyone who was involved in it," she recalls. "And what was happening, clearly, was that he couldn't reach that point. It was impossible. And I thought the fact that he was sending this out in a newsletter was wonderful. I mean, here he was telling the whole right-to-life world that we were not devils."

Like an office memorandum, with checkmarks by the appropriate initials to indicate which staff members had reviewed it, the Circle of Life newsletter worked its way around Reproductive Health Services. Judy kept it long enough to read it and sign off at the top, *JW*, before turning her attention back to work; she had a wary fondness for Sam too, and he was right, she didn't believe he was a reactionary right-wing terrorist. But it made her sigh, reading phrases like "compassion for the unborn" and "vacuum machines of death." People like Sam must see the world in black and white, it seemed to Judy, as though every gray in the moral landscape could be wished away by a simple act of will. "Sincere, but totally out of touch with reality," Maureen McCarthy hand-wrote on an end-of-year report to NARAL, describing the St. Louis meetings with Sam and her pithy assessment of his character; and that was how Judy felt about him too, that if he would just touch down to earth for a while he might walk through her clinic and see what she saw instead of the vacuum machines of death.

She saw women. She saw women she no longer had to direct onto interstate airlines or pull into the station wagon from St. Louis street corners so that they could come home and abort into her toilet. On procedure days the receptionist who opened up the clinic at seven-thirty always found women waiting in the hallway, some of them settled patiently on the bare beige floor with their backs against the walls; women found their way to Reproductive Health Services from all over Missouri, and some of them also came down from Illinois or up from Arkansas or all the way over from Kentucky, two states away. They came because they needed what Judy had built, and she wished sometimes that she could jam the right-to-life people into their seats, like unruly schoolchildren, and force them to listen over and over to the voices of the women they were trying to put back on street corners. Judy was a veteran of the public abortion debates and she knew what people like Sam and Ann O'Donnell said when you flung this at them, but all you had to do was spend an afternoon in the Reproductive Health Services waiting room to understand how wrong they were: making it illegal again was *not* going to change these women's minds, it was not going to make them want the baby, it was not going to do anything except shut down the first generation of American medical facilities to treat elective abortion like a doctor's procedure instead of a secret.

Judy was fierce about Reproductive Health Services, and she knew it. Outside the clinic, her old life was dropping away from her piece by piece and she was terrified sometimes, although this would have startled her own staff, the picture of Judy terrified; in the clinic she was supposed to be the rock, the cheerleader, the brassy voice in the dizzying central office where new people would show up to be interviewed and come away thinking that it was like Grand Central Station, Judy laughing and sizing them up and answering interruptions at the door and then saying suddenly, "Excuse me, I have to take this call from Washington, D.C." She worked long hours, she came in on Saturdays, and when she went home to the suburban house she and Art had bought after Reproductive Health Services opened, she saw that even the street name had begun to grate on her: Country Club Court. It seemed to Judy that some years back she had stopped being this person, the good Mrs. Widdicombe on Country Club Court. She no longer knew how to talk to the man she had married; he was still gentle and attentive, a kind man, a fine father; but the change in Judy made him anxious and dependent, and she imagined sometimes that when she marched out into the world Art was standing sorrowfully at their front door, watching, wishing he could call her back.

In January 1979, four months before her forty-first birthday, Judy told Art she wanted a divorce. Afterward she would say it was the New Year's party that pushed her the rest of the way. The two of them had been invited someplace fancy, where the clock hit midnight and all at once the room was full of couples, men and women in elegant clothing, all kissing and singing "Auld Lang Syne." Judy had gone home crying, and after that she told Art that she didn't want to work on it any more, she didn't want to go to counseling, she just wanted to move out. She loved their sons, but they were grown and nearly grown, nineteen and fifteen; the fifteen-year-old could stay at home with his father while he finished out high school, Judy wasn't going to displace the boy or put him through a custody fight, and Art was home now more than she was anyway.

She took an apartment in the West End, down the street from Reproductive Health Services. At first she had some furniture from the Country Club Court house in her apartment, but after a while even the sight of the familiar things began to bother her, and she called a decorator she had met and asked her to clear the old belongings away and start over. Judy had never in her life felt so detached from the place that was supposed to be her residence; she lived at the clinic, really, or in the hotel rooms and friends' guest bedrooms she occupied when she was traveling, and she sometimes traveled now for weeks at a time: to China, where she toured clinics and factories as part of a visiting delegation of American professionals; to Bangladesh, where she worked with an American physician on a foundation-funded family planning teaching tour of Bangladeshi medical colleges; or to Washington, D.C., where she made regular visits to the NARAL headquarters to hear the latest anxious dispatches about the efforts of Antis around the country.

That was the jargon at the NARAL offices, *Antis*, pronounced "ant-eyes," short for "Anti-choice." The vocabulary was cleaving so thoroughly that there appeared sometimes to be almost no overlap at all. The movement literature and newsletters that arrived on Judy's desk were full of references to Choice and Rights and Freedom and Individual Decision-Making, all of which were described as undergoing relentless assault by persons who referred to themselves, inaccurately, as "pro-life."

Who are the so-called "pro-life" people?

read a fund-raising advertisement by the Abortion Rights Alliance, the NARAL affiliate for the state of Missouri.

> They are the COMPULSORY PREGNANCY people, and that's what they should be called. . . . Your legislators are under attack by the COMPULSORY PREGNANCY people. The poorest women are being denied State and Federal funding for abortion. And that's not all. Clinics, where clean, safe conditions are assured, have been bombed and set afire. Fanatics inflamed by the COMPULSORY PREGNANCY propaganda are picketing and attacking abortion clinics and terrorizing women.

It was not incidental that NARAL had changed the organizational name that filled out its acronym. The National Association for Repeal of Abortion Laws had become the National Abortion Rights Action League, once the old abortion laws were repealed, and the dispiriting part of this vigorous-sounding new label was the sheer volume of Action that still faced them. Judy served on NARAL's executive committee and received so many alarming missives from Washington that she covered clinic bulletin boards with political alerts and made sure the bulletin boards were affixed where patients and Significant Others were more or less forced to notice them. "YOU MADE A CHOICE," began one flyer that was handed post-abortion to every Reproductive Health Services patient during the fall of 1978, and the paragraphs that followed were blunt and frankly political:

> You chose what to do about a pregnancy. Just now you may feel that you will never need the right to choose again and would just as soon forget about it. But you or someone you care about may need it again, a sister, friend, mother or a daughter.

> THE RIGHT TO A SAFE, LEGAL ABORTION IS A BASIC HUMAN RIGHT GUARANTEED BY OUR CONSTITUTION

> In 1973 the Supreme Court legalized abortion and said a woman has the constitutional right to choose, along with her doctor, whether or not to terminate a pregnancy. Freedom of choice means no one is forced one way or the other. It means you can decide for yourself about preg-

nancy and childbearing based on your own situation and your personal beliefs.

YOUR RIGHT TO THIS PERSONAL DECISION IS IN TROUBLE

In June, 1977, the U.S. Supreme Court said that states don't have to pay the cost of an elective abortion for poor women on Medicaid if they don't want to. In other words, the Court said the states can "encourage childbearing" while paying the cost of birth under Medicaid while paying nothing for an elective abortion for a poor woman! Anti-abortion groups are pushing for a "Human Rights Amendment" to the Constitution which would make abortion illegal by declaring personhood from the moment of conception. Calls for a Constitutional Convention have the same goal, and 13 states have voted in favor of the Convention call—35 are needed.

YOU CAN LET YOUR LEGISLATOR KNOW HOW YOU FEEL ABOUT FREEDOM OF CHOICE RIGHT NOW

We have petitions, postcards, addresses and sample letters for you to use if you wish to send a message right now in support of freedom of choice.

It was a controversial decision Judy had made, confronting people with politics when they were at their most vulnerable; her own clinic counselors sometimes objected to the indelicacy of her timing. But the timing was deliberate. Sylvia Hampton had been hired under the title "Health Rights Advocate," and Judy was unrelenting about this; it wasn't right for a patient to whisk in and out of Reproductive Health Services as though her own abortion had nothing to do with law or public policy or the rights of the women around her. Sometimes the counselors would sink into conference-room chairs at the end of the day and trade war stories about patients who had proceeded through the entire abortion process while insisting from start to finish that they were pro-life. The counselors were either volunteers or getting by on lousy Reproductive Health Services pay because they believed in what they were doing, and these particular women frustrated them intensely, the protestors who had come in from the antiabortion ranks long enough to take care of their urgent little personal problem and then go right back out to the picket lines. What were the counselors supposed to say? Suddenly they were like civil libertarians defending the free speech of Nazis: if you believed in every woman's right to have an abortion, then presumably that right extended to the selectively shortsighted too, but it was awfully hard to sound compassionate when your patient was sitting there explaining that her situation was so *very*

grave and special that she had to have an abortion, even though she didn't
believe abortion should be legal in the first place. "One of them was in
parochial school," recalls Bertie Passanante, who at fifty-nine was the
grandmotherly senior member of the counseling staff,

> And I cannot tell you the disdain she had for everybody who was in
> the waiting room. But it was okay for *her* to be there. I can practically
> see her. She had protested, and gone to meetings, and I think she signed
> something in church. But it would kill her parents if she were pregnant,
> and her relationship with the boy was over, and so forth and so on, and
> she knew she would have to live with this for the rest of her life. But she
> was going to go ahead and do it. It was important for her to be on record
> as anti-abortion. And she didn't take it very well when I suggested to her
> that maybe everyone sitting in the waiting room would prefer not to be
> there. She just had in her mind that all the other people out there were
> using abortion as a method of birth control, and their reasons were not
> as good as hers. That made me sick.

Certainly the counselors were instructed to listen for cues, to be
attentive to ambivalence, to understand that they ought to talk more
emphatically about alternatives when a woman began using words like
"murder" and "unborn child" to describe her prospective abortion. But no
Reproductive Health Services counselor was supposed to slam the door on
somebody simply because of some vivid inconsistencies in the patient's
own moral code. If a woman was insistent, if she refused guidance about
raising the baby or placing it for adoption, then it was the counselor's job
to show her to a procedure room and hold her hand—literally, if that was
what the patient wanted—while the doctor took care of her. (Judy had
been known to hold private procedure-room baptisms, when she thought
a patient might be comforted by the ritual; the counselors would stand
aside and Judy would lean over the surgical pan and recite, just loud
enough for the patient to hear her voice, what she hoped was a passable
prayer.) And there was no provision for whistle-blowing, either, no matter
how strong the temptation to call up some *Post-Dispatch* reporter and
whisper *Guess who just brought her own daughter in for an abortion.*
Without a guarantee of patient confidentiality the clinic stood to lose
everybody, not just the occasional hypocrite, and Judy's survival instincts
were far too finely honed for that.

She worked on the defense, guarding her territory, alert to the next
blow. Half a decade of legality had given them a Yellow Pages listing
but no promises, no certainty, no protective public embrace. When the
fund-raising letters went out from NARAL's national headquarters, the
return envelopes would pile up on the Washington, D.C., office work

tables, some of these envelopes stuffed with checks or five-dollar bills and handwritten notes of encouragement, and some with expletives scrawled across the NARAL letterhead, or pornographic pictures ripped from magazines, or razor blades that fell out into the hand of the staff member ripping open the seal. "I am against everything NARAL stands for," wrote one man who had attached to his note both a razor blade and a small coat hanger fashioned from a paperclip. "With all the protective measures there are to keep from getting pregnant, you all should be stuck with the poor fuckin kid. I hereby enclose one Handy Dandy Self Help Abortion Kit, so as not to be a burden on the tax paying public. PAX."

At NARAL they kept Anti letters in a file, archiving, storing the sound of the enemy's voice. "May your husbands—if at all you have one —be STERILE or so productive he gets your best friend pregnant." "I had a baby and I paid for him, so let the girls pay for their own mistakes as I did. Taxes are high enough." "You lesbian-bitches suck!"

I cannot tell you how thankful I am that you were not around when our adopted son was conceived. You would have had him murdered.

I think abortion promotes loose behavior. They can screw around all they want and there are no consequences—I give no money so that people may act without thinking and then someone else picks up the tab.

I can hardly wait for the time when that most vicious of all crimes —the murder of the unborn—is again illegal. As for safe abortions— this should be as expensive, dangerous and difficult as possible. If the tramps want to murder their children—let them pay for it. If a female doesn't want a birth (other than rape) all she has to do is say *no*—and that's for *free*.

Those women whose life is in danger will always have a right to an abortion, don't worry; but as for whores who want the pleasures but don't want to face the consequences, I couldn't care less about their problems. (P.S Hope they defeat you.)

The letters came directly to Reproductive Health Services, too, carried in with the laundry bills and the invoices from medical supply companies. At first they unnerved Judy, especially the unsigned notes in tiny single-spaced penmanship that contained many references to JESUS and MURDERERS and BURNING IN HELL, but after a while she learned to glance over the letters with a kind of medical detachment, even when the paper was smeared with blood and wrapped around cutout Willke brochure pictures or the severed feet of plastic dolls. Sometimes the writers enclosed items they had evidently made themselves, cloth bibs or crocheted baby booties. Viewpoint Mail Opposing, Judy printed coolly on the outside of the file folder in which she saved these mailings, the letters and their dismal props, and every time she opened the file drawer to tuck away a new one she recalculated by reflex the steadiness with which the Opposing file was thickening, and how slender by contrast the file labeled Viewpoint Mail Supporting appeared alongside.

Where *were* they? All those women. All those abortion patients and their mothers, their roommates, their boyfriends, their husbands, their older sisters, their married lovers, all slipping quietly back down into the anonymity of North Euclid Avenue at the end of their clinic day. Did they imagine that it was so simple? Did they imagine that the clinic sat there solely for their own convenience, that it was permanently affixed in the landscape, that its future was as solid and steady as the great hospital complex up the street? It had taken the Missouri General Assembly only one session, after the 1973 rulings in *Roe* v. *Wade* and *Doe* v. *Bolton*, to introduce a lengthy restrictions bill: no saline abortions, no abortions without husband's consent, no abortions on minors without parental consent. And Judy had hauled her roadshow down to the statehouse again, just like the years before *Roe*, two hours to Jefferson City, testify before the committees, lay out the statistics, argue the arguments, keep the voice firm and clear while the gentlemen crack their knuckles and gaze off out the window and excuse themselves to conduct some pressing business elsewhere. Up and down the statehouse halls the Missouri Catholic Conference and Missouri Citizens for Life lobbyists worked that bill, steadily, efficiently, directing the busloads of ladies from the parishes, coordinating the letterwriting campaigns, distributing plastic roses so that the flowers could be hand-delivered one by one to the legislators' office people; sometimes there were whole boxes of roses, dozens piled up at a single legislator's desk, each rose tagged with the name of a different constituent.

And what did the defenders of legal abortion have, there inside the statehouse for the first big test of the new post-*Roe* era? They had Judy and her clinic statistics. They had the *St. Louis Post-Dispatch*, always an earnest ally in its opinion pages, editorializing against mandatory spousal consent. They had a young St. Louis woman explaining in the newspaper that her Women's Abortion Action Committee had stopped holding regular meetings over the months since January 1973: "Abortion is a real feminist issue, involving a woman's right to free choice," the young woman said. "But since the Supreme Court decision, people just aren't interested any more. They think it's settled."

The 1974 bill was approved, in the state House of Representatives, by a vote of 148 to 3. In the Senate the right-to-life lobbyists' triumph was only slightly less lopsided, 27 to 5, and the Republican governor Christopher Bond signed the bill while wondering aloud whether it was entirely constitutional, and within three days Frank Susman had lined up his plaintiffs—Reproductive Health Services' Michael Freiman, a Columbia obste-trician-gynecologist named David Hall, and the Columbia Planned Parenthood clinic—and challenged nearly every provision of the new law in U.S. District Court. This was how the ritual was going to work, Judy supposed for a while: the legislators were going to accommodate the ladies with the roses, and Frank Susman was going to take them to court, and if Frank was right the courts were going to throw out as unconstitutional everything the legislators had just done. *Planned Parenthood* v. *Danforth* followed the script precisely, all the way to the U.S. Supreme Court; in July 1976, three months after Susman and Missouri attorney general John C. Danforth argued the case in Washington, D.C., the Court handed down a lengthy ruling that overturned every one of the provisions the right-to-life lobbyists had worked hardest to promote.

No state could require abortion patients to obtain their husbands' permission first, the Court ruled—that decision-making power, under the guidelines of *Roe* and *Doe,* belonged only to the patients and their physi-cians. No state could issue a blanket prohibition on saline abortions, either, the Court ruled; since saline infusion was the standard method for second-trimester abortions, such a prohibition amounted to an "unreasonable or arbitrary regulation" on abortion after twelve weeks of pregnancy. And the Court found Missouri's parental consent requirement too inflexi-ble: "Minors, as well as adults, are protected by the Constitution and pos-sess constitutional rights," wrote Justice Harry Blackmun in the majority opinion.

There were portions of the 1974 law that survived *Danforth;* the Court approved a requirement that each abortion patient give her consent in writing, for example, even though Missouri law did not require written consent before any other form of surgery. But as an affirmation of the constitutional right—of every woman's right to obtain a legal abortion without excessive interference from the state—*Danforth* had proved so resounding a victory that the casual reader of newspaper headlines might nod her head and go on to the rest of the paper. "Court Strips State Abortion Law." "Husbands, Parents Can't Block Abortion." Surely *now* it was settled.

But of course it was not, and at Reproductive Health Services new bulletins arrived every week from NARAL or the National Organization for Women or one of the other organizations that had energetic young recruits tracking worrisome developments around the states. It was a full-time job simply trying to keep up with the creative efforts of the Antis, whose ranks now included attorneys and professors and physicians and congressmen and the hundreds of county-by-county or parish-by-parish volunteer groups that could apparently be deployed at will. They were pressuring the Congress, pushing for support for a constitutional amend-ment to make all abortions illegal. They were lobbying at the regional level for state resolutions calling upon Congress to convene a Constitutional

Convention. They had run a 1976 candidate for president, the New York Democrat Ellen McCormack, and placed her single-issue candidacy on the ballot in more than a dozen state primaries. "Our opposition clearly intends to be around for a long time," NARAL executive director Karen Mulhauser wrote in the summer of 1976, in one of the regular Dear Friends newsletters shipped out to places like Reproductive Health Services:

> Some of the leaders may admit that they may not be as effective as they wish in influencing the '76 elections, but they have extensive plans for the future. They have begun voter surveys in targeted states to determine their areas of strength and plan to have abortion referendums in many states in the 1978 off-year elections. If they are successful in getting "their" voters out in '78 for the referendums they will use these polls of voter opinions against the 1980 candidates who support legal abortion. *They are increasing their efforts and dedication* as we seem to be becoming more complacent and take for granted that the law is on our side.

The warning was prescient; the first really bad news in half a decade was on its way. In August 1976, after weeks of public infighting, the Congress approved its annual Labor-HEW Appropriations bill with an attached amendment that explicitly prohibited the use of any federal funds for abortion. Nicknamed in homage to Representative Henry Hyde, the Illinois Republican who had first introduced the idea of a federal funding ban, the Hyde Amendment was finally adopted with certain faintly discernible qualifications built in: federal money could be used "where the life of the mother would be endangered if the fetus were carried to term" and for "medical procedures . . . for the treatment of rape or incest victims." Outside those categories the funds were to stop, and suddenly the bulletins from Washington were full of capsule funding explanations and figures that might be brandished in debate.

Medicaid, the joint federal-state program funding medical services for the poor, paid for approximately three hundred thousand abortions per year around the country—a total of about fifty million dollars. *"The costs, in both health and fiscal terms, would be prohibitive,"* declared a NARAL anti–Hyde Amendment argument summary that included some of the big-number extrapolations being passed along in defense of federal abortion funding. "*150–250 deaths would result from self-induced abortions. Up to 25,000 cases involving serious medical complications* from self-induced abortions would result. . . . Since most of those women denied Medicaid for abortions would be forced to carry unwanted pregnancy to term, *the cost to the government* for the first year after birth for medical care and public assistance *would be between $450 and $565 million"* (emphasis in original).

This last line of argument was understood to be perilous in public debate; it left too obvious an opening for the debate opponent to shoot back with Oh, fine, let's save government money by killing poor children off so we don't have to pay for their *upkeep.* And when the lawsuits were filed—the Hyde Amendment was challenged, immediately, in a set of U.S. District Court cases brought by the New York–based Center for Constitutional Rights and the American Civil Liberties Union—the legal briefs concentrated instead on questions of fairness and constitutional protection. The Hyde Amendment unconstitutionally enforced one religious view as law, the briefs argued; the Hyde Amendment violated equal protection guarantees by singling abortion out from all the other still-funded medical services; the Hyde Amendment rendered the right to abortion meaningless for every woman too poor to pay for an abortion herself.

Much of this argument was already working its way through the appellate court system, because some individual regions—including St. Louis—had tried early on to cut off public abortion support at the local level. Ever since *Roe v. Wade,* the two St. Louis city hospitals had refused as a matter of policy to permit on the premises any but life-saving abortions. By tradition the hospitals' obstetrics and gynecology residents came through the medical school at St. Louis University, which was Catholic, and Mayor John Poelker had also announced in the weeks after *Roe* that he had no intention of allowing elective abortions into city-run facilities. "It's a moral issue with me," Poelker declared in the newspapers, and within three months Frank Susman had his lawsuit underway: an indigent woman, a married "Jane Doe" mother of two, who had been turned down for an abortion at one of the free city hospitals. Jane Doe finally had her abortion at Reproductive Health Services, where the clinic's hardship-cases fund covered the costs of the procedure, but *Poelker v. Doe* pushed steadily through the federal courts, its fortunes shifting as it rose; the U.S. District Court ruled in favor of the city, but the Eighth Circuit Court of Appeals took Susman's side. St. Louis officials had no right to try to coerce women into bearing children, the Eighth Circuit held:

> Stripped of all rhetoric, the city here, through its policy and staffing procedure, is simply telling indigent women, like Doe, that if they choose to carry their pregnancies to term, the city will provide physicians and medical facilities for full maternity care; but if they choose to exercise their constitutionally protected right to determine that they wish to terminate the pregnancy, the city will not provide physicians and facilities for the abortion procedure, even though it is probably safer than going through a full pregnancy and childbirth.

By January 1977, Susman was before the Supreme Court again, arguing *Poelker* v. *Doe* as the Court also considered two public funding cases out of Connecticut and Pennsylvania. Both states had tried on their own to drop abortion from their Medicaid plans—to declare, even before the Hyde Amendment was approved by Congress, that Medicaid women living inside their state borders could not obtain elective abortions at public expense. And in June 1977, with the Hyde Amendment lawsuit still proceeding through the lower courts, the Supreme Court handed down on a single morning its rulings in all three regional cases—*Poelker* v. *Doe, Maher* v. *Roe,* and *Beal* v. *Doe.*

They were terrible rulings: Frank Susman telephoned Judy, the first typed notices in his hand, and translated for her the legalese of the majority opinions. Over vehement dissents by Justices Harry Blackmun, Thurgood Marshall, and William Brennan, the three strongest legal abortion defenders on the Court, the majority had declared that guaranteeing the *right* to abortion was not the same thing as guaranteeing the abortion *itself.* Even if officials in a particular state chose to cut off public funding or hospital service, the Court ruled, poor women and more affluent women still shared the same constitutional right to seek out legal abortion in a private facility and pay for it themselves. "We certainly are not unsympathetic to the plight of an indigent woman who desires an abortion," Justice Lewis Powell wrote for the majority in *Maher* v. *Roe,* the Connecticut case, "but"—and here Powell quoted from a seven-year-old Supreme Court opinion unrelated to abortion—" 'the Constitution does not provide judicial remedies for every social and economic ill.' "

It made Judy furious. *Sorry, ladies, life is unfair.* If men got pregnant you could bet the Constitution would damn well provide a judicial remedy for *this* social and economic ill. For four years Judy had been trying to subsidize poor women at Reproductive Health Services, sliding money over from the full-fee patients to cover the Medicaid women; in Missouri, the state had been kicking in seventy-five dollars per abortion, less than half the price the clinic charged private patients. When the 1977 funding rulings were announced, even those grudging state payments came immediately to a halt. Missouri would pay "only where an abortion is medically indicated," the new regulation read, and "medically indicated" was not to be a flexible concept: ". . . where the attending physician, in the exercise of his best clinical, medical judgment, believes a full-term pregnancy and childbirth would cause cessation of the mother's life."

That was when Judy Widdicombe hired Sylvia Hampton to help stir things up at the clinic—to write up the political information sheets, nudge the Significant Others a little, put together a fifty-page *Health Rights Advocacy Program Guide* for other clinics interested in working some civics teaching into their medical services. The Supreme Court was no longer the dependable champion it had seemed a few months earlier, and Missouri's track record as an antiabortion state was by now so widely recognized that every new venture of the Antis seemed to pop up within weeks, neatly repackaged for local distribution, from the prolific headquarters of the Missouri Catholic Conference or Missouri Citizens for Life. "AND NOW, THE ABORTION REGULATION ORDINANCE," read a distraught 1979 flyer from NARAL's Washington office, warning affiliates around the country of the next menace likely to appear in their midst. But Judy didn't need the capital letters to catch her attention; in Missouri, predictably, the Abortion Regulation Ordinance had already arrived.

For purposes of the Missouri General Assembly the ordinance had a legislative title, House Bill 902, and a genial-sounding nickname, the Informed Consent Act. The bill made its statehouse debut with sixty-five co-sponsors, as though the representatives had been falling over each other in their eagerness to affix their names to the next assault on 100 North Euclid Avenue, and there was no uncertainty at all about the political future of H.B. 902: almost everybody was going to vote for it; hardly anybody was going to vote against it; and the bill was going to arrive for signing at the desk of Joseph P. Teasdale, the most recently elected governor, who had publicly described abortion as "murder" and had commemorated the fifth anniversary of *Roe* v. *Wade* by proclaiming a state day of mourning "in memoriam for these unborn children." More work for Frank Susman, that was all; a nearly identical version of the Abortion Regulation Ordinance was already under legal challenge in Ohio, but that was not going to slow down the Missouri General Assembly, not when there was a chance for a fine headline like "Bill to Toughen Abortion Laws Advances in House." Such speechmaking H.B. 902 inspired! "What we have before this body today is life and death!" cried a Democrat from Kansas City, declaiming vigorously before one of the Senate votes. "I hope you all here have the guts to vote for this! The Supreme Court has no guts!"

The actual text of H.B. 902, which was appearing in variously modified form in a dozen jurisdictions around the country, was modeled on a city ordinance that had been composed the previous year by right-to-life activists in the Ohio city of Akron. The ordinance had been approved by the Akron city council, by a one-vote margin and after a citywide debate so passionate that the local newspaper temporarily expanded the letters-to-the-editor section to make room for the mail. At the political heart of the Akron ordinance, and setting off the most emotional of the arguments, was a requirement that every abortion doctor within the Akron city limits instruct his patient, orally, before the procedure,

> that the unborn child is a human life from the moment of conception, and that there has been described in detail the anatomical and physiological characteristics of the particular unborn child at the gestational point of development at which time the abortion is to be performed, including, but not limited to, appearance, mobility, tactile sensitivity, including pain,

perception or response, brain and heart function, the presence of internal organs and the presence of external members.

The doctor was also required to inform his patient that abortion was a "major surgical procedure," and that it could result in "serious complications" or "severe emotional disturbances." Every patient was then obliged by law to wait twenty-four hours, presumably digesting this distressing information, before proceeding with her abortion. There was a minors' parental consent section, too, and a requirement that fetal remains be disposed of in a "humane and sanitary manner," and two weeks before the scheduled start-up date the American Civil Liberties Union had challenged nearly the entire ordinance as an unconstitutional infringement of the privacy rights established by *Roe* v. *Wade* and *Doe* v. *Bolton.* "If this legislation is not overturned," the Cleveland ACLU director wrote to Judy, in a letter seeking financial help with the Akron case, "it will affect every patient, every clinic, and will infringe on how every physician in the country practices medicine. The battle belongs to us all."

But by that time the battle had already opened a second front inside Judy's own state capitol, where the Missouri Catholic Conference lobbyists had drafted H.B. 902 after studying copies of the Akron ordinance, and the new Louisiana law that was based on the Akron ordinance, and the fill-in-the-blanks model statute ("to protect and promote the State of _____'s interest in unborn human life and maternal health") that a National Right to Life Committee lawyer had written up for anybody else who might be interested. Missouri's version of the Akron legislation doubled to forty-eight hours the mandatory waiting time between a woman's "informed consent" and her abortion, and most of the other provisions were woven around the existing Missouri statute, or what remained of it in the wake of the *Danforth* ruling: all second-trimester abortions to be performed in hospitals only, no clinics or doctors' offices; both parents' written permission to be required for all abortion patients under the age of eighteen, with a provision allowing teenagers to request court orders overruling parental vetoes. "Unborn child" to be formally defined, for purposes of Missouri law, as "the offspring of human beings from the moment of conception until birth and at every state of its biological development, including the human conceptus, zygote, morula, blastocyst, embryo, and fetus."

And as H.B. 902 proceeded slowly through the Jefferson City committee process, the right-to-life people were leaving nothing to chance. NARAL would send out suggestion sheets from Washington, *What You Can Do If an Abortion Regulation Ordinance Comes to Your Town or State,* but they were pale little flutters against the determined and precisely coordinated advance of the Catholic Conference and Missouri Citizens for Life. The Missouri Citizens for Life handouts explained the entire state legislative process and singled out exactly which House committee members should be contacted, and when; the Catholic Conference handouts offered five-point highlighted instructions to pastors, parish committee chairmen, and regional organization leaders. "Please return this information to the Missouri Catholic Conference by March 26, 1979," began one record-keeping sheet sent out in multiple copies all over the state:

> Write in the name of each State Representative to whom you sent letters or urged others to do so. In Column (1) insert the number of letters that you *know* were sent. In column (2) insert your *estimate* of the *additional* letters that were *probably* sent due to your urging. This is a pre-addressed self-mailer and requires no envelope. [emphasis in original]

It was preposterous, trying to compete with organization like this; no right-to-abortion group had the resources, the mailing lists, the captive ear of men and women lined up in pews at Mass. Once a newspaper writer asked Sylvia Hampton what their greatest problem was, and Sylvia answered straightaway, "The Catholic Church," and waited for the reporter to write that down in her notebook. "I can't put that in the newspaper," the reporter said. Sylvia asked her why not and the reporter just looked at her, as though astonished that a person could be so thick; it was offensive, it would anger the Catholics, it was like putting a racist remark into print. Sylvia wanted to say, See, they've even got a stranglehold on *you*, but she held her tongue; there was no point antagonizing the Catholics any more than she already did simply by showing up in the morning every day for work. She felt helpless against the Church, they all did, and when she would go down to Jefferson City to lobby against H.B. 902, even the sympathetic legislators would sit her down and sort of pat her on the hand while they explained that their hearts were with her, really. " 'But these little ladies from the parish,' " Sylvia Hampton recalls, mimicking the patient voice of a St. Louis assemblyman she had known since they were both in high school. " 'They just worked their heads off to get me elected, and it was for the purpose of getting my vote on this little tiny insignificant issue. You're not going to ask me to go against my friends from St. Anne's Parish, on this little tiny issue, when you know I'm with you on family planning?' "

House Bill 902 passed, in the Missouri House of Representatives, by a vote of 139 to 18. In the Senate, the vote was 28 to 4. An exasperated senator from St. Louis described H.B. 902 as the Frank Susman Income Maintenance Bill, just before casting one of the four No votes, and within two weeks Susman had persuaded a federal judge in Kansas City to issue a temporary restraining order blocking enforcement of most of the new statute while his lawsuit was readied for trial. The case was called *Planned*

Parenthood v. *Ashcroft,* for the Kansas City Planned Parenthood clinic and the new state attorney general John Ashcroft. The first plaintiffs' witness, called to the stand in October 1979 to describe for the judge and the newspaper reporters the hardships that would be visited upon women if H.B. 902 was allowed to take effect, was Mrs. Judith A. Widdicombe.

JUDY FLEW to Kansas City for her *Planned Parenthood* v. *Ashcroft* testimony, her briefcase loaded up with flyers and Reproductive Health Services informational material she thought might be useful for their case. The statistical data she carried in her head. Eight thousand two hundred abortions per year, at Reproductive Health Services, Inc.; twenty-two percent of them on patients from outside Missouri. Percentage of patients under the age of twenty, at RHS and the smaller Kansas City clinic acting as co-plaintiff in their lawsuit: one third. Number of hospitals offering second-trimester abortion procedures in the St. Louis region: zero.

She had a great deal to say. Reproductive Health Services was the busiest abortion clinic in the state; indeed, by 1979 Judy was executive director to one of the busiest abortion clinics in the Midwest, and her resume as expert witness was lengthening with each passing year. She was president of the National Abortion Federation, a three-year-old providers' organization that Judy had helped organize. She was management consultant to the Planned Parenthood Federation of America, which was rapidly multiplying the number of Planned Parenthood clinics offering abortions along with more conventional family planning services. Judy was in and out of a half dozen clinics a month, assessing equipment and technique and potential patient populations, and it was her abiding belief that the authors of H.B. 902 might just as well have gotten right to the point and sent trucks out to the clinics to pile up big crates in front of the doorways.

A woman could climb over these crates, but she would have to work at it. She would have to be a person of considerable tenacity, single-minded in her determination to have an abortion no matter what discouragements were placed before her. It had been a clever bit of labeling, abbreviating as "Informed Consent" the provisions of the Akron ordinance and its Missouri offshoot, but Judy had debated enough Ann O'Donnells to understand that the objective here was really neither information nor consent. The objective was keeping women out of abortion clinics; or failing that, making sure that each woman felt as tormented as possible about what she intended to do once she was inside.

What genuine good was served by requiring every physician to inform his potential patient of "the probable anatomical and physiological characteristics of the unborn child"? The final version of the Missouri legislation had left out precisely which characteristics the physician was obliged to mention, but in a sense the Akron ordinance had already written out the list for them. *Appearance, mobility, tactile sensitivity, including pain, perception or response, brain and heart function, the presence of internal organs and the presence of external members.* You could think of it as the Fingers and Toes subsection: that was what the Antis really wanted, after all, a statutory requirement that every single abortion patient be confronted by a white-coated physician shouting "IT HAS FINGERS AND TOES" at such volume that even clapping hands over the ears would not shut out the sound. And they could all fall to arguing over whether it did have fingers and toes (Are embryonic protuberances that look like fingers and toes the same thing as actual fingers and toes? How can anybody be sure about tactile sensitivity? Where's the scientific evidence on pain response?)—but really the content of the "anatomical and physiological characteristics" information was far less important than the fact that H.B. 902 required every abortion doctor, under penalty of law, to repeat it. The doctor had to say the words aloud and the woman—it was impossible to write such a requirement into a statute, of course, but that was the obvious implication—had to *listen* to them.

As though she didn't already know, Judy thought to herself the first time she read the new Informed Consent language: as though every pregnant woman at the doorway of an abortion clinic had somehow missed the part about human embryos turning into babies. The Antis liked to suggest that the clinics all slid right over that little complication, that counselors went around waving their hands and making airy references to the "blob of tissue" that came out during the abortion process; but Judy had been one of those counselors and she knew that the process of listening to a prospective abortion patient was infinitely more complicated than that. Did the woman *want* information about the physical appearance of the fetus she was about to abort? Did she want to know it had "brain and heart function," or the "presence of external members"? At Reproductive Health Services they made that information available, as any good clinic ought to; there were pictures and science books on the shelves, and the counselors were instructed to answer each question in a way that was at once gentle and straightforward. Every patient had a right to full information, Judy had believed that since her early nursing days. But in an abortion clinic every patient had a right to denial, too: a right to refrain from asking, a right to shut her eyes, a right to select the vocabulary with which she was going to think about what she had decided to do.

Some women *liked* "blob of tissue" as a working description of what was sucked into the collection jar; it made them feel better. No statehouse full of men had any business removing by legislative fiat this protection that some women extended to themselves—this right to denial. It was probably true that a state-mandated Fingers and Toes speech could inspire a few clinic patients to turn around and bolt; it might even be true that a few of those patients would be glad afterward that they had bolted. But most of the patients were going to push right on because they *would not have this baby*, they had already made the decision, they had to have an abortion except that now they would do it with a special raging pit-of-the-stomach guilt, courtesy of the Missouri General Assembly.

And that was one paragraph of H.B. 902. There were fifty-nine others, every one of them, it was probably safe to assume, composed by somebody who had never tried to manage a clinic full of real women with real unwanted pregnancies. On paper, Judy supposed it sounded perfectly sensible to require that the doctor provide the indicated information "not less than forty hours prior to her consent to the abortion." As a general goal Judy happened to think waiting a few days was a good idea; Reproductive Health Services had long since developed its own informal waiting period by asking women to come in for a pre-counseling session a few days before they scheduled their appointment for the abortion.

But demanding that waiting period, by order of the state, was just shortsighted and cruel. What about the women who drove all the way from Springfield, three and a half hours southwest of St. Louis, or came down on the bus from the farm country up near the Iowa border? Were they supposed to go sit out their two days alone and pregnant at a Quality Inn, spending money they didn't have while enjoying their pleasant stay in St. Louis? Or did the legislators who had voted for this particular provision expect these women to spend twenty hours on the road instead of ten, driving back and forth and back and forth in deference to Section 188.039 (1) of the Missouri criminal code? Sometimes it seemed to Judy that nobody in the sheltered remove of a statehouse chamber should be able to vote on anything that affected abortion clinics, not until they had sat in her waiting room for a month or two and watched the human anguish come and go. Parental consent—there was another idea that sounded reasonable until you studied some of the family dramas that dragged themselves across the Reproductive Health Services threshold, the alcoholic mothers, the abusive fathers, the fourteen-year-olds whose parents had threatened to throw them out of the house. You *encouraged* parental consent; you didn't *mandate* parental consent. It didn't *work* to run abortion clinics from Jefferson City, or from Baton Rouge or the Akron City Hall; an abortion clinic was a medical facility, subject to the same public health standards as any other medical facility, and it was nonsense to pass off as "regulation" these multipaged lists of rules and prohibitions made up by people who thought no woman should be able to have an abortion in the first place.

That there were shoddy clinics, unscrupulous clinics, savagely commercial operations that shoved women through without double-checking the pregnancy tests or sterilizing the instruments properly—Judy knew this too, all the clinic directors did. Some months earlier the *Chicago Sun-Times* had spent two weeks running a series of appalling stories about the worst Chicago clinics, several of which had been infiltrated by reporters and local civic investigators posing as patients or new employees. "Life on

the Abortion Assembly Line: Grim, Grisly, and Greedy," one of the open-
ing headlines read, and that set the tone for what was to come. "12 Dead
After Abortions in State's Walk-in Clinics." "Abortion-Mill Bosses Cut
Corners on Care." "Patient Recalls: 'I Was Just a Guinea Pig.' " "Nurse to
Aide: 'Fake That Pulse!' "

The *Sun-Times* series was rich with detail: screaming patients told to
shut up and hold still, abortions performed on women who were not
pregnant, "counselors" paid on commission and coached in the hard sell.
("Don't answer too many questions," one of the investigators reported
having been instructed, "because the patient gets too nervous, and the next
thing you know they'll be out the door.") Judy had heard enough private
bad clinic stories already to know that the *Sun-Times* reports were almost
certainly true; Chicago had developed a reputation for extraordinarily
avaricious competition among its abortion facilities, and Judy had already
spent nearly a full decade watching states across the country produce a
charmless assortment of businessmen who appeared to see legal abortion
mostly as a quick route to wealth.

They were not all bad owners. Some of them ran decent abortion
clinics. Some of them had medical degrees and did abortions themselves
and were technically efficient at it, good results, no malpractice suits,
satisfied customers. Expensive automobiles, also, you saw that quite a bit,
and the gold and diamond jewelry, and the rapid-fire delivery of the salesman smoothly pitching his product—little men with big egos, Judy used
to mutter, rolling her eyes as she said it. In later years a public health
journal was to publish a study of California physicians who made legal
abortion their principal practice between 1967 and 1972, and the author, a
University of California public health professor named Michael S.
Goldstein, used proper academic phrasing and charts to describe a phenomenon Judy had come to know exceedingly well:

Table 2 summarizes a variety of financial sentiments expressed by
the physicians. Obviously, money was an important motivation. It was
cited by over three-fourths of the sample, including all of the entrepreneurs and workers as well as most community doctors and two academics.
In most instances, money was spontaneously offered by the respondent
as the first and major motivating factor. Typical responses were: "It was
a bonanza, my prime—everyone's prime motivation...I loved the
money!... it was fabulous, impossible to resist"; "Everything I touched
turned to gold"; "Yes, greed was what it was all about"; "Look, by November of '69, I was making over $22,000 a month ... nothing compares
to that"; "Money was the only motive"; "I'm a big spender, an entrepreneur ... money is what I need most and this was the way to get it."

Almost one-fourth of the sample specified that money was their sole
motivation for doing abortions. These physicians were most likely to be

community physicians who did not particularly want to perform abortions but who felt they needed the money badly or entrepreneurs who saw themselves as businessmen providing a service with little regard for what the service was. As one entrepreneur put it: "I never should have been a doctor, I'm not interested in medicine. I'm a businessman, always was, always will be." Another said, "Medicine had nothing to do with it. I'm an entrepreneur. If it wasn't this, it would have been something else."

The National Abortion Federation (NAF) itself had been in part an early effort at sifting good abortion providers from the bad ones, and its own membership directory hinted at the increasing complexity of the new competition for patients. Freestanding Abortion Providers, Federal-Tax-Exempt Abortion Providers, Hospital-Based Abortion Providers, Feminist Health Center Abortion Providers—NAF's first organizers, merging two separate groups that had been operating at cross-purposes, had tried to come up with neutral-sounding titles for what was actually an assortment of men and women of such wildly different outlooks that sitting them in a room together was a daunting exercise in social diplomacy. "You'd come to board meetings and there would be guys in sharkskin suits, you know, with watch fobs, and black Cadillac limousines, ready to take people out to dinner and spend like crazy," recalls Frances Kissling, NAF's first executive director. "And you'd have radical feminists who were there basically arguing for women's rights, just sort of sixties counterculture, the *least* concern for how they looked. And you'd have buttondown Planned Parenthood types, very nice ladies in plaid skirts and white blouses, and you'd have people like Judy, professional dress and gold jewelry. Sometimes I felt like what's-his-name the lion tamer, Gunther, with the whips, trying to keep the peace."

The NAF convention agendas were weighty with professional workshops and papers: "Physician-Nurse Teamwork," "Gross v. Microscopic Diagnosis of Fetal Tissues," "Rh Immune Globulin in the Abortion Patient." Some of the grittier practicalities were in evidence as well (here was "Inventory Control," at the 1978 convention, or "Advertising and Making the Media Work for You"), but it was important to Judy that their gatherings look like medical functions, not trade association meetings or Women's Liberation organizing sessions. This was not a universally admired decision, during the contentious first decade of what right-to-life groups had begun referring to as the Abortion Industry; since the earliest collaboration efforts, the feminists and the entrepreneurs, the hospital doctors and the medically oriented nonprofits had been bickering about the symbolic and practical purpose such an alliance was meant to serve. Should they be arranging medical symposia or negotiating group discounts on office supplies? Were they refining a gynecological subspecialty or marketing a product? The very vocabulary of the arguments made some of the teaching hospital doctors queasy: *client, provider,* as though they were no longer physicians attending patients but instead were suddenly supposed to think of themselves as competently trained auto mechanics.

As a group the feminists appeared to mistrust most of the hospital doctors anyway, enmeshed as the doctors were in the very system the feminist health centers had set out to defy. The feminists mistrusted big nonprofits like Reproductive Health Services, too, for ignoring their consciousness-raising obligations and deferring too meekly to organized medicine; the nonprofit directors mistrusted the entrepreneurs, for making money off abortion and wearing sharkskin suits; and the entrepreneurs resented the nonprofit directors, for acting lofty about profit and failing to appreciate the way private operators had stuck their necks out to bring abortion into some communities when no one else would do it.

"You get all these people in the room and it is a miracle that the organization survived," Frances Kissling recalls. "These were all very strong people. The people who would go out and do these services and run these facilities were mavericks, very highly individualized thinkers, with very strong opinions, and with *very* big egos. You'd have impassioned discussions. Each of these people genuinely believed they knew better than anybody else, that their way was *the* right way."

The *Sun-Times* stories, which appeared every day beside a small drawing of a man in scrubs glaring imperiously down from behind his surgical mask, had set off urgent alarms inside abortion facilities all over the country. "We were all talking to each other on the telephone and the phone at NAF was ringing off the hook," recalls Uta Landy, who had just replaced Kissling as NAF's executive director when the newspaper series began. "I think that in general everybody felt it was an attack on *them,* on everybody, that the media was using this opportunity to discredit abortion clinics, period. Everybody was really aware that there were some clinics that were lacking in certain standards, and we were relieved that none of these clinics from the articles were members of NAF. But we also realized that they could have been, as well, because we didn't have a very stringent screening process."

Rapidly, mindful of the television stations and antiabortion groups that were rushing to capitalize on the *Sun-Times* stories, Landy and her NAF committees began trying to tighten up the screening process, adding details and categories to the written questionnaires sent out to all prospective NAF members. Landy set up a toll-free telephone service, so that callers might pass on recommendations to good abortion facilities and collect complaints about bad ones, and she composed for national distribution a *How to Choose an Abortion Facility* brochure loaded with suggested questions a woman might use to size up a clinic before making her appointment. "What tests are done before the abortion is performed?" "What qualifications does the doctor have?" "How long do you *have* to stay in the recovery room? How long *can* you stay?"

There was no warning to patients in Landy's brochure, no advisory that said, *By the way, there are some real pigs calling themselves abortion doctors out there.* There was no sure method of checking to see whether individual administrators were telling the truth about their clinics, either, and nobody from NAF made unannounced quality inspection visits; the Federation was a membership organization, not a licensing body, and within the membership a circle-the-wagons approach was taking hold as legislation plainly hostile to their interests appeared in one state after another. The NAF members themselves could argue over the definition of good abortion service (they had bruising arguments, in fact, about what kind of performance standards NAF ought to impose), but this was a matter for *them* to resolve, the members themselves, clinic directors, and abortion doctors. City ordinances and state statutes had no business here. At the New York–based national office of the American Civil Liberties Union a young staff attorney named Janet Benshoof was keeping track of nearly a hundred abortion-related lawsuits around the United States, and most of those lawsuits were challenges to legislative harassment—for surely that was the appropriate label for it—that ranged in scale from the local township ordinance to the coast-to-coast prohibitions of the Hyde Amendment.

"Nationwide class action challenging the legality on statutory and constitutional grounds of 'Hyde Amendments.'" A thick packet of summaries, updated regularly and distributed by Benshoof's Reproductive Freedom Project, condensed into single pages the lawsuits case by case. "Class action suit for declaratory and injunctive relief against legislation recently passed in Louisiana." "Plaintiff, an abortion clinic, challenges a Youngstown City ordinance which regulates and licenses abortion clinics." "Plaintiff, a clinic which performs first-trimester abortions, challenges the Town Board of Grand Chute's health ordinance which virtually prohibits first trimester abortions by imposition of impossible standards of operation on the clinic."

The roster of legal counsel was familiar by now, ACLU attorneys and women's rights attorneys and ill-paid young men and women from the regional Legal Services offices, and just as they had during the years before *Roe*, they shipped sample briefs and constitutional theories back and forth, talking directly to their clients—the clinic directors and physicians who had agreed to serve as plaintiffs—only when there was some dramatic development in the lawsuits that bore their names. Judy had grown accustomed to this role, and she knew how to take her legal cues from Frank Susman; it was Frank who had convinced both Judy and the NARAL organizer Maureen McCarthy that since some version of H.B. 902 was clearly destined for approval in the Missouri General Assembly, they ought to hope it would pass in its most apparently *un*constitutional form, thus ensuring its own eventual demise in the courts. Judy had no idea whether Frank was right about this; it seemed plausible enough, but then

Frank Susman went to work every day in a nice tenth-floor law firm in Clayton and was not actually going to be the one who had to sit with some pregnant woman reciting the Fingers and Toes speech because state law made it a crime not to. Judy had no real backup plan, if *Planned Parenthood* v. *Ashcroft* lost in the courts; Illinois required no parental consent for minors, so she imagined she could refer the occasional teenager over to the clinic in Granite City, on the Illinois side of the Mississippi River. She could hand the addresses of cheap motels to out-of-town women forced into the two-day wait. She could go on turning away the second-trimester patients, directing them off to New York or Kansas City, hoping they were going to be able to come up with the plane fare or the bus fare before they were so far along that nobody would take them at all.

Her preparations were for victory, though, not for defeat. There was a gamble written into the opening legal complaint of *Planned Parenthood* v. *Ashcroft:* Frank Susman had challenged more than just the 1979 legislation, the new Missouri abortion requirements modeled on the Akron ordinance. He had also gone after part of the preexisting state abortion statute. Until *Ashcroft,* nobody in Missouri had challenged the state's second-trimester hospitalization requirement; for public relations purposes it would have been foolhardy to do so, inviting by the challenge a line of court testimony that would probably make the staunchest abortion supporters blanch. More to the point, medical practice itself had argued against such a challenge during the first years after *Roe:* later abortions belonged in hospitals, that was the thinking, and even abortion doctors would say so on the stand.

But medical practice had changed, and Judy had been studying the change, and she wanted more out of *Planned Parenthood* v. *Ashcroft* than a simple repudiation of the newest efforts of the Antis. She wanted her clinic to expand its medical reach. It was grim, it was technically ambitious, and there were aspects of its execution that were almost certainly going to provoke intense reactions within her own staff; but Judy wanted second-trimester abortion at Reproductive Health Services. She was ready to try D and E's.

THE SECOND-TRIMESTER language of the challenged Missouri law took up a single sentence: "Every abortion performed subsequent to the first twelve weeks of pregnancy shall be performed in a hospital." Considering the authorship, this was remarkably low-key phrasing. There was a great deal more attention-grabbing terminology that an inventive draftsman might have used. By the second trimester of pregnancy (or so it was believed throughout the early years of vacuum aspiration), a human fetus was too big—its limbs too long, its connective tissue too strong, its bones too hard and brittle—to pass safely through a cannula and vacuum tube.

Generations of gynecologists and skilled dilation and curettage abortionists had also drawn a safety line at roughly the end of the first trimester; much past twelve weeks, according to the conventional wisdom, the uterine lining was too soft and the parts too big to pull all the tissue through the cervix with instruments.

Before the late 1960s, that left a small and dismal array of options for a woman who was past her third month and determined not to give birth to the child. If the pregnancy was going to put her in real physical danger, or she managed in some other fashion to talk her way past a hospital therapeutic abortion committee, then she could undergo the major surgery of a hysterotomy, the cesarean section intended to pull out a fetus too premature to survive. If she happened to find an abortionist who used the "packing" method (often a midwife of uncertain training, like St. Louis' Mrs. Vinyard), then she could pay a substantial amount of cash to have her cervix stuffed with the irritant—clean gauze, if the abortionist took some pride in technique—that was supposed to set off a miscarriage some-time during the next few days. If she lacked the money or the contacts required to hire an abortionist, she could try, generally with calamitous consequences, to abort herself.

Or if she was wealthy enough for the round-trip ticket, she could get on an airplane and fly overseas, where physicians in a few other countries —first Japan and then some parts of Europe—had begun refining newer methods for abortion past what was thought to be the first-trimester D&C deadline. But the American medical community was profoundly uninterested in keeping pace with this particular area of research, and it was not until the first 1960s efforts at state abortion law reform that a very few physicians, initially isolated from one another in separate corners of the country, began experimenting with different techniques for pushing up the time limit without having to cut open a woman's abdomen.

The first to achieve some modest medical celebrity for his work was a Hungarian-born obstetrician-gynecologist named Thomas Kerenyi. Ker-enyi lived in New York, where he and his uncle, a well-known research physician with a special interest in reproductive physiology, had almost unintentionally found themselves studying deliberately induced miscar-riage. In 1960, while Kerenyi was still completing his hospital internship, the two of them had been experimenting after-hours with rabbits, investi-gating techniques that might be used to stop premature labor in women who wanted to give birth. But they had to start the rabbits' premature labor in order to try stopping it, and they found that injecting a concen-trated salt-and-water solution into the animals' uterine cavities did just that: the amniotic infusions brought on contractions that eventually ex-pelled the rabbits' litters.

Kerenyi and his uncle, Arpad Csapo, were not the first physicians to see the connection between saline infusion and abortion; a Romanian physician had experimented with saline as early as 1939, and Japanese doctors had used saline for abortions during the half decade after World War II. But the Japanese had abandoned it as unsafe, and in the United States saline was unheard of for therapeutic abortions, so that when Kerenyi suggested applying their research to a human patient at his hospital, he had to convince his skeptical obstetrics chairman first. A seventeen-year-old girl had just been transferred to the gynecology wing with committee-approved orders for a therapeutic abortion, Kerenyi recalls; apparently she had convinced the committee that she genuinely intended to kill herself if she had to give birth to the child. But she was twenty weeks along, which made her a candidate for hysterotomy only, and the young Kerenyi pleaded for permission to try avoiding the surgery—which would have left the young woman with a long vertical scar and a permanently weakened uterus—by transferring the technique he and his uncle had so far used only on rabbits.

"I couldn't believe they were going to use hysterotomy and make somebody an obstetrical cripple," Kerenyi recalls. They were given a twenty-four-hour deadline and rushed, disastrously, to meet it; their first bottle of saline exploded in the sterilizer and they had to start all over again. But it worked: the second bottle survived, the young woman's amniotic fluid was partly withdrawn through a long needle, and the saline solution was then run directly through the needle into the amniotic sac that surrounded her fetus. "Sure enough, she aborted at thirty-six hours, and we were the heroes," Kerenyi recalls. "And there was our first case."

They did fifteen more over the next year, both committee-approved therapeutics and some spontaneous miscarriages that needed medical intervention to complete the emptying of the uterus. And by 1970, when the New York law changed, the Kerenyi-model saline abortion was widely understood to be the second-trimester method of choice. It was not without its complications—even Kerenyi, who was convinced that saline's hazards could be avoided with proper technique, advised that women be watched carefully for fever, uterine injuries, and signs of potentially life-threatening sodium buildup in the blood—but it was considerably safer than hysterotomy and required no general anesthetic. And although the saline itself had to be injected by a doctor (in theory a doctor trained to know the correct concentrations and precisely where to aim the solution), a woman waiting for her saline to take effect required only a hospital bed and a nearby nurse, which made hospital saline units an economically attractive proposition during the freewheeling early years of legal abortion in New York and California. By the time Judy Widdicombe was scouting out the territory for her Clergy Service referrals, Manhattan alone offered dozens of different hospitals with saline units, many of these small older hospitals like the one where Kerenyi—by then a full-time private physician and medical professor—had taken on a busy second shift to help run the saline abortions.

As a percentage of the city's new abortion cases, their numbers were modest: in New York City, where the health department kept close watch on statistics during the early years of legalization, only one sixth of the abortion patients were showing up late enough in pregnancy to warrant salines. But over two years alone, 1970 to 1972, one sixth of New York's abortion patients was approximately forty-five thousand women. Kerenyi and his saline-trained colleagues were busy men. "There were days when there were twenty lined up and we went from one patient to the next," Kerenyi recalls, "putting in the needles, draining off the amniotic fluid, and then infusing the hypertonic saline."

At that point in the process—needle inserted, saline infused—the doctor, as far as the saline patient was able to perceive, disappeared down the hall. The solution took a while to kick in. Then, often after a long lone wait in her saline unit hospital bed, the patient essentially went into labor —painful, drawn-out labor, with the writhing and the vomiting and the crying out, except that at the end of this labor the woman pushed from her body not a newborn baby but a dead fetus. If she managed the timing right, a nurse stood by to help her with the final expulsion; if the nurse was busy elsewhere, then the woman pushed the fetus out alone, in her hospital bed, and waited for someone to come and clean up the mess. It was gruesome, everybody understood that it was gruesome; Kerenyi himself wondered aloud in later years what it must have been like for the miserable young women who had been under the impression that all they would have to do was receive an injection and wait for miscarriage. In Kansas City, Kansas, where doctors at the university medical center set up a four-bed saline unit after the Kansas law changed, they used to make sorrowful black-humor remarks about a local amusement park's advertising slogan and suggest that perhaps the same words should be printed on a banner over the entrance to the unit: DO WE HAVE A DAY PLANNED FOR YOU!

And there was another problem with saline, an infrequent problem, but one of such unfortunate proportions that on the rare occasions when it did occur it inspired a medical vocabulary of some delicacy. "Fetal demise," as the clinical jargon sometimes phrased it, was not always achieved in a saline abortion: the fetus, in other words, could under certain circumstances come out alive. The saline itself was supposed to kill the fetus, but when it was administered badly or in inadequate concentration, the attending nurses could find themselves confronted with a moving fetus that still registered a heartbeat—a *live baby*, as one of the unpleasant articles in the lay press was to shout in italics, before proceeding into some of the incidents that made physicians like Kerenyi wince: Two-and-a-half-pound male born after saline abortion in Omaha, left on the drainboard of a sink, expired within hours. Four-and-a-half-pound infant born after saline abortion in Bakersfield, California, given oxygen against the attending doctor's orders, later placed for adoption. Two-pound fourteen-ounce female born after saline abortion in Westminster, California, expired after breathing stopped, leading to the physician's criminal trial one year later on charges of murdering the infant by pressing his hand to her neck.

Only a few of these incidents reached the public eye in that formal a fashion, with eyewitness testimony in a court of law, and none resulted in a final conviction.* And bungled saline installation was not the only culprit in live-birth complications. Hysterotomy operations could in some instances produce fetuses that showed signs of life, and for a while physicians were experimenting with second-trimester methods that introduced other substances into the uterus but left open the possibility for occasional live births; during the 1970s there was a surge of medical interest in prostaglandins, which were effective at setting off labor but produced serious side effects and did not directly kill the fetus.

For all its drawbacks, then, saline infusion was the second-trimester method of choice well into the mid-1970s. Even the Supreme Court had endorsed it, cautiously, when the Court found in *Planned Parenthood* v. *Danforth* that the state of Missouri had no right to forbid salines. "The State . . . would prohibit the use of a method which the record shows is the one most commonly used nationally by physicians after the first trimester," Justice Harry Blackmun wrote in the *Danforth* majority opinion, "and which is safer, with respect to maternal mortality, than even continuation of the pregnancy until normal childbirth." That was what a clinic director like Judy Widdicombe had to go on: it was safer than other things and it was what everybody did, but when she first went to New York or Kansas City to watch the saline units in operation, she found she had to work to contain her own revulsion. This was an *advance?* Ashen women in hospital gowns pushing IV poles down the hall and stopping long enough to grab their bellies and double over in pain—it was as though someone had hospital-gowned the women she and Art used to pick up on street corners during the illegal days, except that saline offered one final bit of brutality that the illegal packing abortionists had not: most hospitals would not attempt a saline abortion before the woman was sixteen weeks pregnant, because Kerenyi and other researchers had found that when the pregnancy was smaller, it was too easy to bungle the abortion by misdirecting the infusion outside the uterine cavity. Since both vacuum aspiration and D&C were thought unwise after twelve weeks, and other late abortion

---

* The doctor charged in the 1977 Westminster case, William Waddill, Jr., was cleared after two trials ended in hung juries (Waddill had testified that the infant was already in a "death struggle" by the time he was called to the scene). The 1975 murder case against South Carolina physician Jesse J. Floyd, who had administered an abortion installation that produced a live infant who later died in the hospital, was finally dropped after four years of legal proceedings. Kenneth Edelin, the Boston physician who was charged with manslaughter in 1975 for allegedly clamping off the blood supply of a one-and-a-half-pound fetus during a hysterotomy abortion, was found guilty by a jury, but the Massachusetts Supreme Court subsequently overturned Edelin's conviction.

techniques like hysterotomy and prostaglandin injection had largely been abandoned, that meant the earliest of the second-trimester patients—thirteen through fifteen weeks—were told they had to wait, stay pregnant and wait, until they were advanced enough to abort in the saline units.

Judy hated it. She liked Tom Kerenyi, who was intense and academic and clearly working hard to minimize the complications, but she hated saline. She hated putting women on a bus or an airplane and knowing that was where they were going to end up, in one of those wards with the long wait and the IV pole, and she knew in any case that even if she could stop hating saline quite so much, she was not going to be able to bring it into Reproductive Health Services. State law prohibited it, to begin with (RHS was not a hospital, and saline was of course "performed subsequent to the first twelve weeks of pregnancy"). But everything Judy had learned about saline suggested that the inductions belonged in a hospital setting anyway; no outpatient clinic or doctor's office could safely accommodate women who might spend all night in labor.

And Judy wanted second-trimester abortion at her clinic. There was no hospital in the eastern half of Missouri that would take an elective abortion patient past twelve weeks; Kansas City had its extremely modest saline facilities, five hours' drive to the west, and apart from that it was mostly TWA to New York City, just like the years before *Roe*. In a way the whole arrangement was perversely backward, Judy thought; of all the women who found their way to the waiting room at RHS, it was the second-trimester patients who least belonged on an airplane or a cross-state bus trip. These were the poorest patients, the patients in denial, the substance abusers too lost to their own addictions to think clearly about what was happening to them, the teenagers too frightened to say anything until their bluejeans no longer fit, the women in their forties who had mistaken early pregnancy for menopause. Women with malformed fetuses also tended to need their abortions well into the second trimester, since most gross deformities and chromosomal damage were not detectable in early pregnancy, and these especially were women who needed to be welcomed and comforted, Judy thought, not exiled to another city. The counselors told Judy about patients who appeared to have been deliberately misled about being pregnant, too, some right-to-life doctor trying to hold them off until it was too late, and in the staff meetings they would pass these case histories around as though in somber appreciation of their allegorical value: here's a lone pregnant woman, beaten up by the system, and all we can do is send her out of town for help.

But in fact there was something else they could do—if the law were to permit them to do it.

It was not going to be easy, what Judy had in mind. There was some technical expertise involved, and an emotional leap, and a certain steady-eyed conviction that most of her counselors and nurses had probably not had the time or experience to cultivate: none of them, as far as

Judy knew, had been called upon to examine fetuses in their bathtubs during the illegal days. But Judy had been reading the medical articles and listening closely at the NAF seminars, and by the fall of 1979 she knew that if the U.S. District Court were to strike down the hospitals-only requirement for abortions after twelve weeks, if just this part of *Planned Parenthood* v. *Ashcroft* went their way, then she and Michael Freiman could make Reproductive Health Services the first clinic in Missouri to begin offering dilation and evacuation, which was always abbreviated on second reference, as though hyphenated into a single word: D-and-E.

In 1979 only a few doctors really knew how to do D and E, or at least how to do it well, so that it was careful, planned in advance, not the panicked bloody instrumentation of a doctor discovering too late that his first-trimester case was actually a fifteen-weeker. It was this deliberation, this planning in advance, that defined the medically acceptable D and E: unlike *curettage*, a medical term understood to mean the scraping of the lining of a body cavity, *evacuation* suggested that the doctor commenced his procedure understanding that he was going to encounter a fetus of sufficient size that his primary mission was to extract it piece by piece, using forceps that were properly suited to grasping and pulling tissue the doctor was not actually able to see until he got it out.

Michael Freiman was going to be good at this. Judy was sure of it. She and Freiman had a volatile relationship, six years into the running of the clinic; Judy thought Michael was emotionally erratic, and Michael thought Judy was manipulative and domineering, and their shouting matches in Judy's office were sometimes audible two rooms away. But both of them knew Freiman had the touch. He had never lost his wonderful feel for the precision of a good abortion, his ability to visualize solely with fingers and instruments, and even when she was angry at him Judy thought he was the best medical teacher she had ever known. And he was intensely curious about each new aspect of medical technique, so that when Judy rounded up Freiman and another clinic doctor named Jack Klein for a hands-on D and E lesson in Washington, D.C., Freiman was eager to make the trip. He had some familiarity with the work of William Peterson, the Washington obstetrician-gynecologist who was doing some of the country's most closely monitored research in D and E, and when the three of them arrived at Peterson's clinic at Washington Hospital Center, Freiman listened carefully as Peterson explained some of the conclusions emerging from their studies. The freestanding clinic was proving to be an appropriate D and E setting, Peterson said. The thirteen- to sixteen-week moratorium was no longer an issue; their largest number of patients were in the fourteen- and fifteen-week range. They were also finding that in some cases they could reduce patient discomfort and cervical tearing by using natural dilators called laminaria, which were made from absorbent seaweed and expanded the cervix gently over a period of hours before the abortion itself was to begin.

It was all very interesting, Freiman thought: very sober, very responsible, very attentively presented. Peterson invited them into his procedure room, where a patient lay draped and ready. They gathered around Bill Peterson to watch. Freiman stood over Peterson's left shoulder, studying the doctor's technique, listening to his moment-by-moment explanation of what he was doing and how best to accomplish it, and then a small arm with a hand on it dropped into the surgical pan.

A *hand.* It had fingers. Freiman looked up quickly, trying to catch the eye of Judy or Jack Klein, but they appeared to be entirely absorbed in Peterson's work. Freiman instructed himself to remain where he was, standing respectfully by the side of the physician midprocedure, but he took an involuntary half-step back; he felt momentarily short of air, as though someone had punched him hard in the stomach. Something more was coming now, set lightly with the forceps into the surgical pan, a leg or the other arm or something else that Freiman did not want to see; and when the abortion was finished and they walked away from the procedure table, he found that all he could remember was that first single hand. He felt sick around his midsection still, although he was sure now that he was not going to throw up or otherwise embarrass them all, and he listened in wonder as Judy and Bill Peterson reviewed in their intelligent medical way some of the fine points of D and E. What was wrong with him? Why was he suddenly thinking about Nazi Germany? Freiman had been the chief abortion doctor at RHS for six years, and he knew he had observed his share of bloodied parts; he had once taken a series of post-abortion close-up photographs to study the correlation between fetal foot size and length of gestation. But never before had his work at the clinic required him to consider so plainly the mechanics of dismemberment. When Freiman did his first-trimester abortions he sat facing the opened legs of the patient before him, his back to the illuminated petri dish into which the aspirator contents were emptied. At some point during the procedure he would glance over his shoulder at the petri dish, and what Freiman always saw— he was sure of this, even when he and Judy were fighting after the Washington visit and Judy said Oh come on, you're telling yourself stories to make yourself feel better—was a mush of mangled tissue that looked, if one took the time to think about it, like something a doctor was supposed to take out. Possibly if he had been obliged to examine this tissue closely at the conclusion of every procedure, the way the clinic technicians were instructed to, Freiman might have encountered discrete pieces that made him feel the way he was feeling now—but he was *not* obliged to examine this tissue closely, that was not the nature of his work, and he had never been instructed before on the most efficient wrist maneuver for using forceps to reach inside a woman's uterus and twist off a fetal arm.

Michael turned to Judy, once Bill Peterson had finished chatting with them, and said, "You know, I don't know whether I can do this."

Actually he did know. The medical literature, as it began to accumulate, was going to come right to the point on this aspect of D and E. "The visual impact of the fetus compels some to withdraw from participation." "It is the physician who experiences much of the anguish of the actual performance of the abortion." "Physicians reported the D and E procedures to be emotionally difficult, which may limit the adoption of this procedure despite its value to patients." Easier on the patient, harder on the doctor: that was the shorthand on second-trimester D and E, and when Judy began to understand what Freiman was saying to her, she fought back with everything she had. We're here for the *women*, Judy cried, look how humane this is for *them*. They lie down, they get local anesthesia, they get up a while later—we're the *professionals*, Judy argued, they're just lone women in a crisis, we should be willing to endure the hardest part so they don't have to.

Michael would listen to her and nod: she was right, D and E was plainly kinder to the patient than saline infusion, and the literature had begun to suggest that when skillfully managed it was safer as well. But he would not do it. It was an emotional decision, he would say, not a rational one. Emotionally, psychologically, he was not capable of doing what they had watched Bill Peterson do.

And Judy would press him, her frustration mounting: what *sense* did that make? How could he be emotionally capable of doing it at twelve weeks but not at sixteen weeks, as though a woman's future ought to depend on exactly what it was that Michael Freiman was forced to look at in the surgical pan? Hadn't he seen that the second-trimester patients were often the women who needed Reproductive Health Services the most?

Even this Michael agreed with: it was true, the clinic ought to be taking care of the second-trimester women on site. He would back Judy up on the preparations for second trimester, and he would back her up on D and E's, too—he would assist in emergencies or help with postoperative care. He didn't think D and E was wrong in some absolute moral way; he was not a deeply religious man, but he had always understood Jewish teaching to define the fetus as an organic part of its mother until it emerges to become a separate person. If Judy wanted, he would defend her second-trimester plans publicly, he would sign things, he would make agreeable statements to reporters, he would go to court and testify. But he was *not going to do the abortions*.

Finally Judy gave up on him. Michael Freiman drove her crazy sometimes; he was so good in an abortion and so exhausting in an argument, and this line he had drawn was just self-protective and cheap, somebody else can do it but not me. Really the clinic had no one else to do it. Freiman was the working medical director, the on-site teacher, the best hands at RHS; for more than a year now, all the way into her *Planned Parenthood* v. *Ashcroft* testimony, Judy had assumed that he would be the one to develop what was apparently the considerable skill required to perform D and E's properly. Now she had to regroup. *Ashcroft* was under submission, awaiting a ruling by a federal judge in Kansas City, and Judy wanted to be medically prepared in case the judge lifted the hospitals-only requirement for second-trimester abortions.

So she picked up the telephone and called Robert Crist.

Bob Crist worked out of the abortion facility at the University of Kansas Medical Center, which stood on the Kansas side of the two-state metropolitan area that makes up Kansas City. At K.U. the latest second-trimester patients still checked into the saline unit, but Crist had spent a decade developing a reputation as one of the nation's accomplished D and E doctors, and Judy had watched him at the NAF meetings as he stood with his presentations during the Mid-Trimester Technology seminars. He was an interesting character at those meetings, tall and handsome in a craggy-faced Gary Cooper sort of way, and he could talk with extraordinary calm and detachment about precisely the mechanical details that had made Michael Freiman feel sick. "Clinical" was an adjective sometimes used to describe Crist, both in admiration and dismay: he carried a complicated image among his fellow physicians, who trusted his technical expertise but wondered if he was as aloof and brusque with patients as he sometimes seemed to be with his colleagues.

But Crist was very good at D and E. That he would do it at all set him in rarefied company, and when he talked about it later, after his credentials were better known and people pressed him for explanation, he would tell them about the scorecard and the cancer operations. The scorecard was from Crist's residency, 1965 to 1967, at the University of Kansas Medical Center. Crist started it himself, and it was an actual piece of paper, with actual pencil marks, each one representing an emergency-room patient brought in after botched abortion. Forty-five pencil marks per month, that was the average tally over one full year, and although the scorecard included no written information about the degree of fever or the type of catheter hanging out of the uterus, Crist had the kind of memory that retains detail, and could recall both the color of the most popular catheter (red) and the wording of the mimeographed insert that came with slippery elm bark, which was supposed to be jammed up the vagina and at that time could be purchased in neighborhoods where the poorest people lived. "Feel up inside and feel for the hole of the doughnut." Twenty-five years later Crist could still recite the slippery elm bark directions, verbatim, and their oddly childlike description of the round opening of the cervix: "Peel the bark off the elm. Put the tip in the hole of the doughnut, and push."

Crist was on a traditional schedule in those residency years, thirty-six hours on and twelve hours off, so that sometimes it was three o'clock in the morning when he or one of his fellow residents was shaken awake to patch up an infected abortion patient in the emergency room. At the time there seemed to them something especially stupid about this, the women so damaged and sick and the young doctors losing sleep to take care of them; one of Crist's fellow residents went on both to help legally challenge the Kansas abortion statute and to set up an active abortion practice of his own once the law had changed. And Crist went on to legal elective abortions too, not as a full-time practice, but as a routine part of his gynecological work; he was an Army doctor after his residency and did them in Washington, D.C., for the servicewomen, the military dependents, the generals' daughters. He did them at ten weeks, he did them at twelve weeks; they had no ultrasound then to size the pregnancies exactly and so sometimes he found himself doing them at fifteen weeks and he just proceeded, with the forceps, learning how to pull out parts without hurting his patient.

It wasn't so hard to get through. Cancer surgery was hard to get through, those were the operations that gave Crist bad nights afterward; a woman came in sick and he was supposed to carve out twenty pounds of what should have been healthy female anatomy. When he did his oncology procedures Crist had to take out vaginas, bladders, small intestine sections, whole swaths across the pelvic area; he had to leave women maimed and still terrified of cancer. When he did his abortions the women were relieved, they were whole, they went back to their families or their military careers or their high school cheerleading squads, they sent him Christmas cards at the end of the year and told him in their own handwriting how grateful they were. Crist didn't much like abortions, nobody liked doing abortions, but a second-trimester termination never battered him the way a vaginectomy did, and at the NAF conferences he tried to give his little speeches about taking suffering away from the patient and handing it over to the doctor instead. He had his tiny coterie of D and E allies, the maverick doctors from Washington and Boulder and Minneapolis and San Francisco (Warren Hern, whose Boulder clinic was taking in D and E patients through twenty-four weeks, used to dismiss saline induction, contemptuously, as "stick 'em and split"), and at the professional meetings they saw that the other abortion doctors were looking at them the way the rest of the medical community still looked at abortion doctors: coolly and with considerable distaste, as though they had walked into the room with bloodstains on their shirts.

Still the referrals kept coming, all the patients sent over for D and E by the doctors willing to refer for it without doing it themselves; and in January 1980, Judy Widdicombe offered Bob Crist the second-trimester work at Reproductive Health Services—if the federal court would only clear the way for them. Crist would not have to abandon his obstetrics and gynecology practice in Kansas City; the commuter flight to St. Louis took less than an hour, and Judy was proposing to fly Crist in on a regular basis, the circuit-riding doctor come to do the work no local wanted to touch.

All right, Crist said: he would come. But they needed a favorable ruling in *Planned Parenthood* v. *Ashcroft* first.

On January 23, 1980, they got it. From Kansas City, U.S. District

Court Judge Elmo Hunter issued a ten-part ruling that sided with the clinics in nearly every one of their principal challenges to the 1979 Missouri abortion law. The parental consent provision was unconstitutional, Hunter ruled, since it handed too much veto power to parents or judges. Mandating a forty-eight-hour waiting period was unconstitutional, Hunter ruled, since it required women to make two separate visits to a clinic or abortion doctor and thus had a "detrimental effect on the health interests of women." The philosophical cornerstone of the "informed consent" requirements—obliging doctors to present their patients with information like the "anatomical and physiological characteristics of the unborn child"—was unconstitutional as well, Hunter ruled:

> It is clear that many physicians believe that a mandatory presentation on fetal anatomy and physiology is not in the best health interests of many, if not all, of their patients seeking abortions. There was evidence that most women seeking an abortion do not desire such anatomical information and that the giving of such information may have adverse effects in the form of emotional tension, increased anxiety and fright. It is also clear that many physicians believe that there are no long-term physical or psychological effects of abortion or are unaware of such effects.

And Missouri's second-trimester hospitalization requirement was no longer constitutional, Hunter ruled—precisely because of the emergence of D and E. Trial testimony and published medical studies had convinced him that D and E was now the safest second-trimester abortion method, Hunter wrote, but there was only one hospital in Missouri that would even permit a D and E to be conducted on its premises. That hospital was in Kansas City, on the western edge of the state, which meant that if second-trimester abortion continued to be confined to hospitals, the rest of Missouri's women were effectively denied access to what appeared to be the most medically advisable technique. "As a practical matter," Hunter wrote, "it may force a woman and her doctor to terminate her pregnancy by a method more dangerous to her health than the method made unavailable."

As judicial reasoning this was perhaps somewhat convoluted (would the judge have left the law intact if hospitals around the state had been more receptive to D and E?), but neither Judy nor Frank Susman wanted to quibble with it: Judge Hunter had given them their nod. On Valentine's Day, three weeks after *Ashcroft*, Bob Crist arrived at Reproductive Health Services with one nurse from Kansas City, two counselors from Kansas City, and a cardboard box containing a set of cervical dilators and a ring forceps. He did seven second-trimester patients in a single afternoon.

•   •   •

THE STAFF NURSES gathered outside the procedure room, the day Bob Crist began D and E's at RHS; from where he stood near the examining table he could see them, clustered in the hallway, as though they had come as close as they intended to. "It was almost as though if they even came into the room, something was going to jump out and get on them," Crist recalls. "It was, hands off, I'm the bad guy, how in the world could you ever come to do this? Judy tried in every way—she was committed to this, and she didn't want me pushed away."

He let them be. For a while he continued to bring his own nurses with him from Kansas City. At the clinic he tried to stay around sometimes with the staff, answering their questions, relaxing a little at the end of the day; once a clinic nurse learned to work with him, he would check her periodically during the procedure to make sure she was managing all right. "He could read you like a book," recalls Carlean Turner, who had been with RHS since it opened and now found that she had a difficult time sleeping the night before the second-trimester procedures. "Sometimes he'd look and say, 'Are you okay?' And I'd say, 'Yeah. Sure.' And he'd say, 'Okay.' And later on he'd say: 'I thought I was going to have to carry you out of there, you were looking so odd.' "

Because in fact Carlean Turner was not okay, not entirely, and she learned to propel herself through the D and E's by standing next to the doctor, focusing closely on the doctor, fixing her eyes directly on the doctor or perhaps directly on the patient, and never looking down into the pan. "*Never,*" she recalls. "I would *never* look down. Some of the nurses watched as he removed the tissue, but I never looked. If I looked, I would never be able to work there again."

Among the clinic staff members it had been a difficult process, preparing for later abortions, and in group meetings they had talked about it, working it back and forth, trying to make sense of their feelings. At Reproductive Health Services the only men on staff were the physicians; all the rest of them were women, and many had children of their own. The right-to-life people liked to suggest that there was something mystical and supremely fulfilling about pregnancy, as though any woman who had given birth must surely understand in some part of her being how all abortion was the killing of human lives, and nearly every abortion clinic in the country counted among its staff members at least one woman who understood what nonsense that was: how it was entirely possible to be two or three months pregnant, happily if queasily pregnant, and still fiercely convinced that it was *nobody else's business* whether or not this pregnancy went on to turn into a baby.

But four months, five months pregnant: this was a different thing. The women with children had felt them, at four or five months; there was that first jumping flutter, the touch, the shove from inside that was supposed to mean *I'm here*. "Quickening," the historical abortion laws had called this first sensation of movement, codifying the unmeasurable but universally understood moment at which a baby was perceived to have declared itself an independent being (the earliest written penalties had recognized this dividing line in their harsher punishment for post-quickening abortions). As a scientific or legal verb, "quicken" had vanished from the vocabulary many generations back, but the dictionary still saluted its historical meaning, *to become alive, to receive life,* and in the conference room at Reproductive Health Services the counselors and nurses listened to each other talk about this, the receiving of life, the taking of life, the passionate fidelity to lives already underway. "It went both ways," recalls a former RHS counselor named Leigh Pratter. "There were people who were upset about it, but we had struggled so hard with so many of the patients who were just beyond the first-trimester line. And those situations were usually so very difficult. So there was great relief as well."

In the conference-room meetings they talked about their dreams, which were agitated and intense. One of the counselors had a twelve-year-old daughter and dreamed they were both second-trimester pregnant and both needed abortions, but only one could have it done, there were dismembered legs, she woke up sweating and afraid. Where were their loyalties, as Advocates for Health? What was their obligation, if they believed abortion was every woman's right? Sylvia Hampton's worst abortion dream had come long before the second-trimester discussions began; she had dreamed about going into labor alone on a high open hill, and looking down to see that what was emerging from her body was not a baby but twisted pieces of metal, like something from a junkyard. In the dream she was frantic and when she awoke, her heart pounding, she began to think that the dream was about the whole idea of abortion, the shadowed side of pregnancy, the personal sorrows that enveloped them every working day at the clinic. Sylvia had been pregnant five times—three children, two miscarriages—and the first time somebody at RHS told her they had a two-inch intact fetus in the lab, an eleven- or twelve-weeker, she had run back to see it—literally running down the hall, worried they might take it away before she had a chance to look. "Judy said, 'Are you sure?' " Sylvia Hampton recalls. "But I had to know. 'Don't try to stop me.' I was going to the lab because I *had to see*."

When she saw it she thought it was like something from a public television science program, curled on a cloth, little sockets for the eyes, and Sylvia's stomach lurched enough for her to say to herself: I wish I hadn't seen it. But after a while she decided she was glad she had. She had learned to live with ambivalence the week she arrived at RHS; all of them did, unless they were lying to themselves and pretending that clinic abortion was easy or simple or had nothing to do with developing human life. It was better than the alternative, that was all; it was better than the

catheters and the slippery elm bark, and it was better than forcing women
into childbirth, and even when it was hard to look at it was still the better
of two bad options—for the women who needed it done.

So she learned to talk about D and E, too, when the Significant Others
asked about it in the group discussions, and she hoped she was doing so in
the most careful way she could, using vocabulary that might describe
without shocking. "We would use words like 'expel the fetus from the
uterus, using instruments,' rather than saying, 'They go in and crush it
and yank it out,' " she recalls. "That would be accurate without being
inflammatory." But most of the time she could see that people didn't want
to know, they asked only the basic questions or no questions at all, they
had already walked past the picket signs out front and they had learned
very quickly the art of staring straight ahead without looking at the pictures or the phrases the picketers waved around. It was self-protective, this
tunnel vision, and Sylvia tried not to begrudge it to anybody; she had
called upon it herself from time to time, it was either that or stand out
there debating them every Saturday morning, and after a while a working
person just had to *proceed*, get *on* with it, stop slogging through the
uncertainty and rearguing all the arguments.

"After a while I was without ambivalence," Sylvia Hampton recalls.
"Let's face it. *Nobody* wants to do it. We do it because it's necessary. You
just could not let these cases go by and send these women off to have kids
under these conditions."

The meetings between Sylvia Hampton and Sam Lee ended, cordially,
without resolution. No one yielded. Sam thanked Sylvia for her time.

Judy ordered new brochures, and an updated listing in the National
Abortion Federation directory: 1st Trimester Suction Aspiration, $175. 2nd
Trimester D&E to 17 weeks, $300.

Then in the spring of 1980—perhaps it had started the summer
before, afterward nobody could recall exactly when it began, except that
Bertie Passanante remembered the way three of them walked into the
reception room one morning and lowered themselves one by one into seats
directly beside waiting clinic patients—Sam Lee began bringing sit-in
demonstrators right up the elevator and through the Reproductive Health
Services front door.

TEN

A House Afire

1980

NOW WHEN THEY PRAYED before a sit-in, out in the clinic parking lot or perhaps a block away so they might preserve the element of surprise, Sam knew more than the others about some of the people they were praying for. Generally they would hold hands in a circle and ask the Lord to touch the hearts of the women in the clinic, to reach with His mercy both the clinic workers and the mothers preparing to abort, and when it came his turn to pray, Sam would think sometimes about Sylvia Hampton and also about Judy, although he believed he had come to know Sylvia better.

"Loving Father, help us today," Sam would begin. He always improvised the prayer, as he had at his charismatic meetings. His voice was measured and low. He was cutting his own hair and beard but attended to them only rarely, so that after one of his arrests his friends saw his picture in the newspaper and called to tell him he was looking like Charles Manson, maybe he should use a comb more often. He was still very thin. He ate what other people cooked for him and paid no particular attention to food. "Give us courage as we go into the clinic today," Sam might say, pressing on with his prayer, holding lightly the hands of the protestors beside him, "and bless the hearts of the clinic personnel and doctors who are scheduled to perform abortions today, and bless those women who are contemplating having these abortions. Turn them around so that they might make a life-giving decision for themselves and for their unborn children. Amen."

Then they would enter the clinic, trying not to draw attention to themselves as they approached. This was easier at first, when people at the clinics didn't recognize them from the newspapers or previous sit-ins; one of the women in the sit-in group would have called ahead and made an appointment under a false name, so they could be sure they were arriving on a day when abortions were actually being performed, and they would try to walk into the waiting room by ones and twos, so as not to alert the receptionist until they were all inside. At Reproductive Health Services the patients reached the second floor by elevator, and Sam showed the others how they could avoid arousing suspicion by taking elevators to different floors and then walking down the stairs to gather quietly in the second-floor stairwell. That way when the police came they were already inside, cross-legged on the floor or sitting with their backs against the door between the waiting room and the procedure area, and if they had coordinated it well they could buy an hour or a morning or possibly even a whole day of life for a child who would otherwise have been dead by midafternoon.

The sit-in pamphlets had been instructive in this regard, imploring protestors not to succumb to despair when they saw that most of the women at the clinic could not be convinced to change their minds. "Sometimes, in fact most of the time, a sit-in does not prevent abortions, but only delays them," advised the inspirational Harvard leaflet by John Cavanaugh-O'Keefe:

> This is no reason for discouragement. An hour, or even a minute, may be precious to a tiny child. We know little about the perception of time that very young people have, and can only speculate about it. . . . It is not at all unlikely that for an unborn child a few minutes are equivalent in terms of growth and new experiences to a few years in the life of an older person. In every sit-in, we hope to save children from being killed, give them a chance to grow up and be born and grow up some more. But if we fail in that, if we only delay the death, that may not be insignificant. Wholly new experiences come very rapidly one after the other in the early days of human life.

Certainly it was gratifying when the women did listen, when one or two agreed to accept a ride over to Birthright, the volunteer organization that had been set up to aid women who had decided to have their babies even under difficult circumstances. Sam sometimes wished he were better at this part of the sit-in, the conversations with the mothers in the clinic waiting room, but when he tried to talk to them directly he felt tongue-tied and hesitant and he saw that the women protestors had an easier time of it. His job was a quieter one. If they were singing he would sing with them, "Amazing Grace" or one of the other religious songs, but mostly he just sat with his back against a door and tried to become the Peaceful Presence that Cavanaugh-O'Keefe had written about in his pamphlet.

He *did* feel peaceful. He was twenty-three years old and understood exactly what he was supposed to be doing in the world. All Sam's uncertainty about the priesthood had fallen away in the terrible immediacy of the abortion protests; if he was meant to serve God this was how he would do it, blocking the doors to abortion clinics, and later he would pick up his academic work toward the seminary again, when the protests were no longer necessary.

He stopped going to university classes. He ran out of money to pay his apartment rent. He lived with friends, a spare room here, a borrowed couch there; for a while Sam occupied the basement of a right-to-life family from the St. Louis suburbs and aggravated the mother of the house by leaving file boxes and legal writings and dirty ashtrays all over the basement floor. The mother would complain that it was like having another teenager in the house, Sam sweet-talking her into buying him new clothes when his old ones grew too shabby, but then they would all sit up together until three o'clock in the morning immersed in Bible study and compelling long talks about morality or self-sacrifice or the implications of the Nuremberg trials.

Sam read, he wrote, he went to meetings, he protested, he got arrested —that was his life. When his younger sister Emily came to visit him for the weekend, she went with Sam to a sit-in and was startled, watching her brother; she had never seen him in a leadership role before, but he was somehow guiding the men and women around him without raising his voice or even talking very much. It was as though they looked to him when they were not sure what to do next, Emily thought, and this was more striking still because Sam appeared to have no real idea about what was going to happen each time one of the sit-ins got underway. He just felt his way through them, working on the Peaceful part. In the midst of the Reproductive Health Services sit-ins Sylvia Hampton would come up to him sometimes, walk in close and look into Sam's face as though she were trying to find something there that she knew and could speak to in private, but he kept his expression calm without really looking back at her the way she seemed to want; he knew Sylvia must feel betrayed, but he was doing what he had to and he had already warned her that he would.

Vince Petersen, the young Franciscan who had organized the first St. Louis sit-ins, watched Sam at a Reproductive Health Services protest and felt humbled by what he saw. For some months Petersen had been staying away from the sit-ins, preparing for his novitiate year as a Franciscan brother, and when Sam asked him to come, Petersen felt guilty about having allowed the burden of protest to fall on other people's shoulders.

Petersen himself had never been arrested at an abortion demonstration, which he supposed was an embarrassment by now, and at the RHS sit-in he felt a small thrill of anxiety when the police matter-of-factly cuffed his hands behind his back and began pulling him toward the elevator. He looked over at Sam, who had already allowed his legs and handcuffed arms to go loose in the posture of full passivity—we do not resist, Sam had reminded Petersen, but neither do we cooperate—and Petersen tried to make himself lie limp in the grasp of the police officers, the way Sam had.

"He knew how to go about the whole thing," Petersen recalls. "I remember getting my strength from him." In the paddy wagon they talked about procedure, how long they would be in a holding cell, who would come with the bail money, which of the defense attorneys was on call that weekend. Sam had the routine down cold. "It was like: *piece of cake*," Petersen recalls. Sam even went to the trouble of warning Petersen about the complicated feelings that might come over him in his jail cell—how he would probably be afraid at first and then find the fear giving way to a satisfaction so deep that it had about it a disturbing hint of smugness. He was right about that part, too.

MONTH BY MONTH, over 1979 and the first half of 1980, the sit-in arrests mounted up: six arrested, twelve arrested, nineteen arrested, twenty-two arrested. The charge against them was criminal trespass: "No person, without lawful authority, or without the express or implied consent of the owner or his agent, shall enter any building or enter any enclosed or improved real estate, lot or parcel of ground. . . ."

In St. Louis criminal trespass was an "infraction," a form of low-end misdemeanor, punishable by up to three months' jail time and a five-hundred-dollar fine. The defense attorneys were volunteers, private practitioners, pulled in by word of mouth and Pro-Life Defense Fund leaflets available in some of the Catholic churches. "At first I didn't even know what we were defending," recalls Andrew Puzder, who was one year out of the Washington University law school and working for a firm in St. Louis when he attended his initial meeting of Defense Fund volunteers. "And then they started talking about these sit-ins. And I thought: *Kent State*. I'm back at Kent State again. I *know* about this. This is something I can *relate* to."

Andy Puzder was twenty-nine in 1979, making him the youngest member of the sit-in demonstrators' defense team, and as a right-to-life volunteer he was both a novice and something of a wild card. He had voted for George McGovern, marched in Washington against the Vietnam War, played electric guitar in a rock and roll band. In his suburban Ohio high school, forty miles outside Cleveland, Andy had been suspended three times for refusing to cut his hair. By 1960s standards it was timid hair, a

Beatles-style over-the-collar shag; but when Andy told the story later it became a small parable about rebellion and authority (in which authority ultimately triumphed, to Andy's chagrin—with a crew-cut truant officer promising to administer the haircut himself, Andy agreed to go to the barber). He went on to college in the state university system, and the campus he selected was located an hour's drive from home, in the north-eastern Ohio town of Kent.

Andy was no antiwar leader at Kent State, but he hovered closely enough around the periphery to absorb the excitement as students mounted what was to escalate into some of the nation's most turbulent campus protest against the Vietnam War. In his freshman year he joined a committee to defend students who had been expelled for their antiwar activity, and when the National Guard was called to Kent State in May 1970, after protestors set fire to the ROTC building, Andy was in the crowd as the Guardsmen began firing and killed four students. Andy ran away too fast to see the shootings, and he hadn't known any of the dead students personally; still it was a sobering experience for a nineteen-year-old, and he eventually left Kent State and earned his college degree by attending school part time in Cleveland while he worked at a music store. He married, had two children, kept playing in rock bands. He had no fantasy of turning into a famous stage musician; since childhood Andy had imagined himself growing up to be a trial lawyer, ably arguing legal cases in court. He enrolled in law school.

In his second semester of law school Andy Puzder moved to St. Louis, transferring into the Washington University School of Law, and it was at Washington University, in the fall of 1976, that Andy took his obligatory course in constitutional law. The professor was an acerbic tweed-jacketed gentleman named Jules Gerard. Gerard was reputed to harbor deeply con-servative ideas about law and the proper role of the Supreme Court, but Andy liked him very much, the Socratic elder prodding his students into intelligent argument, and when they got to *Roe* v. *Wade* Andy studied the ruling at home, as he had been taught to do, so that he could summarize it in a written brief before the students and professor discussed the case together in class.

For a long time afterward Andy Puzder would recall in some detail this one evening of study, the battered wooden desk in his apartment bedroom, the bracket shelves stacked with law books, the small pool of light from the gooseneck lamp. He read the entire majority opinion once, then blinked and shook his head and tried reading it again. *We need not resolve the difficult question of when life begins.* What had he missed? Was the Supreme Court dodging the central question because *philosophers* couldn't agree as to its answer? Philosophers couldn't agree on *anything*. If you put enough people "trained in the respective disciplines of medicine, philosophy and theology" in the same room, you could probably work up a rigorous argument about whether the room could be irrefutably proven to exist in the first place. "My mental reaction was: this was an opinion that starts with both feet in the air," Andy Puzder recalls. "I wanted to know, reading this opinion—what was the evidence on when human life begins? What were the facts? What made it a philosophical question? We're changing the Constitution here, and *I can't figure out why.*"

Andy Puzder thought of himself then as an emphatically lapsed Catholic: public schools, childhood prayers he never memorized, Sunday Mass only at the insistence of his parents. In 1973 he had paid the most minimal attention to the *Roe* v. *Wade* headlines, registered briefly the news that abortion was now legal, thought: Well, women's choice, privacy, sounds good to me. And he tried now to brief *Roe* the way his law professors had instructed him to, lay out the reasoning, point by point, one two three. But he could not figure out how to do it. "This one went *one two four five six,*" Puzder recalls. "I couldn't fill in the three. And I thought: 'I'm wrong. I'm going to go in tomorrow, and Jules Gerard is going to explain this in a way that makes me feel like an idiot.' "

But Gerard, who had been teaching law at Washington University for fourteen years, was of the opinion that as judicial scholarship *Roe* v. *Wade* was indefensible. In class he invited argument and tried to explain in professorial tones some of the sharpest critiques of the opinion; in private his voice rose and the pejoratives piled up one atop the other. "Preposterous," Professor Gerard would say heatedly, making his way paragraph by paragraph through the majority opinion in *Roe.* "*Colossally* preposterous . . . ridiculous . . . clearly unpersuasive. . . . *Incredibly bad.*"

Every time he taught *Roe* v. *Wade,* Jules Gerard would find himself in the same argument with some of his students: Why are you so opposed to abortion, Professor? And Gerard would say: I'm not opposed to abortion. I'm opposed to *Roe* v. *Wade.* Before 1973, Gerard had once composed some written testimony for the Missouri state legislature, when somebody asked him to argue in favor of one of the bills loosening Missouri's criminal abortion law; he liked the idea, thought the old law too restrictive. The state-by-state reform movement had seemed to him a promising development. "The debates were proceeding exactly the way you'd hope they would in a democratic society," Gerard recalls. "People were presenting their opinions, it was a rational process, nobody was emotional—or at least not *that* emotional, some people would feel very strongly about protecting life, they would present their opinions. Legislatures would talk, and compromise, as legislatures always do. Bills would be passed, and people were willing to accept it, whether they were happy or not, the way we should in a democracy."

Roe and *Doe* had derailed in a single day what Gerard regarded as that laudable democratic exercise, but what really galled him was the reasoning the opinions used to do it. "The essence of *Roe against Wade,* the square holding of *Roe against Wade,* is that the Fourteenth Amendment forbids the state to define the word 'person' so as to grant protection to a fetus," Gerard says, recalling his first astonished reading of the 1973 rulings. "And there is *no way on earth* you can read the language of the Fourteenth Amendment to mean that. It simply can't be done, except under some kind of Alice in Wonderland game where words mean anything you want them to mean. It *can't* mean that."

As a legal critic of the *Roe* and *Doe* opinions Jules Gerard was somewhat more acid than many of his colleages—but not all of them, by any means. Regardless of their personal views as to the results of *Roe* v. *Wade*, it was difficult in the half decade after 1973 to find a legal scholar willing to put up an enthusiastic defense of the opinion itself—as scholarship, as legal reasoning, as the kind of solid, persuasive, precedent-by-precedent explanation that is supposed to buttress a major ruling by the U.S. Supreme Court. In the nation's law journals, writers argued with each other over just how flawed an opinion it was. The sharpest denunciations came from Catholic law schools like New York's Fordham, where a professor named Robert M. Byrn, already a familiar legal figure in pre-*Roe* right-to-life circles, entitled his article "An American Tragedy" and argued that the justices had grossly misread history, the common law, and the Fourteenth Amendment: "The veneer of scholarship in the Wade opinion is only that and nothing more," Byrn wrote. "Beneath the surface, there is little that is not error."

But attacks had also come from less predictable sources, like the Yale constitutional law professor John Hart Ely, who declared in his much-discussed 1973 *Yale Law Journal* critique that the detailed trimester rules laid out in *Roe* v. *Wade* would probably have made admirable and humane legislation—and that if he were a legislator, he would have voted for them. It was as constitutional law, Ely argued, that the ruling collapsed. "*Roe* v. *Wade* seems like a durable decision," he wrote. "It is, nevertheless, a very bad decision. Not because it will perceptively weaken the Court—it won't; and not because it conflicts with either my idea of progress or what the evidence suggests is society's—it doesn't. It is bad because it is bad constitutional law, or rather because it is *not* constitutional law and gives almost no sense of an obligation to try to be."

Ely's article, which condensed in particularly trenchant prose many of the complaints that were to be directed at *Roe* over the coming years, was entitled "The Wages of Crying Wolf." "I do wish 'Wolf!' hadn't been cried so often," Ely wrote, referring to the fine old tradition of sounding scholarly alarms that the Supreme Court had overstepped its authority in one ruling or another. But this time, Ely argued, the alarms were warranted: in *Roe* the justices had abandoned their role as neutral interpreters of the Constitution and had proceeded instead into the business of imposing their own views upon state legislation, a job the judiciary was explicitly not supposed to do. "The problem with *Roe* is not so much that it bungles the question it sets itself," Ely wrote, "but rather that it sets itself a question the Constitution has not made the Court's business."

In his essay Ely used an improvised verb, "Lochnering," that for the nonlawyer demanded some legal history by way of explanation. More than a half century earlier, in a 1905 case entitled *Lochner v. New York,* the Supreme Court had disgraced itself—that was how a later generation of critics would characterize it—by overturning a New York labor law that had limited the working hours of bakery employees. The law capped each employee's work schedule at ten hours per day and sixty hours per week, and a bakery owner named Lochner had been convicted of violating its provisions. When the case reached the Supreme Court, a five-to-four majority ruled that the New York law was unconstitutional, that it violated what is referred to in the law as the "due process clause" of the Fourteenth Amendment.

The Fourteenth Amendment, of course, contains no language about bakeries. Nor does it make any reference to work hours, places of employment, or the relationship between owners and the workers they hire. The entire due process clause consists of seventeen words in the middle of one paragraph:

> All persons born or naturalized in the United States and subject to the jurisdiction thereof, are citizens of the United States and of the State wherein they reside. No State shall make or enforce any law which shall abridge the privileges or immunities of citizens of the United States; *nor shall any State deprive any person of life, liberty, or property, without due process of law;* nor deny to any person within its jurisdiction the equal protection of the laws. [emphasis added]

The Fourteenth Amendment had been ratified in 1868, two years after the Congress approved a civil rights bill that promised the "full and equal benefit of all laws" to American citizens of every race. In the contentious climate of the post–Civil War United States, the most ardent backers of that bill argued that simply adopting it as federal legislation left too politically vulnerable its protections for the freed slaves. After many months of debate and electoral battle, the Constitution itself was finally amended, for the fourteenth time. The five paragraphs of new language that made up the Fourteenth Amendment contained several provisions aimed squarely and punitively at the Confederacy (for example, government was to pay back no debt "incurred in aid of insurrection"), but Section 1 gave constitutional authority to the new civil rights bill and promised, in effect, that the federal government would protect every citizen against unjust treatment by the laws of his own state.

This profoundly important addition to the Constitution was only two sentences long, however, and like the rest of the venerated document, it was passed on to subsequent generations without detailed instructions as to precisely what it was supposed to mean. Plainly it did not mean that a state could *never* deprive its citizens of life, liberty, or property; if the Fourteenth Amendment had said that, then no state could ever collect taxes or confine criminals to jail or impose the death penalty. Instead states were prohibited from undertaking these deprivations "without due process of law."

And what was due process of law? Judges and legal scholars and crusading trial attorneys were still arguing about that more than a century after the Fourteenth Amendment was ratified. Had a person received his "due process" if the system treated him fairly—if the rules were clearly spelled out, if he was offered the same chance as everyone else to defend himself, if there was no great injustice evident in the *procedures* by which he was being accused and tried? Or could "due process" be understood to apply to the *content of the laws themselves*—to the overriding merits, for example, of a state law that restricted the hours of bakery workers?

The Supreme Court's 1905 *Lochner* ruling, which was the most famous of a series of philosophically similar rulings stretching into the 1930s, dug into the due process clause and found that New York's bakery workers law had violated the "liberty" rights of both owners and employees—that "liberty" now included the right to purchase or sell labor, and that statutes of this nature constituted "mere meddlesome interference with the rights of the individual." The political subtext was evident to many critics and later to historians of the period: the early twentieth century was producing a great deal of reform-minded economic legislation, like minimum-wage and maximum-work-week laws, and the Court was striking these new laws down not because they denied citizens the fair procedures of due process but because the justices disapproved of the *substance* of the laws. The conservative gentlemen of the Court, it was argued later, had been so personally offended by the idea of state intervention in matters of business that the justices invented new meanings for the constitutional ideas of "liberty" and "due process" in order to strike down economic regulations they found unpalatable.

By the middle of the century the Supreme Court itself had repudiated these earlier rulings, and in constitutional law classes "Lochnerism" became a term of derision. Deciding what was good for the country and then distorting the Fourteenth Amendment to force that view of things upon state legislatures that might see otherwise—this was the essence of Lochnering, in the modern view, and in his *Yale Law Review* article John Hart Ely argued that this was precisely what the Supreme Court appeared to have done in *Roe* v. *Wade*. The opinion offered no satisfactory answers, Ely wrote, to the grave questions it raised: Why should "viability" be given constitutional significance? Why should an individual woman's right to abortion be accorded more weight than a state's desire to protect a fetus? Why should it *matter* whether or not fetuses were, as the *Roe* opinion phrased it, "persons in the whole sense"? ("Dogs are not 'persons in the whole sense,' nor have they constitutional rights," Ely wrote, "but that does not mean the state cannot prohibit killing them in the exercise of the First Amendment right of political protest.")* And where was the *legal* explanation—as opposed to the moral, sociological, or philosophical explanation—for deciding that the right to privacy was, in *Roe's* words, "broad enough" to include the right to abortion?

Since by tradition Supreme Court justices generally do not defend or even discuss their legal rulings in public, the most pointed rebuttal to Ely's criticisms came in the form of another law journal article. "The Court's opinion in *Roe* is amply justified," wrote Philip B. Heymann, a Harvard law professor, and Douglas E. Barzelay, a young attorney who had just completed Harvard Law School, in a *Boston University Law Review* essay published some months later. Legal precedents and long-standing principles backed the Court's conclusions, Heymann and Barzelay argued, even if the opinion itself was not worded as persuasively as it might have been: "The language of the Court's opinion in *Roe* too often obscures the full strength of the four-step argument that underlies its decision."

Numbering them, one brief paragraph after another, Heymann and Barzelay summed up the steps.

One: Even though federal judges are generally supposed to defer to state legislatures, and to refrain from imposing their own economic or social beliefs as they examine a particular law for its guarantees of due process, everyone agrees that there is one area of rights—"fundamental" rights—that states may not curtail unless they prove an especially compelling reason for doing so. When a federal judge decides that a state law threatens to limit one of these fundamental rights, then the judge *must* look closely at the substance of the law, to help decide whether the state has proved its compelling reason for taking that right away.

Two: "Fundamental" rights include certain rights explicitly mentioned elsewhere in the Constitution (for example, the right to speak and assemble freely)—*but are not exclusively limited to those.*

Three: Included in these "nonenumerated fundamental rights"—those not written directly into the Constitution but found by successive Supreme Courts to be implied by its wording—are the personal marriage and procreative rights joined together under the general heading of "privacy."

Four: The right to abortion fits "squarely within this long-established area of special judicial concern." If personal decisions about contraception are protected as part of the right to privacy, the authors wrote, then surely

---

* In his 1990 book *Abortion: The Clash of Absolutes,* Harvard Law professor Laurence Tribe took on this argument by suggesting that the analogy was unfair: protestors are certainly able to exercise their First Amendment right to free speech without destroying dogs or anything else. "Nobody *has* to kill animals or mutilate government property in order to exercise the right to engage in freedom of speech," Tribe wrote. But a woman having an abortion *must* kill the fetus, he wrote; that is the definition of a completed abortion. "That is simply an *outcome* that cannot be avoided prior to fetal viability, if the woman is to exercise the right to terminate her pregnancy" (emphasis in original).

personal decision about abortion can readily be protected too. In *Roe* the real challenge that had faced the Court was not deciding whether the right to privacy properly encompassed the right to abortion, Heymann and Barzelay argued; rather, the challenge—and it was a formidable one, they wrote—lay in deciding whether states could ever override that claim by asserting that the developing fetus' life had taken on competing and protectable rights of its own.

"Striking the necessary balance plainly required an agonizingly difficult decision, involving as it did drawing the line where protection of life may begin, a subject on which public opinion was stridently divided," Heymann and Barzelay wrote. "It is, of course, important that a sharp line be drawn to show where human life begins and ends if we are to maintain a respect for life without regard to differences in intelligence, age, capacity and experience. One may fault the Court for not having drawn such a line with sufficient clarity, but surely it was right that the line can safely be drawn well after the emergence of a fertilized egg. . . . What is crucial is the correctness of the Court's determination that there is an early stage at which the potential of the embryo or fetus does *not* justify overriding the right of the woman to decide whether she will bear a child."

At the Washington University law school Jules Gerard followed some of the scholarly debate, as articles like these argued back and forth in law journals around the country, but nothing he read in *Roe*'s defense convinced the professor that the opinion had any merit at all. And by the time Andy Puzder's 1976 constitutional law class had dispensed with abortion, moving on to other subjects in the big blue casebook, Andy was left with the distinct understanding that for legal logic and internal consistency *Roe* v. *Wade* had proved an unqualified disaster. As a law student he found this a remarkable and unsettling piece of information, but in short order it was buried beneath De Facto Segregation and Discriminatory Restraint on Expression, and Andy finished his next two years of law school without giving more than passing attention to abortion or the ruling that had legalized it.

The truth was, Andy would tell people later, he had still not thought through his position very carefully by the time he heard about the Pro-Life Defense Fund. He was a practicing lawyer by then, the newest associate in the law firm of the veteran St. Louis trial attorney Morris Shenker. The firm had hired Andy for summer clerkships while he was still in law school, and even though Shenker himself had a controversial reputation around the state—his client list included some organized crime figures, and FBI and IRS investigators kept trying, without success, to have him indicted *—Puzder loved the firm and turned down other offers so that he

---

* In February 1989, five years after Puzder left the Shenker firm, Morris Shenker was indicted by a grand jury in Las Vegas, where he owned a casino and kept a second home, on charges of concealing income from the IRS and bankruptcy creditors. He died six months later, before being brought to trial.

could work there full time when he graduated. Andy knew that some of the best lawyers in town had received their early training at Shenker's firm; around the office it was standard practice for Shenker to take a junior associate aside and say: Go get some courtroom experience, but not in the big complicated cases our clients hire us to try. Take some *pro bono* work, Shenker instructed Andy: find somebody who needs a volunteer defense attorney, and the firm will cover your expenses.

So Andy picked up a Pro-Life Defense Fund leaflet, one morning at Sunday Mass, and stuck it in his pocket on the way out the door. Some months earlier he had commenced a tentative return to the Catholic Church; near his family's house was a nice old church run by a kindly parish priest, and Andy had begun bringing his sons to Mass, trying to carry on his own family's tradition and at the same time introduce a modest element of religion into the boys' public school upbringing. The church Andy had chosen sponsored a dutiful but unimaginative parish right-to-life committee, and even after he signed on as a Defense Fund volunteer, it seemed to Andy that he was interested more in the potential drama of the courtroom work than in the right-to-life cause itself. Andrew Puzder, Counsel to the Accused—he could see himself pacing before the jury members, orating his way through some stunning defense summation, the way Perry Mason did in the television programs Andy had watched as a child.

The problem was that Perry Mason's clients were never guilty, and the Pro-Life Legal Defense Fund's were. They *had* trespassed. They knew they had trespassed, the prosecutors knew they had trespassed, anybody who bothered to study the write-ups in the *Post-Dispatch* could read the written statements in which they announced that they intended to continue trespassing. ("We hope the police will agree with us and allow us to keep our peaceful presence, but if asked to leave and abandon our unborn brothers and sisters, we will refuse to do so.") They made for a perplexing day's work, Sam Lee and his St. Louis University comrades, and the first wave of volunteer defense attorneys had already cast about for arguments that might sound plausible in court. "At the beginning we were challenging the applicability of *Roe* v. *Wade*," recalls David Danis, one of the two St. Louis attorneys who had been representing Sam and the students charged with him since the initial sit-in arrests in March 1978. "Even though the Supreme Court had said a doctor or mother couldn't be prosecuted, because of the woman's 'right to privacy,' all it said was that the government couldn't try to interfere. It didn't say *we* couldn't walk in and try to stop it."

But that was a stretch, and the lawyers knew it; their clients were charged with criminal trespass, not with violating the dictates of *Roe*. So they had looked through the criminal law texts and lit upon the "necessity" defense, a venerable if rarely successful legal strategy that asks judge and jury to decide that in the case before them, breaking the law was *justified*, that extraordinary circumstances made it the right thing to do.

By midsummer 1979 the defense lawyers had already convinced one jury to accept this reasoning as a basis for acquitting five clinic protestors, including Sam Lee, and as a new Defense Fund volunteer Andy was sent out to the law libraries to help research the next batch of cases—to see what he could find on the shelves that might keep convincing St. Louis juries to let the protestors off without fines or jail time.

After work, alone amid the bookcases of the Shenker office law library, Andy pulled out law volumes and began taking notes. He was intrigued by the necessity defense. "Justification" and "excuse" have different meanings in criminal law, and the difference was critical to what seemed to him the clinic protestors' grander mission. "Excuse" was for the client who had done something bad, everybody understood that it was bad, but special circumstances called for indulgence by the court: the client's cousin was pointing a gun at him, say, forcing him to steal the neighbor's television set. But "justification" suggested that your man had done something *admirable:* he had made a moral decision, had done what society would have preferred even though a law had been broken in the process. Breaking a stranger's window to stamp out a kitchen fire—that was a "justification" sort of crime, in the unlikely event that a person were to be prosecuted for it, and as Andy flipped through the Missouri statute books he was intrigued to find that only months earlier a formal defense called Necessity had been written directly into the criminal code. Subsection by subsection, the state legislature had been revising and updating the entire criminal code for the state of Missouri, and here, by splendid coincidence of timing, was the new Section 563.026:

> Conduct which would otherwise constitute any crime other than a class A felony or murder is justifiable and not criminal when it is necessary as an emergency measure to avoid an imminent public or private injury which is about to occur.

In other words it was state law now, not simply a hit-or-miss legal tactic passed down from the British, but formally set into the statutes to which Missouri judges were supposed to defer. For strategic and theatrical purposes this was perfect; the very phrasing of the statute suggested the kind of evidence and argument that Defense Fund lawyers might brandish in the trespass trials. How were the clinic sit-in protestors going to keep convincing juries that they were acting to ward off "imminent public or private injury"? They were going to talk about abortion. In a court of law, before twelve of their peers, and presumably with newspaper people taking notes, the protestors were going to be given the chance to explain why abortion was the killing of human life—and why protecting this life was more important, more "necessary," than respecting trespass laws.

Andy was not sure he actually believed this himself, that abortion was the killing of human life, but as he saw it his own beliefs were beside the point. The *protestors* believed it. The first time he met them, the young Peaceful Presence volunteers talking intently among themselves in Dave Danis' big suburban living room, Andy felt a wave of nostalgia so sudden and sharp that he could imagine himself turning to Danis like a kind of simultaneous interpreter: I've been here, I know this, when I did it we wore black armbands but our hair was the same and also the expression on our faces, the luminous conviction of purpose. When he listened to the protestors Andy saw that there was no hyperbole or legal gimmickry in offering necessity as the defense argument; from their perspective it was true, they had done what was necessary, and when they described what had taken place at the abortion clinics, the words they used were insistently the vocabulary of the rescuer perceiving imminent disaster. "We were there saving babies," they would say. "We were there saving life."

The most eloquent among them was Sam Lee, who looked like Andy's younger brother except that Sam was still in his Charles Manson mode and Andy dressed like a downtown lawyer, with his straight dark hair trimmed short and combed neatly across his forehead. Both Sam and Andy had brown eyes and stood about six feet tall, and when they met the first time Andy felt right away that they were going to be friends. Andy talked to Sam about Kent State. Sam talked to Andy about Gandhi and Martin Luther King. A decade earlier Sam would have risen swiftly and be-atifically into the leadership of the antiwar movement, Andy could see that about him, and Andy proceeded with real seriousness now into the Defense Fund work, a single practical goal shining up amid all the broader and more ambitious exhortations against abortion. He was going to keep Sam Lee out of jail.

And as he worked up his defense materials, interviewing the pediatrician and obstetrician and anesthetist who had volunteered to submit affidavits on his defendants' behalf, Andy began listening to his experts in a way that was no longer entirely lawyerly. *Human life begins with human conception:* they all said the same thing and they all explained it the same way, nothing spiritual about it at all; the explanations lay not in the grand intent of God but in the observable multiplication of human cells from conception through birth and beyond. Once the embryonic cells began this process, each new cluster was a separate person, an individual genetically different from the mother and father who gave it life. Here was the heartbeat, here the sex characteristics, here the brain waves, here the beginnings of organ development—it made such sense to Andy, it seemed so plain when you thought about it this way, and now as he put in his after-hours Defense Fund time Andy told himself the story his father used to tell, about witnessing the liberation of the World War II Nazi death camps at Dachau. Andy's father had been an American soldier, and he would say to his family, in wonder and contempt: of *course* the German people knew, they *must* have known, it was nonsense to imagine otherwise. But how could they have known and done nothing? What fables must they have invented for each other as they willed themselves not to see?

SAM MADE A little money selling mail-order books for a company owned by a brother of one of the defense lawyers. These were holistic medicine books, popular in health-food stores, and the company office was close to St. Louis University, so Sam could walk over from a meeting or a research session at the campus library. He worked the evening shift answering telephones at a Catholic women's retreat center, too, and eventually he had enough cash for books of his own and coin-operated copy machines and sandwiches at the delicatessen. He rented a room in a boardinghouse for a while, fifteen dollars a week, with a hot plate on the dresser and a bathroom down the hall.

He bought a court suit at the discount store: brown corduroy, plain but respectable.

Andy Puzder began bringing him home for dinner, often accompanied by a St. Louis University Law School friend of Sam's named Tim Finnegan. Andy's wife Lisa took one look at Sam, with the hair and the wire-rimmed glasses that had replaced his heavy black frames, and thought *the sixties are finished, get over it,* but Sam and Tim told funny stories and exclaimed over Lisa's cooking and eventually explained that they were savage Scrabble players, so that Lisa and the two young visitors would stay up until three in the morning, chain-smoking and arguing noisily and insulting the veracity of each other's words. Sometimes they made middle-of-the-night telephone calls to friends with *Oxford English Dictionaries,* insisting that particular words be verified on the spot. When Sam was visiting, Lisa walked around with hiccups from laughing so hard. He knew she believed that abortion was morally complex but ought to remain legal, and she liked the way he kept his sense of humor around her, never preached or pontificated, listened seriously to her when they talked about it. "We'd be doing something or other, the topic would come up, and I would kind of groan," recalls Lisa Fierstein (she and Andy divorced in 1987). "I'd say, 'Oh God, here we go again, *abooortion.*' " And Sam would mock her tone, grinning, his eyes bright behind the glasses: "Yep, here we go," he would say. "We have to talk about *abooortion.*"

On Saturdays Sam got up early, sometimes before it was light outside, and went out to one of the clinics for a sit-in. They were gathering big crowds by now, one or two hundred people on the sidewalks and several dozen more willing to commit the actual trespass on the premises, and Loretto Wagner, the indefatigable public relations chairman of Missouri Citizens for Life, had made good her resolution to join the blockaders and

offer herself for arrest. During the week Loretto still worked the telephones and wrote her MCL press releases, but on Saturdays she was
driving her station wagon out to the Escobedo clinic, which into the spring
of 1980 was the agreed-upon target for most of the weekend sit-ins, and
placing her large middle-aged self amid the young people in the doorways.
Loretto had watched a few of the sit-ins slide too close to hysteria, some
of the women demonstrators weeping uncontrollably or shouting at the
police and the approaching clinic patients, and she and Sam tried to work
out with the supervising police officers a more sensible choreographing of
each event: let the demonstrators talk to every approaching patient all the
way from the patient's car to the clinic front door, for example, as long as
the demonstrators promised no physical contact. Then when the woman
reached a blocked doorway, some linked-arm demonstrators gazing up at
her from what was supposed to be an entrance to the medical area, a
police officer would stand over the demonstrators and recite the prescribed
warning: you have to leave or you will be arrested, you have to leave or
you will be arrested, you have to leave or you will be arrested.

   After the third You Have to Leave, police would drag the limp demonstrators away, the patient would step inside the clinic, and new demonstrators would move into the doorway. There was a certain rhythm and ritual
to it, the singing, the hands clasped together, the ardent pleas to the
advancing pregnant women, and Loretto found with a twinge of uneasiness
that she was beginning to look forward to the sit-ins, as though anticipat-
ing a particularly pleasurable event in church. "It gets to be a heady
experience, going out there," Loretto recalls. "There was a feeling of excitement, almost, when the police would come up with the paddy wagon, and
they'd slap the handcuffs on, and everybody was standing there with tears
in their eyes, isn't it terrible, they're taking that wonderful woman off to
jail. And the next morning in church everybody pats you on the back and
says, 'You're doing such a great job.' "

   By the second or third arrest it was also apparent to Loretto that she
was no longer the only mortgage-holding mother from the suburbs to
have made the decision to walk directly into the trespass area. The arrest
lists themselves, which the newspapers compressed like baseball box scores
at the end of their brief reports on the protests, now included not only
"Loretto Wagner, 45," but also "Valerie Volk, 43," and "Lucille Lavin, 48,"
and "Lillian Goedeker, 51"; there were middle-aged men blocking clinic
doorways, and two diocesan priests; and Nancy Danis, the wife of the
Defense Fund lawyer David Danis, was leaving her five children at home
so that she also could commit civil disobedience for the first time in
her life. At the organizational meetings Nancy and Loretto would worry
together sometimes about their families, about the housework that was
going untended while they were being handcuffed and shoved into paddy
wagons, about the disturbing sensation that their lives were beginning to
split into the plainness of everyday obligations and the full-color drama of

the Saturday arrest. Loretto would listen to a good friend's explanations for declining to join in the next clinic protest, family commitments or pressing appointments or maybe the Girl Scouts were counting on her as a troop leader that day, and for an instant Loretto would want to erupt: *Girl Scouts,* what kind of priorities are *those,* at least those girls are *alive.*

"Jobs, our house, my housework—nothing had any meaning when right down the street there were babies being killed," Nancy Danis recalls. At the height of the protests, in the fall of 1979 and the early spring of 1980, Nancy's name was appearing in the newspaper arrests lists too, and she would find herself in furious argument with her own friends and family members, who had begun to look at Nancy with an expression suggesting that she was becoming something of a public embarrassment. David's brother, Father James Davis, was one of the protesting priests, and that was causing family dissension too, so that at Danis gatherings David and Nancy sometimes stood up at the dinner table, too angry to say anything more, and walked out of the room. Nancy would lie in bed at night, her head throbbing, composing one-line slogans for the picket signs. For a long time after each sit-in she felt that she could still hear the voices of the women who pushed straight past her, determined to walk in there with their babies and walk out without them; she would remember the things the women said, the way some of them smirked and rolled their eyes, especially the teenagers, who had a way of looking so nonchalant sometimes that Nancy wanted to slap them. "It meant *nothing* to them, absolutely nothing, they probably went to the mall after it was done," Nancy Danis recalls.

Word of the sit-ins was spreading, in national right-to-life circles; around the country clinic sit-ins had been flaring up sporadically from one city to the next for more than three years, but the protests were now so clearly and publicly concentrated in St. Louis that a few out-of-state sit-in demonstrators moved to Missouri to join in. Joan Andrews, a thirty-one-year-old Tennessee woman who in later years was to become the most famous arrest martyr of the American right-to-life movement, abandoned her organizing efforts in Delaware and drove west to volunteer in St. Louis. Gerald Mizell, the beleaguered county police major now signing up overtime officers and tactical units for regular Saturday duty at Escobedo's clinic, began keeping notes for *Law Enforcement Bulletin,* the monthly magazine of the FBI. The article that eventually ran, co-authored by Mizell and his superintendent G. H. Kleinknecht, served as a four-page national instruction manual on managing clinic trespass: "To facilitate the large number of arrests, it was necessary to have at the demonstration site prisoner conveyance vehicles, plastic handcuffs, booking equipment, and photographic equipment," Mizell and Kleinknecht wrote. "When the arrested demonstrators reached the prisoner conveyance van, each arrestee was photographed with an identification placard containing name, date, time, and complaint number. . . ."

Some of the police officers were plainly uncomfortable with their Saturday clinic duty, locking into paddy wagons these ladies who might have been sitting two pews away at Mass the previous Sunday. Loretto liked to think they were winning converts that way, too, the officers who let her know with their eyes or their muttered apologies that they were with her, privately, they thought she was doing the right thing. During one of the sit-ins, an airplane circled overhead trailing a STOP ABORTION banner with the Birthright telephone number written out in big bright digits, and for years afterward (although Mizell says he never knew if it was true), the protestors repeated with conviction the story that a county police officer had asked to be relieved from clinic arrest duty and had taken up a departmental collection to charter that airplane himself. The momentum was powerful, the protests were building one Saturday after another, Father Danis got his picture in the paper with his cuffed hands extended serenely before him and the metal grille of the paddy wagon preparing to slide shut. "Two people walk into that room and one walks out," Father Danis declared for the newspaper, repeating with the fervor of the civil disobedient one of the central axioms of the right-to-life movement. "The first time you try to prevent that is an experience you'll never forget."

After that their numbers grew larger still. Father Danis wept before a men's club breakfast at St. Cecilia's Church and was given a standing ovation. Priests began announcing from the pulpit the address and start time of the next planned Saturday protest. Trespass defendants were marched into courtrooms one or two dozen at a time, sometimes accompanied by friends and relations and a separate volunteer attorney for each defendant, and Andy Puzder found that he rather liked watching the judges sigh or put their heads in their hands as the benches before them suddenly crowded end to end with earnest white people in jackets and ties.

"Not the usual City Court crowd," recalls Stephen Kovac, one of the city attorneys charged with prosecuting the trespass cases inside St. Louis city boundaries. Kovac's own family was Catholic, conservative South St. Louis Slovaks, and on more than one occasion the hardworking young prosecutor was obliged to explain to a distressed aunt or grandparent that he was not really defending *abortionists*, that these protestors were breaking the law, that it was his job to prosecute lawbreakers no matter how worthy their cause might seem to someone else. For public relations purposes the necessity defense was proving something of a coup, making parish heroes of the clinic protestors; at the onset of every defense presentation the lawyers would announce their intent to prove justification, and they would offer for submission long footnoted briefs full of legal citations and medical affidavits. ("Dave Danis had a legal brief that must have been a half-inch thick," Steve Kovac recalls; "this is basically a court that operates with no paper, and he has this half-inch brief that he wants to put in, saying they're innocent, he wants this under the necessity defense.") The judges would look at the briefs, and look at the hundred-dollar-an-hour downtown attorneys in courtrooms usually reserved for shoplifters and gamblers and hit-and-run drivers, and the judges found wondrous legal ways to let the protestors off: they hadn't been properly named in the paperwork, their identification was unclear, the prosecutor had failed to prove that the invaded clinic was located inside the municipality in which it was located. *Get these people out of my courtroom,* Andy Puzder could imagine some of the judges imploring under their breath, but as long as the protestors walked out without convictions, the defense attorneys had done their job.

Then in April 1980 the first blow came, and it came, astonishingly, from the leadership of the Catholic Church.

The *archbishop* was attacking them—the newly appointed arch- bishop, John L. May, who had arrived in St. Louis only three weeks earlier from his previous appointment in Mobile, Alabama. Loretto Wagner was so bewildered when she heard the initial reports that she wondered whether there had been some dreadful misunderstanding at the chancery, but Father Danis told her it was true, the archbishop had ordered him to stop getting arrested. There were rumors that the archbishop had gotten up before a priests' conference and denounced the entire sit-in effort as "counterproductive," and then the archbishop talked to the *Post-Dispatch,* repeating "counterproductive" and this time adding "ill-advised," and then during a Saturday sit-in at Escobedo's clinic Loretto looked up from her doorway and there he was, His Eminence himself, walking across the parking lot to meet them.

It was not a jovial encounter. Loretto thought May had come to tell them he had changed his mind; but no, he wanted them to stop. He told them that in Mobile there were right-to-life volunteers who picketed from the sidewalks and tried to counsel women away from abortion, but that they did all that without blocking doorways or getting arrested or being photographed inside paddy wagons. It was unseemly, May suggested to Loretto and the other protestors who had gathered around to listen: all these Catholics breaking municipal laws, as though civil authority had no meaning for them; they were alienating the Protestants, they were threatening ecumenism in the right-to-life movement, they were interfer- ing physically with women who were doing something that was in fact legal, no matter how morally wrong it might be.

"And *what are we supposed to do until the law is changed?*" Loretto had never spoken this way to an archbishop in her life—she had hardly ever spoken to an archbishop at all, she realized, but May had just angered her more than the Jefferson City legislators who voted against her bills. At least the pro-abortion legislators were straightforward about it. That was a harsh thing to think, and Loretto didn't really believe the archbishop condoned abortion, but the sense of betrayal was sickening; the previous archbishop had refrained from public comment on their sit-ins, a silence the protestors liked to think of as tacit support. Now any Catholic who felt called to the sit-ins was going to have to proceed over the disapproval of the highest cleric in St. Louis, and both Loretto and Sam knew that was a very great deal to ask.

In telephone conversations and emotional living-room meetings, the sit-in organizers argued over the proper response to the archbishop's statements. For seven years St. Louis had been one of the great beacons for Catholic right-to-life efforts around the country; the city's Archdiocesan Pro-Life Committee, with its elaborate network of parish programs and its efficient Catholic Conference lobbyists in the capitol, had been so widely admired that the St. Louis work was said to have inspired much of the Pastoral Plan for Pro-Life Activities, an eight-page document issued in 1975 by the National Conference of Catholic Bishops. The Pastoral Plan was a directive, its text unanimously approved at the NCCB's annual meeting in Washington, D.C., instructing "all Church-sponsored or identifiably Catholic national, regional, diocesan, and parochial organizations and agencies" to commence a highly organized campaign against abortion. "The Pastoral Plan seeks to activate the pastoral resources of the Church in three major efforts," the bishops had written:

> 1) An educational-public information effort to inform, clarify, and deepen understanding of the basic issues;
> 2) a pastoral effort addressed to the specific needs of women with problems relating to pregnancy and to those who have had or have taken part in an abortion;
> 3) a public policy effort directed toward the legislative, judicial, and administrative areas so as to insure effective legal protection for the right to life.

The Plan was specific in some of its instructions to the regions—there were details on who was to be appointed to the various committees, and the twelve political objectives that were to be pursued by a national network of bipartisan "citizens' lobbies" not formally linked to the Church. But on the subject of civil disobedience, which back in 1975 had not been under serious discussion as a form of right-to-life protest, the Pastoral Plan was silent. "An intensive longrange education effort leads people to a clearer understanding of the issues, to firm conviction, and to commitment," the bishops had written, and it was language like this that now filled the St. Louis sit-in organizers with such frustration as they watched their weekend protest crowds begin thinning in the wake of Archbishop May's criticisms. What did May suppose they were invoking out there, if not "firm conviction" and "commitment"? What education effort could possibly be more "intensive" than the repeated image of men and women risking jail in defense of their fellow human beings? Had the fear of lawsuit suddenly inspired the Church to demand that faithful Catholics tone down their own moral stance?

The fear of lawsuit was real; Sam and Loretto both knew that. Shortly before May's criticisms were made public, the owner of the building that housed Escobedo's clinic had filed a million-dollar damage suit against twenty-one protestors, the entire arrest list from one of the Saturday sit-ins, and as the volunteer lawyers filed their dismissal motions Sam could see how it was possible that in the next such lawsuit the archdiocese itself might be named—perhaps as a conspirator to the sit-ins. The building owner was a major Republican fund-raiser who told the newspapers that his lawsuit had nothing to do with abortion, he simply wanted to stop the disruption of normal business activities, but as far as Sam and Loretto were concerned a face-off had been called as they were *winning* and the Church had panicked and backed down. "Almost like there was this train," recalls Sam's friend Tim Finnegan. "And the archbishop just reaches out with his hand and stops it and says: 'Whoa—you're not going anywhere.' "

TWO WEEKS after the archbishop's admonishments appeared in the *Post-Dispatch*, Sam Lee and Loretto Wagner were arrested again at Escobedo's abortion clinic. Seventeen others were arrested with them, not a bad day's tally; the Missouri Citizens for Life organizer Ann O'Donnell was on the arrest list this time, her first sit-in arrest, and Loretto told the *St. Louis Review*, the archdiocesan weekly, that seven women had backed away from their scheduled abortions while the protestors were blocking the doors. "Our best sit-in, if there is such a thing," Loretto said gamely to the *Review*, but the damage was already spreading; priests had stopped encouraging them from the pulpit, May repeated his position in his weekly *Review* column, and Loretto was fielding anxious telephone calls from Catholics who up until the archbishop's declaration had been driving out every Saturday to cheer her on. "People were afraid to come," she recalls. "They thought they were maybe breaking a rule of the Church, that this was some matter of dogma, that they had to obey."

And in some chapters of Missouri Citizens for Life—which since its inception had positioned itself as precisely the sort of officially nonsectarian "citizens' lobby" envisioned by the Church's Pastoral Plan—this was cause for considerable relief.

In Jefferson City, for example, much of MCL's day-to-day operation was handled by a thirty-four-year-old volunteer named Carl Landwehr. Landwehr was a paid employee of the Missouri Catholic Conference, where he worked as an office administrator, but in his spare hours he drove and telephoned and canvassed for Missouri Citizens for Life. He was a jacket-and-tie man, master's degree in sociology, quiet and businesslike in his public presence. And Landwehr was appalled by the clinic sit-ins. Over the years he liked to think he had learned a fair amount about the way people look at things, and Landwehr winced every time he read a sit-in story or studied another newspaper picture of Sam standing there like some bearded biblical prophet in wire-rims. It was not going to *work*, Landwehr would say grimly to himself: people like him were out there trying to pull Mr. and Mrs. Average into the right-to-life movement, make them see how universal and mainstream a cause the saving of human life really ought to be, and here were these demonstrators singing hymns and being dragged off in handcuffs and generally managing to make themselves look precisely like the antiabortion religious zealots the media seemed to love putting on display.

Kathy Edwards, the Kansas City homemaker who volunteered as political director for Missouri Citizens for Life, used to mutter that if Sam radiated any more saintliness in public they were all going to witness their first pro-life Ascension. She hated the sit-ins too, the way they fed every stereotype and derailed enthusiasm that might have been put to more practical use; at the MCL meetings it began to seem to Kathy that the sit-in people were becoming mesmerized by their own virtue, and when she telephoned Carl Landwehr in the capital, the two of them would anguish together about the poisonous waste of time. "And we didn't *have* any time," Kathy Edwards recalled in one of a series of interviews before her death from cancer in 1996. "We realized they were creating problems for us. But we could tap into the energy if we could get those folks to do a sit-in on Saturday and a lit drop on Sunday."

For both Kathy Edwards and Carl Landwehr had been working the telephones, tracing progress state by state, and reading the bulletins from National—that was what the local affiliates called the National Right to Life Committee's central headquarters in Washington. By early 1980 it was not yet much of a headquarters, three downtown rooms furnished chiefly in file cabinets and wilting potted plants, but from National a small administrative staff kept track of fifty statewide affiliates, organizing the annual convention and mailing out datelined dispatches in the monthly *National Right to Life News*. And although the *News* still gave over the occasional half-page to accounts of clinic sit-ins, any regional organizer who looked over each issue would find the leadership insisting that the *real* work now before them—the work that merited the full attention of the serious right-to-life volunteer—was politics.

"The fate of this nation and the fate of its children will be decided in the 1980 and 1982 elections," declared the Phoenix internist Carolyn Gerster, then in her second year as National Right to Life Committee president, in her "From the President's Desk" column in the January 1980 *News*. "There is only one way to enact a Human Life Amendment—by electing the men and women to Congress who will vote for us."

The single most important word, in this missive from the president, was "electing." Over the decade since state abortion laws began to change, right-to-life volunteers across the United States had commenced their political education by learning, often by trial and error, the intricacies of statehouse lobbying—how to urge the Yes or No vote from legislators already in office. But even in states of steadiest lobbying success, like Missouri, it was evident toward the close of the 1970s that lobbying alone was not enough to bring back the protections destroyed by *Roe*. Right-to-life organizers were going to have to push directly into the election campaigns themselves—to send hostile legislators a message more emphatic than a sackful of angry mail. "If we can do it in Iowa, you can do it everywhere," the Iowans for Life co-founder Carolyn Thompson exhorted a workshop at the National Right to Life Committee's 1979 convention, six months after Thompson's statewide volunteers attracted national headlines by hurling themselves into the 1978 Senate race between the Democratic incumbent Dick Clark and his Republican challenger Roger Jepsen. "You can *see* the end results of a job like this. It's like if you donate money to a good cause, but you don't really see the children's eyes light up when the turkey basket comes in—education is of course our ultimate and important goal, but in politics you can *see the results*."

Carl Landwehr was part of the audience at that convention workshop, listening closely to Carolyn Thompson, taking notes. By 1979 the Clark-Jepsen story was celebrated in right-to-life circles; the media had given it such a David and Goliath quality that Thompson was able to play for the convention a video consisting solely of snippets from television reports about Fervent Little Single-Interest Group defeating Big Favored Incumbent. Dick Clark was a liberal Democrat and in that 1978 Senate race had been the clear front-runner, personally well-liked in Iowa; Carolyn Thompson, who was a registered nurse with mildly liberal political sympathies, had voted for him herself when he first ran for the Senate in 1972. But Clark's Senate voting record proved to be consistently "pro-abortion," as the right-to-life newsletters usually phrased it: Clark opposed abortion funding cutoffs and announced his opposition to a Human Life Amendment. He would not be satisfactorily lobbied, Carolyn Thompson recalls; she once took a contingent of Iowans for Life colleagues to Washington and watched Clark pull open a file drawer full of letters from distraught right-to-life constituents.

"He admitted to me, in the presence of pro-life people, that his mail ran ten to one our way," Thompson recalls. "But he basically said, 'I have to vote my conscience on this issue.' I said, 'Well, what if your constituents feel differently?' And he said, 'I have to vote my conscience. I feel it's a woman's right to choose.' "

So in 1978, when Clark was challenged for reelection by the conservative former lieutenant governor Roger Jepsen, who *did* support a Human Life Amendment, the Iowans for Life forces began vigorously working the state, as much against Clark as for Jepsen. It was not the first time that regional right-to-life volunteers had organized demonstrations and designed political flyers on behalf of particular candidates (in New York, for example, a group of women had a decade earlier formed the New York Right to Life Party, running candidates for state office; their 1976 gubernatorial candidate, the Long Island homemaker Ellen McCormack, received 130,000 votes in the general election). But when Roger Jepsen won Clark's Senate seat, defying every prediction and confounding Clark's own experienced Democratic pollster, a flurry of national attention suddenly focused on the state of Iowa and what Clark himself came to think of as The Issue. Certainly it had been a boisterous campaign, enlivened by such heat-producing controversies as gun control (Clark for it, Jepsen against it), but voter surveys by the *Des Moines Register* extrapolated that twenty-five thousand votes—many of them, it was assumed, cast by Catholics who were usually loyal Democrats—had gone to Jepsen because of his abortion position. That was only a thousand votes less than Clark's entire margin of defeat, and Clark remained unconvinced by the political analysts who argued afterward that abortion was only one of many factors that might have made the crucial difference. "I'd be willing almost to bet my life that it was The Issue," Clark recalls. "NARAL and these other groups kept saying afterward, 'That wasn't why he got beat.' And then they'd come to me and say, 'Why do you keep saying this? Keep *quiet* about it. Don't give them the satisfaction.' But there was no doubt in my mind at all—and I don't think there was any doubt in anybody's mind."

Right-to-life affiliates all over the United States heard about the Clark-Jepsen race, and when they sent their emissaries to the national convention in Kentucky the following summer, some of them startled Carolyn Thompson by approaching her in the hotel lobby and asking for her autograph. It was as though Iowans for Life had written the newest and most compelling chapter in a rapidly evolving textbook, and as she commenced her convention workshop Thompson tried to joke about the sudden celebrity thrust upon her by Clark's defeat. "How does a girl from East Overshoe, Iowa, make it to the national convention?" Thompson began. (Afterward a reporter hurried up to her to check on the proper spelling of Overshoe.) The workshop was called "Grassroots Organizing," and Thompson explained for an attentive crowd the two critical elements of the Iowa experience: the "lit drop" and the PAC.

The lit drop was not an Iowa invention; American precinct workers had been handing out their candidates' campaign literature since long before abortion was considered an acceptable subject for public debate. But there was a particular inspiration to the Iowa lit drop—for it was the Clark-Jepsen race which set that insider shorthand squarely into the vocabulary of every right-to-life organizer who was trying to master the art of political pressure. On the weekend before the 1978 general election, mobilizing an impromptu army of station wagon–driving volunteers, Thompson and her colleagues had distributed *three hundred thousand* election flyers, each one featuring on its cover a reproduction of the most famous Lennart Nilsson photograph, the eighteen-weeks-of-pregnancy baby with its thumb in its mouth. Superimposed upon the picture was a local clinic's advertisement for abortion services through twenty-four weeks, and beneath this strong graphic, in red and white letters: SEE WHY THIS LITTLE GUY WANTS YOU TO VOTE. No time was wasted hanging the flyers on doorknobs, or waiting in shopping centers for the occasional outstretched hand; from one end of Iowa to the other, all three hundred thousand Little Guy flyers were stuffed on Sunday morning under the windshield wipers of automobiles in church parking lots.

For practicality and economy of effort it looked like a nearly perfect tactical strike—the heart of the right-to-life constituency neatly gathered all at once into physically discrete areas, each one of which could be fully leafleted in fifteen minutes if the volunteer in charge had brought along a few teenagers or a Boy Scout troop to help. The church addresses came straight out of the Yellow Pages, and Thompson had also noticed that churches tend to cluster in distinct neighborhoods, simplifying even further the footwork and station wagon routes.

"It didn't take a rocket scientist to figure it out," Thompson recalls. "It was born out of a lack of funding and the inability to buy media, any type of ads, or anything else. Basically you're just trying to figure out the cheapest thing you can get out, and who you can get it to."

Dick Clark would say later that he knew he was going to lose the election when his own brother, who was Catholic, telephoned that Sunday to describe the windshield leaflets all over his parish parking lot after Mass. Thompson's "legions of foot soldiers," as she liked to call them, had hit every Catholic church in Iowa, and they had tried as well for a wide-scale lit drop on the other denominations, too—no Reform Jews or Unitarians, they were assumed to be a lost cause, but by 1978 there was enough open right-to-life activity in other mainline Protestant denominations, Baptists for Life and Lutherans for Life and so on, that Thompson (who was Catholic and would always point out, when the argumentative religious question was raised, that the Iowans for Life leadership included a Jewish lab technician and a Methodist nurse) directed her legions to avoid only those individual Protestant churches where there was reason to believe the congregation might be hostile. This was a *nonsectarian lit drop,* it was vital that both volunteers and the media receive this message, that right-to-life people from one region to another were linked not by loyalty to the Vatican but rather by a broad belief in God. "We go to *all* churches," Thompson told her audience at the 1979 convention, to applause so loud and sustained that she had to pause for a moment before she could be heard again. "Because as you know, *that's* where our support is. It's *churchgoing* people."

Carl Landwehr and Kathy Edwards grasped at once the political potential for the Iowa-style lit drop—not only for its impact upon the churchgoers who might take the time to read the flyers under their windshield wipers, but also for its mystique, its implicit threat. Was there a politician in the country who could now entirely ignore the possibility of a massive Sunday-before-election-day church parking lot blitz? The pro-abortion people had no equivalent: there were no Yellow Pages listings for the buildings in which tens of thousands of pro-abortion voters might be counted on to gather over a three-hour period exactly two days before an election. And even if there *had* been such sites—if somehow massive numbers of those people took to congregating on schedule at their coffeehouses or their local ACLU headquarters or whatever—the polls kept indicating that their vote was blurrier, it was distracted, it was pulled apart by party loyalties and wide-ranging interests and by the comforting conviction that the Supreme Court had already settled the abortion question anyway.

Right-to-life voters were different—"crusaders," Carolyn Thompson used to say. "Pro-life first, Republican or Democrat second." They would cross party lines, they would push other issues aside, they would hold their noses and vote for candidates they might otherwise have dismissed —they were Special Interest Voters, no other subgroup of the American electorate fit that term more precisely, and when they talked to each other at interstate meetings they sometimes expressed bewilderment at the suggestion that there was something fanatical or undemocratic about voting against a politician solely because of his abortion position. *He supports laws that permit parents to euthanize their unwanted children, but otherwise I agree with him on the issues*—it was lunatic to imagine themselves as anything *but* special interest voters, given what they believed about the nature of abortion. "Yes, as a pro-life person, you have interests in other issues," read a resolute editorial in the Missouri Citizens for Life newspaper:

In that respect you are not a "one-issue" voter, but you can claim the title of a "prerequisite voter." IT IS A PREREQUISITE that the candidate will use the trust of public office to protect the helpless from the powerful who would kill them simply because they are inconvenient or "unwanted."

Defending abortion as the "right" of a woman to provide her "reproductive freedom" is as valid as defending the right to embezzle in order to provide financial freedom. The damage done to the victim of abortion, the dead baby, is far more devastating than the damage done to the victim of embezzlement. Who would *defend* stealing as a legitimate means to achieve financial independence and expect voters to support him for public office?

By the spring of 1980 Missouri Citizens for Life had spent six years courting these "prerequisite voters," its leaders working (not always successfully, in the view of their frustrated opponents) to contain the organization's political activities within the legal boundaries set by the tax code. Like nearly every other American right-to-life group that had survived since the early 1970s, MCL had been assembled as a tax-exempt nonprofit, permitted by law to educate and lobby—but not to intervene directly in political campaigns for office. And over the years stalwarts like Loretto Wagner and Ann O'Donnell had learned how to push those distinctions about as far as they could possibly go. At the onset of every election season, for example, Missouri Citizens for Life distributed a series of Candidate Surveys listing each politician's response to a question about supporting "an amendment which would establish the right to life for all human beings." The candidate's options were Yes, No, No Response, or Altered Response ("candidate altered his or her response making classification impossible"), and every voter studying this list, with its Y's and N's and NR's lined up in column rows, was instructed in print that the flyer was not an endorsement—which would have been impermissible under the tax code—but was presented instead as a "public education service."

For communications and recruiting purposes, the men and women distributing these flyers had always depended on the churches, the Catholic parish system in particular; Carolyn Thompson and her Iowa foot soldiers were not the first people in the country to have grasped the logic of the Sunday morning approach to the congregations. But no one had ever tried a lit drop with the scale and blitzkrieg timing of the Iowa campaign, and in Iowa the right-to-life leaders had managed one additional strategic advantage: they *endorsed*. There was no mincing around about "public education services" or "candidate surveys" in the Clark-Jepsen race; the Little Guy in the Iowa flyers was asking straight out for a Jepsen vote, and in order to design the flyers in that unequivocal fashion, to work directly for Jepsen and against Clark, the leaders of Iowans for Life—which was officially chartered, like Missouri Citizens for Life, as a nonprofit educational and lobbying organization—had learned how to create a PAC.

That was the second part of Carolyn Thompson's message to the states, and it was crucial. The formal name was Political Action Committee, but by 1980 the initials were fusing into a single familiar nickname, so that volunteers around the country began speaking with casual authority about "state PACs" and "federal PACs," as the regional seminars and convention workshops introduced this next advance in their political education. Creating a PAC, the lawyers explained, was like taking out a license to campaign. Whether the goal was donations of money or organized telephone banks or the foot soldiers in the parking lots, any nonprofit right-to-life group that wanted to escalate openly into political campaigning was required by law to form a political committee solely for this purpose, with separate bank accounts and registrations on file at the government election commissions.

It was paperwork, that was all, filling out forms and consulting tax attorneys. But in the wake of the Iowa election, every state political operative was eager to learn it. And by the summer of 1980, with the next general election a single season away, Missouri Citizens for Life had its PAC too—its own license to campaign. "It is tempting to [reprint letters from] people on fixed incomes, young married couples, 'old-stand-bys' who wrote us words of encouragement and sacrificed to help us launch the PAC," wrote Ann O'Donnell, now doing triple duty as the new PAC's chairman, the state delegate to the National Right to Life Committee, and National's vice-president, in a public thank you printed in MCL's September 1980 newspaper. "You understand the most urgent and effective way to stop the killing is to change the law. To change the law, pro-life legislators must be elected to offices in our state."

From a distance these remarks had the ring of a stern lecture to the sit-in people—*change the law*, Ann O'Donnell had repeated, as though in the next breath she meant to add, *and stop this foolishness about breaking it*. But Loretto Wagner, who was still driving down to the clinic sit-ins to join the younger demonstrators in the doorways, had never heard Ann complain aloud about the Saturday protests even though Ann herself had only been arrested the one time, when she and Loretto were both mad at the archbishop. Loretto was beginning to have some serious misgivings about the sit-ins too, but she kept them private and pushed on as though the movement could do everything at once, commit civil disobedience *and* bring court challenges *and* throw hostile politicians out of public office. She was not as enthusiastic as Kathy and Carl about the merits of PACs (she worried about having to endorse one right-to-life candidate over another), but Loretto had her marching orders and from her kitchen telephone she began calling all over the state of Missouri, summoning the local chapter heads, tracking down the church addresses, preparing for November.

It was the biggest election they had ever faced. For governor they had Joseph Teasdale, the Democratic incumbent and ardent right-to-life partisan, against Christopher Bond, the Republican former governor of lackluster right-to-life sentiment (NR on the Candidate Surveys), who had lost to Teasdale in 1976. For senator they had the incumbent Thomas Eagleton, a Democrat who was such an enthusiastic Y that he had personally co-sponsored a Human Life Amendment, against the Republican Gene McNary, who was a Y but appeared to be hedging. (McNary was quoted, with evident disapproval, as having once said, "When the Supreme Court says a woman has a right to abortion, I have to go along with that.") And for president of the United States they had Jimmy Carter, the Democrat who had been frustrating right-to-life leaders since the onset of his administration, against Ronald Reagan, the former California governor who now looked like the first unequivocally right-to-life candidate since 1973 to have a serious chance at the presidency.

The candidacy of Ronald Reagan had set off some turbulent discussion in right-to-life circles, particularly after he named George Bush his vice-presidential running mate. Every right-to-life volunteer with a political memory knew that as governor of California, Ronald Reagan had signed into state law the 1967 Therapeutic Abortion Act, making California the third state in the nation to adopt a Model Penal Code "reform" statute. For years afterward Reagan complained that he had never really approved of the bill, and that California physicians abused its provisions by dispensing so freely the "necessary for mental health" diagnosis; nevertheless it was under Reagan's governorship that California became the pre-*Roe* abortion center for the western United States. And George Bush was a dismal selection as running mate: Bush had already declared publicly, during his own unsuccessful campaign for the 1980 Republican nomination, that he would not support a Human Life Amendment even though he was, as his campaign literature phrased it, "personally opposed to abortion." In right-to-life argot this stance was always referred to, disdainfully, as Personally Opposed But (as in, "I am personally opposed to slavery, but do not wish to impose my views upon others"), and whenever politicians trotted it out in an effort to mollify right-to-life voters, the effect was precisely the reverse. The week George Bush was named as Reagan's running mate, the *National Right to Life News* ran a full page of responses from the leaders of various right-to-life organizations; the word "disappointed" appeared four times, along with "duped" and "wary" and "totally unacceptable."

But from the vantage of state organizers like Kathy Edwards and Loretto Wagner, even a Reagan-Bush ticket was rich with political promise. Bush had agreed to support the Republican platform. The Republican platform called for a Human Life Amendment. The Democratic platform explicitly rejected a Human Life Amendment, expressing instead the conviction that abortion must remain legal; President Jimmy Carter and his running mate, Vice President Walter Mondale, were regarded in right-to-life circles as towering examples of the Personally Opposed But school. "Mr. Carter's administration has been one of the most disastrous in recent history insofar as its effect on the moral ethics of pro-life and pro-family that we treasure so deeply," wrote Jack Willke, some weeks after taking over as president of the National Right to Life Committee, in an October 1980 issue of the *News*. "We shudder at who he might appoint to the U.S. Supreme Court if given the chance, and thank God every night that no opening appeared during his four year term of office."

Ronald Reagan, on the other hand, had taken pains to convince right-to-life leaders that he had evolved over the years into a Personally Opposed *And*—that he did support a Human Life Amendment, that he had learned a great deal about abortion since 1967, and that he was profoundly distressed about what had happened in California. Ronald Reagan received the first and only 1980 presidential endorsement of the newly formed National Right to Life Committee PAC, and to the lead volunteers around the states, surveying the field for candidates who most deserved the loyalty of the prerequisite voter, Reagan succeeded in looking sincere, committed, trustworthy, and electable—the most exciting presidential prospect in the seven years since *Roe* v. *Wade*. "There was just *great* enthusiasm," Loretto Wagner recalls. "The impact of Reagan started to hit us—that this could be a totally pro-life candidate, with the charisma and everything else that he had."

IN URGENT COLUMNS of the Missouri Citizens for Life newspaper, the pace quickened. EDUCATION—REGISTRATION—MOTIVATION—AND HARD WORK NOV. 4, exhorted one headline eight weeks before the election. Loretto Wagner had never worked so hard on a political campaign in her life, hour after hour at her kitchen table with the multipaged lists before her and the telephone crooked into her ear; by nine o'clock at night she would be squinting at her own scribbled handwriting, picking absently at cold Kentucky Fried Chicken pieces, and repeating herself long-distance for the hundredth time. *Can you do it yourself out there? Can you organize it? Can you get somebody else to help you?* They had no computers then; the information was stored in hand-typed lists and manila file folders and Loretto's notes, which tended to crumple and pile up in the kitchen until they became a source of some hilarity among the MCL volunteers. "They used to tease me about my little scraps and scribbles of paper, I'd have it on a million different sheets," Loretto recalls. "I knew exactly where everything was. But God knows what I wrote it on. And these were all pretty dog-eared, all these little markings, by election time."

Loretto Wagner had helped campaign for Gerald Ford during the 1976 presidential election, but that was a halfhearted effort among right-to-life volunteers: Ford was a Republican, the Republican platform had some right-to-life language in it, Jimmy Carter looked like a bad alternative. Nobody in the movement believed Ford was genuinely on their side; his own wife, Betty Ford, was suspected of being a legal abortion sympathizer. But this time the prospect for change was extraordinary. The presidency was within reach of one of their own, a candidate who had declared in public that he truly believed abortion ought to be regarded as the killing of innocent human life. Every state right-to-life newsletter in the country must have quoted from the personal letter Ronald Reagan wrote to the summer 1980 National Right to Life Convention in Los Angeles; the Missouri Citizens for Life paper reprinted Reagan's second paragraph in full: "Never before has the cause you espouse been more important to the future of our country. The critical values of the family and the sanctity of human life that you advocate are being increasingly accepted by our citizens as essential to reestablishing the moral strength of our nation."

The presidency, moreover, was only part of it. This time right-to- life forces had a chance at the U.S. Senate, too. Thirty-four Senate seats were at stake in the November 1980 election, and in more than half those races, according to the minutely annotated tables now accumulating at National, a Human Life Amendment supporter was challenging some incumbent who opposed all serious attempts to pass protective measures for the un- born. Missouri's Thomas Eagleton was not one of those pro-abortion sena- tors, Loretto knew; Missouri Citizens for Life had always been able to count both on Eagleton, who was in most respects a liberal Democrat, and on the Republican incumbent John Danforth. But anybody reading National's dispatches could follow the gathering excitement in Iowa, where the Democratic incumbent John Culver was being challenged by Republi- can representative and Human Life Amendment supporter Charles Grassley; or in New York, where the Republican incumbent Jacob Javits had been defeated in the primary by Alfonse D'Amato, who announced himself as a candidate both of the Republican Party *and* the Right to Life Party. Florida, Indiana, South Dakota, Wisconsin—the races were scattered across the country, one campaign crew after another stacking up flyers with the code words printed under the smiling face of the candidate, *unborn child, Human Life Amendment, restoring the right to life.*

And with all of this at stake, in the fall of 1980—the Senate, the presidency, and by extension the future composition of the U.S. Supreme Court—an entirely new constituency was massing to help decide the vote.

In the resolutely Catholic city of St. Louis, the new presence was at first perceived principally from a distance: the occasional newspaper or magazine article; the television programs broadcast from other states; the radio sermons Kathy Edwards might pick up on KCCV, Kansas City Chris- tian Voice, while driving east across the state. The owner of KCCV was a veteran radio broadcaster named Richard Bott, and like the ministers whose sermons filled much of his air time, Bott identified himself as an "evangelical"—a Christian whose chief allegiance was less to a particular denomination than to a set of theologically conservative religious beliefs all based upon the Message, the Good News, the *evangel.* As a religious label the word "evangelical" was loosely defined, and sometimes argued over by its own adherents; Bott himself attended evangelical Presbyterian churches when he was in Kansas City or St. Louis, and his broadcast ministers included evangelical Baptists, evangelical Methodists, and evan- gelical pastors from the Church of the Nazarene and the Christian Reform Church and the Assemblies of God.

All of these ministers, and presumably the many thousands of other pastors who described themselves as Protestant evangelicals (approxi- mately seventy-five thousand American churches in 1980 belonged either to the National Association of Evangelicals or to the evangelical Southern Baptist Convention), shared certain basic ideas about their own religious faith. They believed the Bible was the original and authoritative Word of

God. They believed they had entered into a personal relationship with Christ—"the present ministry of the Holy Spirit," in the words of the National Association of Evangelicals' seven-point statement of faith. They believed it was their privilege and obligation to carry the Christian Gospel to others, to become messengers, bearers of the News, "evangelists" in the literal sense.

And although Bott's KCCV commentators spent hour after hour discussing Scripture and its lessons for each listener's personal and spiritual growth, they had remained silent, like most public voices of American evangelicalism, when abortion was legalized around the country. Bott himself had conducted a radio interview shortly after *Roe* v. *Wade* with a Kansas physician who helped lead the legalization campaigns; the doctor argued energetically, defending legal abortion and the Supreme Court ruling, and when he was finished Bott leaned into his microphone and urged his listeners to consider what his guest had said. "I remember telling the audience: 'You see, folks, it's a matter of *opinion*,'" Bott recalls. "'Some are very opposed to it. Others are of the opinion of Dr. Kranz. So you see, people have to make up their own minds.'"

It was not that evangelicals were promoting legal abortion, or deferring automatically to the Supreme Court; the National Association of Evangelicals had passed a 1973 resolution deploring *Roe* "in strongest possible terms" and declaring that unborn life, like all other human life, was a sacred gift from God. But even that resolution hedged in its right-to-life commitment, suggesting that a faithful Christian could accept the necessity of some abortions—the same "therapeutic" categories, for example, recognized by the Model Penal Code bills of the 1960s. "We recognize the necessity for therapeutic abortions to safeguard the health or the life of the mother, as in the case of tubular pregnancies," the Association's resolution read. "Other pregnancies, such as those resulting from rape or incest, may require deliberate termination, but the decision should be made only after there has been medical, psychological and religious counseling of the most sensitive kind."

For the next half decade, no great chorus of protest against *Roe* rose from the National Association of Evangelicals' member churches. Evangelical radio and television ministries kept their distance. The Southern Baptist Convention, the largest evangelical denomination in the United States, delegated its formal abortion study to the denomination's national Christian Life Commission, which was run by Foy Valentine, a Texas-born pastor who argued in his writings and speeches that Scripture made no explicit mention of induced abortion, and that people of faith ought to be able to disagree about it without coercion by the government or their fellow Christians. Every year the national Southern Baptist Convention meeting took up the abortion question in debate; and every year through 1979 the Convention's approved resolution read like a guarded endorsement of *Roe:* Southern Baptists, while affirming the "biblical sacredness and dignity of all human life, including fetal life," endorsed the "limited role of government" in all abortion matters—prayerful consideration of the problem, in other words, but no state abortion prohibitions, and certainly no Human Life Amendment to the Constitution.

Foy Valentine genuinely believed in this hands-off policy, and plainly there were others in the Southern Baptist leadership who agreed with him. But there was also a thick strain of anti-Catholicism at work during the early post-*Roe* years. The evangelical magazine *Christianity Today* ran occasional editorials imploring its readers not to dismiss the right-to-life cause simply because Catholics had taken it up, and Richard Bott recalls how readily he and many other evangelicals shrugged off the first decade of abortion controversy as the distant battle of an alien culture—which from the Protestant point of view took an obsessive and irrational position on contraception, too. "When *Roe* v. *Wade* hit, and the Catholics were so opposed to it, you just automatically assumed that it was something to do with their church, like the Eucharist, the way they give communion or absolution," Bott recalls. "As though—if the Catholics believe in it, why, we kind of think they believe strange things anyway. So it was very easy to assume that if *they* believed in it, no one else did."

Then in the autumn of 1979, beginning in Philadelphia and proceeding week by week from one city to the next, a laboriously produced film series aimed directly at Protestant evangelicals began making its way across the United States. The films' two principals and on-camera narrators, Francis A. Schaeffer and C. Everett Koop, were highly respected names in evangelical circles: Schaeffer lived in Switzerland, where he wrote books of religious philosophy and ran an international retreat for Protestant evangelicals; and Koop was the widely published surgeon-in-chief at Children's Hospital of Philadelphia, where he had developed a specialty in the surgical repair of babies born with life-threatening deformities. Koop had already written one book lamenting the perversity of a society that celebrated work like his while permitting other physicians to end life deliberately: "While we struggle to save the life of a three-pound baby in a hospital's newborn intensive care unit," read the opening paragraph in Koop's 1976 book *The Right to Live; the Right to Die,* "obstetricians in the same hospital are destroying similar infants yet unborn."

The Right to Live; the Right to Die was a slim paperback that had sold well over the years; the new film series project had far grander ambitions. In his autobiography Koop recalls visiting Schaeffer and Schaeffer's adult son Franky at their mountain retreat in Switzerland, the three of them beset by the conviction that something must be done to rouse evangelicals from their complacency about abortion, infanticide, and euthanasia—for both Koop and Schaeffer saw "mercy killing" of the unwanted as the logical partner to legal abortion.

"Late that evening," Koop wrote,

We sat in front of his fireplace and scribbled down the scenario for five motion pictures and the outline for a book, the entire project to be known as *Whatever Happened to the Human Race?* Together, the Schaeffers— father and son—and I determined to awaken the evangelical world— and anyone else who would listen—to the Christian imperative to do something to reverse the perilous realignment of American values on these life-and-death issues.

Whatever Happened to the Human Race? moved from city to city like a traveling college seminar, each five-film sequence airing over two or three days in an auditorium where there was room for the panel discussions that were supposed to follow viewings. The films were essentially on-screen lectures, full of somber narrative and imagery—here was Koop bent over an ailing newborn on a surgical table; Schaeffer in a junkyard beside a broken baby carriage; Koop on the shores of the Dead Sea, surrounded by the scattered forms of prostrate baby dolls. The text of both films and book ranged through logic, medicine, history, and philosophy; and when it explored theology, which it did at some length, it used the language of evangelical Protestant preaching:

> The only way we know that people are made in the image of God is through the Bible and the Incarnation of Christ, which we know from the Bible.
>
> If people are not made in the image of God, the pessimistic, realistic humanist is right: The human race is indeed an abnormal wart on the smooth face of a silent and meaningless universe. In this setting, abortion, infanticide, and euthanasia (including the killing of mentally deranged criminals, the severely handicapped, or the elderly who are an economic burden) are completely logical. Any person can be obliterated for what society at one moment thinks of as its own social or economic good. Without the Bible and without the revelation in Christ (which is only told to us in the Bible) there is nothing to stand between us and our children and the eventual acceptance of the monstrous inhumanities of the age.

As an effort at worthy filmmaking, the *Human Race* series received mixed reviews even within the religious press, and *Christianity Today* reported in late 1979 that the twenty-city seminar tour was attracting sparser audiences than Koop and Schaeffer had anticipated. But the message, over the months leading into the 1980 political season, spread far more widely than the films themselves. An evangelical with an attentive ear needn't have sat through all five films and the subsequent discussion workshops to learn that Francis Schaeffer and C. Everett Koop were exhorting Protestants to action, that Roman Catholics had no exclusive claim on this issue, that defense of the unborn could be seen as a mission of *any* Christian person who lived under the Lordship of Christ.

And this message was beginning to appear now in other contexts as well, as though building critical mass as it spread from one evangelical community to the next. War pictures, the graphic photographs that had provided such effective visuals for audiences in the Catholic parishes, were proliferating among the congregations of evangelical churches. The Southern Baptist Convention, its national leadership upended following the 1979 election of a deeply conservative slate of officers, adopted for the first time a resolution supporting a Human Life Amendment. Christian radio stations like Kansas City's KCCV began picking up the three-year-old and increasingly popular program "Focus on the Family," which regularly aired critiques of abortion; the program's host, the California psychologist and evangelical James Dobson, liked to urge his listeners to imagine what the Lord must be feeling as He witnessed the widespread slaughter of innocents.

These new surges of evangelical enthusiasm were evolving by the end of the decade into the first unabashedly religious campaign that right-to-life activists had ever carried to a broad public audience. Unlike their Catholic predecessors, who had worked so hard to convince others that the right-to-life position was rooted in science, logic, and broad moral principles rather than in religious training, the new evangelical recruits looked directly and openly to the Bible as they explained their reasoning to one another and the public around them. Passages from Scripture were cited, examined, set forth as plain evidence of God's intent. Psalm 139:13: *For thou didst form my inward parts; thou didst knit me together in my mother's womb.* Jeremiah 1:5: *Before I formed you in the womb I knew you, and before you were born I consecrated you.* Luke 1:44: *For behold, when the voice of your greeting came to my ears, the babe in my womb leaped for joy.*

It made no difference that the word "abortion" could not be located in the Bible, according to the argument now steadily gaining acceptance in evangelical churches; the teaching of Scripture was clear, that life in the womb was fully human, innocent, and created by God in His image. And Scripture was by definition both true and unchallengeable—"the inspired and only infallible authoritative Word of the Living God," as a National Association of Evangelicals board member wrote in one of the paperback abortion studies circulating around the country, and "therefore the only source from which answers to moral questions can be obtained."

But the most strident of the new evangelical right-to-life champions (and the most intriguing, for the many secular journalists who were suddenly paying very close attention) reserved their finest rhetoric not for scriptural dissection but for political attack—a big-vision, full-blown, righteous preacher attack that *began* with abortion and moved on from there. Abortion-Feminism-Homosexuality-Pornography, that was often how the litany began, as in this excerpt from an interview with the Virginia Baptist minister Jerry Falwell:

> I feel that the dignity of life is a principle we protected in this country until 1973. I think the traditional family, the monogamous husband-wife relationship, is a principle that America has honored until lately. Now we have a 40 percent divorce rate and we accept homosexual marriage, so we are beginning to violate that principle. The principle of moral decency has been honored in this country until lately; pornography is a recent phenomenon. All these principles and many others have been honored in this country, and for that reason God has honored the United States.

Here was the other line of argument drawing evangelicals to the right-to-life cause as the 1980 elections approached: that legal abortion must be stopped not only because it was wrong, morally and biblically wrong, but also because it was part of a hostile campaign to impose liberal ideas upon an unwilling Christian populace. Because this was a political rather than a theological argument, and because the audience to which it was directed was believed to be enormous, general interest reporters had begun writing about it many months earlier; by the onset of 1980, newspapers and magazines had published dozens of articles examining the small group of Washington, D.C., conservatives who had made a calculated decision to rouse Protestant evangelicals into a powerful voting bloc. It was not simply a right-to-life voting bloc these organizers had in mind; the agenda, amid what was now being referred to by both admirers and detractors as the New Right, was vastly more ambitious than that. The magazine *Conservative Digest*, which served as a monthly mouthpiece for New Right enthusiasts, explained in June 1979 how the most visceral issues might agitate an otherwise remote electorate into a broader political view:

> Attention to so-called social issues—abortion, busing, gun rights, pornography, crime—has also become central to the growth of the New Right. But to imagine that the New Right has a fixation on these issues misses the mark. The New Right is looking for issues that people care about, and social issues, at least for the present, fit the bill. As Weyrich puts it, "We talk about issues that people care about, like gun control, abortion, taxes and crime. Yes, they're emotional issues, but that's better than talking about capital formation."

The Weyrich to whom *Conservative Digest* made such familiar reference was Paul Weyrich, the thirty-six-year-old political lobbyist and organizer who was widely recognized as one of the chief architects of the New Right. Weyrich had set up a Washington, D.C., organization called the Committee for the Survival of a Free Congress (he also co-founded, with the Colorado beer brewer Joseph Coors, the conservative Washington research institution called the Heritage Foundation), and by the late 1970s he was part of a Washington direct-mail and political organizing network that attacked not only liberal politics but also what Weyrich and his colleagues portrayed as the flabby, halfhearted compromises of old-line Republicanism.

The list of complaints was lengthy and multijurisdictional: arms limitation treaties (too much giveaway to the Soviets), the Panama Canal (too much giveaway to the Panamanians), Taiwan (too much giveaway to the People's Republic of China), government taxation and regulation (far too much, generally, of both). As the basis for a national movement these were challenging issues, demanding a certain level of opinionated attention to current events, but the tactical insight of the New Right strategists— led by Weyrich, the National Conservative Political Action Committee chairman John (Terry) Dolan, the Conservative Caucus director Howard Phillips, and the right-wing direct-mail specialist Richard Viguerie, who also published *Conservative Digest*—lay in their decision to push at the same time the "so-called social issues" as though the entire roster fit together seamlessly and in perfect continuity, Taiwan and taxes and homosexuality and abortion.

Smaller-scale movements, organized with the same single-minded focus of national right-to-life groups, had already sprung up during the 1970s to take on many of these "social issue" controversies independently; in many state capitols, right-to-life lobbyists worked side by side with the anti-homosexuality lobbyists, the anti–gun control lobbyists, and the lobbyists for Phyllis Schlafly's Eagle Forum, which had built a successful coast-to-coast organization exclusively on opposition to the Equal Rights Amendment. And Weyrich was convinced that if the "social issues" were properly packaged and energetically promoted together, so that they added up to one great shout of protest from conservatives watching cherished American values slip away, they might attract to the New Right a massive influx of votes from within communities that traditionally either voted Democratic or avoided politics entirely.

Working-class Catholics were to make up part of this dream force of new conservatives, but even greater potential numbers appeared to be lurking in the congregations of Protestant evangelicals, many of whom had effectively disenfranchised themselves for years as their pastors urged attention to personal spiritual salvation rather than worldly matters like political campaigns. Exactly how many evangelicals were out there was a source of some confusion, as was the exact nature of their beliefs and political inclinations (one much-quoted 1976 Gallup poll was subsequently extrapolated to suggest that fifty million Americans thought of themselves as "born-again" Christians, a category that overlapped but was not necessarily synonymous with evangelicalism—and that would include such apparently unshakable Democrats as President Carter).* But Weyrich and his New Right colleagues were certain that sizable numbers of Christian evangelicals had the makings of vocal and active political converts: they held conservative and in some cases profoundly conservative religious beliefs; they tended to live in communities of conservative social standards, many concentrated in the southern half of the country; and their pastors appeared as a group to be warmly receptive to the proposition that traditional Christian family life was being threatened simultaneously by the rise of invasive government and the collapse of moral standards.

Better still, for purposes of instant wide-scale communication, evangelicals made up most of the enormous and rapidly growing audience for radio and television ministries. A well-crafted message from a few key pastors might reach a multistate audience numbering into the millions, and nobody appreciated the potential of these demographics more fully than the New Right's political men. According to both Weyrich's recollections and an admiring 1984 biography of the Reverend Jerry Falwell, Weyrich and some of his political colleagues were paying a visit to Falwell's Virginia church offices, trying to persuade the preacher to take up an explicitly political leadership role, when Weyrich mused aloud that a "moral majority of Americans" was waiting around the country for the inspiration a religious man like Falwell might deliver.

"And Jerry Falwell said, 'Stop! Stop!' " Weyrich recalls. "He turned to his people and said, 'That's the name of our organization.' "† Thus was born the Moral Majority, an incorporated triad of organizations (one lobbying group, one tax-exempt foundation, and one PAC) whose principal mission, as Falwell later said, was "to fight together for a pro-life, pro-family, pro-moral, pro-American position." The head of all three Moral Majority divisions was the Reverend Falwell, a Baptist Bible Fellowship preacher whose "Old Time Gospel Hour" television ministry—a weekly broadcast of the standing-room-only Sunday services at Falwell's enormous church in Lynchburg, Virginia—brought in more than thirty million dollars a year in contributions from dedicated viewers. In his preaching Falwell was a fundamentalist, part of the severe, separatist,

---

* The August 1976 poll, asking 1,553 American adults whether they would describe themselves as having had a " 'born again' experience—that is, a turning point in your life when you committed yourself to Christ," found thirty-five percent responding yes.

† The conservative political activist Ed McAteer, then national field director for the Conservative Caucus, was present at this meeting and recalls that *he* was the one who interrupted Weyrich to exclaim that "moral majority" would make a suitable organizational name.

soul-winning form of conservative Christianity that is usually character-ized as one branch of evangelicalism; and from his Lynchburg headquarters Falwell commanded a diversified religious enterprise that included a Chris-tian school, a summer camp, an accredited college, and a nationally broad-cast radio program, which he had already combined with his nationally broadcast television sermons to dispatch messages that had a snappy politi-cal ring to them. The 1976 patriotic rallies Falwell organized around the United States were called "I Love America"; his 1977 college fund-raising sermons were called "Clean Up America"; his church headquarters mailed out occasional polling forms that sought responses to questions like "Do you approve of acknowledged homosexuals teaching in our public schools?"

Jerry Falwell had already done some preaching on abortion, which he liked to call "America's national sin," and he often told reporters that his own despair over the Supreme Court's *Roe* v. *Wade* ruling was what prodded him in the 1970s to begin working political themes into his ser-mons. So it was apparently by mutual consensus, Weyrich and company advising and Falwell seeing the pragmatic and moral wisdom of the plan, that abortion—the subject likeliest to reel in conservative Catholics and disenchanted Democrats (often, but not always, the same people)—was placed at the head of the Moral Majority's sweeping agenda. The nation's other sins were to follow immediately behind, their collective damage made evident to anxious Americans of many religious faiths. Just as the Roman Catholic Church had for generations made abortion into a hyphen-ate of unquestionable wrong, *abortion-and-infanticide*, so now would the Moral Majority and the New Right construct together a new semantic linkage. "God needed voices raised to save the nation from inward moral decay," Falwell reminisced in 1987 in his autobiography, *Strength for the Journey:*

> All across the land people were just as afraid of the dangers that threatened the American family as we evangelical and fundamentalist Christians were. It wasn't necessary to be born again to hate abortion, the drug traffic, pornography, child abuse, and immorality in all its ugly, life-destroying forms. Whatever plan God had for this free nation was being threatened, and we needed to draw together millions of people who agreed on these basic issues to take a stand with us and to turn the nation around.

The Moral Majority churned up a respectable amount of press atten-tion over the following two years, particularly after Ronald Reagan met personally with Falwell during the 1980 Republican convention; when reporters made inquiries at the Lynchburg offices, they were told that the Majority had enrolled between two and three million members, with chapter offices located all over the United States. In retrospect these numbers appear to have been somewhat inflated,* but for public relations purposes they served very well, creating the spectacle of a powerful new force in American politics. Falwell was not the only conservative pastor suddenly rousing his listeners to direct political action; the television evangelists Pat Robertson and James Robison had also begun urging their audiences to the polls, and when Paul Weyrich convened his biweekly political strategy meetings at his offices in Washington, D.C., delegates from evangelical Protestantism now sat side by side with the Moral Majority representative, the National Conservative Political Action Committee representative, and the two dozen other men and women whose names generally made up the *Who's Who* lists of the New Right.

By the fall of 1980 these lists were proliferating—sometimes as graphic who's-linked-to-whom charts, with little circles and arrows—in publications from across the political spectrum. Every list now included the Moral Majority and other organizations generally described as "Christian" or "pro-family," and every list also included at least one right-to-life group, more often two or three. There was the Life Amendment Political Action Committee, which had been set up in Washington, D.C., by a former Kmart manager named Paul Brown; there was the American Life Lobby, founded by Brown's wife Judie, who had quit her job with the National Right to Life Committee to form her new organization with start-up funds from *Conservative Digest* publisher Richard Viguerie. There was a National Pro-Life Political Action Committee as well, and the National Right to Life Committee often sent a representative to Weyrich's biweekly strategy meetings, so that the connection was evident to any outsider studying the lists of names: the New Right was embracing Right to Life, with the state-by-state volunteer networks and the dedicated core of prerequisite voters; and Right to Life was in turn embracing the New Right, with the direct-mail expertise, the money-funneling PACs, and the splendid surge of fresh reinforcements the New Right leaders appeared to have summoned from the ranks of the Protestant evangelicals.

From a distance, it looked like a natural and fearsome alliance. But in the kitchens and parish offices and cramped rental headquarters where right-to-life veterans spent their volunteer time, the courtship of the New Right set off worried debate. They were all Republicans, these New Right warriors; they might make sneering remarks about the Republican Party's timidity before the big issues, but they backed only Republicans and they

---

* Subsequent press accounts, such as Jeffrey K. Hadden and Anson Shupe's *Televangelism: Power and Politics on God's Frontier,* argued that close examination of the newsletter circulation and chapter membership numbers put the Moral Majority's actual membership at substantially less than two million. The image of widespread grassroots activity was enhanced, Hadden and Shupe wrote, because in many communities Falwell was "able simply to wave his wand over preexisting local organizations (many of them church-based) and claim them as Moral Majority chapters."

attacked only Democrats. The right-to-life political rolls had always included a good number of Democrats, some of them considered liberals in
much of their voting record: Missouri's own Thomas Eagleton, a Democrat
and one of the most dependable right-to-life votes in the Senate, was
clearly not destined to receive a cent from the National Conservative
Political Action Committee. "If this situation is allowed to continue, there
is an increasingly grave danger that the right-to-life movement as a whole
will be discredited as a right-wing sham," wrote George C. Higgins, a
Catholic monsignor who had recently left the staff of the National Conference of Catholic Bishops, in an angry cautionary essay printed in the Jesuit
magazine *America:*

> What I am posing is the possibility that in a subtle and sophisticated
> way parts of the prolife movement are being used as a vehicle to promote
> a much broader right-wing agenda. . . . Some have suggested that the
> right-wing connection is a necessary strategy if the right-to-life move-
> ment is to play "political hardball." This is an extraordinarily short-
> sighted attitude. If the Human Life Amendment is to be enacted, it must
> have broad-based support, and cannot be viewed as a vehicle for a small
> group of idealogues to gain political power.

Higgins' September 1980 essay condensed in four eloquent pages the
national in-house argument against selling the soul of the right-to-life
movement to men like Jerry Falwell and Paul Weyrich. And from their
watch in St. Louis and Kansas City, Loretto Wagner and Kathy Edwards
could see the merits of the argument: nobody wanted to be manipulated
by people who thought of abortion as a kind of advertising come-on for a
product too dense or unpleasant to sell on its own. Loretto had some
sympathy with the new "pro-family" speeches she was hearing so much
about; she *was* upset about pornography and drugs and the undermining
of the traditional American family. She agreed with the eloquent Protestant pastors, to a point, when she listened to them assail women's libera-
tionists and urge women to take pride in their God-given role as mothers
and homemakers. But there was a kind of hysteria to their message,
Loretto thought, the way they carried on about God's intent for this and
God's plan for that: they were so frantic about homosexuality, which
Loretto certainly did not approve of but which still seemed to her no reason
to justify discriminating against people; and their spending priorities were
backwards, so much money for defense and none for government services
or the poor. It was as though abortion had caught the right-wingers'
imagination more because it offended their conservative social vision than
because it killed individual human beings. Monsignor Higgins had attacked
these very priorities in his essay, urging Catholics to remember that "sanc-

tity of life" was a principle with many applications, including support for nuclear disarmament, full employment, humane welfare payments, and a long list of other causes in which the New Right was very much the enemy:

> Unfortunately, there are many in the prolife movement who do not share the bishops' broad application of the respect-life principle. Instead they apply the principle selectively—to the unborn child, but not to prisoners on death row, nor to the poverty-stricken family in the inner city. . . . I would hope that prolife Catholics seriously consider the possibility that in collaborating with the right wing on abortion they risk defeat of the overall social justice agenda. Many of the issues on this agenda are an integral part of our concern for the sanctity of human life and are intimately tied, morally and practically, to our opposition to abortion. We cannot abandon our commitment to the social teachings of the church as a trade-off for New Right support on the issue of abortion.

And as Kathy Edwards saw it, here was the problem with the Higgins argument—the "social justice" argument, as the Catholic left liked to phrase it. In their own way, it seemed to Kathy, New Right leaders and social justice Catholics were trying to do precisely the same thing, subsuming right-to-life efforts within a complex set of proposals as to how the world's problems ought to be fixed. If she had been forced to spend a day locked in a room with one group or the other, she would surely have gone with the social justice Catholics, Kathy thought; she found herself growing increasingly irritated with the evangelical conservatives' fixation with Threats to the Family and Man as Head of Household and so on. But neither the Catholic left nor the Protestant right really cared about abortion, Kathy Edwards thought, not the way she did, not with the background and straight-ahead focus of a volunteer heading into her second decade of right-to-life work.

Missouri Citizens for Life *was* single-issue. Kathy wanted to *keep* it single-issue. Once a Human Life Amendment was passed, she and Loretto and all the other state volunteers could go their separate ways and work on gun rights or nuclear disarmament or whatever suited their personal ideas about society; but in the meantime Kathy Edwards was willing to take the New Right organizers for what they had to offer, which was political weaponry with a national reach. These people had a track record. Right-wing PACs had produced notable if probably overhyped showings in the 1978 elections (the National Conservative Political Action Committee had taken credit in Iowa for helping defeat Senator Clark), and for November 1980 they were circulating a widely publicized "hit list" of incumbent liberal Democratic senators selected for focused attack by direct-mail letters and paid newspaper and television advertisements. The advertising copy skipped swiftly through the list of New Right complaints —South Dakota senator George McGovern was accused of hobnobbing with Fidel Castro at taxpayers' expense; California senator Alan Cranston was accused of kowtowing to big labor; Iowa senator John Culver was described in one NCPAC fund-raising letter as "the *most radical* member of the U.S. Senate." Idaho senator Frank Church was on the hit list too, as was Indiana senator Birch Bayh, and for the right-to-life organizer there was one essential fact that bound together all of these incumbents: they were No votes on a Human Life Amendment.

Besides, the New Right was also supporting Ronald Reagan, who had gone out of his way to appeal to conservative Christians, making several campaign stops to speak to gatherings of evangelical pastors. No serious right-to-life voter in the country could really want Jimmy Carter to serve another term as president; if prerequisite voting meant anything at all, it was going to mean voting in November 1980 for at least one Republican and in many cases for a whole slate of Republicans, who represented the only national party that had written endorsement of the Amendment into its political platform. The social justice Catholics might accuse Jack Willke of pandering to conservatives every time he showed up at one of Weyrich's office strategy meetings, but the fact was that conservative Republicans had the potential to tip the right-to-life balance in Washington: if enough of them won their races, and if Reagan won the presidency, then it was only a matter of months—a few years, perhaps, the political machine worked slowly even in victory—before national protection for the unborn was in adopted Amendment form and on its way to ratification by the states.

And the 1980 election returns, as they began to mount up on the afternoon of November 4, were exhilarating. Ronald Reagan won the presidency. Iowa's John Culver lost his Senate seat. Indiana's Birch Bayh was out, South Dakota's George McGovern was out, Idaho's Frank Church was out; the only "hit list" candidate to retain his office was California's Alan Cranston, and conservative Human Life Amendment supporters also helped push out anti-Amendment incumbents in Georgia, North Carolina, Oklahoma, New York. Missouri lost its right-to-life governor, the Democrat Joseph Teasdale, beaten at the polls by the noncommittal former governor Christopher Bond, but that was a minor disappointment before the extraordinary sweep of congressional victories, and in St. Louis the jubilant celebration parties lasted late into the evening as the network television men called one race after another. "We were *ecstatic*," Loretto Wagner recalls. "I knew we'd had a big win—that this was possibly a new day."

The *National Right to Life News* stripped a banner headline across the top half of its front page: "PROLIFE GAIN: PRESIDENT, 10 SENATORS & MORE." An urgent and commanding editorial, written by a National executive committee member, took up the entire bottom half. "*Now* is the time to insist upon the submission of our Human Life Amendment to the American electorate, before the prolife movement loses its victorious edge," the editorial began:

> If we will all cooperate in this effort, I am confident that we can have our Human Life Amendment approved by the Congress and submitted and ratified by the 38 states within 12 months. Let's plan now to use the March for Life, to be held in Washington this January 22, as our kickoff for *ratification* and plan to attend the March for Life event 12 months later for the greatest prolife victory celebration in the history of mankind. ... *We must not now bask in the glory of incomplete victory and allow our complete victory to slip away.*

BUT IT DID slip away, and the damage was irreparable. To this day there are right-to-life activists who believe that what happened next crippled their movement, damaged long-standing friendships, and destroyed the chance for an amendment protecting unborn life in the Constitution— that never again would they come so close.

The infighting began in December, six weeks after the general election, as right-to-life chapter organizers around the states were still explaining for their memberships the wobbly alignment of the promising new Congress. At National's Washington office the vote counts were supposed to be tabulated by the full-time paid lobbyist, a dedicated young Notre Dame graduate named Charles Donovan, and Donovan reported back with a complex assessment of the odds. Any proposed amendment to the U.S. Constitution must be approved by two thirds of the Congress before being sent on to the states for ratification, and in the Senate, which was habitually more hostile to right-to-life causes than was the House of Representatives, the Yes votes appeared to be up by at least ten—but Donovan could still count fewer than fifty sure Yeses, even if every one of the new freshman conservatives made good on his promise to support a Human Life Amendment.

Another twenty senators *might* now be reachable, Donovan reported, pulling the Senate vote up to the mandatory sixty-seven. But it was a very close call. It depended on a lot of unmeasurables, like the degree of panic among the unprincipled No votes—senators who didn't really care one way or the other about abortion and might now see the political wisdom of switching sides.

"I can remember telling people that in terms of strict head counts we weren't there," Donovan recalls, "unless the election returns had so intimidated fifteen to twenty percent of each chamber that they would vote for us out of political considerations, and not out of conviction."

Ann O'Donnell and Kathy Edwards shuttled this news back to the regional chapters in Missouri. Ann was by now the chief conduit to National, flying regularly to Washington or Chicago for the unwieldy fifty-member board meetings (one board member per state, that had always been the policy, it was a Grassroots Movement, although some of the veterans were beginning to wonder testily whether for administrative purposes they had perhaps produced an overabundance of grass). Kathy was the state president, charged with passing along new developments to the loyal lit-droppers and envelope-stuffers around the state, and it was late in January that Ann brought back the latest startling news from Washington: North Carolina senator Jesse Helms, a profoundly conservative Republican with a reputation for marching off in his own direction, had introduced a Human Life *Bill*—not an amendment to the Constitution, but instead a piece of ordinary legislation, requiring for passage nothing more than a majority vote in both houses and the signature of the president.

Inside the National Right to Life Committee's Washington offices —and, within a matter of days, inside the state headquarters of every right-to-life chapter in the country—the Human Life Bill set off an uproar. Over the eight years since *Roe* v. *Wade*, amending the Constitution had been the grand obsession of the right-to-life movement, inspiring every march and every speech and every hard-fought political campaign. Since 1973 the most tireless activists had argued back and forth about the wording of such an amendment, but the principle had remained unscathed: amending the Constitution was as permanent a change as the American system permitted. Once the unborn were explicitly recognized and protected in the Constitution—whether the written reference was to "moment of fertilization," "unborn offspring at every stage of their biological development," or any of the other key phrases that had surfaced in the assortment of competing amendment proposals—then clearly the right to life would be shielded both from the whims of state legislatures and from the justices of the Supreme Court, who were required, after all, to defer to the Constitution in every case that came before them.

But fixing these ideas into a *bill*, a simple piece of legislation, like a tax increase or an interstate trucking regulation—this was revolutionary, in right-to-life circles. By what power could the Congress approve such a bill? Wouldn't the Supreme Court overturn it at once? And how had this proposal managed to spring up out of nowhere, so fully drafted that nobody in the mainstream leadership had seen it until it was being introduced as legislation?

Indeed, part of the problem with the Human Life Bill, quite aside from the legal questions it raised, was its source. The chief propagandist for the bill was a pugnacious and famously noncooperative "anti-abort" named James P. McFadden, who on his own constituted most of the leadership of a Washington organization called the Ad Hoc Committee in Defense of Life. McFadden, a fifty-year-old former writer for William F. Buckley's conservative magazine *National Review,* had made up his committee's title shortly after *Roe* (he deliberately threw some Latin into the name, he recalls, since he knew he must look to fellow Catholics for his initial support), and for the last eight years he had published both a scholarly-appearing journal, the *Human Life Review,* and a cheeky Walter Winchell–style biweekly called *Lifeletter,* half political update and half personal tirade, in which McFadden typed single-space and underlined for emphasis and took obvious pleasure in annoying as many people as possible, including every right-to-life leader who annoyed *him.* "Anti-abort" was a frequent McFaddenism, as were "pro-abort" and "apparat" and a lot of Germanic military terms that seemed to McFadden suitable for describing conflict escalation, as in this excerpt from the *Lifeletter* published immediately after Helms—joined on the House of Representatives side by Illinois Republican Henry Hyde and Kentucky Democrat Romano Mazzoli —introduced the Human Life Bill:

> The basic idea is by no means new: it has been discussed since at least '74. . . . What is new is a political situation that, as we say, might well mean not only that an HLB could pass but also that the Court might actually accept it (or at least see in it an opportunity to back off even further from the now-untenable Abortion Cases?). That is why the Helms-Hyde-Mazzoli apparat swung into action. While "differences of opinion"—chronic in any broad-based movement—were not unexpected (and anyway could be considered later), it was expected that, with such leadership, the vast majority of anti-abortionists would get behind the move (as one member put it: "If you won't follow these guys, who will you follow?"). Alas, not so: there is already plenty of bitter opposition (even from some "experts" who favored the concept when it was "their" idea!). Thus, as we go to press, the whole HLB effort—the blitzkrieg thrust which seemed the perfect strategy at a crucial moment—may be in jeopardy.

It was in the Winter 1981 issue of McFadden's *Human Life Journal,* a quarterly collection of original essays and reprinted articles generally more sober and measured than his own newsletter, that a young Washington attorney named Stephen Galebach had laid out the theoretical framework for protecting unborn life by statute rather than by constitutional amendment. Over twenty-eight pages of heavily footnoted text, Galebach argued that the route to congressional protection lay in the Fourteenth Amendment, which prohibited states from depriving any "person" of life, liberty, or property without due process of law. Since the Supreme Court had refused in *Roe* to decide when life begins, Galebach argued, then

Congress had the power to make that decision instead: by majority vote, the members of the House and Senate could decide that human life begins at conception and that "unborn children," as Galebach wrote, are therefore persons protected by the Fourteenth Amendment.

McFadden himself had recruited Galebach to write his article for the *Human Life Review,* and the hastily introduced Senate and House bills bore so many McFadden handprints that as the arguments grew louder and the warring memos began to circulate, it was obvious to some right-to-life insiders that they were fighting as much about bruised egos as about the substance of the bills. McFadden was a renegade, smoking his pipe and pounding out his little newsletter and using the telephone to work the Washington congressional *apparat* to which he liked to make such secretive reference; the former National Right to Life Committee president Dr. Mildred Jefferson had once accused him of subverting the movement by dividing loyalties and hurting National's fund-raising efforts, and McFadden told her flat out that he didn't care, if they couldn't raise enough money on their own they deserved to be subverted. It was vintage McFadden, staging a blitzkrieg without consulting anybody else; the entire operation left National looking foolish and scrambling to catch up.

"We were sort of in the wilderness," Jack Willke recalls. "We didn't know *what* was going on." National board members exhorted each other to one side or the other, called Chuck Donovan to wheedle insider data from the working lobbyist, studied the legal and pragmatic arguments now emerging at voluminous length from various parts of the country: It was a good idea because it could *work,* they could get a majority vote, the president would sign it, the momentum would be unstoppable, the nation's Congress would be placed on record as defending the right to life. It was a bad idea because the Supreme Court would certainly overturn it, Congress *had* no such authority, it would drain all efforts toward the only permanent remedy, which was still a constitutional amendment. *"To the majority the HLB can appear to be a solution to the abortion problem,"* read one vehemently argued paper entitled "The Case Against the Human Life Bill(s)." *"That is one of its greatest dangers."*

And there was a deeper and more complicated argument about the Human Life Bill as well: its wording, according to some of the right-to-life lawyers now examining the provisions line by line, might have no direct effect at all on private clinics offering privately funded abortions. Even if it were found constitutional, these lawyers argued, the bill would simply clear the way for individual states to pass their own laws permitting or prohibiting abortion—now "informed by Congress's judgment," as Galebach wrote in his article for the *Human Life Review,* "that unborn children are human life and human persons."

In the lexicon of right-to-life activists, there was an eight-year-old shorthand for this approach: "states' rights." It was a term borrowed deliberately from the vocabulary of nineteenth-century slaveowners, because for the activists who now repeated it—distastefully, as though reluctant to resurrect a strategy that had been argued over and largely discredited during the movement's early days—the logic was identical whether the discussion was about slavery or abortion. Should the voting adults of a particular state be permitted to decide who is and is not a full human being?

To the first generation of post-*Roe* activists, the answer to that question had appeared so obvious that there was no point in discussing it seriously. If the unborn were human in Pennsylvania, then they must also be human in New York; the elements of life did not magically rearrange themselves because a pregnant woman had crossed a geographical boundary line. A lot of sobering compromise had been forced upon right-to-life activists since those naive early years Carolyn Gerster described as the Children's Crusade, but surely this at least was inviolable, this fidelity to the personhood of the unborn; surely they had spent the better half of a decade working for something finer and more honorable than the right of every state to sanction the medically sanitized killing of its unborn if that was what its voters chose to do.

Now the states' rights argument was back, as lawyers from the right-to-life ranks upbraided each other in written disputes about what the Helms bill would actually accomplish in law. Then in September 1981, with the movement already in philosophical disarray over whether and how to support the Human Life Bill, the Utah Republican senator Orrin Hatch added a wholly new complication by introducing for congressional consideration a states' rights *amendment*—a fresh proposal for changing the Constitution, with the attendant requirements for two-thirds approval by Congress and ratification by the states.

Loretto Wagner learned about Senator Hatch's amendment proposal in the Sunday morning newspaper, which she was reading in the passenger seat of the family station wagon as her husband drove the Wagners back from church. Loretto was state vice-president of Missouri Citizens for Life by then, third rung down in the MCL hierarchy, and she was astonished that nobody had told her the states' rights approach was being hauled out again. She considered it now with a veteran's eye, wondering whether the movement's national leaders had been right back in 1973 to insist that the entire debate was a repetition of the nineteenth-century slavery arguments. The slavery arguments had ended in civil war, after all; maybe it did make sense to give up a little moral ground, to place some faith in local democracy, to trust each state's voters to do the right thing. Missouri was as good an example as any: the people of Missouri had outlawed abortion before *Roe*, and they would surely do so again after *Roe*—which was what this amendment appeared to be offering, a national life after *Roe*.

And as Loretto studied line by line the newspaper description of Senate Joint Resolution 110, which was the formal name for Senator Hatch's amendment proposal, she found herself thinking, *Yes. This is right.*

This will *work*, it will *pass*, it will go into the *Constitution*, and the Supreme Court will not be able to do a thing about it.

The text of Senate Joint Resolution 110 read:

> A right to abortion is not secured by this Constitution. The Congress and the several States shall have the concurrent power to restrict and prohibit abortions: *Provided*, That a law of a State which is more restrictive than a law of Congress shall govern.

Loretto telephoned Ann O'Donnell as soon as she got home. "Did you see the paper?" she asked. She was very excited. "Did you see what Hatch has done?"

Ann said she had seen the paper, and then—Loretto was taken aback, she had known Ann for years and never heard her sound like this—Ann exploded into anger on the other end of the telephone. She thought it was a *terrible* idea. It was stupid, it was divisive, it was settling for the unacceptable—Ann liked the bill that Senator Helms had introduced, she thought it was a limited but workable way to prime the country for a serious constitutional amendment protecting the unborn of every state, and now the Hatch forces had apparently negotiated some kind of backroom Washington deal placing the Helms bill on hold while Senator Hatch held hearings to attract a lot of attention to *his* amendment, which if it passed would affix the states' rights approach so permanently in the Constitution that they would all be stuck with it forever.

It was the bitterest argument Ann and Loretto had ever had. It went on for the next four months; the more Loretto thought about it, the more convinced she was that Orrin Hatch, the devout Utah Mormon whose right-to-life convictions no one dared impugn, had offered up a constitutional amendment realistic enough to be approved by two thirds of both bodies of Congress and ratified by the states. *Of course* it was a compromise, Loretto wasn't blind, she wanted full constitutional protection as much as Ann did, but compromise was better than going down in saintly flames, for heaven's sake. No serious Washington lobbyist believed they had the votes for an amendment to protect all human life; it simply *could not be done*; it didn't matter how noble an idea anybody thought it was, there were too many abortion apologists out there ready to start holding forth about "privacy" or doctor-patient relations or Roman Catholic dogma masquerading as law. But an amendment that simply reversed *Roe*, that left the follow-up act to the democratically expressed will of the people— this was manageable, it seemed to Loretto; this could be sold even to elected officials who weren't personally convinced that abortion ought to be prohibited at all.

And Ann would fire back at Loretto, her voice rising: you're wrong, you don't know Washington the way I do, you don't *understand* this. Promoting the Hatch Amendment was like raising a giant white flag: Okay, we give up, let's fight together for passage of an amendment that gives states the right to keep abortion *legal*. Let's write it into the *Constitution*, let's actually place the words "right to abortion" into the great document itself, let's take the one shot we have (for the movement was never going to get a second chance to amend the Constitution, Ann was adamant about that) and use it for wording that suggests that there *is* such a right and that states have only to decide for themselves whether to recognize it. *Nothing* about the humanity of the unborn. Nothing about the right to *life*. Was Loretto out of her mind? How in the world did she imagine that they were going to convince hardworking volunteers around the states to spend their valuable time defending a proposal like this?

And Loretto would answer, knowing it was a cruel thing to say, but meaning it: what price was Ann willing to pay for those untarnished principles? Was Ann ready to assume responsibility for all the babies who were going to be aborted while she and her compatriots held out for the perfect life-begins-at-conception amendment? Loretto and Ann shouted at each other over the telephone, Ann would slam down the receiver without saying good-bye, Loretto's children would find her crying at the kitchen table. "This is *not a game*, Ann," Loretto would hiss, before the line went dead because Ann had hung up on her again, and on that much they agreed: the entire right-to-life leadership was at war over the tactical decision about which approach to support, and from all over the country incendiary memoranda and heartfelt position papers began demanding allegiance to one position or the other:

You were pro-Hatch, meaning that your immediate goal was passage of the Orrin Hatch–sponsored amendment declaring that no "right to abortion" existed in the Constitution.

Or you were pro-Helms, meaning that your immediate goal was passage of the Jesse Helms–sponsored Human Life Bill, followed directly thereafter by the *only acceptable change to the Constitution*, which was an amendment explicitly protecting all human life throughout every stage of pregnancy.

Or—here was a late-entry complication to muddy the allegiances even further—you were an advocate of the wildly optimistic Two Step Plan, according to which the right-to-life movement would push for passage of the Hatch Amendment and then, once Hatch had been approved by two thirds of both houses and ratified by the states, the movement would regroup at once to push for passage of a *second* constitutional amendment protecting life from conception on.

By the time the Two Step Plan surfaced, to jeers and catcalls from its many detractors, the infighting over which strategy to pursue in the Congress had grown so vicious that the worst of all right-to-life slander was beginning to spread between warring camps: they support this extremely bad idea, one faction would suggest about another, because *they really aren't pro-life at all.* When the National Council of Catholic Bishops came out for the Hatch Amendment, the Council's president declaring in public testimony that the Hatch approach had the distinct advantage of appearing "achievable," anti-Hatch forces denounced the bishops as unconscionable traitors. "The Catholic *apparat* deliberately caused a new division that, at *best,* can paralyze action," James McFadden wrote to a Washington colleague in a personal letter that reflected some of the venomous spirit of the moment. "The *apparat* will do anything to prevent the [Human Life] Bill from passing. Why? In my judgement, the reason is as clear as it is detestable: the Bill would keep the abortion issue alive, whereas the *apparat* wants to abort it. The great Abortion Debate has crippled the Washington Catholic bureaucracy's real agenda, which is symbolized at the moment by El Salvador, but has, for many years, involved as *primary* concerns a host of 'liberal' political causes and concerns."

A Kansas right-to-life newsletter described the Hatch Amendment, the insult fully capitalized for emphasis, as a "PRO-CHOICE AMENDMENT." ("It simply transfers the CHOICE for abortion from women and doctors to Congressmen and state legislators.") A Pennsylvania right-to-life newsletter filled twelve pages of tiny-print typing with passionate defenses of the Hatch Amendment and the moral demand for compromise. *("Who among us is willing to watch another million babies die while we wait for the 'perfect' amendment?")* A pro-Hatch open letter accused the pro-Helms side of "bare-knuckles lobbying," "distortion and bias," and "creating the most serious division in the pro-life movement since it began"; a pro-Helms open letter accused the pro-Hatch side of "destructive activism," and demanded to know how people could claim to be pro-life while sabotaging legislation they ought to be championing; a *National Catholic Register* columnist compared the pro-Hatch forces—Catholic bishops and all—to lemmings on the march, swarming blindly and stupidly ahead without any regard for the devastation they leave in their wake.

In a way, some right-to-life veterans reminded each other as the hostilities intensified, milder variants of this argument had been bedeviling the movement for nearly a decade already. When so much of the work consisted simply of proving to the American public that the unborn child was as human as the born child, there was profound philosophical danger in *any* concession that in another context might look like level-headed pragmatism. Consider the long-standing discussions about abortions for rape and incest victims, for example, or abortions for fetal deformity, or abortions to "preserve the mental health of the mother"—what were commonly referred to, in movement jargon, as the Exceptions for Hard Cases.

Within the right-to-life movement, among women and men who were serious about their commitment and had tried hard to think this matter through, it was very difficult to find *believing* advocates of Hard Cases Exceptions—people who sincerely believed that the law ought to permit abortion for women who fell into these categories. Believing such a thing undermined the entire central premise. The law does not permit a woman to have her five-year-old put to death in a medical clinic because the child was conceived during a rape, or because the child has cerebral palsy, or because the woman will be made mentally ill by the child's continued existence. If it truly was a child that grew from conception on inside the pregnant woman's body, then it was a child regardless of the unhappy circumstances of the pregnancy; if right-to-life people defended Hard Cases Exceptions as ethically appropriate, then by definition, logically, those same right-to-life people were declaring either that killing children *was* occasionally acceptable, or that the unborn were not really children after all.

Thus during the formative years of the first regional and national right-to-life organizations, nearly all the in-house literature that raised the Hard Cases questions answered them succinctly, in the manner of the early, pre-*Roe* edition of Jack and Barbara Willke's *Handbook on Abortion*, which referred to fetal deformity abortions as "pre-natal euthanasia" and printed its final remarks on rape abortions in the format of a poem, or a protest placard:

> *Isn't it a twisted logic*
> *that would kill an innocent*
> *unborn baby for the crime*
> *of his father!*

But the repeated setbacks of the 1970s had convinced some right-to-life leaders that this principled, internally consistent opposition to Hard Cases Exceptions was a kind of moral luxury reserved for those who had taken the time to consider the issue at length—and that to make any progress at all, the movement was going to need the support of people who were sympathetic but had not so carefully thought through their views. An ordinary American citizen, contacted over the telephone by a polling company, was not likely to have read and deliberated extensively about abortion and its relationship to the law; that person (a homemaker, say, with no formal affiliation to an abortion group of any kind) was going to be operating on instinct, on what seemed sensible based on her general moral principles and her understanding of the news. And whenever abortion questions were included in national opinion surveys—this had been true since the early 1970s, and every serious right-to-life activist knew it —a majority of those polled said they believed Hard Cases abortions ought to be legal.

Those survey results were often summarized triumphantly by legal abortion advocates, who liked to insist in their speeches and press releases that polling proved again and again that "choice" was the position of the American majority. But anybody who examined both the response breakdowns and the survey questions themselves could see that the polling in fact suggested something more complicated than that. Here was a fairly typical Gallup poll, for example, conducted in early 1979 and including abortion as part of a broader national opinion survey:

Question: Now, here are some questions dealing with the subject of abortion. Do you think abortions should be legal under any circumstances, legal only under certain circumstances, or illegal in all circumstances?

Responses:
Legal under any circumstances 22%
Legal only under certain circumstances 54%
Illegal in all circumstances 19%
No Opinion 5%

Even if a reader put aside for a moment the vocabulary used in that question (for poll studiers had learned this, too, that people tended to answer abortion questions differently—sometimes the very same people, directly contradicting themselves within a single survey—if the questions were reworded to include phrases like "protecting unborn human life"), there were a variety of ways to do the arithmetic on those Gallup poll numbers.* It was accurate, certainly, to say that seventy-six percent had agreed that abortion should not be outlawed. But it was also accurate to say that seventy-three percent had agreed that abortions *should* be outlawed unless they fell into the category of "under certain circumstances." And when the surveys included more detailed questions as to what those "certain circumstances" might be, the lists showed with some consistency

---

* A *New York Times*/CBS News Poll, for example, in August 1980 polled a random sample of Americans about their abortion views. When the respondents were asked, "Do you think there should be an amendment to the Constitution prohibiting abortions, or shouldn't there be such an amendment?," twenty-nine percent of those surveyed were in favor, sixty-two percent opposed, and the rest uncertain. When the *same people*, later in the survey, were asked, "Do you believe there should be an amendment to the Constitution protecting the life of the unborn child, or shouldn't there be such an amendment?," fifty percent were in favor, thirty-nine percent opposed, and the rest undecided. Wrote E. J. Dionne, Jr., in the *New York Times* article recounting the poll results: "Fully one-third of those who opposed the amendment when it was presented as 'prohibiting abortions' supported it when it was presented as 'protecting the life of the unborn child.' "

that a majority supported Hard Cases abortions *and did not support most of the others*—categories like "If the family cannot afford to have the child," or "If she is unmarried and does not want to marry the man."

Thus as a right-to-life person read the results of American surveys on abortion, a majority of Americans could reasonably be said to believe that *most* abortions should be illegal—for everybody understood how few women genuinely fit into the Hard Cases categories—but that some should be permitted by law. And even before the Hatch Amendment fights, friends and allies in right-to-life organizations had argued with each other about whether to capitalize on this information or to brush it aside, the way they imagined an abolitionist might have brushed aside the results of nineteenth-century national opinion polls about the rights that ought be granted to black people. Should legislation be written to satisfy the less committed majority, even if doing so meant inscribing into law those philosophically perilous exceptions for Hard Case abortions? What if that was the only way *any* protections might result? As the Pennsylvania right-to-life newsletter observed amidst its twelve-page defense of the compromises inherent in the Hatch Amendment, the movement had in a sense been preparing for years for the battle that now consumed it; the clearest legislative victory of the 1970s, the newsletter reminded its readers, had been the repeated congressional passage of the Hyde amendments, which annually cut off federal abortion funding for all but Hard Cases abortions.

"Do you remember what we settled for in the earlier Hyde amendments?" the newsletter asked:

> No funding *except* for rape, incest and physical health. Was that a compromise? Maybe, but it *did* save some babies! . . .*
>
> Reservations about rape or incest or physical health of the mother, while not acceptable to the right-to-life movement, are concerns expressed by many on the periphery of the movement. *We cannot wait until we convince them of our position to gather their needed support.* The Hatch offers them something they can work for without holding back for exceptions we cannot accept.

Not since the devastating news of January 1973, urged the newsletter, had so important a subject confronted the right-to-life movement. Yet divisiveness now threatened to undermine them all:

---

* From 1977 until 1980, Hyde Amendment annual funding cutoffs exempted abortions for rape, incest, and—added on in 1978—"severe and long-lasting physical health damage to the mother." In 1981, the cutoff rules were tightened to allow exemptions only for life of the mother; that version was to be adopted every year until 1993, when rape and incest were reincluded as exemptions.

We have it within our grasp to stop the holocaust. We cannot let this moment slip away. Those who have wrapped all their hopes and efforts in the Human Life Bill to the exclusion and detriment of the Hatch proposal ignore reality. While we cannot "impose our morality" on them, we can and do object to the sniping at Senator Hatch and the outrageous charges circulating in Washington implying that the Roman Catholic bishops are somehow backing off on their opposition to abortion by endorsing the Hatch. . . . Can those who call this a compromise sacrifice the lives of babies who will not be born if we don't use the means we have at our fingertips to stop legal abortion *now?*

By late 1981 the casualties were accumulating and the internal warfare had spilled out into the general press. The National Right to Life Committee lobbyist Chuck Donovan had grown so frustrated by the increasingly bitter divisions among board members that he had resigned from his job. New Right organizers pleaded publicly and privately for a truce, warning leaders of the various factions that if they could not unite behind one approach they were destined to lose everything—the bill, the amendments, and what should have been the accumulated goodwill of eight years' work on Capitol Hill. ("The right-to-life movement will have pulled the trigger on its own heart," declared Connie Marshner, the New Right policy analyst who was chairing the biweekly Washington strategy meetings, in *The Washington Post.* "The average politician will throw up his hands and say, 'A pox on both your houses—don't any of you ask me to do anything for you ever again.' ") Internal memoranda were leaked to the newspapers, senators received telegrams insisting that any vote for the Hatch Amendment was unequivocally "anti-life," and in Kansas City Kathy Edwards' handwritten notes turned dark and anxious as she wrote down snatches of telephone conversations with Washington and St. Louis:

"Critical decisions on direction of the movement."
"Pull victory out of anything."
"Are we sowing the seeds of our own destruction?"

A thousand miles from Washington and the fractious National board meetings, Kathy found the entire Hatch-Helms fight both depressing and a spectacular waste of effort. Kathy was a pragmatist, a lobbyist had to be, and philosophically she was inclined to follow Loretto: a states' rights amendment was possible, a life-begins-at-conception amendment was not. But it was also evident that in Washington and around the country the battle had turned vindictive and that men and women who were supposed

to be allies were beating each other up over power and politics and intraof-
fice rivalries that had nothing to do with abortion law. The University of
California professor John T. Noonan, who had spent more than a decade
establishing his credentials as one of the movement's legal scholars, wrote
from his office in Berkeley that the right-to-life leadership was beginning
to look like squabbling relatives in front of a three-alarm fire.

"We are in the situation of a family whose house is on fire and some
of whose children are inside," Noonan wrote at the close of a long analysis
of competing amendment proposals. "With the passage of every hour a
child is killed. Meanwhile we stand outside arguing with each other and
with the fire department as to the hoses they should use. What is needed
is to turn on the water and start pumping."

In December 1981, Kathy Edwards sat down at her dining-room table
and wrote out by hand the longest and most emotional memorandum she
had ever prepared for the board members of Missouri Citizens for Life. In
a week the fifty-member National Right to Life Committee board was to
vote in Chicago on endorsing or rejecting the Hatch Amendment, and it
was up to the Missouri board to prepare voting instructions for Ann
O'Donnell, who was still Missouri's delegate to the National board—and
still vehemently opposed to the Hatch Amendment.

Page after page, in her clear, loopy handwriting, Kathy poured out
her arguments for making peace as a movement and uniting behind Senator Hatch. "It has been argued that the Hatch proposal compromises our
principles," Kathy wrote:

> Hatch does not *address* the principles. . . . Hatch does not address
> "personhood," but as a practical matter, the average person doesn't under-
> stand the term or really care. What they oppose is the killing of unborn
> babies. If you stop and think why you became involved, it was more than
> likely because you opposed the killing. Our understanding of society's
> moral corruption came second. We must continue our work to both stop
> the killing and to restore the moral principles.
>
> Ours is a killing society. It took many years to sink this low and it's
> going to take many years, probably several generations, to restore respect
> for all human life. There is no magic solution to the problem. Each piece
> of legislation, each amendment proposal provides an opportunity to move
> forward. No legislative or amendment formula is going to mean the end
> of our work.

Two days later Ann O'Donnell sent off a heated countermemo: even
if the movement were to abandon its most cherished ideals and vote for
Hatch's wretched proposal, she wrote, it still appeared to have no real
chance for passage in the Senate. "I am truly heartsick about what is

happening to the right to life movement," Ann wrote. "The rush to support a measure that is the antithesis of what we stand for AND a loser at that, is a curiosity I am unable to understand."

On December 12, with Ann O'Donnell waiting in Chicago for her orders from home, the Missouri Citizens for Life board of directors met in a hotel conference room in the mid-Missouri city of Columbia and approved by voice vote a resolution endorsing "the passage and the ratification of Senate Joint Resolution 110, the legislative authorization amendment known as the Hatch Amendment to the United States Constitution." Somebody got up to call Chicago, and it was not until later that afternoon that they heard what happened when Ann O'Donnell received her voting instructions from Columbia. Ann cast her vote, waited for the full-board National vote to be tallied, and saw that by a narrow margin the pro-Hatch forces had won. Then she quit the National Right to Life Committee. "Very calm voice," recalls a National board member who watched Ann O'Donnell walk out of the room. "Something like: 'This is my formal notice. I'm tendering my resignation. I shall not be serving on this board.' "

Six weeks later a dozen National board members, all close allies of Ann O'Donnell in the Hatch-Helms fight, sent out a letter demanding Jack Willke's resignation as president of the National Right to Life Committee. The letter accused Willke of maneuvering National into "a compromise of the very principle upon which we were founded—that the right to life is inalienable," and as copies fanned out around the country, a veteran Midwestern board member responded by sending out a desperate and mournful plea for calm. "We have hit rock bottom," wrote Anthony J. Lauinger, chairman of Oklahomans for Life:

> What are we doing to ourselves? . . . For the past year the pro-life movement has been like a blind man stumbling in circles, being flogged ceaselessly by others—not on the pro-abortion side—but within the movement itself. This internal turmoil hasn't been the fault of Jack Willke. He has been but one of a cast of many characters. It is the fault of all of us. We have to realize—each of us individually—that if we continue to seek to cannibalize each other, we do so not only to the irrevocable and inexcusable detriment of the unborn child, but also that we do so [sic] at our own peril. For how could such self-destruction—self-destruction based on pride—as our movement has been engulfed in for the past year possibly be excused or forgiven (if it continues into the future) by Him whose work we do?

AT THE SATURDAY clinic protests, sometimes sitting in now with only four or five other protestors for company, Sam Lee understood how very, very

far he was from the fighting over national strategy—how small and contained an arena he had chosen instead.

Sam could see for himself the newspaper headlines, ANTIABORTION
MOVEMENT BADLY SPLIT OVER TACTICS, and he listened from one side of the
room as the tension rose between Loretto Wagner and Ann O'Donnell
during planning meetings, and he knew from the anxious hallway talk
afterward that the St. Louis stalwarts were bewildered themselves about
which of their strong-willed leaders to follow. Instinctively Sam felt that
Ann's arguments made more sense, that it was repugnant to support a
constitutional amendment that left states the option of legalizing abortion;
but it was also plain to Sam that in the real world around them neither
Jesse Helms' bill nor Orrin Hatch's amendment was going to get in the
way of a single abortion any time soon.

That was Sam's job. He got in the way of abortions. By the spring of
1982 Sam had lost count of his total number of arrests, maybe thirty or
forty; the records were piling up in Andy Puzder's law office, and when
Andy needed extra help now he sent Sam out to look up the case law
himself. Some months earlier Andy and Lisa Puzder had invited Sam to
move in with them, since they could see that he was shuttling between
boardinghouses and obliging friends' couches, and Sam had settled into
the third-floor Puzder guest room like a beloved and exasperating younger
brother, leaving his dirty clothes on the floor and taking out the garbage
and walking the children to the park when Lisa was too busy to do it.
Andy was traveling quite a bit on his private cases, and when he was gone
Lisa would jokingly introduce Sam as "the man I'm living with"; she still
thought Sam was probably wrong about abortion law, but she felt deeply
affectionate toward him and had come to believe that he might be the most
trustworthy and ethically consistent person she had ever met. It moved
her to watch him come downstairs in the morning, his hair still damp
from the shower, and head out into a daily life that appeared to be based
upon nothing but moral principle.

"Gotta go out and get arrested today," Sam would say, smiling a little,
but quiet and entirely serious. Ever since the archbishop condemned the
sit-ins, their support crowds had been dropping away, and Sam had misgivings now every time he started off for one of the clinics. Loretto Wagner
had confided that the sit-ins were beginning to trouble her so much that
she did not think she could keep participating much longer; there was an
obsessive quality to the whole Saturday exercise, Loretto said, as though
it were an addiction, or a cult, and sometimes when he looked around the
clinic waiting rooms Sam could see what was making her nervous. The
sit-ins were losing their Lorettos, the good suburban ladies whose very
respectability lent a certain order to the group. Only the most determined
stayed on Saturday after Saturday, the women and men whose field of
vision was as focused as Sam's, and there were days when he could feel
the frustration and sorrow swell so palpably inside a crowded clinic waiting
room that he knew he was supposed to contain it somehow, to say some-

thing that would help people understand how Gandhi or Martin Luther King would have wanted them to behave. "We have to take on the role of the unborn child," Sam would say. And the others would grow impatient with him, crying, Yes, yes, but that doesn't mean you have to be completely passive, there are things we can *do.*

In the police holding cells, late into the afternoon, with the prostitutes and the drunk drivers slumped into plastic chairs at the other end of the room, Sam and the others would smoke cigarettes and prod each other, jousting, vying for the high moral ground. After Loretto stopped showing up so often the senior sit-in protestor was usually Ann O'Brien, a grandmotherly-looking St. Louis woman who was partial to emphatic movement vocabulary like "abortuaries" and "killing rooms"; there was also a young man named John Ryan, who told people that his father had explained abortion to him when he was still a teenager by bringing him along to right-to-life meetings; and a black-bearded Southern evangelical pastor-in-training named Mike Chastain, who had helped stir up right-to-life protest at the St. Louis Presbyterian seminary where Chastain was a divinity student; and the increasingly famous Joan Andrews, who had a plain face and a small whispery voice and so fevered and indomitable a resolve that as she traveled from city to city, chaining herself to clinic doorways and refusing to cooperate in her own arrests, right-to-life locals had taken to calling her Saint Joan.

Once Joan Andrews went through a window at Reproductive Health Services, right in the midst of a sit-in. It was an inside window, one of those sliding-glass partitions between waiting room and medical area, so it wasn't as though Joan crashed through glass or came swinging down like a cat burglar; still there was something thrilling about the invasion, the idea of advancing directly toward the killing rooms instead of just sitting there waiting for the usual arrest routine, and when Sam said afterward that he hated the window stunt, the others assailed him for being so docile. Wasn't it a good idea to unplug the abortion machines —disarming, at least momentarily, the weapons of death? What about unplugging the abortion machines and then pulling out some wires in the back, or wrapping the cord around your body (Joan Andrews tried this one afternoon, and then berated herself afterward because it was such a feeble gesture), so that anybody who tried to drag you out would topple over the machine? What about unplugging the abortion machines and then messing things up in a selective sort of way, dirtying the sterilizers, maybe urinating on the instruments (that idea came from Mike Chastain, who also pointed out that spitting on them or throwing them on the floor would achieve the same purpose), so that nobody could shove you out and then pick them all up to begin killing again?

They were crossing a line, perhaps they had already crossed it, and Sam couldn't pull them back. He could feel Vince Petersen's model of peaceful Franciscan witness sliding irretrievably away. When John Ryan shouted about murder, or made his baby-crying noises as the police were dragging him through a clinic doorway (John was awfully good at that, falsettoing his voice so it sounded like an infant crying for its mother, it gave Mike Chastain the chills to hear it), Sam would recoil and close his eyes, praying, wondering what Gandhi would do, trying to make the calm more powerful than the noise. Sam's strength had always come from his quiet. He had never been an organizer, not in the conventional sense; Mike Chastain, who had graduated from the South Carolina military college The Citadel, sometimes thought of Sam as a *disorganizer*. Mike liked discipline, order, action; he wanted to lead the troops into battle with the scriptures held high. Sam tested Mike's patience, with his beatific follow-your-own-light docility. One morning some hostile people came up to both of them outside a sit-in and said, *You believe in capital punishment, don't you, how can you be such hypocrites?* And Sam the pacifist Catholic, who did not believe in capital punishment, looked at Mike and said, "You see, they've got you in a moral dilemma here."

But Mike had that worked out and was disturbed that Sam couldn't see it. "There is no moral dilemma," he said. "There are instructions from God as to how a life ought to be taken. Hitler's life should have been taken —through due process."

Sam wondered why Mike didn't just march into the clinics and start killing all the abortionists. Weren't they supposed to be *defending* human life? How was Mike going to make his moral distinctions once he had justified for himself the deliberate, preplanned taking of life—not self-defense against a threatening attacker, but instead the rational decision that a particular human life merits extinction?

"As a Christian I don't have the power of the sword," Mike said. "Only the state has the power of the sword." Mike would quote Scripture to Sam, whenever they began arguing like this: he liked Genesis 9:6, for example, *Whoso sheddeth man's blood, by man shall his blood be shed: for in the image of God made he man.* Capital punishment was *sanctioned* by Scripture, Mike would insist, as long as the punishment was justly applied to the evildoer and not to the innocent. Sam would keep invoking Gandhi, or Martin Luther King; once he absently threw in the abolitionist John Brown, too, and Mike turned to Sam in triumph and cried, "John Brown was a *murderer!*" Slaveowners died during Brown's violent revolts, Mike reminded Sam, and Yankee legend and song had made a hero of Brown anyway. Couldn't Sam see that it was wrong sometimes to be too passive in the face of evil, that disarming the abortion machines was a small but manifestly moral act? Wasn't it Sam himself who sometimes likened their work to defending a child from attack on the street? If one of them happened upon such an attack in any other context—a man with an upraised club, say, preparing to strike the next blow—would they hesitate for an instant before taking the club away and breaking it to pieces?

And Sam would shake his head *no*, it was not the same, abortion was not like any other context. Two years of Saturday mornings at the clinics had taught Sam something about the fragile relationship among them all —protestor, abortionist, mother, unborn child. It was not really possible to prevent the abortion without winning the heart of the mother, and it was no use to anyone—Sam knew this, he was sure of it, he had sat down and talked to these people—to think of clinic workers and abortionists as deliberate complicitors to murder.

Every morning of every protest Saturday, well into the summer of 1982, Sam worried about the role he ought to play—as leader, as pacifist, as civil disobedient. There were practical dilemmas on top of the moral ones: the clinics all had lawyers now, and one by one the half dozen abortion facilities in greater St. Louis were obtaining restraining orders that specifically, by name, prohibited each of the best-known sit-in protestors from entering the premises. Sam Lee was named in every one of the restraining orders, and when the Pro-Life Legal Defense Fund lawyers explained the possible consequences of violating a legal injunction of this nature, their warnings were sober and sharp: simple trespass was no longer the issue. A restraining order was the direct command of the court, the personal instruction of the issuing judge, and flouting such an order was like a deliberate insult, inviting serious criminal charges and lengthy jail time—perhaps a six-month sentence, the lawyers warned, for each offending sit-in.

And what good was an abortion protestor locked up in jail for six months at a time? How useful was martyrdom, as a practical matter? Every philosophical treatise on the early clinic sit-ins had directed the participants to remember that their primary job was literal rather than symbolic; they were to place their bodies between unborn children and abortionists, and the protest inherent in the trespass, the grand gesture of public defiance, was supposed to remain secondary. "The reason for sitting in is to save lives, not to test the law," the influential 1978 pamphlet *A Peaceful Presence* had instructed.

But if they walked into clinics now in the face of injunctions explicitly ordering them not to—if they pasted the restraining orders on cardboard and hung them insolently around their necks, as Mike Chastain once suggested—then what plausible defense could their attorneys possibly present? Wouldn't a "necessity" defense simply add one more layer to the insult, as though the accused were saying, *Well, Your Honor, your little injunction just doesn't matter as much as my principles?*

For a while Sam and his sit-in companions dodged the restraining orders by moving the protests from site to site: as soon as one clinic obtained its injunction, the sit-ins changed venues. But by midsummer 1982 they had run out of clinics. Reproductive Health Services; Bolivar Escobedo's clinic; the street-level clinic operation called the Ladies' Center; the women's health facility run by two osteopaths out in St. Louis County —every building was now off limits, by judicial injunction, to the small group of men and women persistent enough to keep pushing their way inside. And as of June there was no longer any question about trying to work up some stepped-up version of a necessity defense, either, because the Missouri Court of Appeals put a stop to it in *all* the trespass cases. Abortion was a "legally protected activity," a three-judge panel held in an appeal of a set of trespass arrests dating back to one of the early Reproductive Health Services sit-ins. Thus it made no sense, the Court of Appeals ruled, to claim that the protestors were only trying to prevent some grievous injury when they broke the law:

> In *Roe* v. *Wade*, the Supreme Court of the United States recognized that a woman's decision to abort her pregnancy is protected by her constitutional right to privacy. Since abortions, like those in issue here, are constitutionally protected activity and, therefore, legal, their occurrence *cannot be* a public or private injury. Thus, defendants cannot rely on the statutory defense of necessity.

It was a very bad blow, this Court of Appeals ruling. There was a little comfort in the sympathetic objections of the lone dissenting judge, who wrote about the "extensive studies" indicating that life begins at conception, and who brought his dissent to a stirring close by quoting Shakespeare's *Hamlet* on the sanctity of human life. (" 'What a piece of work is man!' " the judge wrote, just over his Respectfully Dissenting signoff. " 'In action how like an angel, in apprehension how like a god!' ") But the dissenting judge was only one out of three, and it made no difference how many more Missouri juries might have agreed with him; the Court of Appeals had effectively forbidden defense lawyers even from presenting the argument in court.

Planned defiance—deliberately breaking one of the injunctions, even as the lawyers warned them not to do it—might now be the only option the protestors had left. It was that or walk away, give up the interventions altogether, and even Sam could see that they had come too far for that. Civil disobedience had its consequences, Sam had understood that the first time he volunteered for arrest, and there was some deep cowardice in backing off the minute the consequences began to look real. Joan Andrews and John Ryan argued that the movement needed their defiance anyway; perhaps the threat of long jail sentences would attract people's attention, bring the support crowds back. Joan was so enthusiastic about the idea that she promised to cut short her trip to the East Coast, where she had some door-blocking sit-ins planned for Delaware and Maryland, to come back to St. Louis and defy the judiciary. "I thought that breaking the injunction would be the only way to ignite the movement," Andrews was to write in her autobiography some years later. "I thought we should show that we had only contempt for such abuses of law, such blatant tyranny as those injunctions."

But Sam had something else on his mind too, and it was complicating every decision he made—pushing him toward heroism and making him wonder in an entirely new way whether this particular kind of heroism was really worth the risk. Sam Lee was twenty-five years old in the summer of 1982, a lapsing pre-seminarian with dubious career prospects and one brown suit purchased solely so that he could wear it to his own arraignments, and for the first time in his life he had fallen in love.

SAM LEE met Gloria Fahey on a blind date arranged by a friend who grew tired of trying to convince Sam to go introduce himself to Gloria on his own. This is a wonderful woman, the friend kept saying; she's intelligent, she's soft-spoken, she has a sense of humor, she's pro-life, you'll find a lot to talk about. Gloria had a master's degree in counseling and a job with the Social Security department, but after work and on weekends she volunteered at Our Lady's Inn, the women's residential facility that Loretto Wagner had helped set up inside an abandoned convent. Everybody was proud of the facility, which was remodeled and furnished by volunteers and offered shelter to poor women who wanted to give birth to their babies, but Sam never did barge in there looking for Gloria the way his friend kept urging him to. Sam was going to be a priest someday; he was less and less certain how he was ever going to settle back into his university classes to prepare for the seminary, but he had never left off believing that he was meant to serve God fully, in a priestly way, without the distractions of romance and family. Around St. Louis he had friends who were women; he'd blocked clinic doorways with women and gone to jail with women, but no one really felt like a *girlfriend*, someone to make him lose his bearings.

Gloria Fahey made Sam lose his bearings. On their blind date, tagging along awkwardly with their friend and his date, they went to a movie and then sat in a nearby popcorn-and-peanuts bar afterward to talk. The movie was *Chariots of Fire,* in which a Christian athlete misses an Olympic race because he refuses to compete on the Sabbath, so they were able to make interesting conversation about religious faith and moral principle. Gloria had a wry half-smile and there was something quiet and powerful about her, Sam couldn't stop thinking about it afterward; she was third-generation St. Louis Catholic and it was as though they had arrived in each other's company by some arrangement grander and more momentous than a set-up double date. Like Sam, Gloria had been caught up in charismatic Catholicism; like Sam, she had lived for a while amid Franciscans; like Sam, she worried deeply about the spiritual and physical toll that legal abortion appeared to be taking—upon the mothers, Gloria remarked, as well as the aborted babies.

A week later Sam drove down to Our Lady's Inn and asked Gloria to come have dinner with him. After that they went on a picnic in the woods, and Gloria showed Sam the working-class parish street where most of her family still lived, and Sam scraped together enough money to buy a great bouquet of helium balloons and hand it to Gloria right at the front door of Our Lady's Inn, with the young pregnant women watching from the windows and whistling and applauding until Gloria blushed furiously and took the balloons inside. Sam showed up at midday at Gloria's Social Security office, too, driving the red van that belonged to the mail-order company whose books he sold part time. When Gloria walked out of the building with him she saw that Sam had unloaded a portable picnic table into the middle of the Social Security parking lot. Over the table he had spread a red checkered tablecloth, with plates of fruit and good delicatessen food and a small cassette recorder that began playing accordion music, the kind Gloria imagined you might hear at a Paris bistro, as soon as Sam reached over and pushed the right button. "Are you ready for lunch?" Sam asked.

They were engaged on the Fourth of July. There was no dramatic down-on-one-knee tableau for Sam and Gloria to reminisce about afterward; they just realized one day that they were talking about children and values and the shape of a life together. Gently, irretrievably, the priesthood vanished from the imaginable future of Samuel Lee, and he felt only the smallest twinge of regret at watching it go; he was happy in a rich and private way that was utterly novel to him, and at the house where he was living, Lisa Puzder could see that Sam suddenly looked as though someone were shining a brilliant light on his face all the time. When he brought Gloria home to meet Lisa and Andy, Sam tried to make a flourishing presentation of it, telling Gloria (a little insensitively, in retrospect, Sam was new at this) how extraordinary Lisa's cooking was, how much Gloria was going to love whatever Lisa whipped up. That night Andy had to leave town on work and Lisa, not expecting company, served frozen fish sticks and Campbell's soup. She and Gloria looked at each other over the fish sticks, and when Lisa said, "What do you think, Gloria, aren't these *yummy*," and Gloria burst out laughing, Lisa knew it was going to be all right, that Sam really had found a partner.

The wedding was planned for the following summer. They would marry in the church in which Gloria Fahey had gone to Mass every morning of her childhood. Gloria's parents would be there, and her five brothers and sisters, and the assorted children of her five brothers and sisters, and the grandmother whose house the Faheys had lived in when Gloria was small (four children in one bedroom, the grandmother upstairs in the other, the parents on the foldout couch in the living room). It was a thick and powerfully interconnected St. Louis family that Sam was marrying into, and when he argued with himself now about whether to break one of the clinic injunctions, the weight of new responsibility bewildered him. Sam and Gloria had spent many hours talking over the moral imperative of the sit-ins; Gloria had never been arrested herself, but she had lettered signs of her own—ABORTION STILLS A HUMAN HEART, that sort of thing—and gone out on Saturdays to hold her sign quietly and watch the demonstrations from the sidewalk. It grieved her to stand there, all those women hurrying past them and right into the clinic, and after a while she had stopped showing up at the sit-ins, feeling that she was happier and more useful at Our Lady's Inn. But Gloria understood Sam's idea about nonviolent blockades, and she respected him for it. Sam should follow his own conscience, Gloria told him; it was not Sam's civil disobedience that made her wonder on occasion whether she was making the right decision by agreeing to marry him.

The argument—it was a serious one, and it escalated after they were formally engaged—was instead about how they ought to live once they began a family of their own. Sam had very pronounced ideas about this, now that he was contemplating marriage and children for the first time in his adult life. He believed mothers should stay home with their children, he told Gloria; he did not want Gloria working at all after their first child was born, and eventually Sam thought he would like to move to the country and live in a self-sufficient rural Christian community something like the one he had visited in upstate New York when he was a teenager. He had a picture in his mind about how peaceful this life would be, Sam would say, and how healthy for the children: a vegetable garden, a simple wood house, Gloria baking bread while the children played outside on a tire swing under the apple tree.

Gloria had heard Sam's stories about that upstate rural Christian community, where the women were expected to wear ankle-length skirts as they went about their dawn-to-dusk farm chores, and she told Sam she had no intention of living that way. She was a twentieth-century urban Catholic woman, Gloria said, and she planned to go on behaving like one after she was married; she wanted to live in the city, close to her family and friends and church. And she did believe raising children was the most satisfying and important work a woman could do, she wanted to stay home for at least a while after each baby was born, but she was not going to promise Sam that she would never go back to work; Gloria had worked hard earning her master's degree and she intended to continue putting it to use. Besides, where was Sam in this rosy picture of life in the country? Using the family car to drive back and forth to clinic sit-ins in St. Louis? Had he figured in the monthly payments on the house with the apple tree? Did he imagine that they were going to live on no income at all?

Sam knew he did not have the details worked out, and he was shaken by Gloria's adamance as they talked; he had assumed the two of them would share by God's grace a single vision of a household in which father and mother dedicated themselves—in some traditional fashion Sam had supposed would come into focus as time went on—to the raising of children. We'll have eight or ten children, Sam would say merrily, half teasing but half not, and Gloria would say, "Yeah, *right*," in a voice that told Sam to back off, and then she would smile a little and say, Three or four, Sam, maybe five. But she dug in her heels about staying at home full time for the next fifteen years. This was not the life she wanted to live, Gloria told Sam, and if Sam would take a good look at himself and his current earning capacity he would see that this was not the life *he* wanted to live, either. Sam was going to have to make some large decisions about what he planned to do and how he planned to do it: a traditional stay-at-home wife was not really a luxury a civil disobedient could afford.

Finally they agreed to wait, to give it time and see what happened once they were married the following year. In August 1982, Sam drove Gloria to Chicago, troubled by their disagreements, still not knowing what he was going to do about violating the injunctions at the clinics.

They were separating for the autumn, as they had anticipated they would; since long before they met, Gloria had been saving money for an ambitious vacation across Europe, her first trip out of the United States. She wanted to be in Rome for the two hundredth anniversary of the death of St. Francis of Assisi, and she and a friend had planned to low-budget backpack together through every country they could. Gloria promised to write to Sam every day, to tell him everything she had seen and keep talking to him in her letters about her ideas on marriage and children, and they made out a list of American Express offices where Sam could send all the letters *he* planned to write. Maybe they could talk by telephone once in a while, Sam said, and he would send her cassette tapes too, so that he could talk to her in a more relaxed way, tell her how much he was missing her.

Sam and Gloria spent three days with friends in Chicago, and then Sam left Gloria off at O'Hare Airport, feeling wistful. While they were gone, an abortion doctor outside St. Louis was kidnapped and locked into an abandoned ammunition bunker for eight days straight.

THE ABDUCTION of Dr. Hector Zevallos took place in a southern Illinois town called Edwardsville, just across the Mississippi River and twenty miles northeast of St. Louis. Zevallos ran the Hope Clinic for Women, which was near Edwardsville in Granite City, also on the Illinois side of the river; the Hope Clinic had its own set of demonstrators, and when the sheriff's department and the FBI began investigating, they worked both lists—Illinois names and Missouri names—until the summons from the authorities began to take on a grim cachet. Who was important enough to be questioned as a possible abortionist snatcher? Zevallos and his wife had disappeared from their house on a Friday morning in August, leaving the lights on and both cars in the garage; a few days later a long ransom letter surfaced, warning of the deceits of Satan and threatening to execute

Zevallos unless President Reagan "put an end to legalized abortion in 48 hours, even if need be that you put everyone who supports unborn baby killing to death."

The ransom letter's authors identified themselves as "the Army of God," and all over St. Louis telephones in right-to-life households began ringing the morning excerpts from the letter appeared in the newspapers. Nobody had ever heard of the Army of God, not the sit-in protestors, not the Missouri Citizens for Life stalwarts, not the lecturers and organizers who traveled out of state; the story went that when Ann O'Donnell got her call from the FBI, she found the entire scenario so preposterous—the notion that serious right-to-life activists would deliberately destroy their own credibility by masterminding something like this—that she held the receiver a foot away and yelled off to one side, as though her husband had just walked into the room, "Hey, Eddie, you got those tacos for the people in the basement?" Zevallos had been kidnapped by a crazy person, anybody could see that. The membership of Missouri Citizens for Life was not responsible for every crazy person in the United Sates. But the headlines and television stories managed to lump them together as though they all sat around chatting in their weekly planning meetings: here's the lobbying committee, here's the protest committee, here's the kidnap committee.

Sam Lee was interviewed by the FBI in an office building in downtown St. Louis and came away feeling queasy and afraid, knowing he was probably off the suspect list—they seemed to believe him when he explained that he had been in Chicago the day of the abduction—but wondering privately whether some civil disobedient from St. Louis or Granite City had taken it upon himself to dispense with nonviolence and move on to the next step. If that was true, they were finished, the sit-in movement was done for, Sam had lost his fragile hold on the rest of them, and when outsiders called them terrorists or fanatics the outsiders would finally be right. Even after it was over—after the Zevalloses had been let out of a car near their house, and the FBI had announced that the doctor and his wife were physically unhurt but had suffered an "emotional ordeal," and warrants were issued for a trio of Texas men who appeared to have no connection to any right-to-life group Sam knew*—Sam could hear, in the

---

* The chief suspect, a forty-two-year-old former businessman named Don Benny Anderson, who in his Army of God letters had also taken credit for abortion clinic firebombings in Virginia and Florida, was convicted in January 1983 and sentenced to forty-two years in federal prison for conspiracy and extortion in connection with the Zevallos kidnappings. Two young men, eighteen- and twenty-year-old brothers whose family had known Anderson's in Texas, were convicted of helping with the abduction and received sentences of under ten years each. In later years other right-to-life and radical right-wing activists would also use the name "Army of God," but the lead federal prosecutor in the Zevallos kidnappings says Anderson and his two younger associates appeared to have no formal connection to any other group or individuals.

late-night arguments of his friends and fellow protestors, the familiar and perilous ricocheting back and forth between morality and pragmatism, between service to higher principles and daily effectiveness here in the world.

What this so-called Army of God had done was wrong; everybody agreed on that. But what exactly was wrong about it? Was it wrong because it was stupid, because it would have no effect on abortion, because the headlines had hurt them all? Was it wrong because Dr. Zevallos deserved to be treated like a human being, even if he refused to extend that basic dignity to the unborn in his own clinic? What if he had not been blindfolded into a bunker and fed Fritos and peanut butter, but had instead been locked up someplace comfortable and forced to read a hundred essays in defense of the unborn? What was the definition of "violence" when innocent human lives were at stake? Where *was* this critical line that the nonviolent abortion protestor was never supposed to cross?

The clinic injunctions were still in place. Sam wrote to Gloria, long emotional letters, his pages filled with small and carefully punctuated printing. "We could be found in contempt of court and the judge might get very nasty," he wrote as the summer ended. "Please pray for wisdom for me in these matters. The obligation is upon me primarily to decide what is to be done, and thus I am responsible for many people, both born and unborn."

Two WEEKS into September, Senator Orrin Hatch's abortion amendment and Senator Jesse Helms' abortion bill collapsed, nearly simultaneously, in Washington. The Hatch Amendment appeared to lack the votes for passage, Senator Hatch announced, and thus was now formally withdrawn from Senate discussion for the rest of the congressional year. The Helms bill was "tabled," removing it from further consideration, by a 47–46 Senate vote that followed three failed attempts to break an opposition filibuster.

"The prospects are bleak," Sam wrote in a dispirited letter that reached Gloria Fahey in Ireland:

> Both pieces of legislation would have protected unborn children if they had been passed. After the 1980 elections, we have had the strongest pro-life representation in the Senate and Congress since 1973. Many thought that 1982 was *the* year for pro-life. But with the defeat of this legislation, and the 1982 elections just over a month away. . . .
>
> During an off-year election like 1982, when the party not represented in the presidency (that is the Democrats) normally gains an increase of congressmen, we can expect it to lose some pro-lifers and gain

some pro-abortionists in Washington. Thus, some people are saying we should concentrate more on local affairs concerning the abortion issue.

The Zevallos kidnap had been a devastating blow, Sam wrote, and had left the media "looking for the next violent move on our part":

> Considering all that pro-life folks have done—opened Our Lady's Inn, taken women into their homes, Birthright, confronted the sexual suicide in today's society—what the movement is being accused of seems absurd. But alas, the man in the street tends to accept the lies of others if told over and over again that lies are the truth. . . .
>
> The Gospels, and adherents of non-violent principles for social change like Gandhi, King, and Chavez, all point to the importance of personal suffering to bring salvation and change. We clearly see this in Jesus' death on the Cross, and the indignities and beatings the blacks and whites suffered in the civil rights movement. . . . When people see other people just like themselves being arrested, taken to court, fined and some-times jailed, they must consider whether unborn children are really humans. We only make sacrifices for things of real value, and most people recognize this fact. If we make large, even heroic sacrifices to stop abortion, then folks are going to believe that abortion should be stopped.

Standing on a street corner outside the Dublin American Express office, Sam's twelve-page letter refolded into her backpack, Gloria Fahey understood what her fiance was trying to prepare her for. The lawyers' warnings were not enough, the threat of a full year's jail sentence was not enough, it was not enough that Sam and Gloria had already planned out their wedding day and what were supposed to have been the first happy months of married life together. He had decided to break the injunctions.

PART THREE

The Seventh Question

ELEVEN

The Litmus Test

1986

ON A JANUARY MORNING in 1986, some days before the annual Jefferson City convening of the Missouri state legislature, a lobbyist named Mary Bryant climbed three flights of stairs inside the capitol building and proceeded down the hallway toward the office of the chief clerk for the House of Representatives. Mary could hear her own heels clicking against the marble floors; the corridors were empty, the great House and Senate chambers silent, and soft yellow light shone through the mottled-glass window in each closed office door. The doors were mahogany, worn and dark. Burnished brass gleamed from the knobs and the hinges, the drinking fountains were made of brass and Carthage marble, and between the wrought-iron light fixtures overhead were round hallway clocks that hung suspended from the ceiling by double chains, like fine men's pocketwatches.

Mary Bryant was about to commence her fourth legislative session as the official lobbyist for the Planned Parenthood Affiliates of Missouri, Inc., and during the month before every session's start-up she made this circuit once a week: office of the House clerk, office of the Senate secretary, and then back to her own modest office—two rented rooms, grandly identified on her business cards as "Suite B," over a restaurant near the statehouse. Mary liked the solemnity of the capitol before the legislators arrived; most of Jefferson City is dull and flat, spread gracelessly out along the banks of the Missouri River, but the statehouse is a massive neoclassical wonder, and in the weeks after Christmas a person walking the length of the third-floor corridors could study without distraction the murals and the sculptures and the stained-glass windows of the high central rotunda. Brief phrases of civic inspiration were chiseled into the granite walls. NOT TO BE SERVED BUT TO SERVE. NOTHING IS POLITICALLY RIGHT THAT IS MORALLY WRONG. AS THE STRUCTURE OF A GOVERNMENT GIVES FORCE TO PUBLIC OPINION—this last epigram stretched all the way around the rotunda walls, so that some fragment of its uplifting admonishment was legible from every vantage—IT IS ESSENTIAL THAT PUBLIC OPINION SHOULD BE ENLIGHTENED IN PROPORTION.

Mary Bryant pushed open the door to Room 307, exchanged morning pleasantries with the women typing side by side at their close-quarters desks, and set down her briefcase to look through the bill bin. There was no real organization to this bin, which was in fact just a cardboard box shoved into a corner by the doorway; every few days the stapled copies of new, early-filed House bills piled up inside the box, and Mary had learned to sift quickly through the piles with an eye for the words or phrases that might signal potential trouble for Planned Parenthood. *Relating to grain dealers and warehouses.* No. *Relating to the practice of chiropractic.* No. *Relating to the use of public funds*—well, maybe. Mary straightened up, stapled bill in hand, and began to read.

A LONG TIME LATER, after the fifteenth or twentieth reporter had asked her to describe her feelings at the moment of this reading, Mary Bryant would find that she was developing a stock answer. "Sick," she would say. "Sick, and cold." Sometimes she would add: "Incredulous." There was a feeling at the back of her neck, also, the flushing sensation that pushes up when a terrible thing has happened and it is somehow your fault—even though this was not true, it was not her fault. But she hadn't expected it. And when she saw what was being proposed, she guessed at once that a coordinated attack was under way, that there was more waiting for her on the Senate side of the statehouse.

HOUSE BILL NO. 1596

Relating to the use of public funds.

BE IT ENACTED BY THE GENERAL ASSEMBLY OF THE STATE OF MISSOURI, AS FOLLOWS:

Section 1. As used in sections 1 to 4 of this act, the following terms mean:

(1) "Public employee," any person employed or compensated by this state or any agency or subdivision thereof;

(2) "Public funds," any funds received or controlled by this state or any agency or political subdivision thereof, including, but not limited to, funds derived from federal, state or local taxes, gifts or grants from any source, public or private, federal grants or payments, or intergovernmental transfers.

Section 2. It shall be unlawful for any public funds to be expended for the purpose of performing or assisting an abortion, not necessary to save the life of the mother, or advocating or counseling in favor of an abortion.

Section 3. It shall be unlawful for any public employee within the scope of his employment to perform or assist an abortion, not necessary to save the life of the mother, or to advocate or counsel in favor of an abortion.

Unlawful for *any* public funds to be expended? Unlawful to *counsel?* What could these people be thinking of? Every state-salaried nurse or physician who believed in the right to elective abortion was now to be officially silenced in front of patients—or commanded to tell lies instead?

Mary jammed a copy of House Bill 1596 into her briefcase and hurried down the corridor to the offices of the Senate. House bills and Senate bills were supposed to be filed independently, each bill carrying on paper the illusion of authorship by its lead sponsor, but Mary had been in the statehouse long enough to recognize instantly the handiwork of Missouri Citizens for Life and the Missouri Catholic Conference. Representative Jim Barnes, the Kansas City Democrat whose name led off the sponsors of House Bill 1596, was not known to be particularly interested in abortion; Mary knew Barnes cared intensely about adoption, something about Barnes having been adopted himself, but if his was the primary name on an abortion bill, it meant somebody else had written the bill out line by line and persuaded Barnes to carry it.

And if such a bill had reached the House, there was surely a Senate version, too—and here it was, Senate Bill 766, typed and stapled in the files of the Senate bill room. *Relating to certain uses of public funds, facilities, or employees.*

S.B. 766 was wordier than H.B. 1596, two double-spaced pages instead of one, but its central mandate was identical—"no public employee within the scope of his employment," etcetera etcetera, ". . . nor shall any public facility be used directly or indirectly for advocating or counseling in favor of an abortion." What was *indirect* use of a public facility, for God's sake? Mary scanned the printed list of early-filed Senate bills, looking for the next bad news, and when she found it she kept her voice level as she asked the clerks if they would kindly copy for her this one additional bill. You do not let them see you sweat, Mary had been working for three years to teach herself that; in the capitol you are always on public display and you never slap your forehead or curse aloud or burst into tears where somebody else might see. Businesslike. "S.B. 728, please." *Relating to unborn children and abortion.*

S.B. 728 was sixteen pages long.

Its opening paragraph repealed eight sections of the Missouri criminal code. Its second paragraph began the language with which those sections were to be replaced:

> 1.205. 1. The general assembly of this state finds that;
> (1) The life of each human being begins at conception;
> (2) Unborn children have protectable interests in life, health, and well-being;
> (3) The natural parents of unborn children have protectable interests in the life, health, and well-being of their unborn child.
> 2. The laws of this state shall be interpreted and construed to acknowledge on behalf of the unborn child at every stage of development from conception until capable of independent existence, all the rights, privileges, and immunities available to other persons, citizens, and residents of this state, subject only to the Constitution of the United States, and decisional interpretations thereof by the United States Supreme Court and specific provisions to the contrary in the statutes and constitution of this state.
> 3. As used in this section, the term "unborn children" or "unborn child" shall include all unborn children or the offspring of human beings from the moment of conception until birth at every stage of biological development.

Mary Bryant put S.B. 728 into her briefcase with the others and left the capitol to read all sixteen pages in the privacy of her office. The opening paragraphs made her so angry that she had to look away for a moment to compose herself. Fertilized eggs—particles so small that if they lay in petri dishes they would be unrecognizable without the aid of a microscope —were now supposed to be granted citizenship. "Rights, privileges, and immunities" for *eggs*. It would have been ludicrous if it had not been typed onto legal paper with a dozen senators' signatures on the cover page. The top signature on the list read "Schneider," in bold scrawl, which suggested to Mary that S.B. 728 was being introduced with a lead sponsor who was entirely serious about pushing this legislation through the General Assembly: John Schneider was a Catholic Democrat from St. Louis, a liberal vote on many social issues, but so vehemently opposed to abortion that he might have written S.B. 728 himself.

Or perhaps Schneider had written only some of it, Mary guessed, as she made her way page by page through the muddy subsections of S.B. 728; perhaps he had gathered up a collection of random antiabortion ideas and numbered them into a single bill. On its face, when she reread it more carefully, the bill's opening passage made no sense. "Unborn children" were to be granted the "right to life," evidently the right to be treated as American citizens,

> to the full extent permitted by the Constitution of the United States, decisions of the United States Supreme Court, and federal statutes.

In other words the state of Missouri was simultaneously proclaiming this right to life, according to S.B. 728, and acknowledging that abortion was legal anyway. What exactly was the point here? Was the entire passage purely for show? Was it nothing more than eye-catching legal bait—an invitation to one more round of litigation over Missouri abortion law?

The rest of the bill was only marginally more coherent. There was some language requiring that all abortions after sixteen weeks be performed in hospitals—very nearly the same language already found unconstitutional by federal courts. There was a requirement that every minor obtain written parental consent or a court order before receiving an abortion. There was a long passage making it unlawful for employers, labor organizations, or educational institutions to penalize any individual for refusing to "participate in abortion," a term that was defined at such voluminous length that S.B. 728 made it sound as though a network of passive resistance was now to be formally protected by the state: a telephone workman could refuse to repair the lines at an abortion clinic; a bank teller could refuse to accept the clinic's deposits; a hotel clerk could refuse to register a visiting abortion doctor.

And what was this near the end, as though slipped in amid the fine print? "Section C, 188.017. 2: A person who performs or induces an abortion on another not necessary to save the life of the mother, which results in the death of the unborn child, or which results in the death of the child aborted alive, shall be guilty of a homicide offense and, upon conviction, shall be punished as provided by law." Had Schneider decided as an afterthought to use his little Senate bill to try recriminalizing abortion in the state of Missouri? No, here was the explanation, two pages on, down in the bill's final paragraphs:

> The provisions of section C shall become effective upon a decision of the United States Supreme Court, ratification of an amendment to the United States Constitution, or the enactment of a federal statute:
>
> 1) Returning to the states the right to regulate abortion as is prescribed in section 188.017; or
>
> 2) Recognizing that the right to life set forth in Section 1, Amendment XIV of the United States extends to unborn children.

So they were planning ahead, right here in their Senate bill, readying themselves in print for the Supreme Court decision or the constitutional amendment that would give states the authority to make abortion illegal again. For a moment Mary Bryant felt a giddy wave of relief, reading one more time the language of the bills that now promised to consume her 1986 session in the Missouri state capitol: ". . . at every stage of development from conception until capable of independent existence." Surely this time they had overdone it, Mary thought; surely this time the Antis had come up with legislation so offensive that even the Missouri General Assembly would turn them away. "I felt: *They can't get away with this,*" Mary Bryant recalls. "*People will see how bad this is. The good people will finally rise up. And they will stamp this thing out.*"

Mary picked up the telephone to call Kansas City and St. Louis. She was going to need a lot of help.

DURING HER FIRST MONTHS as a Planned Parenthood lobbyist, Mary Bryant had walked the statehouse corridors feeling that she carried some communicable disease and everyone in the capitol knew it. "Lobbyists would scatter, literally," she recalls. "I could walk down a hall, or walk into the library, and people would be busy, or they would find a way to excuse themselves. And I'm talking about male lobbyists who I knew *shared* my views. But they wanted nothing to do with the issue. There were times when I felt like the Antichrist, I think—there would be priests in the halls, and clergy, and the Catholic Conference counsel Lou DeFeo—and there would be the evangelicals and Pentecostals that came in at various times. It was a devastating experience."

Mary Bryant had gone to work for Planned Parenthood in the fall of 1982, not as a lobbyist, but as an administrative assistant—someone to run the small office in Jefferson City while the director of public affairs attended to the statehouse lobbying and press releases. The office work seemed entirely suitable to Mary, who had been restless in her previous job as a departmental administrator at a local college; she had never campaigned publicly for legal abortion or birth control, but she was forty-two, raised in the conservative surroundings of southwest Missouri, and she still remembered the stories from her teenage years about girls who had vanished long enough to undergo something secret and dreadful, illegal abortion or illegitimate birth. Mary had three children and had divorced twice before marrying the man who was now her husband, so she had at intervals been a single mother herself, and at one point she had worked as an alcoholism counselor and spent long afternoons in an office out by the sorry little trailers and cabins around the Lake of the Ozarks, listening to beaten-down women who appeared to have had no chance at a life of their own. Mary thought the word "choice" was precisely the right one for the feminists and NARAL activists to have written onto their banners, and she was pleased that Planned Parenthood had hired her, pleased to be working for an organization with what seemed to be such admirable goals.

Then two weeks after Mary began her Planned Parenthood job, the public affairs director quit. The legislative session was about to start up, and Missouri Planned Parenthood wanted its lobbyist; around the country Planned Parenthood had been steadily increasing its visibility as a public defender of legal abortion, and in Missouri the state board of directors had decided to follow other states' example, setting up an office next to the capitol so that at least one person on the payroll could learn the nuances of statehouse politics. Mary Bryant had never in her life worked a committee hearing or deciphered a legislative calendar, but she offered to do what she could, and when the General Assembly convened for its 1983 session Mary set off for her inaugural season as the Antichrist. She was an extremely presentable-looking woman, attractive, trim, her graying hair cut short and wavy above her good dress blazers, and when she introduced herself to legislators or their aides she would watch the faces change expression the instant Mary identified her employer.

"You have to understand that the words 'Planned Parenthood' were so highly charged that there was this incredible silence after I would say that," Mary Bryant recalls. "Incredible silence. And then: 'Well. You don't need to talk to me. I'm pro-life.' No one was ever hateful. I think they simply dismissed me. *Very* uninterested that I was there."

For one season Planned Parenthood hired a professional lobbyist to work alongside Mary in the capitol, but after that she was on her own, and she found that when she set out for the statehouse she always brought along her briefcase, an authoritative-looking oxblood leather satchel that she would fill with printed reports and pending legislation until the weight of the loaded case strained her shoulder when she walked. Mary thought of her briefcase as a talisman and she carried it everywhere, imagining that something about its bulk and sobriety might make people regard her seriously, but after a while she saw that she could have driven an Army tank through the halls and it would have made no difference; they would still look right through her and go busily about their work. She had arrived a decade too late. Missouri Citizens for Life and the Missouri Catholic Conference had turned the statehouse into a closed shop. The St. Louis delegation was in thrall to the Church; the rural representatives came from conservative farming communities; the southern half of the state was ceded broadly to the Southern Baptists and the evangelical Assemblies of God. There were brave little pockets of enlightened thinking in Kansas City and the university town of Columbia, and once in a while the occasional renegade would rise from St. Louis and shrug off the insistent urgings of the Antis; but every time an abortion vote was tallied the numbers were so comically lopsided that a press corps wag once described Missouri's hapless band of pro-choice legislators, the men and women who could be relied upon to cast a public vote in defense of legal abortion, as the Kamikaze Squad.

The name stuck. There are 197 legislators in the Missouri House and Senate, and on a good day—a *good* day, when nobody was out sick or hiding in a back office until the troublesome vote was over—Mary could count on twenty-two, maybe twenty-three votes. They were capable speechmakers, some of these Kamikazes; Representative Karen McCarthy, the young liberal Democrat from Kansas City, always got to her feet to give her speech about privacy, the state's obligations to poor women, the primacy of the doctor-patient relationship. Karen McCarthy would stand at the microphone and push on gamely while the legislators yawned and wandered out of the chamber to get coffee or return their telephone calls; there were some lively office backgammon games during the abortion speeches, Mary knew that too, and when it was time for the votes the boys would file back in and press whichever button the Antis had directed them to, Yes or No. It didn't much matter either way, as long as it placated the busloads from the parishes and fell in line with everybody else. " 'Wayull, you *know* how it is, Mary, the votes are *there,* girl, cain't you *count?*' " (A decade later Mary could still mimic the dismissive drawl of the men who would bother talking to her at all.) "They would say—this would make me so angry—'Get me a close margin and I'll vote with you. I'm not going to hang myself out there to dry when you've only got five votes.' "

This was the part that really galled Mary Bryant, as session by session she began to understand the intricacies of Missouri statehouse politics: most of the legislators didn't care about abortion at all. The Kamikazes cared; they were genuinely convinced that abortion ought to be legal and unencumbered by governmental restrictions, and at the opposite end of the spectrum was a second informal coalition, similarly modest in numbers, that might be thought of as the True Believers—legislators like John Schneider, who gave the impression of unimpeachable sincerity every time he orated about the sanctity of human life.

But the rest of them—a numerical majority, Mary was certain of it, even though this would have been nearly impossible to prove—wished nothing more than to have the matter go away and leave them alone. They didn't want to talk about it, they didn't want to read about it, they didn't want their legislative calendar kept hostage while the Karen McCarthys and John Schneiders held forth on the House and Senate floors. But they didn't want their names turning up in the wrong column of a Missouri Citizens for Life windshield flyer, either, and even Mary Bryant could see how straightforward the voting decision must be once you dispensed with the idea that convictions or moral principle had anything to do with it. "Litmus test" was a term Mary had heard more than once, referring to the ritual of the annual abortion vote; she would corner some pleasant legislator from the suburbs, an intelligent fellow, a reasonable prospect, and he would listen to her arguments and nod thoughtfully, agreeing with her, such a courteous man. "And I would talk to him and talk to him and talk to him. And he'd be: 'Yeah. Yeah. Yeah.' Things like: 'This is a really difficult issue.' And: 'I want women to have services.' Then he'd vote with *them.*"

In her sessions at the legislature Mary learned a great deal about lobbying by sleight of hand. She came to think of a good day's work as running her traps, an expression she had heard around the statehouse; Missouri was a big hunting state and Mary assumed she was metaphorically pacing from one spring-loaded device to the next, looking for bodies and stolen bait and visual cues recognizable only to the expert: "You have different levels of contacts. Maybe when you first hit the capitol in the morning—if you're smart you get there a little before eight, and you have coffee in the cafeteria, and that's when you can catch legislators when they have their least amount of armor on. They're half awake, they're not really into it yet, and you can sort of graze and schmooze and pick up the mood of the capitol. The capitol has different moods on different days. On some days everybody's just mean as hell, you know going in that it's going to be one of those days. So you go into a couple of hearing rooms to see what lobbyists are out, what's really happening that day.

"Then you run your operatives, other lobbyists, and secretaries. Secretaries are absolutely essential, because they watch the comings and goings. And sometimes lobbying is—well, some of the greatest, most stupendous scoops or breaks I ever had were overheard conversations, in elevators. Because after you're there a while you get to know all the faces, and all the nuances, and you can see two people walking across the rotunda, and you know from the pace they're walking, or how close they're walking, whether they're exchanging information or intend to exchange information when they go around the corner. Body posture. I've done it myself. I'd see someone coming, and we'd know we've got to talk, but you cannot be seen exchanging information. Say for instance you're attending a hearing. There's a bill you've sort of been watching, and you're sitting in the back of the room. A staff person gets up and walks out into the hall and looks back to where you are. It's just the look. And you go out and you get a drink, go down to the water fountain or walk to the Coke machine. And sometimes there's a conversation, maybe there's a word or two, or it's just body language: *There's trouble here, and it's brewing.*"

In mid-January 1986 the legislative session started up in earnest, and Mary Bryant set out at once to begin running her traps. She met with the Kamikaze Squad, singly and in conspiratorial groups. She rounded up witnesses willing to testify at House and Senate hearings. She typed multipage statements, each objection highlighted for quick review, and distributed them up and down the hallways to the legislative staff: THIS BILL IS AN ATTEMPT TO LEGALLY RESTRICT AND ENCOURAGE HARASSMENT OF PROVIDERS OF REPRODUCTIVE HEALTH CARE AND EDUCATION—AND THOSE WHO SEEK THESE SERVICES.

Subsection by subsection, Mary laid out what seemed to her the most dramatic hazards in the House and Senate bills, starting with the notion of "protectable interests in life, health, and well-being" from fertilization on. "Imposes one religious and philosophical viewpoint on all Missourians," Mary wrote. "Could change legislative district boundaries based on population."

And: "Definition of 'public funds' is so broad it could possibly prohibit abortions in all Missouri hospitals, public and private."

And: "Denies physicians and other health professionals First Amendment rights to counsel patients concerning the option of abortion and frustrates their ability to adhere to standard medical practices—thereby subjecting them to malpractice liability."

And: "Is AFDC paid to fetuses? Are taxpayers allowed deductions for fetuses? How is date of fertilization proven?"

It made no difference; it didn't matter what she wrote. Mary had a full-time statehouse assistant working with her by now, a young woman she had hired into the Jefferson City Planned Parenthood office, but the two of them were not even slowing the process down. The three bills merged under the heading of House Bill 1596 and Mary was stunned by the swiftness of the passage through committee and toward a full floor vote; it was as though someone had latched a diesel engine to H.B. 1596 and then leaped aside to get out of the way. Mary would sit down with legislators and indicate this sentence or that, pointing out that H.B. 1596 was unconstitutional, that they were preparing to vote for legislation that violated the United States Constitution, and the legislators would smile at her and nod their heads until she began to understand that *that was why they were voting for it,* they *knew* it was unconstitutional, they knew Frank Susman would show up with a court challenge the day they signed off on it. " 'Now, Mary, this just isn't going to go anywhere. It's going to pass, but the courts are going to throw it out.' "

Some of them would sort of wink at her as they said this: Simmer down, Mary. It's just the litmus test.

She wanted to shout at them. At the end of the day she drove home exhausted and furious and thought about what would happen when the courts did receive the legal challenge. Would they overturn H.B. 1596 the way they had overturned so much antichoice legislation already? What if they didn't? What if the Antis had something grander in mind, another push all the way to the Supreme Court? Mary had spent enough time around antiabortion lobbyists to know that they were wholly serious about their mission in the statehouse; there was Lou DeFeo, the humorless white-haired Missouri Catholic Conference lawyer who fought Mary on everything she did, and there was the bearded younger man, Sam Lee, who worked for Missouri Citizens for Life. Actually Mary rather liked Sam, who seemed somehow to radiate integrity as he walked quietly through the statehouse halls; she knew he had served a very long jail sentence for blocking clinic doorways, but she had read in the newspapers about his pacifist philosophy, and when she watched him work she found this easy to believe, that Sam was a genuine pacifist. All that language in the legislation about state-protected passive resistance to abortion—that sounded like Sam.

But Mary had no illusions about Sam Lee's broader intentions; if he had the power to do so he would outlaw all abortion, and in Anti legislation that was always the distant target, the end of abortion as a legal alternative for women. Why was she having such a hard time making people understand this? She convinced Frank Susman to drive to Jefferson City and testify; she pulled some Planned Parenthood colleagues in from Kansas City; she had some statehouse help from lobbyists who worked for the American Civil Liberties Union and the multidenominational organization called the Religious Coalition for Abortion Rights. But she had no busloads and she had no parking lot windshield campaigns and she had no home district volunteer troops, not the way the other side did. In 1986 the Speaker of the House was a politically astute Democrat named Bob Griffin, and when Mary tried to approach him for support Griffin was cordial, in his own gruff way, but uninterested in either her arguments or her polling data. "He says—I'm paraphrasing—'Goddamnit, don't show me polls,' " Mary Bryant recalls. " 'Show me warm bodies in a room.' "

In March Mary drove to Bob Griffin's district, up northeast of Kansas City, to find him some warm bodies in a room. Griffin lived in Cameron, a one-main-street town amid depressed farming country, and for a few days Mary went from one place to another looking for people to talk to about abortion and the importance of reproductive freedom. She talked to school guidance counselors; she visited the home of a local newspaper editor; she heard stories about bankruptcies and failing farms and thirteen-year-olds being ferried down to Kansas City for their abortions and their birth control pills. At night Mary stayed in a motel by the highway, listening to the March wind bang against the window, and when she had trouble sleeping she read the paperback she had brought with her, the Margaret Atwood novel *The Handmaid's Tale*, about a society in which women are used solely for sexual pleasure and the bearing of children. The motel-room heater worked badly and the room, like the book, gave her chills.

BOB GRIFFIN'S abortion epiphany had come to him in the hallway outside the basement community meeting room of the courthouse in Plattsburg, which was one of the county seats in the district Griffin represented. It was midsession 1985 when this had occurred, still late winter in northwest Missouri. The right-to-life people had a bill in the House, the way they always did, and Griffin, who as Speaker of the House had the power to stash legislation more or less wherever he wished, had stashed this bill in a committee run by one of the Kamikazes. The particular Kamikaze Griffin selected was known for his keen sense of justice, and Griffin had been of the opinion that under this gentleman's committee stewardship the right-to-life bill would probably vanish gracefully and without much public notice.

In later years Griffin would have difficulty distinguishing that 1985 right-to-life bill from all the others that had come before it. Maybe it was the bill refusing abortion insurance coverage for state employees, maybe it was the bill prohibiting abortions undertaken because the baby was the wrong sex—Griffin lost track, after a while; it pained him to have to yield good legislative time to abortion arguments, and for years he had voted the way the right-to-life organizations asked him to, automatically and without long soul-searching about the rights of women or the nature of embryonic life. The people with the roses lobbied him relentlessly, the other side never did, and it was only during the last session or two that a few legal abortion advocates had begun approaching Griffin, pleading for help. After that Griffin had tried to even the balance a little, quietly, by using his power as Speaker to slow up some of the antiabortion legislation.

These were the sort of moves that should have gone by smoothly, blurred amid the confusion of legislative procedure: the occasional bill ushered into an unreceptive committee, or stuck into the House calendar quite a long way down. But the right-to-life people got wind of it this time, the lobbyists must have spread the word, and Griffin was asked to come to a meeting in the Plattsburg courthouse to explain himself. He drove over from Cameron and walked into the courthouse building and then stopped near the top of the stairs, looking down at the hallway before him: it was crowded with people, men and women packed side by side and several deep along both walls, spillovers from the large, already filled community room inside. The faces were intensely unfriendly. Griffin was going to have to run a kind of gauntlet just to reach the space they had cleared for him at the front of the community room. It occurred to him to wonder about his own physical safety; Griffin had been a state representative for fifteen years and never before seen a constituent meeting this crowded or tense. Once when he was a county prosecutor people had packed angrily into the courthouse for a hearing on taxes, but Griffin had been the hero then, arguing on behalf of county taxpayers against a state-imposed tax increase. Now he was the enemy; he felt as though he had walked by accident into the wrong neighborhood, and nobody in the crowd appeared placated when Griffin reviewed for them his fine pro-life voting record. They wanted their *bill*, they wanted it moved from the hostile committee, and they had learned enough about legislative procedure to know exactly what the Speaker of the House was able to do to take care of it.

Griffin took care of it. When he got back to Jefferson City, he pulled the bill and sent it to another committee, where it was whisked straight through for approval on a floor vote.

Right-to-life bills were different from all the others: that was Bob Griffin's epiphany. They came with propulsion of their own. Other bills rose and fell with the passing of the session; there was bartering, dealmaking, private conversation that everybody understood was to remain within the walls of the capitol. But when the right-to-life people had a bill in their sights, it was as though a battery of invisible sentries had been dispatched to the statehouse to shove the bill along no matter what some faithless legislator might do. Jim Barnes never had much enthusiasm for H.B. 1596; Griffin knew that, Barnes would say it privately, legislators who didn't know Barnes very well would watch him shrug when they asked about his new abortion legislation. Barnes was a big athletic-looking man who had gone to a Jesuit college but maintained uneven relations with the Missouri Catholic Conference; he had been having trouble securing the Catholics' help on some adoption legislation he was carrying, a bill facilitating searches between adopted children and their birth parents, and when Lou DeFeo asked him to sponsor a one-page abortion bill during the 1986 session, Barnes—who was moderately interested in discouraging abortion, since he was gratified that his own birth mother had adopted him out instead of finding an abortionist—had seen the potential merits of doing the Catholic Conference a favor.

It was simply a small public funding bill, Barnes recalls DeFeo assuring him—a logical extension of earlier Supreme Court decisions that permitted states to stop paying for abortions. "It was just kind of a cleanup matter," Barnes recalls DeFeo telling him. "He says, 'Of course, we'll be sued. But we're going to make it because it's constitutional. We've already got precedent on it.' "

So Barnes agreed to carry DeFeo's bill. "I didn't know this thing had a nuclear reactor tied to it," he says.

H.B. 1596 swelled and changed and was augmented piece by piece as the legislative session proceeded, and every new development took Jim Barnes by surprise until he began to feel that the charade of his sponsorship was somewhat ridiculous. Hearings he had never organized seemed to materialize on their own; witnesses he had never called showed up to testify; passages he had never approved appeared on updated copies of the legislation until H.B. 1596 was suddenly the Big Abortion Bill and the newspapers were writing about it and everybody was arguing back and forth about what would happen when its constitutionality was tested in the courts.

Then Barnes began hearing rumors that a constitutional test was the *point*, that the various authors of the now-blended bill had actually set out to load this year's abortion legislation with provocative language, that the right-to-life people had managed to affix Jim Barnes' name to the next missile aimed at the Supreme Court.

Now Barnes was flat-out angry, or so he recalled years afterward—feeling used and misled. At one point he turned to a colleague and said, "You know, I feel like punching John Schneider in the mouth." Barnes held no great admiration for *Roe* v. *Wade;* he believed that abortion had become too casual since the Supreme Court legalized it everywhere, and that *Roe* had abruptly cut off what ought to have been a grave and ongoing state-by-state debate. But all the rhetoric wore him out. It wasn't his battle. Barnes had other business to attend to; he didn't think abortion belonged in the court system in the first place, and if he had known he was being invited to attach his name to a test case, he would have said no and passed the headache off to someone else. As it was, legislative decorum required him to make a public show of defending his bill aloud in the House chamber, responding for the record to questions put to him about H.B. 1596, and when the time came Barnes stood and gave his perfunctory responses, knowing they were all simply going through the motions, knowing the bill was going to pass, knowing Lou DeFeo and Sam Lee had won and Mary Bryant had lost and that when the bill was sent to the governor Barnes was going to skip the signing ceremony entirely, find something else to do that hour.

The final version of H.B. 1596 was adopted by the Missouri General Assembly in two votes, the first in the Senate and the second in the House, during the week between April 17 and April 25, 1986. The combined vote total was 142 to 41. By a strict accounting of the numbers these were better results than Mary Bryant had ever seen, but the accounting made no difference; the Antis had beaten her again, and during the final House of Representatives vote Mary watched the lights blink on one by one down the big vote tally board on the wall, green for Yes, red for No, the greens aligned up and down in long straight rows. When Mary told the H.B. 1596 story some years later, after she knew what would happen next, how much bigger it was going to become, this was the part that made her cry, recalling the defeat illuminated there before her in the splendid high-ceilinged chamber with the reporters crowded into the spectators' balcony to watch: she was so *angry* at herself, unreasonably, understanding that the odds against her were insurmountable but feeling nonetheless that there must have been something more she could have done, some opportunity she must have squandered, some No vote she could have wrested from the Yeses if she had worked even harder or been cleverer about statehouse persuasion. But on the day of the big House vote she permitted her face to register none of these emotions. Mary stood silently at the back of the spectators' balcony, watching the vote-count board, waiting for the last light-bulb indicator to come up green. Then she picked up her briefcase and went out to talk to the reporters.

On June 26, Governor John Ashcroft signed H.B. 1596 into law. Reporters and photographers were invited into the governor's office for the signing, during which the governor sat at his desk and a semicircle of lobbyists and legislators stood behind him, smiling for the cameras. Representative Jim Barnes was not in attendance.

On July 14, one month before H.B. 1596 was scheduled to take effect, Frank Susman filed his lawsuit.

FRANK SUSMAN began the lawsuit against H.B. 1596 the way he had begun so many others over the last thirteen years: he went to Judy Widdicombe and asked Reproductive Health Services to back him.

And for the first time in their long and contentious friendship, Judy thought seriously about turning him down.

Judy Widdicombe had never liked reading printed legislation, all the Notwithstandings and In Lieu Thereofs, but when Frank Susman brought her the summary of the final version of H.B. 1596, she listened to find out just what he and Mary Bryant were so worked up about. Why was Frank convinced they had to challenge this in court? And was Reproductive Health Services supposed to foot the bill again? Sometimes it seemed to Judy that she was constantly in the midst of paying for one legal challenge or another; RHS made very good money for a nonprofit, so generally there were funds in reserve, and Judy knew the lawsuits advanced arguments that might ultimately benefit both the clinic and the women who used it. But every case ran up substantial bills in legal expenses and fees —for Frank Susman's meter kept running as the legal work proceeded. Judy remembered writing one check for twenty thousand dollars, another for thirty-six thousand dollars. Why was it always Frank? Why was it always RHS? Couldn't somebody else carry the torch for a while?

The truth was Judy was losing her drive. She knew it and Frank knew it. At the clinic Judy was restless, distracted, overlooking things that would have consumed her attention a decade earlier. She was on so many national lists that she traveled regularly, to NARAL meetings or National Abortion Federation symposia or consultations in other states; international organizations had begun inviting Judy overseas to advise on birth control services in developing countries, and she had by now made a second trip to Bangladesh, and one to Colombia. Whenever Judy came back from these trips it took her some weeks to readjust, and she no longer roamed the clinic halls the way she had in the beginning, straightening pictures and opening drawers to make sure all the medical instruments were in place.

She had other people to take care of that. Reproductive Health Services was an admirable clinic, everybody who knew abortion facilities said so; in the thirteen years since its opening the clinic had quadrupled its size, spreading out across the entire second floor at 100 North Euclid Avenue. The counseling and Significant Other programs were national models. Out in the suburbs RHS ran a branch clinic now, a smaller, elegantly decorated office in the West St. Louis County building where the frequently picketed

Bolivar Escobedo ran his practice. Judy had pushed on with that expansion despite misgivings by staff members who worried that the West County branch was a waste of money, but Judy was insistent: affluent women from the suburbs also needed abortions, and RHS ought to be ready to market more strategically, to offer these women medical care within their own community.

So for a while there was West County to worry about—stocking it, furnishing it, rearranging schedules to staff both clinics at once. Then that was done too, and by 1985 Judy could feel herself stiffening with routine, there at her walnut desk with the prints and posters and appreciation plaques on the wall. She was better at building things than maintaining them. When Judy was in St. Louis her life had a pleasant, airless regularity to it: all day in one clinic or the other, maybe out in the evening for a dinner date or a movie, home to the two-story brick house she had bought in University City, just outside the St. Louis city boundaries.

She had a friendly relationship with her ex-husband, whose new wife was a woman Judy liked very much. Her sons had grown into fine young men, one a banker, one in charge of his own construction business. She played bridge with friends once a week, more for the companionship than the challenges of the game; she practiced piano, which she had studied with such seriousness back when she thought she would become a professional singer; she shopped the department stores for tailored clothing and good gold jewelry. She experimented with her hair, which had been styled into its smooth short cut for so long that Judy tried reddening the blonde and curling it into a permanent. (Not a success, Judy decided; she felt like Raggedy Ann and for a month she started every time she glanced into a mirror.) She wished she could lose weight.

It was not that running an abortion clinic had become a placid line of work—the Antis were never going to permit that to happen, as far as Judy could see. The National Abortion Federation, which Judy had envisioned at its founding as a sober professional organization dedicated to the enhancement of medical practice, now included as part of its standard information gathering a national tally on bombing and arson attacks at abortion clinics: fifteen during 1985, thirty the year before that. At every NAF meeting the clinic directors compared notes on security systems, hand-scrawled bomb threats, and raspy-voiced midnight telephone calls, and Judy could never quite decide whether she found these anxious huddles comforting in their camaraderie or profoundly depressing. Reproductive Health Services had been using security guards for five years already, one for the downstairs lobby and one for the hallway outside the front door, and it still made Judy's shoulders sag to see the entrance to her medical facility being patrolled like a prison corridor. She no longer saw the picketers down on North Euclid, even when she walked right past them; she had learned to glance in their direction without registering their presence in any but the most abstract and automatic way, and when the receptionist buzzed Judy's office to tell her that demonstrators had made it past the security guards and into the waiting room, Judy would swear quietly and sometimes just sit in her desk chair for a while before walking out to see what they were doing this time. She could attach names to most of their faces when she was obliged to do so, and when she thought it would make any difference she would address them personally, unless they were singing or praying so loudly that she would have had to shout to make herself heard. Sam Lee usually said hello to Judy without engaging her in conversation or coming over to shake her hand, and Judy realized she felt the same way he did, that there was not much point in trying to talk about it any more.

At her house she had security floodlights installed at every corner, so that at dusk the property was automatically bathed in glaring light that must have horrified the neighbors, although nobody ever called to complain. She tried to vary her routes to the clinic, the police had suggested that after Hector Zevallos was abducted, but Judy knew she was too casual about it and most mornings she got into her Mazda and drove straight past Forest Park, the direct point-to-point route any kidnapper would have expected her to take. The telephoned clinic bomb threats no longer flustered her the way they used to; she had learned after a while to debrief the receptionists mechanically, When did the call come, Exactly what was said, Was it a vague threat or a specific one, as in: ten minutes from now a bomb will explode in your waiting room. The police had taught Judy that, how she was to clear the clinic at once when the threats were specific, get the patients off the procedure tables, send them down to the first-floor lobby in their surgical gowns, wait for the squad car and then follow the bomb-sniffing dogs through every room of the clinic, looking for anything out of the ordinary, a potted plant moved to one side, a wastepaper basket out of place. Judy had been powerfully frightened during the first few evacuations, darting this way and that, hurrying everybody along, but so far nobody had found an actual bomb; the callers were only telephone heavy-breathers and Judy had begun to loathe them in a dull and irritated sort of way, resenting the effort it took to summon any emotion about them at all.

Every few weeks another National Abortion Federation alert arrived in the mail. "On October 27 at about 3:00 a.m., the River City Women's Clinic in Baton Rouge, LA, was destroyed by fire." "On Monday, December 2, the Portland Feminist Women's Health Center in Portland, Oregon received a package bomb through the mail." "On December 25 at approximately 3:30 a.m. explosions damaged three facilities in Pensacola, Florida." The more decorous Antis hurriedly disassociated themselves from each new incident, often in injured tones that suggested how unseemly it was to have imagined any connection; one of the standard Anti responses to reporters' post-violence inquiries, Judy knew, was to suggest that clinic owners ("the financially lucrative abortion industry," as one sit-in protestor had put it, darkly, in newspaper remarks about arson reports) were setting off the bombs and fires themselves, either to collect on insurance or to build public sympathy for the clinics. Sometimes the very huffiness of the right-to-life leaders' denials made Judy snort as she read them: what did they suppose they were provoking, after ten years of encouraging their rank-and-file to call abortion doctors babykillers, if not a certain rogue strain of eye-for-an-eye fanaticism?

But even Judy could see that most of the Antis really did have very little use for firebombing, which from their point of view must look disastrous for public relations purposes. Besides, even a successful firebomb closed only the random single abortion clinic—and temporarily at that. Sophisticated Antis had so many other more far-reaching tactics to select from and refine: the intimidation of politicians, the sidewalk harassment of women, the group coercion in church. Indeed, as time went on the right-to-life people had developed a way of mounting surprise assaults from unexpected directions, so that defensive regrouping had become something of an annual ritual for NARAL and Planned Parenthood and the various other organizations working to keep abortion legal. The early months of 1985 had provided as good an example as any: here were the pro-choice stalwarts deep into the work of their second decade, the lawyers preparing their legal arguments, the fund-raisers writing their donor letters, the political organizers building their voter lists—and suddenly the issue of the moment was a *movie*, Bernard Nathanson's movie, the twenty-eight-minute low-budget documentary that Ronald Reagan was reported to have watched at the White House.

For two or three months, from its release in December 1984 and on through the Antis' annual January march on Washington, this movie appeared to be pushing its way into every newspaper and magazine and talk show in the United States. Judy would open her *Post-Dispatch* or her *National NOW Times* and there it would be again, Bernie's movie, with the word "Scream" in boldface type. It was the best antiabortion propaganda Judy had seen in years. The film was entitled *The Silent Scream*, and for public relations purposes that alone was enough to carry it some distance; it was an arresting title, like the cover of a paperback thriller, and long before Judy actually saw the film she had heard about the gothic central premise from which Nathanson had drawn this phrase. Apparently Bernard Nathanson—NARAL's Bernard Nathanson, the New York obstetrician-gynecologist who had helped found the National Association for Repeal of Abortion Laws and for nineteen months ran New York's Women's Services, the busiest pre-*Roe* abortion clinic in the United States—had decided to make a film in which to expound upon his defection to the Antis. He had gotten hold of an ultrasound videotape that showed one complete abortion as it looked inside the uterus, the entire procedure recorded in the fuzzy black-and-white visuals of ultrasound imagery, and it was Nathanson's contention that as this videotape proceeded it was possible to watch the twelve-week fetus first struggling to escape the suction cannula ("the child is extremely agitated," Nathanson's somber narration declared, "and moving in a violent manner"), and then opening its mouth, an image that Nathanson suspended by momentarily stopping the videotape.

His own face off camera, the tip of his pointer tapping at the blurred light and dark of the videotape, Nathanson described for the viewer the specifics of this portion of the ultrasound:

> Once again we see the child's mouth wide open in a silent scream in this particular freeze frame. This is the silent scream of a child threatened immediately with extinction.
>
> Now the heart rate has speeded up dramatically, and the child's movements are violent at this point. It does sense aggression in its sanctuary. It is moving away; one can see it moving to the left side of the uterus in an attempt, a pathetic attempt, to escape the inexorable instruments which the abortionist is using to extinguish its life.

Poor Bernie. Judy thought she had come to know Nathanson reasonably well in the pre-*Roe* years, when she would fly to New York for abortion conferences and he would be striding around telling amusing stories and making energetic pronouncements about the hypocrisy of abortion law. He had a flamboyance to him in those days, as though he had discovered in the abortion movement something gloriously invigorating; Arlene Carmen, the New York church administrator who had kept track of the interstate underground referral networks for the Clergy Consultation Service, once privately described Bernard Nathanson as a great mass of fervor trolling eagerly for a cause. "Unguided missile" was how Arlene put it, and in retrospect Judy decided Arlene was right: Bernie was a zealot no matter which way he was pointed. If he really had changed his mind about the morality of abortion—if Nathanson had genuinely come to believe, as he wrote in a 1974 *New England Journal of Medicine* essay, that his directorship at an abortion clinic amounted to having "presided over 60,000 deaths"—then it was evidently not in him to withdraw quietly from the leadership and let his former allies carry on. He had to excoriate himself in public and then turn up as a propagandist for the other side.

He was good at it, too. When she finally saw *The Silent Scream*, Judy realized just how clever Nathanson had been, constructing his film around a single seven-minute ultrasound; since an ultrasound picture is not a photograph, but rather an image assembled from the recorded echoes of sound waves, the visual that finally presents itself is so grainy and indistinct that the lay person requires some help seeing anything in it at all. Judy played *The Silent Scream* on a video monitor at RHS, a roomful of wary abortion activists gathered around her, and the women who had never watched an ultrasound before were peering closely at the screen, cocking their heads this way and that, trying to understand exactly what they were looking at as Nathanson spoke. It was like one of those psychologists' inkblot tests, Judy thought; if you stared at it long enough your imagination invented the pictures for you, and for the sympathetically inclined Nathanson's narrative was so tragic and rich with drama that he could have held an audience spellbound while moving his pointer across a map of Montana. *The child senses the most mortal danger imaginable . . . the child is being tugged back and forth . . . the body is now being systematically torn from the head. . . .* It was brilliant. Nathanson had seized upon the grim central reality of an abortion procedure—something with movement and a heartbeat was being sucked into a vacuum tube or dismembered by surgical instruments, that was in fact what happened— and embellished it with personality and evil intent until it had swollen into a kind of on-screen morality play, the murderous abortionist stalking the innocent child.

The *Silent Scream* was introduced into the *Congressional Record,* in transcript form, by the New Hampshire Republican senator Gordon Humphrey, who announced that Nathanson's film was destined to change American history. President Reagan said he prayed that every member of Congress would watch it. Judy had learned a great deal over the years about ignoring the creative output of the Antis, but it was obvious that someone was going to have to respond to *The Silent Scream;* the title alone was certain to reach many more Americans than would ever see the film itself, and within weeks of the film's release Planned Parenthood's national office was mailing out around the country a ten-page rebuttal pamphlet, its own title stripped in large red capital letters across the cover: *THE FACTS SPEAK LOUDER.*

It was a good solid pamphlet, the text laid out on big sheets of heavyweight glossy paper, but when Judy studied her copy it seemed to her that the Planned Parenthood people were allowing Nathanson to lure them into the wrong arguments. Planned Parenthood had lined up a panel of impressively credentialed physicians to assail *The Silent Scream* point by point: a twelve-week fetus is much smaller than the televised pictures in the film; a twelve-week fetus is not developed enough to experience pain or fear or to make any purposeful movement beyond basic reflex response; a twelve-week fetus cannot scream. "The mouth of the fetus cannot be identified in the ultrasound image with certainty," read the pamphlet. "The statement that the screen identifies the open mouth of the fetus is a subjective and misleading interpretation by Dr. Nathanson. His conclusion is not supportable."

Wrong argument. Maybe the fetus *did* open a mouth, or something that would have turned into a mouth if it had been left to develop on its own. Point taken: the human fetus was more than some superfluous body part, Judy knew that, her clinic workers knew that, she wouldn't have been reciting baptisms in the procedure rooms if she believed the Blob of Tissue label sufficed for every pregnancy termination. The great deception in *The Silent Scream* lay less in the images on the screen than in the cast of characters Nathanson had chosen to include: one fetus, one abortionist, one propagandist. No *patient*. Nathanson had left the woman out of his abortion movie, as though every termination was performed upon a disembodied torso, and in the media flurry that accompanied *The Silent Scream*, the clinic patients also seemed to have been dropped entirely from the discussion. Nanette Falkenberg, the young political organizer then in her third year as executive director of NARAL, lost her composure at one point during a newspaper interview and cried out in frustration, "Do they think the fetus is housed in a Tupperware jar?"

Nanette Falkenberg said later that she was mortified the instant that comment was out of her mouth, but the syndicated columnist Ellen Goodman repeated it in print some days afterward and people began calling Falkenberg to tell her how much they liked it. Visual imagery was an art NARAL organizers had never really mastered. At the Washington headquarters of NARAL most of the staff members were Falkenberg's contemporaries, women in their twenties and thirties who had still been in high school or college when *Roe* was handed down; they were ardent in their commitment and could write and speak passionately about "choice" and "reproductive freedom" and "the rights of all women," but these were not ideas that fit readily into single photographic images. How do you photograph a life reclaimed, a mistake contained, a private sorrow thoughtfully and painfully resolved?

"We had *always* failed in the visual war," says Falkenberg, recalling the distraught weeks at NARAL amid the clamor over *The Silent Scream*. "How do you portray the complexity of a woman's life? How do you portray that this is a single woman who has three kids, let's not even make her battered, she's just a working-class woman with a minimum-wage job and she has three kids and a boyfriend and he leaves. How do you portray that visually? We always tried to keep the rational high ground: they were emotional, we were thoughtful, we told the truth, we went to Capitol Hill with facts. But it was so clear that you had to respond emotionally on this."

Among NARAL veterans the visuals dilemma was more than a decade old by the time Nathanson's movie set it off again. Antiabortion fetus pictures had begun circulating in 1970, during the state-by-state legalization efforts, and for fifteen years leaders at NARAL and the adjunct abortion advocacy organizations had been arguing with each other about how to respond publicly to these pictures: how to characterize them, how to counter them, what to say about them when confronted by reporters or debate opponents.

No one had come up with an entirely satisfactory response. The pictures were "inflammatory," or the pictures were "shock tactics," or the pictures were "misleading" because they overemphasized late abortion and magnified things that in reality were extremely small. Lawrence Lader, the New York writer who co-founded the National Association for Repeal of Abortion Laws, used to make angry references in his NARAL memoranda to "Catholic 'horror' fetus pictures," and in fact the element of horror— the blood, the mangled tissue, the dismembered arms and legs—had in its own way provided a certain shield against the content of the photographs themselves. Daily newspapers and television programs were often reluctant to display such gruesome images, and many of the pictures prompted such immediate aversion of the eyes that it was easier to dismiss them all without acknowledging what was in them, without having to give voice to the uncomfortable core of legal abortion advocacy, without ever being obliged to talk about the pictures while looking directly and steadily at what they showed. *Yes—it's a severed foot in a surgical pan. Isn't it dreadful? It has to be legal anyway.*

At intervals the "dead woman" option had been advanced, uneasily, a tactic Lawrence Lader described as "meeting shock with shock": since the pre-*Roe* years the office files at NARAL had included a small collection of coroners' photographs of adult women lying dead on the floor, their buttocks and vaginal areas bloodied by botched criminal or self-induced abortion. In a horrible-for-horrible contest the dead-woman photographs could plausibly hold their own against the slide collections of the Antis, and the coroners had also included in the photographs some useful atmospheric details: the soiled rags crumpled up beneath the body, the sheets of newspaper spread across the floor. In one of the pictures a rubber catheter lay near the dead woman's pale splayed legs, and the legs themselves protruded through a bathroom doorway, as though the woman had been crawling toward the toilet when she collapsed. NARAL once printed the catheter picture in a pamphlet, some weeks after *Roe*, with the caption: A DEAD WOMAN ON THE BATHROOM FLOOR? OR LEGAL ABORTION? WHICH DO YOU PREFER?

As a persuasive tactic this was controversial even then, and the passage of time had only escalated the arguments. "Inside the movement there was a lot of controversy about dead-woman pictures," Nanette Falkenberg recalls. "Some people thought it was exploitative. And others—I think I would put myself in this category—felt by the mid-1980s that abortion had been legal so long that they didn't believe the pictures anymore. It wasn't even legality so much; it was the improvement in medical procedures, the idea that doctors wouldn't let that happen to women in emergency rooms now, that women might find illegal abortionists, but they wouldn't do *that*, they'd be safe."

The other drawback to the dead-woman pictures was implicit in the hate mail that still arrived at NARAL, daily, hand-addressed, twelve years after *Roe v. Wade*. "Whores." "Tramps." "Females that love to intercourse."

"You who have no ability to control yourselves." "If these women want to play with fire, let them suffer the consequences." To anyone who opened the envelopes, or listened to the fiercest shouts from the sidewalk protests, it was apparent that for some perhaps sizable portion of the American populace, a photograph of a naked woman's corpse evoked not compassionate resolve but something uglier and more complicated. The women looked sexual even though they were dead. It was this very sexuality that seemed to rouse certain people to rage, as though they were far less interested in protecting the fetus than in punishing the woman who had the audacity to have conceived it, and in most of the dead-woman pictures the faces were turned away from the camera, making the women not only sexual but also faceless. It seemed to Nanette Falkenberg that this was the essence of the visual image problem, the weakness Nathanson had exploited so deftly with his dangerous movie: in the abortion fights *the women had no faces*. They were anonymous, theoretical, blurred amid the sloganeering, so that the letter-writer from New Paltz, New York, could manage with no evident twinges of uncertainty or conscience to describe abortion patients as "immoral" women who demanded readily available clinic abortions "so that they can enjoy all the illegal sex they want."

Would it have made a difference if someone had led Mr. New Paltz up and down the streets of his own hometown, introducing him to ordinary women who had made the decision to have an abortion? What if he had known about the schoolteacher, the clerk at his grocery store, the middle-aged woman at the bus stop, the English major home from her freshman year at Cornell? In a way, the abortion movement had become a victim of its own public stories; every chance at a microphone was used to talk about the dramatic and the heartrending, figures of extraordinary tragedy, not the kind of woman everybody knows. "It was always the thirteen-year-old who'd been raped," Nanette Falkenberg recalls. "Or the mother with nine children who was going to have a baby with Down Syndrome."

More than a decade earlier, with no real orchestration or formal declaration of intent, abortion activists around the country had set out for the first time to remake the vocabulary with which Americans talked about abortion—to replace the phrase "pro-abortion," which in the pre-*Roe* years had served as the standard label for a person in favor of legalization, with the loftier and less combative "pro-choice." The marketing of "choice" as a central principle had always infuriated abortion opponents, who expounded in many speeches and articles on the logic of granting Americans the "choice" to kill other people; in Anti publications the label was still "pro-abortion" (or, in moments of particular vehemence, "pro-death"). Many big-city newspapers, like the *Post-Dispatch* and *The New York Times*, tried to steer a neutral course by using "pro-choice" only inside quotation marks, but even inside those quotation marks both the phrase and its subtext were repeated so insistently—our cause is not *abortion*, it is the *right to choose* abortion—that by 1985 a political demonstrator could hold up a sign reading I'M FOR CHOICE and leave no confusion at all as to the meaning of the words. That much abortion activists had achieved solely by phrasing things a certain way over and over again: they had territorialized new language, so that each side had equal access to the one-word distillation of the morally complex.

But they still needed visuals. Bernard Nathanson had just taught them that all over again. And if dead-woman pictures were wrong, and abortion-advocate-giving-stump-speech pictures were wrong, and black-on-white graphics of coat hangers were wrong (too strident, too antiquated —"I always thought the coat hangers were dreadful," Falkenberg recalls), then perhaps there were other ways to force visual imagery into the public discourse, to put faces on the women in abortion clinic waiting rooms. "Change the *message* of the pro-choice movement," NARAL's Washington office instructed in a resolute thirty-page memorandum mailed in February 1985 to every affiliate in the United States:

> THEMES you should begin to use are:
> We are your mothers, your sisters, your daughters, your friends, and we have chosen abortion. (Ordinary women that everyone knows.)
> Every case is an "exception." (Every woman choosing abortion has a compelling reason.)
> We are decent, caring, intelligent women making responsible choices.

In considerable detail, with a formal launch date and instructions on month-by-month implementation, the NARAL memorandum announced plans for an orchestrated national campaign entitled—in direct rebuff to Bernard Nathanson—"Abortion Rights: Silent No More." At every state affiliate, local directors and volunteers were to begin working these themes into their written materials and their public appearances: *ordinary* women, *decent* women, the *women you know*. Letters were to be solicited from around each state—letters of *ordinary, decent* people who wished to write about their experience with abortion. At the climax of the campaign, each state NARAL affiliate would hold a series of "speakouts," public gatherings at which women and men (ordinary, decent women and men) would tell their personal stories and read aloud from the collected letters, presumably with reporters on hand to "provide a wave of state and local publicity," as the memorandum said, "that emphasizes the refocusing of the debate about abortion to include women."

It was a campaign of exuberant ambition, and for some months into 1985 it was possible to pick up a newspaper in many parts of the country and find writers leading off their copy the way one columnist did during the spring: "It was an evening in the summer of 1964. She was 21 and scared. She was to wear 'something green' and have the $500 in cash. . . ." In St. Louis the local NARAL organizers took up their mandate with vigor, composing righteous "Dear Friend" letters—

Are you tired of . . .
. . . Hearing about "The Silent Scream"?
. . . Women's lives being left out of the abortion debate?

—and printing their Speak Out invitations on eye-catching heavy yellow stock, Theme Number One worked smartly into the salutation. *We are your mothers, your daughters, your sisters, your friends, and we are speaking out for our right to choose.* The Missouri NARAL office was a third-floor walk-up not far from Reproductive Health Services, and by April the two young women who ran it had learned to brace themselves every morning for the postman's delivery of the handwritten letters with their far-flung Missouri postmarks: Rolla, Joplin, Kirksville, St. Joseph. "I remember every day we'd go for the mail and we'd sit there and read them," recalls Lucia Miller, who worked as the state volunteers' coordinator. "And some would be so moving that we'd have our eyes watering." The Missouri NARAL director Laura Cohen, who had been hired in 1983 and had so far spent most of her working hours trying to rouse voters or build the membership lists, now found herself reading abortion story letters written by women she had met personally, many of these older women who wrote in brutal, impassioned detail about experiences that Laura— who was thirty-two then, and had never undergone an abortion herself— understood only in the abstract.

"A really close friend of mine wrote an incredible story about having an illegal abortion in Philadelphia," Cohen recalls. "She ended up in Puerto Rico. She was nineteen, with her boyfriend. And I'd never known it. That kept happening, over and over. A friend of mine who was in her seventies writing about her *mother* having an abortion, I couldn't believe it. Unbelievable stories. It was like a hidden history coming out."

The St. Louis Speak Out was held on a Sunday afternoon in May, in the basement of the First Unitarian Church. There were reporters in the back of the room and picketers outside (miniature coffins fashioned from Styrofoam coolers, Lucia had never seen that prop before), and crowded together in the folding chairs of the big church meeting room, 150 women and men who took turns standing up to speak or read aloud from the letters that had arrived at the Missouri NARAL office. The local television stations put it on the news that night. The newspapers wrote it up. *The New York Times* ran an admirably sized Washington, D.C., Speak Out story, with an opening paragraph and accompanying photograph that served the grander purpose precisely—here was the lovely Rabbi Shira Stern, hand-in-hand with her rabbi husband, describing the abortion she had undergone when they learned that their fetus carried a fatal birth defect. Rabbi Shira had a *face*, resolute and gazing directly into the camera under her clipped blond hair, and in her *New York Times* picture the backdrop even included the banner with the big theme lettered along its length: WE ARE YOUR MOTHERS, YOUR DAUGHTERS, YOUR SISTERS, YOUR FRIENDS . . .

The Planned Parenthood Federation of America, which was in the midst of its own national advertising campaign, ran some full-page newspaper ads in which two serious, dignified-looking women, each identified by her real name and hometown, described their illegal pre-*Roe* abortions. (". . . The man had a whiskey glass in one hand and a knife in the other. . . ." ". . . I drove secretly to a Pennsylvania coal-mining town and was led to a secret location. . . .") Clinics and NARAL affiliate offices across the country began receiving copies of NARAL's "We Are Your Mothers, Your Daughters" poster, with its attractive, racially balanced foursome of women drawn large beneath the slogan. Lucia Miller and Laura Cohen thought the picture was so wholesome as to be simpering and said it looked like a Tampax ad, but they stuck the poster up on a wall somewhere; at least now they had a visual that would not make anybody flinch. It was a good campaign; it had pushed Bernard Nathanson out of the spotlight, it had swelled NARAL's membership rolls, it had helped propel the movement through one more otherwise anxious year. "We continue to receive letters almost daily at the Missouri NARAL office, telling stories of abortions or stating support for the right to choose," wrote Cohen and Miller in a group thank-you letter as the Silent No More campaign wound down. "Together, St. Louis pro-choice activists have built a strong base for abortion rights work in this city."

JUDY WIDDICOMBE wished she could believe it. She had tried to be a useful partner in the NARAL campaign; she had made room for the Speak Out invitations in the Reproductive Health Services waiting area, and the wholesome poster and EVERYWOMAN'S RIGHT TO CHOOSE bumper stickers now decorated various high-visibility bulletin boards along corridors inside the clinic. But it seemed so futile, trying to win back the image wars. They had lost those battles a long time ago, Judy thought; the Antis had the pictures, the Antis had the don't-bother-me-with-the-real-world moral simplicity, the Antis even had the language, as hard as the NARAL and Planned Parenthood people had tried. "Choice" was a respectable start; "decision" was a nice sober word; "ordinary" was fine, all the ordinary women amid their ordinary tragedies. But if you were trying to wave phrases at a distracted legislator, none of the NARAL vocabulary was any match at all for "life," "unborn babies," the "right to life."

It was just not going to get any easier, that was all. For three years now Judy had been hearing occasional reports from Europe about an experimental French pharmaceutical called RU-486, which some enthusiastic early accounts were describing as an "abortion pill," a medication women could swallow during very early pregnancy to bring on a miscarriage without aspirators or surgical intervention. In the anecdotal descriptions of RU-486—the name was an abbreviation of the drug's five-digit identification label during its development at the French pharmaceutical company Roussel Uclaf—the pills were made to sound like a medical advance as revolutionary as the vacuum aspirator had been some fifteen years earlier: if RU-486 worked, the stories suggested, pregnant women might someday be able to report to their local physicians for medication handed over as quietly and matter-of-factly as a packet of antibiotics. No surgical gowns, no metal instruments, no electric whirring of the suction machine —no *clinics*, if the speculation was carried to its logical conclusion. At National Abortion Federation meetings Judy had heard clinic directors trade RU-486 rumors not with eagerness but with deep anxiety, imagining a time when their specialized offices and costly vacuum equipment might be seen as relics of a less enlightened era.

And Judy wished she could believe that, too—that the directors were right to be anxious, that abortion-by-medication might someday antiquate the clinic system and in doing so correct what turned out to have been the colossal flaw in modern American abortion practice. Had any of them really understood what they were doing when they began fifteen years earlier to set up independent abortion facilities, separated by label and by physical distance from the offices and hospitals where "conventional" medical care took place? Tidy little bull's-eyes they had built, magnets for the Antis, medical ghettos whose thresholds a physician could so easily decline to cross. It was tempting to imagine the subversive potential in the new French pills, the means of abortion tucked into the cabinetry of every drugstore in the land, the abortions themselves taking place inside private and anonymous bedrooms and bathrooms, the Antis left with nothing to picket except the entire roster of American gynecologists and family physicians and pharmacists.

But this was nonsense. The RU-486 reports themselves argued against it. In the first place, the pills were apparently useful only during the very first weeks of pregnancy; they worked by shutting off the progesterone receptors that signal a uterus to accept a fertilized egg, and most of the clinical trials were accepting women whose first missed menstrual period was only one to three weeks overdue. For the woman composed enough both to discover her pregnancy and resolve to end it so rapidly, this gave RU-486 a distinct advantage over vacuum aspiration (suction was somewhat unreliable during very early pregnancy, and many clinics were reluctant to schedule a vacuum aspiration until the patient was at least seven or eight weeks pregnant), but Judy had spent enough years at an abortion clinic to know that no quick discovery abortion method was going to empty out her waiting room. There was too much confusion out there, too much cataclysmic life trouble, too many women who had required elaborate moral and practical preparation just to get themselves and $250 up to North Euclid Avenue before the first trimester was *over*.

Besides, aborting by RU-486 was laborious, as Judy understood it, requiring multiple visits to the physician (one for the physical examination; one for ingestion of the pills; one for a follow-up prostaglandin dose, by injection or vaginal suppository, to enhance the pills' effectiveness; and one for the final checkup). It was not cheap. It was not painless. It was private only in the sense that the woman bled out more or less on her own, no physician pushing cannulas into her and flipping a switch. It would find its place in the practical array of abortion techniques—it deserved to, if the early reports proved true; Judy sometimes thought that if she had been offered an abortion pill back in 1970 she might have swallowed it instead of waiting so many weeks and then having to lie passively on a procedure table while someone else scraped around inside her.

But the pills would end nothing, they would quiet no one, the Antis would hate them as much as they had hated dilation and curettage once and they hated suction now. Probably they would call them Babykilling Pills or something, Judy could picture the headlines on the newsletters; they would have no gruesome visual aids to help them along, since what RU-486 expelled was a tiny cell mass that no amount of magnification could turn into severed hands and feet, but Judy had a feeling that this would not even slow the Antis down. They were *never going to go away;* the picketers would *always* be out there; the clinics were *never* going to be secure the way a hospital or a dental building or for God's sake a veterinary clinic was secure. The Missouri Catholic Conference and Missouri Citizens for Life were going to keep restaging the annual spectacle over and over again, drafting their single-minded legislation, lining up their obedient legislators, rolling up their sleeves for the litigation that was sure to come.

And as Frank Susman stood in Judy's Reproductive Health Services office in April 1986, explaining one more time why he and RHS together shared the moral obligation to take the state of Missouri to court, Judy realized that she was tired of it. She didn't want to do it any more. She didn't want to front the legal fees, she didn't want to be accommodating for the reporters, she didn't want to gather up all her statistics and take the commuter flight to Kansas City and sit in a hard-backed witness stand chair for four hours at a time while some unctuous cross-examining lawyer made a big public show of the fact that she was a nurse and not a doctor. Couldn't they just live with H.B. 1596? Wasn't it smarter once in a while to see the Antis' legal bait for what it was—to walk away instead of risking so much?

But this law declared that in the state of Missouri human life began at *conception*, Frank said. Was Judy ready to walk away from that? Was

Judy going to let the Antis make it unlawful for a public health nurse to tell a pregnant woman that she could end her pregnancy safely at Reproductive Health Services? A lot of people were depending on their legal challenge, Frank argued; half the legislators had probably voted Yes on the bill because they knew it would be knocked down immediately in federal court.

Judy had a lousy feeling about this. She had known Frank Susman for fifteen years and had fought with him many times before; Judy had never forgotten that Frank was one of the people who had refused to support her as the inaugural director of RHS. But it had never before been so obvious to Judy that Frank's ambition was driving him as hard as his convictions, that he hadn't thought through either the expense or the inherent risk in the legal challenge he was now asking her to back. The entire process had turned into a reflex action for Frank, Judy thought, an automatic defense of his own professional turf: they pass laws, I do the lawsuit.

Judy wanted out. We're not paying for it, she said to Frank, not this time.

Frank reminded her that Reproductive Health Services could lose its late-second-trimester cases if H.B. 1596 went unchallenged—that the legislation, as one of its apparently random assortment of feints and jabs at abortion practice, required all post-sixteen-week abortions to be performed in hospitals rather than clinics. The Supreme Court had already decided second-trimester-hospitalization requirements were unconstitutional, but those 1983 rulings—one of them had been Frank's own case, *Planned Parenthood* v. *Ashcroft*, Judy had testified for him herself—were based on laws requiring hospitalization after *twelve* weeks, not sixteen. The only way to overturn this new sixteen-week clinic cutoff was to go back into court, Frank said.

Look hard at the damage the Antis were trying to do here, he urged Judy; except for the sixteen-week-hospitalization requirement, its provisions might not affect RHS directly, but they promised to present real difficulties for some Missouri women. Out in Columbia, for example, the publicly funded medical center at the University of Missouri was the only hospital in the region that allowed second-trimester abortions to be performed on the premises; even though most of those abortions were privately paid for, the patients paying in full both the doctors' fees and the operating-room costs, the hospital was nonetheless a "public facility" and would thus be required under the new law to prohibit any but lifesaving abortions on the premises. That would force patients to turn to other facilities for help, but there *were* no other second-trimester facilities in Columbia; both the private hospitals and the principal local clinic refused to accept second-trimester patients.

And the law made it illegal, Frank added, for any public employee who could plausibly be defined as a health professional—doctor, nurse, social worker, counselor, or "persons of similar occupation"—to "counsel a woman to have an abortion not necessary to save her life." Even the lobbyists who pushed H.B. 1596 seemed uncertain as to exactly what was meant by that counseling phrase; perhaps the public employee was forbidden to utter the word "abortion" in the presence of a pregnant patient; perhaps the public employee was allowed to say "abortion," but never in a voice that suggested abortion might under the circumstances be a good idea. In any case the public employee was almost certainly prohibited by law from speaking or writing the words "Reproductive Health Services, One Hundred North Euclid Avenue, reputable place, here's the telephone number," while in consultation with a pregnant woman who wished to know what her options were.

And there was one *more* thing, Frank Susman said: the law required every doctor performing an abortion on a woman at least twenty weeks pregnant to precede the abortion with tests—"such medical examinations and tests as are necessary"—to determine whether the fetus was viable. This last provision was either simple harassment or cleverly rigged legal strategy, depending on how its wording was ultimately interpreted by the courts. It appeared to require all second-trimester abortion doctors to separate out their post-nineteen-week patients and then subject them to some special set of viability tests; the law did not specify what those tests were supposed to be, but said only that they must include those tests "necessary to make a finding of the gestational age, weight, and lung maturity of the unborn child." Loosely interpreted, that might require tests as routine as pre-abortion ultrasound, a practice Reproductive Health Services had commenced some years back as part of the standard protocols for all second-trimester abortion. More literally interpreted, the provision might require every post-nineteen-week patient to undergo amniocentesis, the extraction of amniotic fluid with a long hollow needle, since amniocentesis was the medical test physicians used (generally in high-risk obstetrical situations, when a woman who *wanted* to give birth was threatening premature delivery) to help assess the maturity of fetal lungs.*

Amniocentesis was an expensive and uncomfortable procedure; at the going rate, an obligatory amnio might add $250 to the cost of the abortion. But there was a subtler and more important problem lurking in the viability-test requirements of H.B. 1596; any attorney familiar with abortion rulings could see that this language had been written as an affront to one of the key holdings in *Roe* v. *Wade*. The *Roe* majority opinion had constructed its instructions to the states around trimesters of pregnancy: first trimester, no state interference allowed; third trimester, states allowed to ban all but those abortions deemed necessary for the woman's life or

---

* Measuring the ratio of certain fetal lung by-products found in the amniotic fluid can help physicians decide whether the developing lungs are mature enough to reduce the infant's chances of fetal respiratory syndrome after birth.

health; second trimester, states allowed to regulate abortion *only in those ways that protect the health of the abortion patients themselves.**

The H.B. 1596 viability-testing requirement was obviously not written to protect the health of women who wanted to abort; its aim, indisputably, was the protection of "unborn children." And because it specified that the viability testing must take place at twenty weeks, during the second trimester, the new Missouri statute was effectively defying the Supreme Court by regulating abortion in a way that *Roe* had prohibited. And if *that* regulation was left unchallenged, then other second-trimester regulations —regulations written not to protect the pregnant woman's health, but instead to protect the fetus and thus make the abortion more cumbersome for the woman and the abortion doctor—would logically follow close behind. For that matter, why hold the logic to twenty-week abortions? Why not begin imposing "protection of the unborn child" restrictions for eighteen-week abortions, or for sixteen-week abortions? There was a reason Frank and Judy had always challenged these abortion bills together, Frank argued: the bills were *bad legislation.* It was *right* to take them on in court.

Still Judy was dubious. Frank back-pedaled, easing off the pressure a little; what about offering a more limited form of support? Would Judy at least allow Reproductive Health Services to act as a plaintiff in this case— to be formally named in the list of agencies and individuals claiming that H.B. 1596 violated their constitutional rights?

All right, Judy said. He could use the clinic's name. Some of Frank's arguments had sounded reasonable as she listened; she did care about second-trimester, Susman was right about that, and she could see that the viability-testing language had been written mostly to discourage both patients and doctors, that nobody really believed amniocentesis was appropriate for women preparing for abortions. Judy was willing to say as much in court, if Susman needed her testimony again, and he could name his case as he liked, Reproductive Health Services versus Antis or whatever, she would back him that much. But she was not giving him the money for it, Judy said. Frank was going to have to find somebody else to pay for his lawsuit.

• • •

* The majority opinion in *Roe* never specified the exact weeks at which the second and third trimesters were supposed to be divided. Instead the opinion described the three stages of pregnancy as: 1) "the first trimester"; 2) "the stage subsequent to approximately the end of the first trimester"; and 3) "the stage subsequent to viability." Because the "viability" of a fetus depends on many factors, this third trimester—the stage at which *Roe* permitted states to assert their interest in the potentiality of human life—thus had no fixed starting point; the beginning of the third trimester might vary from one pregnancy to the next. But twenty weeks' gestation, under the *Roe* guidelines, was clearly second-trimester; neonatologists consider twenty weeks too early for any fetus to be considered truly "viable," or able to survive outside the womb.

So FRANK SUSMAN went to the regional office of the American Civil Liberties Union, where the preliminaries for an H.B. 1596 lawsuit were already underway; all the ACLU needed was a lawyer to take the case.

Susman had a complicated relationship with the ACLU. At the local level, he was a veteran board member and part of the regular roster of attorneys who volunteered their time handling ACLU-sponsored cases. But the ACLU's full-time abortion experts were in Manhattan, where the old brick national headquarters building housed the special in-house litigation unit called the Reproductive Freedom Project. As a physical setting the Project was unmemorable, some office furniture and a wall of scratched file cabinets crowded into one corner of the seventh floor, but by 1986 those file cabinets contained the nation's most complete archive of legal records from abortion cases: scores of cases, in state courts and federal courts, in which Reproductive Freedom Project attorneys had acted as lead counsel, or submitted friend-of-the-court briefs, or provided local attorneys with advice and sample briefs.

And Frank Susman was not a beloved figure within the offices of the Reproductive Freedom Project. For nearly a decade the Project's lawyers had worked under the directorship of the ACLU staff attorney Janet Benshoof, and Benshoof and Susman had barely been on speaking terms since squaring off three years earlier over *Akron* v. *Akron Center for Reproductive Health*, the major ACLU-backed abortion case that led the Supreme Court to overturn the restrictive abortion ordinance adopted by the city of Akron, Ohio. Susman had played no direct role in the arguing of *Akron*; at the time he was preparing one of his own Supreme Court cases, *Planned Parenthood* v. *Ashcroft*, which was scheduled for oral argument the same day. But he knew that a local law professor named Stephan Landsman had handled the ACLU start-up work on the *Akron* case, and that Janet Benshoof had provided Reproductive Freedom Project help on the briefs and the appeals. When Benshoof proposed over Landsman's objections that *she* conduct the oral argument at the Supreme Court, Frank Susman—who knew personally the owners of the Akron abortion clinic named as lead plaintiff in the suit—began making telephone calls from St. Louis, intervening in what was rapidly becoming an ugly ACLU tug-of-war, to urge the clinic plaintiffs to insist that Landsman present the argument instead.

Frank Susman had never been adept at organizational diplomacy; he was quick and impatient and tended to sound curt even when he was trying to be cooperative. He and Janet Benshoof had not been especially fond of each other before the *Akron* conflict; as an abortion litigator Frank had accumulated a stellar resume (two dozen cases, six Supreme Court rulings; Susman was filing the *Rodgers* v. *Danforth* papers while Janet Benshoof was still in law school), but within the small community of post-*Roe* dedicated American abortion lawyers, Susman's credentials had always been regarded as faintly suspect. He lived in St. Louis, a long way from the insular circles of East Coast attorneys who shuttled between New York and Washington, D.C. He worked for a corporate firm, not a law school or a social advocacy nonprofit. He had a reputation for forging along by himself, paying only minimal attention to the advice and legal theories of attorneys who might logically have served as his abortion law colleagues. That he was male made matters worse; almost all the full-time reproductive rights lawyers were women, the men's names generally relegated to a roster of respected second-tier associates, and after Janet Benshoof finally lost her intense campaign to argue *Akron* before the Supreme Court, Susman heard from ACLU associates how deeply angry she was, and how convinced she remained that Susman had intervened principally because he wanted to deny *her* the argument.

Frank Susman and Janet Benshoof never confronted each other directly about the *Akron* fight. For three years the two of them avoided one another whenever possible, Susman explaining to anybody who would listen that he had meddled in the *Akron* dispute only because he believed local ACLU attorneys had the right to expect a chance at Supreme Court argument in cases they had started. And when Susman now agreed to challenge H.B. 1596 on behalf of the Eastern Region of the Missouri ACLU —the ACLU would pay the court costs and other expenses incurred in preparing the suit, but Susman would be expected to work on a volunteer basis—there was no collegial embrace from the Reproductive Freedom Project office in New York. Susman commenced his lawsuit on his own, accompanied from the early weeks by a single partner from his law firm, a younger attorney named Tom Blumenthal, who was to write his memories of the case into an unpublished book-length manuscript:

> I looked up from whatever I was doing just in time to duck the projectile which Frank had hurled at my head. It was a copy of Senate Substitute for H.B. 1596.
>
> "Look through this and tell me what you think. We need to draft an injunction and file it in front of Judge Wright before the bill goes into effect."
>
> Freely translated this means: draft a complaint. Write it to be filed in the Western District of Missouri federal court, where we have jurisdiction over both the State and the Attorney General, both necessary parties. In this way we will be assured of drawing Judge Scott O. Wright, who gets all cases filed in the Central Division for the Western District of Missouri, who has no patience for the State's inanity, and who is very familiar with the constitutional questions.

Tom Blumenthal was thirty-four years old that spring, ten years younger than Susman, with unkempt curly hair, a cherubic, clean-shaven face, and a mildly wiseass demeanor that emerged when he was feeling shy. Susman's own thick hair and mustache had gone prematurely white, so that when Blumenthal stood beside him they might have been taken at first glance for father and son, and during Blumenthal's three years at Susman's firm the two of them had in fact learned to communicate in a kind of familial shorthand, making each other smile with what Blumenthal once referred to as "our mutually sick senses of humor." At the firm the abortion work now fell to both of them, the RHS telephone calls handed off to Blumenthal when Susman was busy, and within a week after the final General Assembly votes on H.B. 1596, Blumenthal was piecing together the earliest draft of a class-action suit.

This is a civil action pusuant to 42 U.S.C. 1983 for judgment declaring Missouri Senate Committee Substitute for House Bill No. 1596 (hereafter the "Act"), which regulates the performance of an abortion, assisting abortion, and abortion counseling, both in the public and private sector, to be unconstitutional. Blumenthal had filed class actions before, and he knew the start-up routine by heart: find your plaintiffs, explain why the offending law affects them directly, explain why they can properly be found to represent large classes of similarly affected persons or institutions. Reproductive Health Services was the first name on the list, but others would need to follow immediately behind, and by the end of May, Blumenthal and Susman had a seven-name plaintiffs' list to lead off their complaint: RHS; three physicians; one public health nurse; one social worker; and the Planned Parenthood clinic in Kansas City, which accepted patients through the fifteenth week of pregnancy and had become, as the complaint phrased it, "the second largest abortion provider within the State of Missouri."

Planned Parenthood of Kansas City had initially intended to bring its own lawsuit against H.B. 1596, bypassing Frank Susman and soliciting the help of the Planned Parenthood Federation of America's newly aggressive legal department in New York. For eight years now, Planned Parenthood's national offices had been operating under the direction of the famously energetic Faye Wattleton, a registered nurse and former clinic administrator with extraordinary presence and public speaking ability, and in the early 1980s Wattleton had persuaded her board to support a new litigation unit that might do for Planned Parenthood what the Reproductive Freedom Project appeared to be doing for the ACLU—that might actively patrol the states for legal conflicts significant enough to merit outside help. If this venture worked well, the thinking went, it would surely advance both the cause of reproductive freedom and Planned Parenthood's public visibility and fund-raising efforts. "They wanted to get *out* there," recalls Roger Evans, the New York attorney Planned Parenthood hired in 1983 as the litigation unit's first director. "They wanted me to kind of walk in there on Monday morning, and by Friday afternoon identify three incredibly sexy cases I was going to bring by the next week—they wanted to get *visible*."

Roger Evans and his staff had brought or assisted in about a dozen cases since then, and Evans had been monitoring Missouri for a potential H.B. 1596 lawsuit ever since the first long-distance distress calls from

Mary Bryant in Jefferson City. But it made no sense to fall into some undeclared competition with a nearly identical lawsuit in the same federal court, so by late spring Evans and Susman had arrived at an uneasy agreement—for Evans knew Susman's reputation on the abortion law circuit, and Susman and Blumenthal knew Evans was going to be weighing in from two thousand miles away—to work jointly on a single suit.

It was not a particularly momentous lawsuit, in those opening weeks of first-draft work: one more case, one more punitive regional abortion statute to dismantle in federal court. *REPRODUCTIVE HEALTH SER-VICES, et al., Plaintiffs,* v. *WILLIAM L. WEBSTER, Attorney General of the State of Missouri, and THE STATE OF MISSOURI.* No asterisk marked the case name as numbered draft paragraphs of legal complaint passed back and forth between St. Louis, Kansas City, and New York; it was not inconceivable that *RHS* v. *Webster* might find its way to the Supreme Court, but that was true of almost every serious abortion law challenge in the country, and neither Evans nor Susman and Blumenthal saw reason for exceptional worry as they settled on their principal points of legal attack. Missouri's new abortion law impermissibly impeded the work of clinics and doctors, they wrote:

> Named Plaintiffs and the medical provider class members are se-verely inhibited, restrained, deterred and precluded from offering medical counseling services in their finest, wisest and most healthful manner by the present, actual and threatened enforcement, effect, application and impact of the Act.

H.B. 1596 defied previous Supreme Court holdings, they wrote, by requiring that all post-sixteen-week abortions take place in hospitals. H.B. 1596 created "situations of gross professional malpractice," they wrote, by making it illegal for public employees to offer their patients abortion advice and counseling. H.B. 1596's opening passage—the Missouri General Assembly's declaration that "the life of each human being begins at con-ception"—violated the First, Fifth, Ninth, and Fourteenth Amendments to the Constitution:

> The Act on its face expressly violates the prior decisions of the United States Supreme Court in *Roe* v. *Wade,* and in *Doe* v. *Bolton,* and their progeny. . . .

On July 14, 1986, Tom Blumenthal and Frank Susman flew together to Kansas City to meet Roger Evans and deliver the complaint to the U.S. District Court.

TWELVE

The Count

1983

ON THE DAY Sam Lee was admitted to the Adult Correctional Institution in Chesterfield, Missouri, he wore a pair of bluejeans, a white cotton shirt, and a small cross he had carved from a piece of Israeli olive wood and suspended from a thong around his neck. On the back of the cross Sam had etched in Hebrew letters both his name and Gloria's. He had carved an identical cross for Gloria, and it made him feel better to think about his wife wearing her cross while he wore his; Sam kept his gold wedding ring on, too, even though he had been instructed to leave it at home. If prison officials took the ring away there was nothing he could do about it, Sam supposed, but he had decided that he would not make this any easier for them by pulling it off himself.

Sam was delivered to the Adult Correctional Institution on June 8, 1983, in the back of a paddy wagon that was also carrying John Ryan, Ann O'Brien, and Joan Andrews. The women were routed toward a separate part of the jail, and Sam and John waited at the intake office while admitting guards made written inventories of the garments and personal items each prisoner possessed upon arrival. Someone handed Sam an orange jumpsuit and told him to take off his street clothes. The warden came out of his office and spoke to Sam and John in a friendly way, saying he was sorry that he had to put them in jail, that he did not believe they deserved to be there, that he would try to watch out for them while they did their time. A prison guard twisted the wedding ring off Sam's left hand, sealed it into an envelope, and locked it away.

Sam Lee had been convicted on multiple counts of contempt of court —six separate violations of court orders barring him from the premises of certain abortion clinics—and sentenced to serve 314 days in jail. While his conviction and sentence were on appeal he had married Gloria Fahey, honeymooned in rural Wisconsin (one of the defense lawyers lent them a cabin on Lake Michigan), and moved with his new wife into an apartment that took up the first floor of a narrow two-story house in South St. Louis. It was not a very big apartment, three rooms and a bathroom, but it had a marble fireplace and good hardwood floors and a real stained-glass window. Sam and Gloria decorated the walls with their wedding photographs and a framed picture of Our Lady of Guadalupe; in the bedroom they hung the heirloom crucifix Gloria's mother had given them, and for eighteen days —that was how long they had together, from the day their honeymoon ended until the day Sam's appeal was denied—Sam woke every morning, for the first time in his life, to the ordinary pleasures of a married man inside his own home. He made coffee in the kitchen. He shopped at the farmers' market for vegetables and cheese. He cooked dinners for Gloria, who was working full time at Our Lady's Inn, and it made Sam happy to surprise her with meals of alarming ambition: once he overspent the food budget on a live lobster and left it waving its claws in the bathtub while Sam prepared drawn butter and put the water on to boil.

He was deeply in love. In the spring, before his wedding, Sam had written out a seven-page proposal for a pamphlet that was supposed to be entitled *Why I Am in Jail for Saving Lives;* there was to be no whining in this pamphlet, which was envisioned as lucid and inspirational, a dispatch from the jail cell in which Sam assumed he would find himself sooner or later. But after he was married Sam could see that the prospect of imprisonment had been easier to imagine when he was sleeping on other people's couches or basement cots. He knew his sentence was likely to be upheld, and he tried to make himself feel calmer by thinking about Gandhi and imprisonment, how Gandhi had written that the nonviolent protestor should accept without complaint the terms of his punishment. "Be a model prisoner," Gandhi had said, Sam had read that somewhere. "A model prisoner, beyond reproach."

In June a newspaper reporter telephoned the apartment to tell Sam the appeal had been denied. All four contempt of court sentences—Ann O'Brien had been sentenced to 314 days, just like Sam, and John Ryan and Joan Andrews had received sentences of 225 days each—had been upheld by the Missouri Court of Appeals. The *Globe-Democrat* put their picture in the paper, the convicts en route to jail: in the photograph John Ryan still had his tie knotted loosely under his suit jacket, and Sam stood tall and bearded with his hand atop the cardboard box that held his underwear and religious books, and Ann O'Brien wore a matronly print dress with a nice neck bow, her arms around a paper shopping bag of her own, as though she had stepped into the college dormitory long enough to settle her boys in for the semester. " 'We're planning on appealing to the Missouri Supreme Court,' said Samuel H. Lee," read the second paragraph of the *Globe-Democrat* story. " 'In the meantime, we'll go to jail.' "

SAM LEE and John Ryan were housed separately in long concrete rooms with stainless-steel urinals and bars on the windows and single beds lined up in rows against the walls. After three days both Sam and John were ordered to pack up for a transfer and the paddy wagon came for them again and drove them the half hour back to the St. Louis County work-release jail in Clayton. This jail looked more like an army barracks than a prison lockup, televisions in the bunkrooms and no bars on the windows, but guards stood at the entrance and every prisoner had permission to travel only to and from work, no detours. On paper Sam still had his job with the mail-order company; it was not what Sam would have called his *work*, his work was at the abortion clinics, but the mail-order company gave him a place to go and the red company van to drive there in, and in the mornings Sam would drive over to the office and put in a few hours' work and then hurry over to South St. Louis to meet Gloria at their apartment.

He was violating the terms of his work release, but no one ever caught him at it. By six o'clock Sam was always back in his quarters at the work-release jail, eating institutional suppers off plastic plates and reading until the lights went out. He was a model prisoner, beyond reproach. When he broke the rules he did it quietly, in the borrowed red van, and after a little while Sam also began leaving the mail-order company at midday so that he could walk up to the library at St. Louis University. Sam had a powerful feeling about this library because it was here that he had first studied abortion carefully enough to understand the direction in which his life was supposed to proceed, but it was not the theology and philosophy shelves that interested him now; Sam no longer needed help clarifying the moral nature of abortion. He wanted to read Supreme Court cases this time. He wanted to think about law.

Dear Sam,

We heard about you on the Pro-Life Action League's Hotline. God bless you. We know this is terrible unfair. I cannot even pretend to know the pain you are going through, but I hurt for you. Just know that you are a shineing example for us, and we are all working twice as hard for life because of you.

Dear Sam,

I have read that you are sentenced to 1 year in Jail for picketing an Abortion Clinic. How unjust that you should be spending time in Jail for trying to *stop* the killing. 10 years ago it would have been the abortionist in Jail for killing.

Dear Samuel, John, Ann and Joan,

It is very late in the evening as I write this letter to all of you. My precious four year old daughter (who will be 5 in November) is snug in her little bed. I sit here unable to sleep thinking about the four of you and then I wonder if you are resting well in your cells.

Dear Mr. Lee:

Please know that your sacrifice is known. Your bravery is respected. Your mistreatment is noted. We thank you for your willingness to stand up to injustice in the face of the consequences. We know that it is such devotion to principle which ennobles the human race.

May God bless you and may you soon be free.

In the work-release jail the lights and television sets went on at five o'clock in the morning, so that Sam woke to simultaneous glare and noise in a roomful of drug dealers and burglars and check bouncers and drunk drivers. Many of these convicted men were serving sentences of three to six months, approximately half the length of Sam's, and when he sat on his bunk to write back to the people who had written him—the letters were coming from around Missouri and from other states too, from Illinois and Maryland and New York—Sam tried to write about his sit-ins in a way that would cheer up the intimidated, the women and men who apologized in their letters because they could never sacrifice the way Sam and the others had. "You can do whatever God calls you to," Sam would write.

He worked out ways of putting things, trying to make them more personal and immediate. "If God calls you to raise those three children to be good Christians, that is your pro-life work—as important as going to jail," Sam would write. "If God calls you to do canvassing from door to door in the evenings after work for a pro-life candidate, and that is all you can do, *that* is as important as going to jail."

Sam brought into the work-release jail some copies of Dr. Willke's *Handbook on Abortion*, which had been expanded over the years and was by now in its twenty-first printing, and offered them to prisoners to examine on their own. He and John Ryan sat together on one of their bunkbeds and prayed, their voices nearly inaudible amid the clamor of the televisions, mentioning aloud in their prayers the jailers, the abortionists, the women at the clinics, and the Catholic judge who had ordered the

contempt sentences. In the letters Sam wrote he suggested that his corre-
spondents might want to meditate upon Proverbs 24:10–11:

> *If you remain indifferent in time of adversity,*
> *your strength will depart from you.*
> *Rescue those who are being dragged to death,*
> *and from those tottering to execution withdraw not.*

But privately Sam was having trouble seeing exactly what that meant
—seeing how he was supposed to go on rescuing those who were being
dragged to death. If judges had jailed him for one year, they might readily
jail him for two. Sam wanted to raise a family with Gloria. He didn't want
his children growing up with a father who could visit them only during
work-release hours. And even if Sam had been willing to strike such a
bargain, to offer up his own family life as the price of trying to protect
others from death, how many would be ready to join him? What was the
point of the four-person sit-in? Without a decent crowd the clinic sit-in
lost its impact, its visual image of a community roused to anger, and Sam
and John argued back and forth about how many men and women it took
to constitute a decent crowd. They settled on one hundred, and in the jail
they made up a posterboard declaration for people on the outside to sign,
a written promise that each signer would join new clinic blockades once
ninety-nine others made the same promise.

They got some signatures on their posterboard, but not very many,
thirty or forty at the most. John Ryan told Sam that was all right, that he
didn't mind downscaling their definition of a crowd; if they were insistent
enough in their demeanor, John pointed out, two dozen people might be
enough to shut down a single abortion chamber for a day. One clinic was
Auschwitz and another was Dachau, John would say; in the grand scheme
of things it didn't matter which one you chose or how many comrades you
had with you as long as you were out there doing *something*, interfering,
making noise, placing your body in the way.

But Sam was not convinced. He was talking by telephone to Andy
Puzder—after a while, once Sam had been granted weekend passes and
dinner-out privileges, he and Gloria would spend evenings at the Puzders'
house, and Sam would listen intently as Andy talked about the right-to-
life work *he* was doing now that his primary sit-in clients were in jail.

He was working on a municipal law, Andy said, an ordinance that
might help one St. Louis suburb keep a clinic from opening inside the
town boundaries; maybe Sam could help with the research. And there was
a law review article Andy and another lawyer had been preparing together,
the outlines of a legislative proposal, an innovative way to force appellate
court judges to confront the humanity of the unborn. Andy said he had
thought of this proposal one morning while he was jogging in Forest Park.

He had come right home and written it down on the back of an envelope. It was only a few paragraphs, even after he thought about it more and typed it up, but it seemed so intriguing that Andy expected he could make something substantial of it—a principled declaration, a stripping of euphemism, maybe another test case to push the Supreme Court toward reconsidering *Roe* v. *Wade*.

Andy showed Sam the single typewritten page. It began, without preliminaries: *The Legislature of this State finds that: 1. Actual human life exists from conception.*

Line by line, through his three numbered "findings" and his Therefore paragraph, Andy explained to Sam what he had in mind. *Roe* v. *Wade* had explicitly forbidden the state of Texas from "adopting one theory of life"—from declaring what Andy and Sam both knew to be true, that human life began with conception—for purposes of the state abortion law. A state could not invoke such a "theory," the *Roe* majority opinion had read, to "override the rights of the pregnant woman that are at stake." Like the rest of the opinion, this passage in *Roe* v. *Wade* appeared to both Andy and Sam to make no moral or legal sense at all, but never mind that; the point was, Andy explained, that in other areas of the law states were still permitted to act as though the unborn were human and worth society's protection.

Unborn children could inherit property, for example. They were recognized in some personal injury cases as human beings, distinct from their mothers. So if a single state legislature were to distill these ideas into a statute, Andy said, it might force into the open the moral hypocrisy *Roe* v. *Wade* had imposed upon every state: for this purpose they're human, for that purpose they're not. A draft paragraph from the law review article summarized Andy's plan:

> As a matter of strategy, if states choose to enact the legislation and make the findings proposed by this article, both the Supreme Court and the advocates of "pro-choice" will face a potentially untenable dilemma. Advocates of "choice" will predictably oppose the statute since it personifies the fetus and suggests that abortion is immoral. However, by opposing the statute they will be opposing childbirth since the statute is designed to protect only the unborn of women who choose to bear children, as well as the childbearing women themselves. An individual can hardly call himself "pro-choice" if he only advocates abortion. When the Supreme Court ultimately reconsiders *Roe* on its merits, it will squarely encounter the quintessential issue which is ostensibly ignored in *Roe:* When does life begin? The Court will not be able so facilely to disregard this issue when presented with specific legislative findings on the subject; and, assuming enough states enact the proposed legislation, it will not be able to conclude by fiat that there is no consensus of opinion on the issue.

Andy was excited about the possibilities for his state law proposal. *The laws of this state shall be interpreted in such a manner as to grant to unborn children from the moment of conception all the rights, privileges, and immunities available to other persons, citizens, and residents of this State.* Then some language about the Supreme Court, both to keep the legislation from being unconstitutional on its face and to highlight the essential illogic of *Roe: subject only to the Federal Constitution of the United States and decisional interpretations thereof.* Really, appellate court judges could do whatever they wanted with this, Andy said. They could uphold it: no big deal, updates state legislation, advances admirable goals, doesn't interfere with *Roe.* Or they could strike it down: impermissible, prohibited under *Roe,* unborn children *have* no "rights, privileges, and immunities."

In either case, it was bound to end up in court, Andy said; it was too provocative not to. If a state legislature were to adopt language like this as law, then somebody, Frank Susman or one of his counterparts outside Missouri, was going to challenge it. And every challenged abortion case— even a case that merely alluded to abortion without restricting it outright —had a chance at reaching the United States Supreme Court.

And Ronald Reagan, the right-to-life president, was in the third year of his first term—which meant that *if* such a law were challenged, and *if* Reagan were reelected in 1984, and *if* the right number of Supreme Court justices were to retire during the next five years, and *if* Reagan selected his Supreme Court appointments from jurists who found *Roe* v. *Wade* as indefensible as the president liked to say he did, then Andy Puzder's one-page legislative proposal might eventually find its way to a reconstituted Court, a Court that might look at the words *actual human life exists from conception* and declare: Well, yes, in fact, it does.

So MANY pitfalls, it seemed to Sam—so much guessing and counter-guessing, so much waiting, so many maybe-nots, so many pieces of paper to fuss over while the clinics went on about their terminations. For five years Sam had left to other people this cooler, less immediately gratifying part of the right-to-life movement's work: the writing of bills, the shepherding of legislators, the paragraph-by-paragraph repetition of legal arguments and theories that might never advance beyond some local judge's desk. It had the aspect of a massive chess game played with excruciating slowness, and Sam could see that if he were at liberty around St. Louis he would still be blocking clinic doorways instead of arranging footnotes for speculative articles about what *could* happen if only *other* helpful developments took place.

But because Sam was incarcerated (loosely incarcerated, but under close enough watch that a single clinic trespass would have sent him instantly back to the concrete rooms with bars on the windows), he began to think seriously now about the mechanics of the chess game itself. What lines of attack were available to someone taking on abortion through the lawsuit or the regulatory bill? What exactly had the Supreme Court held—not just in *Roe* and *Doe*, nearly everybody in the movement knew that, but in the rulings that came afterward, the rulings that were generally referred to, in exquisitely inappropriate legalese, as the "progeny" of *Roe* and *Doe*?

One by one, the bound *Supreme Court Reporters* stacking up before him on the St. Louis University law library tables, Sam studied the Court's abortion rulings over the decade since 1973. Printing quickly into his yellow legal pads, he wrote down what appeared to be the highlights.

Connecticut v. *Menillo*, 1975. States may require that abortions be performed by licensed physicians.

Planned Parenthood of Central Missouri v. *Danforth*, 1976. States may *not* require the husband's consent before an abortion—unconstitutional, under the holdings of *Roe* and *Doe*. Unconstitutional to prohibit saline infusion as an abortion method. Unconstitutional to impose a blanket parental consent requirement for minors. States *are* permitted to require abortion facilities to keep records and obtain written consent from each patient.

Maher v. *Roe, Beal* v. *Doe, Poelker* v. *Doe*, 1977. States and cities may refuse both to pay for abortion and to offer abortion as a public hospital service.

Bellotti v. *Baird*, 1979. Unconstitutional to require parental consent for minors' abortions unless the consent law gives minors adequate appeal to judges.

Harris v. *McRae*, 1980. "Hyde Amendment," ending federal funds for abortion (with exceptions only for reported rape/incest or threat to the mother's life), upheld as constitutional.

City of Akron v. *Akron Center for Reproductive Health*, 1983. Unconstitutional for government (city of Akron, in this case) to mandate the following: twenty-four-hour waiting period between clinic visit and abortion; hospitalization for all second-trimester procedures; obligatory informed consent process requiring doctors to tell all patients about the fetus' "appearance, mobility, tactile sensitivity, including pain, perception or response, brain and heart function, the presence of internal organs and presence of external members."

Planned Parenthood of Kansas City, Missouri v. *Ashcroft*, 1983; *Simopolous* v. *Virginia*, 1983. Companion rulings to *Akron* (handed down the same day). Unconstitutional for the state of Missouri to require hospitalization for abortions after twelfth week of pregnancy. Constitutional for the state of Virginia to require that second-trimester abortions take place in hospitals *or* licensed outpatient clinics. Other aspects of the Missouri abortion law upheld as constitutional: mandatory pathology reports, a requirement that minors obtain parental or judicial consent, and a rule that a second doctor be present for abortions performed after viability.

Sam had read appellate opinions before, when he was helping Andy with research into the necessity defense, but until the fall of 1983 he had never examined the Supreme Court's abortion rulings one after the other, as though they added up to something bigger and more complicated than *Roe* v. *Wade* alone. For five years right-to-life newsletters had been telling Sam that *Roe* was merely the starting point for an increasingly adamant Court majority—that when a person considered the major rulings as a whole, it became apparent that what Right to Legal Abortion really meant was Right to Abortion Without Impediment: the right of a woman to abort her baby without being obliged to obtain the permission of her husband, or to take an extra twenty-four hours to think about it, or even to be confronted with a description of the baby's physical appearance. The only straightforward victories had come in the public funding cases. But as he made his way through the rulings at the law library, reading not just the majority opinions but also the sometimes vehement objections in each dissent, Sam began to see why Andy was nevertheless optimistic about bringing yet another case to the Supreme Court—how it was possible to find hints of encouragement even amid so many defeats.

The key, Sam could see, was in the numbers. Nine justices sit on the U.S. Supreme Court; all it takes, for a ruling that makes precedent or refines precedent or reverses precedent entirely, is five votes. If five justices could be persuaded that *Roe* v. *Wade* was a bad decision, that it was wrongly decided, or that new evidence or circumstances now justified a reversal, then ultimately it would make no difference that the right-to-life movement had nearly destroyed itself trying to win approval for a Human Life Amendment. For all practical purposes the Amendment was dead; the last big vote had come in June, when Orrin Hatch's still-controversial constitutional proposal (reduced by now to ten words, *A right to abortion is not secured by this Constitution*) had failed to win even a bare majority in the Senate. After the vote there was brave talk from this or that faction, Senator Hatch insisting that his amendment efforts were simply "the first step"; but the moment had passed, everyone knew it, the Congress wanted the wreckage cleared away as efficiently as possible, and Sam knew that within the policymaking branches of the right-to-life movement the central topic of discussion was now the Supreme Court and the complex political and legal calculations that might produce those five crucial votes.

Roe and Doe were both seven-to-two rulings, with the two minority justices, William Rehnquist and Byron White, making it plain in their written dissents that they believed the Court had no authority at all to force states to legalize abortion. Nothing either justice had written since suggested any change of heart. So that was two votes.

Then in 1975, Justice William O. Douglas, who had voted with the Roe majority, resigned from the Court. But his replacement, the former Seventh Circuit Court of Appeals judge John Paul Stevens, proved to be a pro-Roe vote as well—so the Douglas-Stevens switch appeared to leave the voting balance unchanged.

In 1976, the Chief Justice, Warren Burger, joined Rehnquist and White as they dissented, vigorously, in Danforth. Burger wrote no Danforth dissent of his own, but the opinion Burger signed onto was barbed and emphatic as it assailed the majority for overturning state spousal consent requirements. "It is truly surprising that the majority finds in the United States Constitution, as it must in order to justify the result it reaches, a rule that the State must assign a greater value to a mother's decision to cut off a potential human life by abortion than to a father's decision to let it mature into a live child," wrote Justice White in the Danforth dissent. "These are matters which a State should be able to decide free from the suffocating power of the federal judge, purporting to act in the name of the Constitution."

If joining this opinion meant Justice Burger was switching sides—if by aligning himself with White and Rehnquist, Burger was signaling some change of heart on Roe itself—then possibly that was three votes.

Then in 1981 Justice Potter Stewart, who had been part of the seven-vote majority in Roe and Doe, retired from the Supreme Court. His replacement, and Ronald Reagan's first appointment to the Court, was the Arizona state appeals court judge Sandra Day O'Connor. As a Supreme Court nominee O'Connor had been publicly reviled by right-to-life leaders, who found in her background nothing to convince them that this was the "sanctity of human life" justice they had expected from Ronald Reagan; during the 1970s, it seemed, when O'Connor had served three terms as an Arizona state senator, she had on four occasions cast votes that might be construed as favorable to legal abortion.* And during her confirmation hearings, O'Connor declined to answer questions about whether she would vote to overturn Roe.

Sam remembered the uproar over Sandra Day O'Connor; in the

---

* In 1970, according to O'Connor's testimony during her confirmation hearings, she cast a preliminary vote for a bill repealing Arizona's criminal abortion statute. In 1973 she co-sponsored a bill increasing the availability of birth control information—a bill O'Connor testified did not provide for or directly involve abortion. In 1974 O'Connor voted against a resolution calling for a Human Life Amendment (at the time, O'Connor testified, she believed the measure had not received "proper reflection or consideration"); she also voted that year against an amendment banning abortions in some state hospitals.

weeks leading up to her confirmation the *National Right to Life News* had referred to her, in a front-page editorial, as a "pro-abortionist." He remembered also that she had been confirmed by a vote of 99 to 0—even the movement's staunchest Senate colleagues had balked at the idea of voting against a conservative Republican who was also the nation's first female Supreme Court nominee—and that for many months afterward O'Connor had offered no public hint as to her judicial views on legal abortion in general and *Roe* v. *Wade* in particular. Then in June 1983, one week after Sam was sent to jail, the Supreme Court had released its ruling in *City of Akron* v. *Akron Center for Reproductive Health.* The ruling itself was another disappointment, the Court declaring once again that government was essentially prohibited from interfering with abortion in ways that might push a woman to change her mind; but from the confines of jail Sam had read the daily newspapers closely enough to see that the big news in *Akron* was Sandra Day O'Connor's dissent. The new justice, voting for the first time in a Supreme Court abortion case, had sided with Rehnquist and White.

The new justice, moreover, had written the dissent herself, and written it as a lengthy and detailed attack on *Roe* v. *Wade*—on the "fallacy inherent in the *Roe* framework," as O'Connor said in her widely quoted dissent. Nearly every newspaper account in the country referred to the same phrase in O'Connor's *Akron* dissent: the internal logic of *Roe* v. *Wade*, the justice wrote, "is clearly on a collision course with itself."

Now in the law library, the full *Akron* ruling before him, Sam was able to study the O'Connor dissent from beginning to end and to see why right-to-life lawyers had been arguing so excitedly about it. O'Connor had stopped short of calling for *Roe's* reversal. Her dissent did not repeat Justice White's 1973 argument that the Supreme Court was simply bullying the states with "raw judicial power." But she picked apart the reasoning with which the original *Roe* opinion was constructed—the argument that pregnancy can be logically divided into three trimesters, and that each trimester warrants a different level of acceptable government interference. "The decision of the Court today," O'Connor wrote, "graphically illustrates why the trimester approach is a completely unworkable method of accommodating the conflicting personal rights and compelling state interests that are involved in the state abortion context."

Medical practice and technology had changed rapidly over the last decade, O'Connor wrote. "Trimesters" no longer served, if indeed they ever had, as a defensible guide to reviewing state abortion law. Dilation and evacuation was a case in point, O'Connor wrote; in 1973 this kind of instrumental abortion was not supposed to be safe for second-trimester pregnancy. Now, apparently, it was—at least partway into the second trimester. Since according to *Roe* states were permitted to regulate second-trimester abortion only in ways that protected maternal health, O'Connor wrote, the State must continuously and conscientiously study contemporary medical and scientific literature in order to determine whether the effect of a particular regulation is to "depart from accepted medical practice" insofar as particular procedures and particular periods within the trimester are concerned. Assuming that legislative bodies are able to engage in this exacting task, it is difficult to believe that our Constitution *requires* that they do it as a prelude to protecting the health of their citizens. It is even more difficult to believe that this Court, without the resources available to those bodies entrusted with making legislative choices, believes itself competent to make these inquiries and to revise these standards every time the American College of Obstetricians and Gynecologists (ACOG) or similar group revises its views about what is and what is not appropriate medical procedure in this area.

The time had come, O'Connor argued, to abandon the trimester principle entirely. *Roe* declared that states could pass laws to protect "potential life" only during the third trimester or after viability, but as legal doctrine, O'Connor wrote, this really made no sense: "At any stage in pregnancy, there is the *potential* for human life," she wrote. "*Potential* life is no less potential in the first weeks of pregnancy than it is at viability or afterward" (emphasis in original). It was O'Connor's contention that the Supreme Court ought instead to be examining all state abortion laws, regardless of which trimester they affected, under a new and uniform standard. "Undue burden" was the phrase she proposed: Was the regulation in question unduly burdensome to women seeking abortions? If it was, the Court should overturn it; if not, the Court should get out of the way and stop trying to second-guess state legislatures.

And according to this standard, O'Connor wrote, the city of Akron's abortion ordinance was entirely acceptable. Requiring hospitalization for second-trimester abortions was not an undue burden—there was no evidence in this case that Akron hospitals refused abortion patients. Requiring doctors to give specific kinds of pre-abortion information was not an undue burden—the city was only trying to ensure, O'Connor wrote, that each abortion decision was made "in light of that knowledge that the city deems relevant to informed choice." A mandatory twenty-four-hour waiting period was not an undue burden either, and even if it was—even if one assumed for the sake of argument that such a waiting period raised to a burdensome level the price of clinic or hospital visits—then the waiting period was justified anyway, O'Connor wrote, because the state maintained so compelling an interest in "maternal physical and mental health and protection of fetal life."

In right-to-life legal circles the Sandra Day O'Connor *Akron* dissent was rapidly becoming the second most minutely studied opinion ever to emerge from the chambers of the Supreme Court, and Sam tried to read it the way he thought a lawyer probably would, looking for signals between the lines. What was the new justice trying to tell them? The very use of the word "burden" suggested that O'Connor believed some state interference could be viewed as unacceptable, a load too heavy for women to have to bear. But the justice gave no hint as to what she might view as such a burden, and her reference to "potential life . . . in the first weeks of pregnancy" was surely cause for some encouragement. If a case came before Justice O'Connor in which the question was posed in an all-or-nothing way, uphold *Roe* or reverse the ruling and start over again, how was she likely to vote?

At the very least, Sam thought, the movement now appeared to have an ally in Justice O'Connor. Perhaps President Reagan had not betrayed them after all. It was tenuous arithmetic that now sustained them from one case to the next: Justice Burger was still hard to position; he had joined the anti-*Roe* justices in *Danforth*, but then in *Akron* Burger voted with the majority to find the Akron ordinance unconstitutional. Still he might be a vote against *Roe*, if the question were properly framed. And if Justice O'Connor was also a vote against *Roe*, then that was four.

One vote left—one Supreme Court retirement, that might be all they needed, one more vacancy for President Reagan to fill. "The numbers that really count," a Christian Action Council lobbyist had declared in the newspapers, the day the Hatch Amendment failed in the Senate, "are the ages of the Supreme Court justices."

Sam began wondering what it would be like to compose some provocative legislation of his own.

He looked again at Andy Puzder's proposal. Andy had the right idea, it seemed to Sam: the trick was writing the legislation so that it served many purposes at once. You wanted your bill to teach, to inspire public discussion, to force the subject back out into statehouse debate. You wanted to word your bill so that it would serve some practical use even if it did become law, even if nobody challenged it in court. But at the same time you wanted to dangle the bait, see if the lawyers on the other side would take it. Anything might set off a test case if the justices were lined up and ready to vote for reversal of *Roe*.

Sam put some ideas onto his legal pads. *Noncooperation*, he wrote.

He carried his legal pads to the university, brought them back to the jail with him at night, piled them up in the bedroom at the apartment in South St. Louis. *Refusal to participate in abortion*, Sam wrote. He had been reading a great deal about Gandhi; Sam had finished the long Robert Payne biography, and he also dug out his Gene Sharp books, the three-volume *Politics of Nonviolent Action*, with their heavily footnoted chapters on Selective Social Boycotts and Rent Withholding and Refusal of Industrial Assistance.

Unlawful for employers to discharge any individual because of refusal to participate in abortion, Sam wrote.

Sam was improvising here, trying to make his phrases read like some of the printed legislation he had seen over the last few years. He had an uplifting image in mind: Imagine a bank teller who is handed a bundle of money from the day's take at the local abortion clinic—and refuses to accept the deposit. Imagine a repairman who is called to fix the clinic's telephone lines—and refuses to enter the building. If office suppliers refused to rent copy machines to the clinics, and janitors refused to bring their mops and pails inside the clinics, and hotel clerks could look up from their registration desks and say, "I'm sorry, but I am not willing to admit an abortion doctor into this establishment"—if all these actions were protected by state law, perhaps even sanctioned by state law, then the collective force of this noncooperation might create a kind of chokehold on the abortion business itself.

Sam kept his legislation ideas to himself at first, writing phrases one way and then another to see how they looked. Five months into his jail sentence he was transferred again, this time into a county-supervised apartment building to which he was supposed to return every week night by nine o'clock. He shared his apartment with John Ryan and a manslaughter convict, a sad-faced alcoholic who had gotten drunk one night and shot to death a visitor he mistook for a burglar; the manslaughter convict would gaze at Sam from time to time and shake his head, saying, Look at us, I took a life and you tried to save one, you got a longer sentence than I did. Sam would tell him not to worry about it, that Sam was managing pretty well considering that he was still technically a prisoner of the St. Louis County correctional system. He saw friends in the evenings. He gave public talks at St. Louis University. His wife was pregnant and due to have their first child the following summer. Organizations around the Midwest were inviting Sam to come speak to their members, and he was eager to oblige; in January 1984, one day after he was officially released from the St. Louis County jail, Sam got into his old Volkswagen and drove four hundred miles north to address a right-to-life group in Waterloo, Iowa. He carried with him one of his legal pads, the top seven pages filled with prompts for his speech. *One purpose of jail is to reform criminals and to bring about repentance*, Sam had written to himself, outlining the conclusion of his Waterloo talk. *Dept of Justice Services has failed in that task. I'm as pro-life as the day I entered jail.*

FOR A WHILE, for as long as he could bear it, Sam lived something approaching the life of a conventional wage earner. He registered at a temporaries agency and was hired on at Monsanto to conduct legal research for the company's law offices. It was all corporate law, patent infringement cases, but it was interesting work and it paid ten dollars an hour, more money than Sam had ever made before. He and Gloria had an idea that they would start up a secondhand bookstore, and they began driving to garage and estate sales to buy good-quality hardcovers, art books and religious books and cookbooks and novels. They stacked the books in the closets of their apartment, thought up clever bookstore names like Novel Lovers Under the Covers, picked out a lively collegiate street where they imagined they might rent a suitable storefront.

Their daughter Miriam was born at the end of June, in the bedroom of the apartment. A doctor and nurse came to help. Gloria labored all night, Sam timing the contractions, the radio tuned to an oldies station, Gloria walking around the apartment trying to breathe evenly while from the bedroom came loud music of the Supremes and Freddy Fender and the Rolling Stones.

Every month Sam attended the regular meeting of the board of directors of Missouri Citizens for Life, Eastern Region. The meetings dragged on into the evenings and sometimes adjourned after midnight. The agendas proceeded noisily and dutifully through the business of the MCL volunteers: Treasurer's Reports. Reports on Proceeds of Carnation Sales. Legislative Updates. Sample Drawings for Christmas Cards. Membership Mailing Costs. Knights of Columbus Handicapped Children Tootsie Roll Drive. Letter-Writing Campaign Asking IBM to Stop Donations to Planned Parenthood.

With his good job, and his apartment, and his wife at home with the baby, Sam tried to persuade himself that this was enough—that he was doing what God wanted of him.

But he did not really believe it. At the meetings he was earnest and serious and worked at appearing attentive while people sat around Ann O'Donnell's living room discussing the costs of various forms of office telephone service. "Sam was the Bearded Young Man," recalls Ceil Callahan, who was working in Jefferson City for Missouri Citizens for Life. "Many of the right-to-life people were women, most of us were housewives, with time available—we *made* time available. But he looked like he came out of my college days. He'd walk into a right-to-life meeting and you'd go, whoa, long hair! He was very soft-spoken. But no one ever tried to talk Sam out of a position. He was always trying to talk other people into a position. A lot of the time people in the room would be going one direction, and Sam was, no, we need to go *this* way."

In October 1984, nine months out of jail, Sam quit his job at Monsanto. He didn't talk it over at length first with Gloria—that was a sore point between them for a long time afterward, Sam just quitting, looking slightly startled when Gloria asked him how they were going to pay the rent, making it sound so uncomplicated when he explained it to her and their friends. He was going back to full-time right-to-life work, Sam said. He was *called* to full-time right-to-life work. This was his vocation, Sam said, the ministry he was supposed to be conducting in the world; it was considerably more freeform than the ordained priesthood, but Sam was surer about this ministry than he had ever been about the Franciscans. He would raise his own money, Sam said. He would write letters asking for donations and organizational grants. Gloria could return to her work at Our Lady's Inn; she had always said she wanted to go back to work eventually, Sam could see now that she was right about that, and the baby was old enough to go along with her. They would manage one way or another from one month to the next. God would guide them along.

April 1, 1985

Msgr. Joseph W. Baker
3202 Pulaski St.
St. Louis, MO 63111

Dear Msgr. Baker:

I spoke to you about a week and a half ago regarding some possible sources of funding for my full-time ministry in pro-life. Enclosed is a sketch of my family and a resume of what I've been doing since I started full-time in right-to-life last October. Perhaps this can be of some use to you in explaining to others what I am doing.

My wife Gloria and I believe that God has called us to completely dedicate our family to protecting the unborn and aiding their mothers. Despite the long hours and the financial difficulties, we find the work so necessary, and we plan on continuing this commitment to pro-life.

Thank you very much for your interest, and your willingness to help out by contacting others who may consider contributing financially.

In Christ's Love,
Samuel Lee

Sam and Gloria lived on six or seven hundred dollars a month—less when the donation checks slowed. They made another thousand dollars selling off all the secondhand books at a yard sale in front of their apartment. Some Pro-Life Legal Defense Fund attorneys gave Sam an office to use, a spare room in the back of their law office in Clayton, and in the mornings Gloria would buckle the baby into the carseat and drive Sam over to Clayton so that he could work amid his growing stacks of notepads and reprints and legal briefs. He wrote research memos for the Defense Fund attorneys. He wrote editorial copy for right-to-life newspaper advertisements. He wrote press releases for the Pro-Life Direct Action League, which was the organization John Ryan had started up in the fall of 1984; Sam had signed the incorporating papers with John, and as he composed his press releases on the computer the attorneys had lent him, Sam learned how to format boldface headlines and mimic what seemed to him the straightforward, informative voice of newspaper prose.

Pro-Life Activists Plan Large Sit-in For Friday Morning

In what is expected to be one of the larger pro-life sit-ins in recent months, about 24 pro-life activists will enter Reproductive Health Services, Inc., an abortion facility, and attempt to nonviolently stop abortions from being performed. The facility, located at 100 N. Euclid, City of St. Louis, has been the site of similar activities by pro-lifers in recent months. Also expected are a large number of picketers who will peacefully demonstrate and pray on the public sidewalk in front of the clinic.

At the top of these press releases Sam always wrote his own telephone numbers, home and office. FOR MORE IMMEDIATE INFORMATION CONTACT: *Samuel Lee.* But there was something wrong with this phrase, and Sam knew it every time he typed it out: he no longer had the information, not in the way that mattered. The sit-ins were John Ryan's work now. Sam still came along on the occasional Saturday, when they could find a clinic with no legal injunction banning him from the property, but the rules of engagement had changed and Sam could see that the others had lost interest in arguing with him about it. The padlock had become a popular sit-in accessory, people chaining and padlocking themselves into clinic doorways; there were stinkbombs and smokebombs and John Ryan had taken to grasping wildly with his hands as he was being dragged from the clinics, on the chance that he might grab hold of some curtains or a piece of furniture or an electrical cord. John had found that he could set off a noticeable reaction among clinic staff people by repeating the same phrases in a quiet, silky voice, *wash the blood off your hands, wash the blood off your hands;* when he recognized someone he would make it personal, the voice still silky, he had learned how to project from within the paddy wagon so that he could call softly through the bars until whoever he was gazing at turned away from him and went back inside. *Come on Amelia wash the blood off your hands. Come with me now. Wash the blood off your hands. How much does Judy Widdicombe pay you for each baby you kill? Wash the blood off your hands.*

"Organizers of Friday's action deny recent reports that pro-life activists at other sit-ins in the St. Louis area have assaulted clinic personnel, and screamed and yelled at patients," Sam wrote in one of his press releases. But this was not precisely true, and he knew it, and it made him feel dishonest. He knew there was yelling at some of the sit-ins, and he knew there were some dramatically elevated voices that might have been described, by an unsympathetic listener, as screaming. When he thought about what was going wrong, it sometimes seemed to Sam that emotion had taken the place of thoughtful belief, that people like him—pedantic, Jesuitical sorts of people, Sam supposed, with their philosophical way of picking everything apart and worrying over the very best and most moral course of action—had been shoved aside to make room for a new generation that operated principally from the gut. John had passion, Sam had to say that for him; he had a charisma and a forceful drive and people liked singing "Onward, Christian Soldiers" with him, all that martial vigor. Maybe Sam was behaving like a Pharisee again, overestimating his own worthiness—maybe the movement needed people like John; or Joan Andrews, who had taken to "disarming weapons" by cutting apart cords and tubing on the suction machines; or the Chicago activist Joseph Scheidler, who had made himself famous by marching in front of abortion clinics while shouting exhortations through a bullhorn.

One weekend John drove Sam up to Appleton, Wisconsin, where two pastors had invited Direct Action people from around the country to a national convention—more of a family reunion than a convention, Sam thought, a few roomfuls of people in a small motel. Joe Scheidler was there, and Joan Andrews sent her regards from jail, and on the outdoor motel sign some of the Appleton organizers had put up letters reading: WELCOME, HAVE A BLAST! This was a joke: a bomb joke, Direct Action humor. Inside the motel the conference participants wore lapel pins shaped like small sticks of dynamite. The point was not lost on Sam, let's make fun of the public image since the press thinks we're all terrorists anyway, but he declined to put on his lapel pin and when he listened to the conversation around him Sam heard an undercurrent of ambivalence, an uncertainty about exactly how the real bombers—the handful of men around the United States who had so far been convicted of bombing or torching abortion clinics—were supposed to be regarded. No one wanted to condemn them. No one wanted to applaud them, really, not directly and certainly not with filmmakers and reporters in the halls of the motel, but Sam could see that in the moral climate the Direct Action people were creating for themselves, clinic bombers were assuming the role of the frontline hero pushed just a little too far. Do we deplore the man who blows up Dachau? Do we care more about bricks or about babies? The April convictions of Matthew Goldsby and James Simmons had already inspired some troubled, introspective articles in the radically conservative Catholic newspaper *The Wanderer:* Goldsby and Simmons were the young Pentecostal Christians arrested, along with Goldsby's fiancee and Simmons' wife, for the Christmas Day 1984 bombings of three abortion clinics in Pensacola, Florida. "It should be recalled that just war criteria can sometimes apply to intolerable social situations," one *Wanderer* columnist had mused in print a few weeks after the young men's arrest:

If American crematoriums had been incinerating 4,000 Jews or blacks a day for some 12 years, and a militant generation arose (the

Pensacola kids were but six and nine years old respectively when *Roe* v. *Wade* was imposed) to attack the crematoriums, there would be some who would call such attacks terrorism, or anarchy. . . . And perhaps they would be right.

Or perhaps not. What should be remembered, however, is that if such arguments are applicable to the abortion clinic bombers, then they apply as well to any German who might have tried to blow up the train lines that helped transport the Jews to Auschwitz.

Even Don Benny Anderson, the Texas man now serving forty-two years in federal prison for the kidnap of Dr. Hector Zevallos, was a complicated call amid the new Direct Action leadership. Joe Scheidler had publicly referred to the Zevallos kidnap as "a citizen's arrest on a mass murderer," and in Scheidler's new book *Closed: 99 Ways to Stop Abortion*, which was widely admired in Direct Action circles, Scheidler's brief chapter on violence (counterproductive, Scheidler argued, and never more than a temporary solution anyway) included a compatriot's salute both to three convicted clinic arsonists and to Anderson, who appeared at least momentarily to have escalated the tactical options from attacks on property to physical assaults on the abortionists. "While we might respect the zeal that would prompt such activities, we do not condone or recommend them," Scheidler had written.

All four men are dedicated to the belief that unborn children's most basic right—the right to life—is being violated by abortion and that daring actions are needed to awaken Americans to the terrible reality of abortion. But most pro-lifers would say that all four, if guilty, went too far. . . .

We must point out for the sake of proper perspective, however, that no amount of damage to real estate can equal the violence of taking a single human life. . . . It is a sign of the deterioration of our values that much of the national media concentrates on damaged buildings, with pictures of charred real estate, while refusing to present pictures of the human victims who are heartlessly and systematically dismembered and painfully killed inside that real estate.

At the Appleton conference Sam watched Joe Scheidler personally introduce John Ryan to an expectant crowd ("The first time I heard John talk, I really felt like a wimp," Scheidler declared), and when John stood to speak he kept his hands in his pockets, his own oratory skills being more of the guerrilla nature, and looked convincingly embarrassed by the public praise. "We're really at a pivotal point, I think, in the pro-life movement in this country," John said. "I don't think I'm being overly dramatic or melodramatic when I say that." The word "rescue" was very much in evidence that weekend; Sam used to quote it himself from Proverbs but had never thought of it as a movement label before. Now he heard people talking about Rescuers, the Rescue Movement, and the new vocabulary only deepened his uneasiness; it was quite a different image, more aggressive and heroic, as though the old notion of the vulnerable Peaceful Presence had been formally cast aside.

Sam wasn't sure he wanted to be a Rescuer. When he and John drove back from Appleton Sam was sober, thinking through his possibilities. Plainly the Missouri sit-in leadership—the Rescue leadership, Sam imagined he was now supposed to call it—had passed from Sam to John, just as it had passed a half decade earlier from Vince Petersen to Sam. This might be good for the movement or it might not; maybe Sam was wrong about the padlocks and the stinkbombs, maybe John would figure out how to pull in more than two dozen sit-in people, Rescuers, so that his Saturdays at the clinic made a *dent*. But Sam didn't think so. In his public work he was extremely careful about what he said and wrote: he went on composing the press releases; he never criticized John Ryan publicly; he gave speeches in which he defended the legitimacy of sit-ins. *Who in here has young children under two years of age? Let's say that the U.S. Supreme Court ruled today that children under two are not persons. What would you do?*

And once a week or so, when he could get a ride or Gloria could manage overnight without the car, Sam went to Jefferson City to learn how the state legislature works.

It was not that he was abandoning sit-ins entirely, Sam would tell himself during the two-hour drive to the capitol (the vocabulary was not coming easily to him; he could see that he was never going to be able to call them Rescues without sounding unnatural). But history seemed to suggest that sit-ins worked only when they could rouse a community to action. If crowds of the dedicated had been willing to mass around passionate John Ryan, if Sam had been sure that John could hold them all to peaceful witness, if there was any convincing evidence that sit-ins *mattered* in the long run—Sam might as well be candid here, he had believed so intensely, he had given so much of himself, but abortion was just as legal and just as prevalent as it had been the first time he lowered himself into a clinic doorway. For six years Kathy Edwards had been arguing that well-crafted legislation was the best tool the right-to-life movement had, that Sam and his sit-ins were alienating far more people than they attracted; and while Sam wasn't willing to go that far, he had come to see that Kathy was right about legislation as a serious tool—especially now, amid what appeared to be the gradual realignment of the Supreme Court.

The 1985 session of the General Assembly was Sam's Jefferson City apprenticeship. Kathy Edwards and Ceil Callahan led Sam through the halls of the capitol, pointed him toward the library, sat him down with printed primers on legislative procedure. On his first few visits he was bewildered and fascinated and felt entirely out of place; his beard and discount-store tweed jacket made him look like an impoverished scholar among the politicians, and moreover Sam could see at once that the capitol building housed a baroque and private subculture into which he might be admitted only after considerable study. He spent hours at the library tables, reading House and Senate journals, trying to follow the daily entries carefully enough to understand how to phrase his own questions. What was a Senate Committee Substitute? How were amendments attached to existing bills? What was the difference between the formal and informal calendar?

"Sam was truly a quick study," recalls Ceil Callahan, who would often put Sam up overnight at her family's house in Jefferson City. Sam slept on the foldout couch in the Callahans' living room, and when the capitol had closed down for the evening and the Callahans' five children had gone to bed, Sam and Ceil would sit up late into the night, talking about strategies and personalities and the relationship between political work and their own religious faith. Ceil told capitol stories and Sam listened intently, trying to get his lessons right: how the busfuls of people from the parishes and the local MCL chapters had to be properly primed before they walked into the building; how there was always some righteous contingent that seemed to think it was the legislators' duty to show up for the General Assembly's opening day on January 15 and outlaw abortion on January 16; how you had to warn your citizen lobbyists beforehand that they were going to encounter liquor and irreverent statehouse humor and legislators who appeared to be heartily shaking their hands without actually listening to a word they said. ("I used to see how horrified people were by the things going on in the capitol," Ceil Callahan recalls. "I did used to have to say to them, look, abortion is just *one issue* for these folks. Sometimes the people would just get livid and argue with me.")

Ceil told Sam how a young Republican assemblyman had once said to Kathy Edwards, "You-all could throw a dead dog out on the legislature floor and everybody would vote for it," meaning this remark as high flattery, the spirit in which it was received. Ceil told Sam how Bob Griffin had to be watched especially closely because there was something duplicitous about him, he could sweet-talk you during the lobby days and then pull the plug on you when he thought nobody was looking. Ceil told Sam who his allies were, the public leadership, the assemblymen and senators willing to carry right-to-life leadership and defend it vigorously in floor debate. And Ceil knew Sam was sharp, she had been listening to him for years at MCL board meetings, but even she was surprised at how readily Sam worked his way into the Missouri statehouse; he loved it, he was elated by it, he ate dinner with legislators and talked up old capitol hands and went back to St. Louis to rewrite the subsections of his own legislative prose. *Participate in abortion: To receive, obtain, perform, pay for, assist*

in, counsel for, suggest, recommend, refer for, promote, or procure an
abortion; or engage in a course of study promoting, or training for, the
performance of abortions; or

To provide advertising, equipment, materials, bookkeeping, medical
supplies or services, materials, counsel, banking, office space or real estate,
hotel or motel rooms, insurance, delivery or installation of goods, private
guards or security, or any other product or service that furthers the opera-
tion of an abortion facility or assists a physician in the performance of
abortions;

188.105. 1. It shall be unlawful:

IN CHICAGO, at the turn-of-the-century office building that housed the principal legal think tank of the American right-to-life movement, Missouri House Bill 1596 did not begin rousing long-distance interest until the fall of 1986, when Governor John Ashcroft had signed off on it and Frank Susman, obligingly, had hurried it into federal court.

"At *first* it wasn't targeted as The Vehicle," recalls Edward Grant, the attorney who served during the mid-1980s as executive director and general counsel to the Americans United for Life Legal Defense Fund. Operating out of its eighteenth-floor offices a half mile from Lake Michigan, the Legal Defense Fund gathered its resources from a nationally scattered pool of lawyers willing to volunteer their services, or to work full time for the Fund's modest salaries, so that the right-to-life movement could develop a national counterpart to what appeared to be the abundant legal resources of the American Civil Liberties Union and Planned Parenthood. "To meet the challenge of the anti-life legal forces of the above-named organizations," an Americans United for Life mailing described the mission of the Legal Defense Fund, "wherever and whenever they strike."

Americans United for Life (AUL), the parent group from which the Legal Defense Fund had grown, was a venerable old right-to-life organization that had been cranking out bulletins and newsletters since 1971. With its Midwestern mailing address and its resolutely educational focus—no lobbyists in the statehouses, no county-by-county citizens' chapter groups —AUL had for some years positioned itself as the quieter, more genteel cousin to the unruly National Right to Life Committee. In 1972, a year before *Roe* v. *Wade*, AUL backing had produced a hardbound essay collection, *Abortion and Social Justice*, that had a footnoted, scholarly look beside the conversational Q&A format of more popular volumes like the Willkes' *Handbook on Abortion*. There were follow-up AUL collections over subsequent years, each dense with material that never appeared in windshield leaflets or shopping center flyers: "Urological Complications of Legal Abortion." "Fetal Development: A Novel Application of Piaget's Theory of Cognitive Development." "Toward an Understanding of the Abortion Debate: Rhetoric as a Reticulate Structure."

At the core of the AUL leadership, helping edit these books and prod the organization along, was a group of Chicago physicians and attorneys who had gathered during the pre-*Roe* years to galvanize opposition to any change in Illinois state abortion law. The physicians marshaled medical and statistical arguments for their debates and academic papers, and the attorneys—in particular one odd and engaging duo, a Catholic trial lawyer named Dennis Horan and a Jewish law professor named Victor Rosenblum —set directly to work on the law. Horan, who was by many accounts a gruff, charming, intensely literary man, a lover of Irish poetry and a scrapper of apparently inexhaustible energy ("like a football player," recalls his widow, Dolores V. Horan, "who puts his head down and goes crashing into the fray"), was appealing to the courts on behalf of the unborn as early as 1972, writing legal briefs and asking to be appointed *guardian ad litem* for fetuses threatened with abortion. Rosenblum, who taught law and political science at Northwestern University, had contributed to *Abortion and Social Justice* an essay entitled "Coercion in Liberation's Guise," and over the ensuing years had settled in as Horan's professorial-looking older ally, gray-haired and voluble, his office a floor-to-ceiling disarray of books and papers and bound law journals balanced in precarious stacks. Rosenblum kept a picture of Groucho Marx on his Northwestern office wall, with a quote lettered in under Groucho's grinning face: "Great Spirits Have Always Encountered Violent Opposition from Mediocre Minds."

In the 1950s, when Rosenblum was a junior faculty member at Northwestern, he had written a short academic book entitled *Law as a Political Instrument*. Nearly half Rosenblum's book was devoted to an examination of the judicial history of racial segregation in public schools—to the series of cases that led, in gradual steps over more than twenty years, to the unanimous 1954 ruling in *Brown* v. *Board of Education* that segregating schoolchildren by race was a violation of the children's constitutional rights. "The decision was hailed in many quarters as a triumph for the principles of democratic government," the young Rosenblum wrote back in 1955.

That historic decision, as we shall see, was more the culmination of a line of policy developed gradually over the years by the Court than it was a revolutionary declaration of policy emerging full blown in the single case. Several decisions toward the end of the nineteenth century, of which *Plessy v. Ferguson* [upholding Louisiana's right to require racial segregation in passenger trains] became the most notorious, had established the validity of "separate but equal" as a form of racial discrimination. After those early decisions, the Court began gradually to chip away at the foundations of the segregation doctrine. Invalidation of "separate but equal" in the field of public education was the logical outgrowth from this line of cases.

Rosenblum's book made no mention of abortion law, but the metaphor that seemed to suit the segregation cases so well—a morally repugnant set of rulings finally collapsing as its own foundation gave way—endured for nearly thirty years to become the defining image at Americans United for Life. Certainly Rosenblum was not the only right-to-life recruit to perceive parallels between the legal rights of black people and the legal rights of the unborn; for the first decade after *Roe* those parallels were invoked repeatedly, in books and posters and organizational newsletters urging people to action: we must march on Washington, we must commit civil disobedience, we must amend the Constitution. But by 1984 the AUL lawyers were attracting attention from all over the country with a strategic blueprint modeled not simply upon the moral reasoning of the civil rights movement, but specifically and in detail upon the segregation cases themselves. One academic paper, published as a follow-up to a major 1984 conference of five hundred right-to-life leaders from around the United States, summed up the directive plainly in the title of its opening section: "*Brown v. Board of Education* and Its Lessons."

Crucial among those lessons, that paper proposed, was the patience and tenacity of the National Association for the Advancement of Colored People attorneys who brought the desegregation cases. "The N.A.A.C.P. lawyers still put their faith in the capacity of the American judicial system to self-correct," wrote Richard S. Meyers, an AUL volunteer attorney then in private practice in Washington, D.C., describing the long-range strategy civil rights attorneys had taken up as far back as the 1930s:

> They accepted the current state of the law and worked with it. They did not argue immediately for a reversal of *Plessy v. Ferguson*. Rather, they argued that if segregation were to continue, absolute equality between the races was constitutionally mandated.

Then, having convinced the Supreme Court to consider how *unequal* various segregated educational institutions really were, the NAACP lawyers had continued bringing cases to urge the Court to expand its own definition of equality in those institutions—to consider not only the physical facilities, but also intangible factors like access to faculty and the free interplay of ideas. *Brown v. Board of Education* was the logical finale to the Supreme Court's rulings in these cases, Meyers wrote, but only because the NAACP lawyers had spent so many years helping shape that logic through the particular cases they selected and the arguments they chose to make.

Here, then, was the working model for pushing the modern Supreme Court toward a reassessment of its own rulings on abortion—for "Reversing *Roe v. Wade* Through the Courts," as AUL entitled the ambitious

Chicago conference at which Meyers was invited to discuss his paper. The illustration on the cover of the conference brochure was a line drawing of the Supreme Court's front facade, the columns and portico skewed up at a forbidding angle as though viewed from below, through the eyes of a person advancing slowly up the great marble staircase. It was not a subtle image, and in the lectures and conference discussions (all of them conducted by white people, a *New York Times* reporter observed; for all its homage to the NAACP, the conference's five-hundred-person registration list apparently included not a single black participant), the metaphorical vocabulary returned again and again to buildings, edifices, structures with foundations, as though *Roe* v. *Wade* were a construction project to be whacked at with sledgehammers and crowbars. "The collapse and reversal of *Roe* can be accomplished either by a direct frontal assault or by laying a siege and waiting for its inner erosion," wrote Rosenblum and an AUL colleague named Thomas Marzen in a keynote essay entitled "Strategies for Reversing *Roe* v. *Wade*." "In either case, some fundamental precept of the *Roe* doctrine must be destroyed before its edifice falls."

For many months before and after that 1984 conference, the Americans United for Life lawyers tried to function from Chicago as a central command post for the assault upon the edifice. The favored expression was "chipping away at *Roe* v. *Wade*": strike softly, the directive instructed, but methodically and with deliberate aim. Each blow was to fall as part of an orderly sequence of events: a state adopts new abortion legislation that is written carefully, with an eye toward the apparently wavering justices and in particular toward Sandra Day O'Connor's *Akron* dissent; the new law is challenged by one of those ever-vigilant lawyers from Planned Parenthood or the ACLU; the state attorney general's office defends the new law all the way to the Supreme Court. Then pillar by pillar (it was and would continue to be an extremely durable metaphor), the Supreme Court begins pulling away the logical structure that holds up *Roe*. Perhaps the trimester system goes first. Perhaps "viability" is reexamined and discarded as too arbitrary an idea to form the basis for Supreme Court doctrine. Perhaps the Court decides that states can work into their laws just a few prohibitions on those abortions that nearly everybody agrees would be undertaken for reprehensible reasons—women aborting girls, for example, because they want only sons.

This plan depended entirely upon first-stage success at the state level, and a decade of post-*Roe* litigation had already taught national right-to-life organizers certain important lessons about the particular combination of ingredients required to guarantee that first-stage success. No single factor was sufficient on its own: energetic right-to-life volunteers were not enough; shrewd lobbyists were not enough; receptive state legislators were not enough; even large communities of Catholics and Protestant evangelicals were not enough to set the promising lawsuit into motion unless everything else was in place simultaneously, the volunteers *and* the lobbyists *and* the legislators *and* the governor willing to sign the bill *and* the state attorney general willing to mount a vigorous legal defense *and* the pro-abortion attorney who could be counted on to bring the lawsuit that might end before the Supreme Court.

Every one of these ingredients was present in the state of Missouri—and had been since 1973.

For sheer abundance of output—legislation introduced, legislation passed and signed, lawsuits argued in federal court—Missouri's only real competition in the post-*Roe* era was the state of Pennsylvania. A half dozen other states maintained local right-to-life organizations influential enough to have attracted the attention of compatriots across the country (Minnesota Citizens Concerned for Life, for example, was widely admired both for its political organizing and for its voluminous, detailed printed materials), but by midsummer 1986, when law school classes and abortion organizations drew up lists of Supreme Court abortion cases and the states from which these cases had originated, Missouri and Pennsylvania together accounted for nearly half the list. *Planned Parenthood* v. *Danforth* —Missouri. *Beal* v. *Doe*—Pennsylvania. *Poelker* v. *Doe*—Missouri. *Colautti* v. *Franklin*—Pennsylvania.* *Planned Parenthood* v. *Ashcroft*—Missouri.

Scholars of state history and sociology might have ventured certain academic explanations for this pattern: each was a spread-out state, with strong Catholic organization and distant rival cities separated by a wide swath of conservative farming country. Each shared a border with an abortion-friendly state whose very proximity could be understood to lower the stakes for a legislator trying to make up his mind about a right-to-life vote (Missouri women could easily travel to Kansas, should abortion become difficult to obtain in Missouri; Pennsylvania women could easily get to New York).

And each could boast not only of the proper combination of ingredients for a Supreme Court challenge, the volunteers and the willing-to-sign governor and so on, but also—by coincidence of personality—of an exceptionally single-minded and politically adept corps of right-to-life crusaders in the statehouse. In Pennsylvania the political sparkplug was one abrasive, inventive Republican representative, a former FBI agent named Stephen Freind; four years earlier the formidable combined efforts of Freind and his legislative allies had produced a wide-ranging state Abortion Control Act that was signed by Pennsylvania governor Richard Thornburgh, challenged almost immediately in U.S. District Court, and debated through the appellate process until *Thornburgh* v. *American College of Obstetricians and Gynecologists*, the latest legal challenge to originate in the Penn-

---

* *Colautti* v. *Franklin* was a 1979 ruling in which the Supreme Court overturned a Pennsylvania statute that had subjected physicians to criminal penalties if they failed to follow certain directives aimed at protecting the life of fetuses during post-viability abortions. The majority opinion, written by Justice Blackmun, found the statute unconstitutionally vague.

sylvania statehouse, now emerged as the central Supreme Court abortion case for the 1985–86 term.

When the *Thornburgh* ruling was announced, in June 1986, the decision appeared on the face of it to be another disappointment for the right-to-life litigation efforts. The Abortion Control Act, the Pennsylvania bill at issue in *Thornburgh,* had included a long list of familiar post-*Roe* gambits: there was an informed consent section, requiring clinics to provide printed materials describing the fetus; there was a mandatory twenty-four-hour waiting period; there was a parental consent section and a second-trimester-hospitalization section and a requirement that any doctor performing a possible post-viability abortion use the technique "which would provide the best opportunity for the unborn child to be aborted alive." Similar versions of each of these proposals had already reached the Supreme Court in previous litigation and been rejected at least once; in his *Thornburgh* majority opinion, Justice Harry Blackmun wrote what was clearly intended as an angry rebuke to the insistent legislators who kept repeating their efforts to undermine the abortion right first articulated in 1973:

> In the years since this Court's decision in *Roe,* States and municipalities have adopted a number of measures seemingly designed to prevent a woman, with the advice of her physician, from exercising her freedom of choice. . . . States are not free, under the guise of protecting maternal health or potential life, to intimidate women into continuing pregnancies.

Subsection by subsection, in *Thornburgh* v. *A.C.O.G.,* the Supreme Court overturned as unconstitutional every challenged provision of the Pennsylvania Abortion Control Act. But the ruling was only five to four —Justice Sandra Day O'Connor dissented again, even more adamantly than she had in *Akron.* Chief Justice Warren Burger dissented as well, this time declaring outright that the sequence of Supreme Court post-*Roe* rulings was now convincing him that his anti-*Roe* brethren on the Court were right: "If *Danforth* and today's holdings really mean what they seem to say, I agree we should re-examine *Roe,*" Burger wrote.

And for the first time in any Supreme Court abortion case, the United States solicitor general entered the proceedings as an *amicus curiae*—a friend of the Court, not directly involved in the litigation at hand, who files a brief offering further arguments for one side or the other—to ask outright for a reversal of *Roe.** With the implicit backing of the Depart-

---

* In 1982, when the Supreme Court took up *Akron* v. *Akron Center for Reproductive Health,* the challenge to the abortion restrictions adopted by the city of Akron, then–solicitor general Rex E. Lee entered that case as an *amicus curiae* to argue that the Akron restrictions should be upheld. But in his brief, and in his oral argument before the Court, Lee stopped short of calling for a reversal of *Roe.*

ment of Justice and the Reagan administration (the wording on the paperwork read, momentously, *Brief for the United States*), Solicitor General Charles Fried argued in his *Thornburgh* brief that the Supreme Court's thirteen years of abortion rulings amounted to a fatally flawed body of opinion—"not just wrong turns on a generally propitious journey," Fried wrote,

> but indications of an erroneous point of departure. Indeed, the textual, doctrinal, and historical basis for *Roe v. Wade* is so far flawed and, as these cases illustrate, is a source of such instability in the law that this Court should reconsider that decision and on reconsideration abandon it.

Thus in spite of the ruling itself, *Thornburgh* v. *A.C.O.G.* appeared to offer real promise on the Supreme Court front: an apparently unmistakable margin of only one vote, and the chief litigating attorney for the U.S. government arguing directly and publicly for an overturn of *Roe*. Then on June 17, less than a week after the *Thornburgh* ruling, President Reagan announced Chief Justice Burger's impending retirement from the Supreme Court. Justice William Rehnquist had been nominated to replace Burger as Chief, the president announced, and for Rehnquist's replacement Reagan offered up the third Supreme Court nomination of his presidency: Antonin Scalia, a deeply conservative U.S. Court of Appeals judge whose voting record and debate speeches left the strong impression, even though he had never ruled in an abortion case, that Scalia believed *Roe* had been wrongly decided.

The National Abortion Rights Action League issued urgent alarms both about Scalia and about Rehnquist's nomination as Chief; in Senate testimony, NARAL board chair Karen Shields described Rehnquist as "a nineteenth century man wishing to push society backwards." But the protests were futile. In September, Rehnquist's ascension was approved by a vote of sixty-five to thirty-three, and Scalia was confirmed unanimously. It was widely noted that the new justice was only fifty years old, quite a young man by Supreme Court standards. If the extensively publicized suppositions about Scalia's abortion views were correct, President Reagan had just presented the right-to-life movement with a sympathetic justice who could be expected to remain on the Supreme Court for at least twenty years.

But the count still held, at five to four. Everybody knew the count: NARAL knew it; AUL knew it; the National Organization for Women knew it; the National Right to Life Committee knew it. Women and men stuffing envelopes in county right-to-life offices found themselves able for the first time in their lives to recite correctly the names of all nine Supreme Court justices, each surname tagged in the memory with an annotation so fixed that this additional piece of information, the anticipated direction of the next reaffirmation-of-*Roe* vote, might have been carved on the nameplates as the justices sat in their solemn row at the Court.

William Rehnquist: No.
Byron White: No.
Sandra Day O'Connor: No.
Antonin Scalia: No.
Harry Blackmun: Yes.
William Brennan: Yes.
John Paul Stevens: Yes.
Thurgood Marshall: Yes.
Lewis Powell: Yes.

It was about this time that somebody at Americans United for Life began talking by telephone to Jefferson City, Missouri, where the state attorney general's office had an abortion bill to defend in court.

SAM LEE drove across the state to Kansas City for the H.B. 1596 trial, which began in December 1986 and ran for four unenthralling days in a windowless federal courtroom downtown. The case had a long file number and a name that suggested a confrontation of admirable proportions, *REPRODUCTIVE HEALTH SERVICES, et al., Plaintiffs,* v. *William L. WEBSTER, Attorney General for the State of Missouri, and THE STATE OF MISSOURI, Defendants.*

But the proceedings were passionless, a doctor testifying calmly and without apparent emotion, Judy Widdicombe testifying calmly and without apparent emotion, another doctor, another clinic director, another doctor. The only courtroom matters Sam had ever sat through were criminal trespass trials, usually his own, and he was accustomed to a certain level of intensity, people acting as though they genuinely cared about what was taking place around them. This trial seemed almost mechanical by comparison, a few newspaper reporters writing quietly into their notebooks, Frank Susman suggesting by the relaxed expression on his face that he knew he was going to win, and the two men from the Missouri attorney general's office looking mildly chagrined by the entire process, as though their job description sometimes obliged them to defend state laws they didn't particularly care about one way or the other. When court adjourned for the day Sam would drive over to the University of Missouri's Kansas City campus to look through the medical library for articles the lawyers might introduce as evidence, "Analysis of Gestational Interval," "Improved

Survival and Short-Term Outcome of Inborn 'Micro-premies,' " but he could see that the state lawyers did not always appreciate the extra help and wished Sam would go away and let them get on with their business.

Sam knew he was being pushy, but it bothered him, watching the lawyers. It wasn't *their* bill Frank Susman was trying to gut. *They* hadn't carried it by hand up and down one statehouse corridor after another, urging, negotiating, rewriting definitions, courting sponsors, crossing out phrases that troubled this person or that. Sam and Andy Puzder had spent more than two years championing the language that made it into H.B. 1596; Sam had written out his noncooperation paragraphs, and Andy had written out his life-begins-at-conception paragraphs, and at Missouri Citizens for Life a pitched battle had broken out because some board members wanted to sponsor a Puzder-Lee bill, legislation combining both Andy and Sam's ideas, and other board members wanted to try again with an Americans United for Life sex-selection abortion ban, which would have passed the year before in Missouri if a pro-abortion senator hadn't stood in the chamber during the final vote, placed one hand dramatically upon his Bible, and commenced the filibuster that killed the bill. Kathy Edwards was dubious about the new legislation Andy and Sam wanted to push for 1986; she said so plainly at the board meetings, and when it came to a vote and a majority of the board went with Andy and Sam, Kathy said, well, if that was the sentiment of the board she was not going to get in anybody's way, but Sam was going to have to take over the statehouse lobbying more or less on his own.

Thus Sam had been granted a title, rather suddenly, and a full-time job to go with it: State Legislative Chairman, Missouri Citizens for Life. Because the job paid no salary Sam had to keep writing his fund-raising letters, but he was exuberant about working the capitol by himself and he felt as though he had been shoved without warning into Advanced Lobbying, a realm that required bartering and reading between the lines and a remarkable variety of diplomatically phrased implicit political threats. In the capitol Sam studied the manner of the Missouri Catholic Conference lobbyist Lou DeFeo, who was the senior right-to-life man in Jefferson City; Lou was laconic and cool, not much given to fraternal chitchat, and as he cornered legislators in the hallway or sat to one side of committee rooms making notes on his yellow legal pads, it was obvious that Lou was tabulating extensive minutiae of procedural detail and potential strategic advantage, so that when Sam's right-to-life bill was swallowed into Lou's right-to-life bill—when H.B. 1596 became the one complex package for the provocative legislation Andy and Sam had envisioned from the start —Sam was anxious at first, but came to see that Lou and the experienced right-to-life legislators knew just what they were doing, that for tactical purposes they needed a single bill to lobby through the session.

Besides, Sam liked Lou's abortion bill. Lou's bill was simple, it seemed to Sam, and so straightforward that Sam could lobby it as earnestly as if he had written it himself: no public funds for elective abortion, period. No counseling, no advocating, no assisting, no private use of public facilities —Sam worked out quickly his lobbyist's answers to every hostile question that might arise from these provisions, and he thought of an analogy that seemed useful to him, a way to make people see why a public health nurse, for example, should be prohibited by law from giving a woman information about how to get an abortion. What if public school counselors, Sam would argue, were permitted to detail every possible option to depressed teenagers? " 'You know, I think one of your options is suicide,' " Sam would say, playing the part of the counselor, keeping his voice neutral and kind, the way he imagined a counselor might. " 'And here's where you can get some pills to do that.' "

Was this additional cutoff in public involvement going to make any practical difference at all? Sam knew the specifics as well as anybody in Missouri: the St. Louis public hospital already kept abortions off the premises; the Kansas City and Columbia public hospitals had together provided facilities for fewer than a hundred of the state's 1985 abortions. Of the seventeen thousand women who had undergone abortions in Missouri during 1985, nearly all had been accommodated either at Reproductive Health Services, at one of the Planned Parenthood clinics that provided abortions, or at a smaller private clinic or physician's office. In the abstract it was mildly satisfying to imagine these clinics and doctors receiving no help whatever from the state, but in practice they were receiving only the most indirect help now, perhaps a social worker handing over an address, or a printed brochure made available at a public health clinic. When he thought about it Sam doubted that there was any woman in Missouri who would give birth instead of aborting because this information would now have to come from the Yellow Pages, or because the law kept her out of Truman and the University of Missouri Hospital.

Still it was the legal strategy that mattered, the dangling of the bait, and the clear moral value of advancing *any* provision that might dissuade a woman from going through with her abortion—Sam rehearsed these explanations, in Jefferson City and at home in St. Louis, whenever he fell into silent imaginary arguments with John Ryan. Where was the principled Sam, he could imagine John Ryan demanding; where was the door-blocking Sam, the uncompromising Sam, the Sam who understood every abortion to be the killing of a human being? Where was the prisoner of conscience who had prayed for the unborn from a bunkbed in jail?

And Sam would answer: In the halls of the statehouse, John, trying to make something *happen.* Sam still received his regular mailings from the dwindling but aggressively vocal no-compromise wing of the right-to-life movement, particularly the prolific Judie Brown, the former National Right to Life Committee board member who had quit to form her competing American Life League ("Personhood is our absolute; we cannot waver from it"). And nobody doubted Judie's sincerity, her large-circulation magazine and newsletters glowed with piety and scriptural references, but Sam was beginning to see how refusal to compromise could become an exercise in deep self-absorption: it made people feel good without actually having to accomplish anything. Judie Brown considered it a *victory* that the Hatch Amendment had failed. Her autobiography was to refer to the entire Hatch episode with what sounded like genuine pride: "We testified, we prayed, we fasted, and in the end we defeated the Hatch Amendment." How nice for them, Sam would find himself thinking, crossly, as he sat in John Schneider's office rewording sentences about viability testing and sixteen-week hospital abortions: the Judie Browns of the movement can glide along serenely, upright and moral and unsullied by compromise, and in the meantime abortion is still legal in every state.

Sam worked so hard that session, spending his Jefferson City nights on the Callahans' couch, driving back and forth between St. Louis and the capital in Gloria's eight-year-old brown Chevrolet. He had accumulated enough respectable clothing to make him look somewhat like a lobbyist, but his fund-raising was erratic, and he learned after a while just to work hungry sometimes, to forgo lunch because he had no cash that day, to see if there were leftover sandwiches in some legislative office or to wait until he got home to the meal Ceil Callahan might have left for him. In St. Louis Sam and Gloria had given up their apartment and moved to a smaller place in a row house owned by Gloria's grandfather; they paid no rent, but the bathroom was in the back and the pipes kept freezing during the winter and Sam would be fumbling around at dawn with a torch and insulating tape, trying to get the toilet to flush. By the spring of 1986 Gloria was pregnant again, and there were days when she turned to Sam, her patience exhausted and the panic rising in her voice, and cried: "We don't have any *money*. How are we going to buy the *milk*?"

And Sam would say: "I'll get some money." And he would get it. He would write another fund-raising letter, a check or two would arrive in the mail; if things were bad enough he would make some telephone calls or visit some offices in person, maybe some Lawyers for Life, anybody he could think of who was sympathetic and earned enough money to be a potential donor. Sam hated going to people's offices to ask for money in person. He never knew what to say; he knew he could be eloquent about some things but not about this, it was degrading for everybody, and every time he had to do it he would find himself standing outside a closed office door, feeling embarrassed and trying to work up the gumption to knock. Once he ran out of gas while he was driving to the bank to get enough money to buy some gas. Sam left the car where it was and began walking toward the bank, arguing angrily with God as he strode up the street: What am I doing wrong here? Why have You led me to this life? Are You testing me, seeing how much I can manage? Or are You signaling me that it's time to quit?

God did not argue back. Sam had never felt that God smacked him on the head or explained Himself with any clarity the minute Sam railed at Him; Sam's relationship with God had always been slower and more elusive than that, and when the answers seemed to come they were gentle, vague, and all-encompassing, so that Sam never lost his faith, but neither could he consider the debate entirely closed. *You may think you know it all, Sam, but you don't, so you just have to trust Me,* that was what Sam might figure out God was telling him, and when God put it that way Sam supposed that he was pretty much like any human being with a ministry to conduct, he was meant to keep doing it even on the days when it was a mess. He got the gas to his car. He kept milk in the refrigerator. He stood behind the governor's chair when H.B. 1596 was signed into law. His son Nicholas was born in October, squalling and healthy, and the baby was two months old when *Reproductive Health Services* v. *William L. Webster* convened in U.S. District Court and Sam drove across the state to watch.

And this at least was evident to him, as the bill Sam had helped to write passed formally if unceremoniously into the federal judicial system: this much he could know he had accomplished. The entire trial was conducted before a single judge, the veteran U.S. District Court judge Scott O. Wright, and when the proceedings were over, Sam got in his car and drove back east toward St. Louis and wondered whether federal judges sometimes thought of themselves as glorified ladder rungs. Judge Wright had seemed to Sam a genial, no-nonsense sort of person, alternately stern and humorous as he moved matters briskly along, but even he must have understood that everybody in his courtroom regarded him less as a wise arbiter of justice than as an obligatory pause on the ascent to the appellate courts. Whatever Judge Wright's ruling, the loser would appeal, and Sam had by now learned the federal appellate process for a Missouri lawsuit as carefully as if he were preparing for law school exams: from the U.S. District Court to the Eighth Circuit of the U.S. Court of Appeals, and from the Eighth Circuit—should the case be accepted as appropriate for the nation's final appellate review—to the Supreme Court.

All Sam could do now, as much as he might wish otherwise, was watch. H.B. 1596 was in the hands of working attorneys: the state attorney general's office would handle the appeals, Andy Puzder and Lou DeFeo would do their best to urge the state men along, and the national people at Americans United for Life had already begun weighing in from Chicago, offering advice and mailing down to Jefferson City their packets of constitutional research and case summaries from other states. Sam knew from his conversations with Andy that AUL's interest in *R.H.S.* v. *Webster* was only beginning to warm; Missouri was not the only jurisdiction with an abortion lawsuit underway, and in any case H.B. 1596, which the Missouri Citizens for Life people had taken to calling the Kitchen Sink Bill even as Sam was lobbying it, was plainly not the kind of one-small-step-at-a-time legislation AUL had been promoting around the country. There was an awful lot of baggage in this bill, the AUL lawyers had observed: look at all these sections, the no-public-involvement section, the hospitalization-after-sixteen-weeks section, the noncooperation section, this paradoxical preamble about life being protected from conception on except that it was legal to abort it anyway—the AUL theoreticians had in mind some tidier, less confrontational vehicle for their next appellate bid.

Nonetheless it was a federal lawsuit, argued and submitted, and over the early weeks of 1987 Sam knew somebody in Chicago was keeping close track of *R.H.S.* v. *Webster*, if only because federal court litigation now appeared to be the brightest prospect the movement had. Sam still tried to get up every morning feeling optimistic, liking his work, thanking the Lord for his wife and his children, telling himself that most people were good at heart and could eventually be reached or at least moved somehow, but some days the weight of accumulated discouragement settled over him and he found himself realizing that litigation was the *only* prospect the movement had. Every passing year pushed the Human Life Amendment farther from their grasp. In 1986, the Missouri congressman Richard Gephardt had publicly defected from the ranks of Amendment supporters, a turning point of such sorrowful symbolic portent that Loretto Wagner's voice still trembled with emotion when she talked about it: loyal Dick Gephardt, South St. Louis' beloved Democrat, the former alderman from an overwhelmingly Catholic district (although Gephardt himself was Baptist, a detail largely unknown to voters)—Loretto used to call him "our fair-haired boy" and kept rallying the right-to-life volunteers around him because on every measurable Human Life Amendment index, surveys and speeches and co-sponsorship of bills, Gephardt was such a reliable Yes.

But when he began maneuvering for the Democratic presidential nomination, when he needed not just St. Louis but a national Democratic base of support, Gephardt's principles suddenly caved—that was how Loretto saw it, and she told the story over and over around St. Louis, making sure everybody in the movement knew it, how Gephardt hadn't even had the backbone to tell them to their faces, how someone had telephoned her one day in May about a three-sentence item on the front page of the *Wall Street Journal:* "Missouri Rep. Gephardt, a possible Democratic presidential candidate, drops his support for a constitutional amendment banning abortion. He says he personally opposes abortions but now favors the status quo under the Supreme Court's decision. His new posture will sit better with party activists."

Loretto had refused to believe it at first and went straight to Gephardt's district office people, and when they kept hedging and clearing their throats she agreed to a meeting with Gephardt himself, just Loretto and Ceil Callahan and a priest from the archdiocese, all confronting the congressman in a private room at a restaurant in Clayton. Gephardt brought a few of his aides along, but he did his own talking, Loretto had to grant him that; he looked deeply unhappy and his face was sweating and he kept saying, "Yeah but, yeah but," and talking about how he was beating his head against the wall on the Amendment, he had no consensus in Washington at all, he had searched his soul, it just wasn't workable, and so on. And Loretto, who had known Dick Gephardt since he was a twenty-eight-year-old anxious-to-please newcomer on the St. Louis Board of Aldermen, had wondered aloud whether this was the first time in Gephardt's political life that he had ever beaten his head against the wall and whether he didn't think maybe it was his job to *build* consensus, wasn't that what elected leaders *did*, wasn't he supposed to fight for things he believed? Did he or did he not believe unborn children possessed a right to life? Loretto told Gephardt that just at that moment she had more respect for Senator Ted Kennedy than she did for Dick Gephardt; at least as a pro-abortionist Kennedy appeared to stand by his own convictions. As the meeting concluded Gephardt extended his hand and said, "Loretto, I'm so sorry," and Loretto said, "So am I," and she accepted his handshake but they both had tears in their eyes, that was the way Loretto told it. She said to Gephardt: "I feel like I'm losing a friend."

After that Loretto went around talking about vengeance. "There's no vengeance like the vengeance of a movement scorned," Loretto would say, pulling out her media contact lists and her volunteer telephone trees to spread the word on Richard Gephardt: flip-flop artist, can't be trusted, looks like a Boy Scout and then abandons you to political expediency. It was a noisy and reasonably effective publicity campaign the MCL volunteers set into motion, lots of articles with headlines like GEPHARDT'S SWITCH ON ABORTION MAY BE COSTLY, but Gephardt's still-undeclared presidential campaign pressed on anyway, leaving Loretto and her volunteers to absorb the larger and uglier lesson about the direction of the political expediency itself: for a Democrat with national aspirations, support for a Human Life Amendment was now so poisonous that fifteen years of Yeses were apparently to be shoved discreetly aside like some embarrassing lapse in judgment.

At least they still had the president. Ronald Reagan was careful, halfway through his second term, or perhaps more accurately his handlers were careful; the president never showed up in person at the January 22 March for Life in Washington, you wouldn't see him grasping the march leaders' hands on the podium or anything, but he always spoke to the marchers by loudspeaker-broadcast telephone or sent a personal message of support to be read aloud. And Ronald Reagan was gradually delivering to the movement a remade Supreme Court. Already the president had produced Sandra Day O'Connor and Antonin Scalia; already the president had set up as Chief Justice one of the two original No votes on *Roe.* The president's next Supreme Court selection, if one more retirement were to be announced while Reagan was still in office, would surely prove at least as hostile to *Roe* as Justice O'Connor appeared to be. The suspense was in the count: when the time came for the next appointment, would the president be able to tip the count? Which justice would leave the Supreme Court first? Would Ronald Reagan be filling a seat vacated by a No on *Roe* or a Yes on *Roe?*

For some weeks into 1987, no word came from the chambers of Judge Scott O. Wright.

Sam waited, watched the newspapers, telephoned Andy Puzder every few days to see if he had heard anything. In Jefferson City Sam was lobbying again; for the 1987 session his principal Missouri Citizens for Life legislation was an adoption bill, the Special Needs Adoption Tax Credit, mandating state tax reimbursements to parents who incurred expenses adopting children who were handicapped, of a minority race, or otherwise difficult to place through conventional adoption. The word "abortion" appeared nowhere in Sam's 1987 legislation, and Sam believed this was as it ought to be, that H.B. 1596 ought to push through the courts without further distraction from the Missouri statehouse.

On March 17, 1987, Judge Wright delivered his ruling. Every challenged provision of H.B. 1596 was unconstitutional, Wright ruled: the life-begins-at-conception preamble was unconstitutional; the hospitalization-after-sixteen-weeks rule was unconstitutional; the no-public-involvement sections were both unconstitutionally vague and "a significant barrier to a woman's right to consult with her physician and exercise her freedom of choice." Even the section requiring every abortion doctor to inform each patient whether she was or was not pregnant was unconstitutional under *Roe* v. *Wade,* Wright ruled:

> It is clear that some physicians believe that it is in the best health interest of many patients seeking abortions not to be told definite results of pregnancy tests. These are women who suspect they are pregnant and choose to abort, but cannot emotionally deal with the specific issue of being pregnant. Some physicians, in the exercise of their best medical judgment, would choose to respect the woman's evaluation of her emotional status and not foist on her specific information which she does not wish to hear. . . .
>
> It is not the province of this Court to approve one view or another. This Court must approve the principles enunciated in *Roe* v. *Wade* and its progeny: that the right to privacy, the conception of personal liberty, encompasses a woman's right to terminate her pregnancy.

In the news stories the Missouri attorney general was variously described as "expressing disappointment" and "considering an appeal," and when Andy Puzder called the attorney general's office in alarm to say, What do you mean *considering,* what's going on here, the lawyer Andy talked to said, That's what we always say, don't worry about it, of course we'll appeal.

Three months later, Justice Lewis Powell announced his resignation from the U.S. Supreme Court.

THIRTEEN

The Seventh Question

1987–89

JUDY WIDDICOMBE heard the Powell news on the radio and watched the reports on television that night, the justice standing before the cameras, one arm raised in a farewell wave. On television Powell looked elderly and gaunt; he was seventy-nine years old and under treatment for prostate cancer, Judy read that in her *Post-Dispatch* the next morning, along with the local reaction story, DEPARTURE IS HAILED, MOURNED, in which the first person quoted was Sam Lee.

> Samuel Lee, the state legislative chairman for Missouri Citizens for Life, said members of his group were "ecstatic" about the announcement of Powell's resignation on Friday.

Alone in her tidy house in University City, her resignation from Reproductive Health Services six months behind her, Judy read about Justice Powell in one somber article after another. *Supreme Court Turning Point. Powell Won Conciliator Reputation. High Court Vacancy: Despite Disclaimers, Abortion Is Pivotal Issue in Nomination.* If she had been at the clinic Judy imagined she would have launched reflexively into action, convened a meeting or typed out urgent new broadsheets for the hallway bulletin boards, but in the quiet of her kitchen Judy gazed at the headlines and realized that she had already begun calculating how long it would take to reassemble the underground once Sam's people had won. Would it be like 1971, with the TWA abortion flights to Manhattan and the Greyhound buses for the women who couldn't afford the airfare? What about all the medical equipment the clinics had accumulated, the aspirators and the cannulas and the trays full of instruments? Would rebel nurses disperse these items surreptitiously across the state so a new generation of Mrs. Vinyards could go about their work with the proper tools at their disposal? There were efficiently constructed little hand-operated suction devices around now too, plastic syringes as fat as turkey basters; you could draw back the plunger on one of these syringes and create enough suction to empty a uterus in early pregnancy, which made them especially suitable for poor Third World countries where the medical care was primitive and abortion was illegal anyway, so that with the local officials you could make vague reference to "menstrual regulation" and thus manage through euphemism to help women maintain some tiny vestige of control over their own childbearing. *Damn* the Antis. Was this what they wanted? Was there a person among them who genuinely believed that making it illegal again would accomplish *anything* except pushing it back underground? Sam Lee was not a stupid man, Judy had read those newsletter essays and talked to him long enough to know she was right about that; he wasn't blind, he didn't appear to hate women, he had sat there in the RHS waiting room and *watched* them line up for their abortions, watched them step right over his cross-legged praying self and push on into the clinic procedure rooms. Where the hell did he think those women were going to go? Had he truly convinced himself that they were going to have *babies* if he and his brethren managed to shut down RHS for good?

   Lee said the resignation was an "important breakthrough in stopping abortion."
   He said the vacancy gave President Ronald Reagan an opportunity to appoint a new justice firmly opposed to abortion, "thus greatly enhancing the possibility of a reversal of the abortion decision" of 1973.
   The court's 1973 decision, known as *Roe* v. *Wade*, struck down many restrictions on abortion. Powell sided with the majority in the case.

The *Post-Dispatch* ran a cartoon graphic, nine sets of black-robed Supreme Court shoulders with small balloons of information where the heads ought to be. It made Judy want to cry. Lewis F. Powell, Jr., Age: 79. Thurgood Marshall, Age: 78. William J. Brennan, Age: 81. Harry A. Blackmun, Age: 78. Of the *Roe* v. *Wade* supporters still remaining on the Court, only John Paul Stevens was under seventy, and he was sixty-seven, older than Rehnquist, older than O'Connor, older than Scalia, and surely older than whatever abortion-hostile candidate President Reagan was going to put up for confirmation this time. The constitutional right to abortion was being held in place by fragile old men, and when Judy looked at their ages lined up like that she saw how silly it was to have wished for a moment that Powell could just delay his retirement until Ronald Reagan was out of office. It was impossible to fend off: the fifth Yes vote was gone, something momentous was going to happen, and as the days went by and the Supreme Court stories kept appearing Judy found herself thinking first *this is it, it's finished,* and then *this can't be it, they can't give legal abortion to a whole generation of women and then take it away again,* and she would try to imagine the closing of Reproductive Health Services —the literal, physical closing, how would it work, what if the staff people refused to cooperate, would men from the city police department come with trucks and handcuffs?—and found that she was unable to do it. Nothing had closed that clinic down, not the picketers on the sidewalk, not the rosary-praying people in the waiting room, not the urine on the carpets or the stinkbombs in the hallways or that steely-faced woman who climbed through a window to get to the procedure rooms. None of the false bomb threats had ever closed Reproductive Health Services, and when the real call finally came, the clinic had stayed open through that, too.

It was summer 1986 when this happened. The Supreme Court had just reaffirmed *Roe* again in *Thornburgh* v. *A.C.O.G.*, and the call had come after midnight from an RHS employee who said, "West County's on fire," just like that, as though the entire county were aflame. Judy had run to her car and driven breakneck out Highway 40 and crested the hill above the West County clinic, the RHS branch she had wanted so badly for the suburbs. There were flashing lights and fire trucks and by the time Judy got out of her car the flames had subsided, so that when she stepped over the hoses large pieces of her clinic lay wet and broken at her feet. The skylight had blown out, the waiting-room chairs were smoldering, slabs of ceiling tile had crashed to the floor, electrical wiring hung ragged and hissing overhead, and Judy looked from one thing to another like an emergency-room triage nurse: numb, detached, assessing, grim. Four days earlier a pipe bomb had exploded out one wall of a clinic over in Wichita. "Anti-abortion domestic terrorism" was the phrase Judy used, when the reporters pressed around her for comment after the fire at the West County clinic, and she added, hoping the papers would print it, that while the clinic was under repair every scheduled abortion patient would be accommodated at the RHS central clinic in St. Louis. "They will have accomplished nothing," Judy said in her *Post-Dispatch* interview, "except to inconvenience patients."

Nothing. She wanted them to hear that, whoever they were. The morning after the fire, when the interviews were over and she walked into her house and closed the front door behind her, Judy sat down hard in an armchair and put her head in her hands. She had no idea who had thrown a firebomb into her clinic; she didn't think it was any of the protest regulars, it wasn't Sam and it wasn't Loretto Wagner and even the more volatile local sit-in people seemed unlikely prospects for a furtive midnight bombing attack. Probably an unstable person enthralled by the rhetoric, Judy thought; you shout *babykiller* often enough and God only knows what reptilian depths you can tap. Maybe it was a plain womanhater, somebody enraged by the spectacle of women deciding for themselves whether to go on being pregnant; maybe women's lives in the modern era were just too goddamn independent for somebody to bear. She didn't know. She was not going to find out. She was very, very tired of fighting back all the time. Judy wanted to quit; she had no plan, no particular future lined up for herself, but she had been thinking about this for months, how used up she felt, and she was startled to find that the firebombing had not renewed her passion or inspired her back into battle. It sickened her, that was all, it made her sick and sour, and all she wanted to do was repair the West County clinic—rebuild it perfectly, a genteel upraised finger to the midnight arsonist—and get out of there.

She quit in the fall of 1986, three months after the fire. She took some weeks to finish emptying her office, clearing her shelves and desk drawers a little at a time. She lived off her savings. She began seeing a therapist. She worked without pay for the citywide AIDS program, helping the program directors organize volunteers. She read self-help books, curiously at first, and then voraciously, volume by volume, so that her shelves at home began filling with titles that suggested failed love affairs and plummeting self-regard: *Women Who Love Too Much. Love and Addiction. The Dance of Intimacy. Feeding the Hungry Heart. The Betrayal of the Body. If I'm Successful, Why Do I Feel Like a Fake?*

Some of these books Judy read three or four times, circling paragraphs, folding corners of pages, stacking them up by her night table so that she could reach over and reread certain passages as she lay in bed. At the foot of her bed she kept an end table and two small upholstered chairs, and in the evenings she would sit at the table and write in a notebook about how she was feeling, the first time she had ever done such a thing, just sitting there alone writing phrases that looked on the page like poetry. Sometimes Judy found herself going on and on, writing page after page, stopping long enough to cry for a while and then pull herself together and begin writing again.

She was in terrible shape. She had spent the whole of 1986 in terrible shape, she could see that now, and when she thought about it and talked to her therapist Judy began to believe that the personal part of her life, that internal part she was reading so much about, was so withered and wrong that for some years now she had felt herself to be Maverick Clinic Director and Veteran Reproductive Health Expert and nothing else, no person left over once the workday was finished, no real inside Judy at all.

She was still not sorry about divorcing Art, they would have made each other more and more unhappy; but on the day Art remarried, Judy had sat alone in her backyard for a long time, feeling that an imaginary crutch had been yanked from beneath her. After that she had dated at intervals (quite seriously, in one instance; there was a high school social studies teacher Judy thought she might be in love with), but never *wisely,* Judy was coming to see, never easily, never unclenching into that delicious conviction that she loved and was loved and that was all there was to it. She was always feeling shoved around one way or the other, feeling too demanding or too independent, and the therapy and her self-help books were teaching Judy that the problem was probably her feeling about herself.

Judy worked hard on this. Her therapy sessions doubled to twice a week. She underwent hypnosis, talking to the therapist about her memory that her mother had been at once emotionally distant and rigid in her expectations, that as a child Judy had never felt entirely safe or cared for, and that this frightened feeling might account for Judy's adult need to take care of other people in a driven, single-minded, almost bullying way. The therapist taught Judy certain calming techniques, how to enter a meditative state by breathing deeply, how to imagine herself in a hammock with a soft breeze blowing, and for a time Judy kept some of the therapist's tapes in the glove compartment of her car, so that when anxiety overcame her as she was driving (there had been some frightening moments in the car, Judy hyperventilating for no discernible reason; once she had found herself driving up a one-way street the wrong way), she could put a tape into the cassette player and calm herself down.

After a while Judy came to see that she was her own biggest project: she was trying to construct an interior self, a woman who felt satisfied and secure even when she was not fighting excellent battles on behalf of all womankind. She tried to keep reminding herself how difficult this was going to be; but as the months slid by, she began to feel that she was making some progress, that she was relaxing, playing the piano, talking late into the evening with her closest friends, learning to let things go. The clinic was functioning reasonably well without her—although Judy was back-seat-managing more than she ought to; she frequently talked by telephone to the new director, a honey-voiced former RHS counselor named B. J. Isaacson-Jones, and inside the clinic, staff members complained to each other that Judy was neither present nor absent, that you could argue with B.J. and hear Judy's answers coming out of B.J.'s mouth. Judy knew what people said about the line of succession at RHS: "Hard shoes to fill." More than once Judy had said it herself, with a somewhat unsavory mix of sympathy and pride; she had plucked B.J. from the clinic staff and trained her right up to the associate directorship, seeing in the younger woman a certain eloquence and potential for leadership. B.J. was an attractive person, intelligent, thoughtful, with large blue eyes that had a way of fixing on you as though your remark just at that moment was of the *most* intense and consuming interest. But there was something soft and tentative about her, no match for Judy at all (in later years, when she was being honest with herself, Judy would admit that she must have seen that in B.J. too, and known she was picking a successor she could shepherd from a distance). And it was taking some noticeable span of time, right on into the summer of 1987, for Judy to begin allowing herself to disappear from the clinic she had built.

And now *this*—the Lewis Powell resignation, the tipping of the Supreme Court. Nobody called Judy for comment; the reporters went to Frank Susman and NARAL and the ACLU, and Judy watched from a distance, studying her daily newspapers, the drama thickening in Washington day by day. For nearly a week Justice Powell's resignation was in the news every morning: appreciations, analyses, speculation on the future of the Court.

Then, on July 1, President Reagan nominated to the Supreme Court the D.C. Court of Appeals judge Robert H. Bork. Judy had never heard of Robert Bork, but all at once the newspaper pages were full of him—his speeches, his interviews, his Senate testimony, his Court of Appeals opinions—and the more Judy read the worse it looked. Bork was "a committed conservative," the papers said. Bork was "a powerful and restless intellect." Bork was "someone who has a very sharp edge as to how he wants to see this court move"—that was Senator Joseph Biden, the Democratic chairman of the Senate Judiciary Committee, warning that the president's new nominee appeared upon early inspection to hold profoundly conservative ideas about the judiciary's role in interpreting the Constitution.

Judy thought of herself as a liberal Democrat, and every week the dispatches from Washington contained new broad-scale alarms about the president's nominee: Bork as a critic of famous civil rights rulings, Bork as a bluenose on free speech protections for literature and the arts, Bork as a champion of "original intent," the idea that Supreme Court justices must base their modern-day rulings solely upon the original, demonstrable intentions of the men who wrote the Constitution. But the biggest news about Bork was his assessment—his recorded, reprinted, unambiguous assessment—of *Roe* v. *Wade*. Unlike Sandra Day O'Connor and Antonin Scalia, whose views on *Roe* could only be guessed at as their confirmation hearings approached, Bork had left an extensive paper trail containing repeated public denunciations of *Roe*.

During congressional testimony in 1981, the judge had described *Roe* v. *Wade* as "an unconstitutional decision, a serious and wholly unjustified usurpation of state legislative authority," and there was a good deal more than that in Bork's articles and speeches; Judy read the excerpts in one Bork profile after another. If she understood correctly what she was reading, Bork didn't even think the Supreme Court had ruled properly in *Griswold* v. *Connecticut*, the 1965 Connecticut case that laid the groundwork for *Roe* by finding a constitutional privacy right guaranteeing married couples the freedom to purchase and use contraceptives in any state. Here was a right invented by modern judges, Bork seemed to be arguing; the authors of the Constitution intended nothing of the sort, and regardless of one's opinion about the morality of abortion or contraception (Bork himself registered no discernible sentiment either way), the Supreme Court had exceeded its authority when it "created" this privacy right— that was apparently Bork's word, *created*—and used it to force states into changing their own laws.

As a Supreme Court nominee Bork was circumspect about what these convictions might come to mean in practice: no one ever quoted him declaring *And I will vote to overturn* Roe *the first chance I get.* But Judy had learned enough about the Supreme Court to know nominees never say anything like that. On the subject of abortion, Bork appeared to present no mystery at all; he was a No on *Roe,* the surest No Ronald Reagan had yet offered up in nomination, and every time she studied the judge's picture in the newspaper, Judy remembered the way Jack Willke liked to count down the numbers on public debate stages. "Seven–two, six–three, five–four," Willke used to say, in that smooth, self-assured voice that made Judy want to hurl something at him. Those were the *Roe, Akron,* and *Thornburgh* votes, majority to dissenters, 1973 to 1986: seven to two, six to three, five to four. The way Willke said it made the next thing seem as simple and inevitable as a rolling ball's progression down a hill, and Judy had learned enough about appellate law to sense now the beginnings of a terrible premonition about what the next thing might be. She could not even call to mind the full formal name of Frank's lawsuit, but she knew the words Reproductive Health Services were in it and that the system, somehow, was pushing it toward the Supreme Court.

THAT FALL Sam Lee kept the television on in the living room of the South St. Louis apartment, and because the apartment was small, its four rooms lined up one behind another, Gloria could be in the kitchen or the bathroom and still hear the voices of Robert Bork and the witnesses who were testifying for and against Bork before the Senate Judiciary Committee. It was the first confirmation hearing Sam had ever followed day by day, and as he studied the lineup of organizations opposing Bork—it was a massive liberal battalion, it appeared, not just NARAL and NOW but also the NAACP Legal Defense Fund, the Sierra Club, the Association for Retarded Citizens, the Southern Christian Leadership Conference, the AFL-CIO, and scores of other advocacy groups—Sam realized that Bork was not his idea of the perfect Supreme Court justice either. Even if the liberal opponents were distorting Judge Bork's record—and Sam's right-to-life newsletters assured him they were, that "radical special-interest groups," as the

National Right to Life News described them, were exaggerating and quoting selectively as part of an elaborately organized national campaign to rouse opposition to Bork's confirmation—it looked to Sam as though Bork cared more about the *process* of interpreting the Constitution than about the practical effect of his own rulings. If Bork was so set on restricting himself only to what he believed the Constitution's authors had in mind, Sam wondered, then how would he have voted on slavery, or school segregation, or the public welfare benefits of the New Deal?

The National Right to Life Committee was ardent about Bork—for some weeks into September the *News* ran full-page Action Boxes listing the names of each state senator and imploring readers to FLOOD SENATE WITH MESSAGES OF SUPPORT—and Sam found a certain irony in this, since Bork had shown no evidence that he cared about the rights of the unborn, either. His famous 1981 remark about *Roe* being unconstitutional had come in the midst of Senate testimony *against* the Human Life Bill, which Bork also described as unconstitutional. ("The deformation of the Constitution," the Yale Law School professor Robert Bork had gone on to say, "is not properly cured by further deformation.") Nonetheless Bork was plainly the movement's fifth vote against *Roe*, this was what they had elected Ronald Reagan to do, and Sam watched the hearings intently, remembering the case names that went with Jack Willke's familiar countdown. *Roe* v. *Wade*: seven–two. *Akron* v. *City of Akron*: six–three. *Thornburgh* v. *A.C.O.G.*: five–four.

William L. Webster v. *Reproductive Health Services, et al.* was in the Eighth Circuit, awaiting its hearing, en route to the Supreme Court.

Sam knew that must be presumptuous, assuming any case was en route to the Supreme Court—every year the Supreme Court received more than seven thousand requests to review lower court rulings, and the justices accepted for argument fewer than two hundred of those. But *Webster* was right for the Court. Sam was sure of it. The lawyers had told him how good their chances were. Sam could feel the heat rising in the legal offices as the states' attorneys prepared their briefs for the judges of the Eighth Circuit—they had the Americans United for Life attorneys seriously interested now, and there were meetings in Jefferson City, interstate telephone calls, memos issuing back and forth as the attorneys argued about what to appeal and how to word their briefs. If it *was* a realigned Supreme Court that awaited them at the end of the appeals process—if it truly was possible that representatives of the Missouri attorney general's office might become the first state lawyers to bring a serious abortion case to a Court prepared to overturn *Roe* v. *Wade*—then what exactly was the case Missouri ought to bring?

Sam was anxious about these meetings and kept calling the attorney general's office, worrying over the telephone, wondering if this or that aspect of the District Court ruling was going to be included in the appeal to the Eighth Circuit. The state lawyers had developed a brisk don't-call-us-we'll-call-you way of thanking Sam for his observations and hanging up, but Sam kept at them anyway, urging them to put everything in their brief, to push every possible aspect of H.B. 1596 through the appellate process. Sam privately felt a little disappointed that Frank Susman had never bothered to challenge in his lawsuit the noncooperation paragraphs Sam had first composed for H.B. 1596; those paragraphs were state law now, although they had been considerably edited down as the bill proceeded through committee, and nobody in Missouri seemed to be paying much attention to them: there were no reports of plumbers or UPS drivers suddenly refusing to enter the doorways of abortion clinics. Sam could see in hindsight that for his purpose it might have been better if Susman had attacked as unconstitutional every single line of H.B. 1596, so that all of it, from its resounding opening sentence *(The general assembly of this state finds that: Unborn children have protectable interests in life, health, and well-being)* to its final bloodless phrase *(. . . shall have standing to bring suit in a circuit court of proper venue to enforce the provisions of section 1 to 4 of this act),* could be argued about and thus defended, publicly and dramatically, in federal court.

For his purpose. It was the autumn and then the early winter of 1987, nearly one full decade after Samuel Lee had driven his Volkswagen into St. Louis and been guided by God or happenstance into the organizing meeting that rerouted his adult life, and in his quiet moments now—on his way to Jefferson City, for example, all alone with the farmland rolling by—Sam wondered whether he still understood precisely what his purpose was. When he lectured and visited right-to-life meetings he met men and women whose eyes now shone when they talked about the coming reversal of *Roe,* as though it would be over then, as though they were already imagining all the pleasurable ways they would spend their time once the movement no longer required their services. And Sam would resist the impulse to shake them a little: *It won't be over,* he wanted to cry, *it won't be over at all.* A reversal of *Roe* was not the same thing as a constitutional amendment protecting life from conception on, but this distinction appeared to be lost on much of the general public; reversing *Roe* was only a retreat back to 1972, when each state was permitted to fashion its own abortion laws.

Sam knew what would happen when the Supreme Court reversed itself on abortion-as-a-constitutional-right: some states would prohibit it, some states would keep it legal, some states would tinker with their criminal codes so that abortion was legal under certain circumstances but not others, and Sam would lobby and write legislation and beat himself up working Jefferson City until the state of Missouri adopted as complete an abortion ban as could possibly be urged upon the General Assembly. That would take a year, Sam imagined, maybe two.

Then what?

Then abortion would probably still be legal in Kansas. Almost certainly abortion would be legal in New York. Abortion might be legal in Illinois, harder to predict there but it might, and if Illinois and Kansas both stayed legal then the St. Louis women could drive east to the Hope Clinic in Granite City, Illinois (twenty-five minutes, right off Interstate 270), and the Kansas City women could drive west to Comprehensive Health for Women in Kansas City, Kansas (twenty minutes, right off Interstate 435), and what would Sam's exhaustive efforts have produced, finally, except some very nice legislation offering statutory protection to the unborn in certain scattered, modestly populated, far-from-a-useful-state-line counties of central and southern Missouri?

The best we can do. In his lecture notes Sam tried to keep his answers logical, sequential, so they made both practical and moral sense. He had Ethical Climate: in a no-legal-abortions state, some women will give birth instead of aborting because they genuinely believe in abiding by the law. He had Distance: in a no-legal-abortions state, some women will give birth instead of aborting because the nearest legitimate clinic is too far away; they don't hear about it, it isn't listed in their telephone books, it's too complicated to get there. He had Alternatives: in a no-legal-abortions state, some women will give birth instead of aborting because the very legislature that outlawed abortion—here Sam was making a rather dizzying idealistic leap, he knew, but he deeply wanted it to be true—will have augmented its abortion ban by lavishing unprecedented state resources on support for maternity homes and adoption centers and aid to needy mothers.

But when he argued with himself in private—when he heard again the imaginary reproaches of John Ryan, or envisioned himself confronting the calm, single-minded young man Sam Lee had been a half decade earlier—the truest answer Sam found he could articulate was: This is the best we can do. *Anything* that dropped the numbers, that had become the great glittering goal; the numbers were printed listings of the deliberate killings of infants, and anything that lowered them was better than not lowering them at all. Sam knew all the appropriate responses to the women-will-do-it-illegally-and-die argument (for example: we don't make bad things legal just because people do them anyway; and: women are already being maimed by profit-hungry legal abortionists; and: the numbers of pre-*Roe* deaths were wildly exaggerated for years by abortion advocates like Dr. Bernard Nathanson, who after his right-to-life conversion declared publicly that the statistic he and his early NARAL colleagues liked to use— five to ten thousand abortion deaths per year during the pre-legalization era—was fiction, invented for propaganda purposes. These were standard-issue comebacks for any right-to-life speaker on the debate circuit), but the fact was that some women *were* going to do it illegally; it might happen infrequently but it was foolish to pretend that it would not, and once in a while it would happen in a particularly brutal way so that two people were killed, baby and mother, instead of only one. This was sorrowful to contemplate, but constitutionally protected abortion was worse; that was the clearest way Sam could put it to himself. Constitutionally protected abortion was one point six million deaths per year, *one point six million:* at the marches you always saw graphs of that statistic drawn onto signboards, with lines of small white crosses to illustrate the numbers, More Deaths Than Vietnam War, More Deaths Than World War II, and so on.

Sometimes Sam thought about moving his family to Kansas or Illinois, once the Missouri work was finished, and lobbying there; he was only thirty, three or four decades of vigorous work still ahead of him, and perhaps he and others like him could move from place to place, organizing, learning the legislatures, repairing criminal codes one state at a time. Or he might stay in Missouri and become an alternatives-to-abortion expert, quadruple the number of crisis pregnancy centers, help Gloria expand the facility at Our Lady's Inn. Or he might write legislation aimed at out-of-state clinics, keep them from advertising their services inside Missouri, maybe he should look up the old case law on interstate abortion ads. The breadth of his new pragmatism startled Sam and made him feel at once cynical and powerfully engaged—he could see that if the Hatch Amendment fights were breaking out now, he would be rolling up his shirtsleeves with the states' rights people, arguing loudly and passionately for the possible, the limited, the compromise that might protect only a few—and as he watched the confirmation of Robert Bork slide irretrievably toward defeat, Sam thought to himself, coldly: Well, too bad, let's get on with the next nomination.

The U.S. Senate voted fifty-eight to forty-two against the confirmation of Robert Bork, the largest margin of rejection ever recorded for a Supreme Court nominee. Everything Sam read convinced him Bork had been railroaded (Sam's *National Right to Life News* editions were indignant with references to "misrepresentations" and "slanderous attack"), but he was too restless for postmortems, and every morning he turned on the early television news just to see whether the name of the next nominee had been announced yet. *Where was the fifth vote?* The news suggested astonishing disarray there in Washington, and it made Sam testy. He wished they would hurry up; the movement had *work* to do: D.C. Court of Appeals Judge Douglas Ginsburg, a free market conservative and philosophical Bork ally, nominated to the Court. No, Ginsburg nomination withdrawn; it seemed the judge had been seen smoking marijuana with friends during the 1970s. Ninth Circuit Court of Appeals Judge Anthony Kennedy, a California conservative with ties to some of the Reagan administration Californians, nominated to the Court. The reporters naturally rushed out looking for Anthony Kennedy controversy, but there appeared to be none to find: the judge's personal history suggested a life of uninteresting moral rectitude. His judicial record was scholarly and prudent, tilting toward government power and away from individual civil rights claims; he had written no provocative articles and delivered no provocative speeches and gave no evidence of ever having expressed any public opinion whatsoever on abortion or the merits of *Roe* v. *Wade.*

When Anthony Kennedy was nominated, Sam called the National Right to Life Committee office in Washington. What is this, Sam demanded, what are the Reagan people doing, who *is* this guy? And Sam's Washington man said: Listen, we can't tell you how we know this, but we have sources, Kennedy is with us. The insiders say he's okay.

That was good enough for Sam. "Okay" was good enough; "okay" was the fifth vote. Anthony Kennedy's confirmation to the Supreme Court was approved by the Senate on February 3, 1988, and through the spring Sam waited for word from the Eighth Circuit. His life had a certain familiar rhythm to it now; nobody would call it an elegant life, his family still scraped for money every month and Gloria made careful remarks about how nice it would be to live in a house where they weren't on top of each other all the time, but in Jefferson City Sam was a recognized lobbyist, nodded at in hallways, handed cups of coffee in legislators' backrooms, confided in, leaked to, sought out by reporters who needed the Missouri Citizens for Life response to this or that development. He went to board meetings. He spoke to church groups. He had a private office for his work in St. Louis, a single room on the floor above a pawnshop; the rent was $170 a month, which Sam budgeted into his fund-raising, and he had hauled in a secondhand desk and stacked his papers on the floor and hung a single wood crucifix on a bare white wall, so that the cumulative effect was both monastic and disorderly, surroundings in which Sam felt entirely at ease.

He no longer made even the nominal appearance at the sit-ins, the rescues, even though both Sam and Gloria had friends who were still planning their weekends around the regular Saturday arrest. Every few weeks the mailman delivered Sam's copy of the *Direct Action News,* the newsletter John Ryan was assembling from his home in St. Louis, and as Sam read the resolute blockade-and-arrest reports from Georgia and Kansas and Florida, he felt the widening gulf between himself and the men and women who now seemed consistently to refer to themselves as "rescuers." They were becoming a national submovement, these rescuers; they had annual conventions and Plans for Action and a roster of verbal, eager, publicity-wise leaders, the most visibly ascendant of whom, a wiry-haired upstate New York evangelical named Randall Terry, was developing a rapt and multistate following with his biblical exhortations urging Christians to defy civil authority—to "obey God's Word," as one of Terry's early articles put it, "even if it means disobeying the ungodly laws of men."

Sam had met Randall Terry, briefly and unmemorably, during several interstate right-to-life gatherings at which Terry had offered to play his guitar and sing. He had a passable voice and seemed to be paying careful attention to everything around him; that was all Sam could recall. Apparently it was around that time that some fiery sense of purpose had roared up in young Terry—the Lord had challenged him during a church prayer meeting, that was the way Terry would explain it in the training seminars he began conducting and recording on videotape—and this challenge from the Lord had placed into Randall Terry's heart the notion of a vast Operation, an organized series of blockades and rescues at clinics, "abortuaries," around the land. "A 'rescue mission' is a group of God-fearing people saying, 'NO! We're not going to let you kill innocent children,' " Terry had written in the summer of 1987, as though presenting this novel idea for the very first time, "and peacefully but physically placing themselves between the killer and his intended victims."

In his training tapes, which were showing up now at various right-to-life houses around St. Louis, Terry was actually quite effective—articulate, self-deprecating, veering between mischievous humor and lectern-thumping righteousness. He spoke like a rigorous preacher at Bible study class, rattling off scriptural references so fast that nobody in the audience could possibly keep up with him, and when Sam watched the Terry videotapes he felt a curious chill: the man was as repellent as one of the hostile feminist leaders from the National Organization for Women, Sam thought, looking so angry all the time, goading people for not being angry enough. "We have sinned as Christians by letting this holocaust go on," Terry liked to say, and: "We need to act like it's murder." Terry appeared to be laboring under the impression that he had invented the concept of the large-scale clinic sit-in, a bit of puffery that Sam found mildly irritating; but beyond that Sam could see at once how readily Terry lent himself to caricature, how precisely his carriage and language suited every secular cliche of the heavy-breathing right-to-lifer. He had worked as a used car salesman, a detail reporters repeated frequently and with obvious delight; he was partial to apocalyptic statements about "America's future" and "whether we're ultimately going to be an ash-heap and a rubble-heap and live and die like jackals and wild animals"; he was frequently to be seen holding his Bible aloft and segueing feverishly into speeches about totalitarian humanism or bringing prayer back into the public schools.

In private the traditional wings of the right-to-life movement were arguing unhappily about Randall Terry's sit-ins, which were now being organized under the catchy and somewhat militaristic name Operation Rescue. The blockades were attracting more national attention than Sam's St. Louis sit-ins ever had—by early 1988 Terry had summoned people to respectably large clinic blockades in New Jersey and Pennsylvania, and had let it be known that he planned something more ambitious for Atlanta during the summer weeks of the impending Democratic Convention—and the National Right to Life Committee leadership had taken a public stance of official no comment, which insiders understood to mean: We *wish* they would go away. National's lawyers fretted about the possibility of conspiracy charges, the Committee being seen for legal purposes as colluding with Operation Rescue and thus sharing liability in the lawsuits the blockaders were certain to invite. But besides that the rescues were so evangelical, so joyously and noisily religious, that they horrified many of the old-line Catholics who made up much of the first generation of main-stream right-to-life activists. Every time Randall Terry opened his mouth, he undermined twenty years of laborious effort to portray abortion as a secular issue, an evil as universally recognized as rape or murder. All the O.R. blockaders put together added up to a tiny fraction of the right-to-life volunteers across the United States, but the media loved them; a rescuer transported by prayer made a snappier front-page photograph than a law-yer dictating reply briefs or a middle-aged homemaker folding flyers into envelopes. In the newspaper stories about O.R. sit-ins no one ever "held" a Bible or a rosary: these items were always "clutched," as though everyone present had fallen into amusing spasms of holy-roller frenzy, and Sam now found himself reading the Operation Rescue write-ups with the same exasperation he knew more seasoned right-to-life organizers had once felt toward him.

Bless them, Lord, Sam would think to himself, but maybe while You're at it, You could to a certain extent shut them up. He felt no embarrassment about his own history, it was right for the time; did the lunch-counter sit-in demonstrators in the South pretend they had never defied the law, once they went on to get their law degrees or whatever? But he felt no real pull to go back, either; he hardly ever found himself longing for the snap of the handcuff or the sated camaraderie of the police wagon ride to jail, and when Sam updated his resume he was vague about the specifics of his early years as a movement volunteer. *Samuel Lee has been active in the pro-life movement since 1978 and has worked full time in the right-to-life cause since October 1984.*

Every few days he telephoned Andy Puzder's office, wondering whether Andy had heard anything from the Eighth Circuit yet. Sam made a list of the names of Supreme Court reporters at major newspapers; when the Eighth Circuit ruled, the reporters would want to know about it, collect some background information for their articles. He wrote down their ad-dresses and made sure he had their names spelled right: Linda Greenhouse, *The New York Times*; David G. Savage, *Los Angeles Times*; Al Kamen, *The Washington Post*; Richard Carelli, Associated Press. Sam went through his piles of Missouri newspaper clippings and pulled out what looked like a good sampling dating back to January 1986, when H.B. 1596 was first introduced (*Anti-abortionists Try New Route*, Sam liked that one very much), and then he laid them out carefully onto eight-by-eleven sheets and copied them one or two articles to a page, so the reporters would find them more convenient to read quickly.

The summer heat came up and in his office Sam turned on the old air-conditioning unit, which was wedged into the window and roared when it was working. He wrote out draft paragraphs for press releases, deleted them, and tried writing them again. In the evenings he took his children to the playground, with clean diapers tucked into an old vinyl bag. He smoked too many cigarettes.

On July 13, the three-judge Eighth Circuit Court of Appeals panel reviewing the appeal in *William L. Webster* v. *Reproductive Health Services* released its ruling, which with one small exception upheld the District Court on every challenged point. "8th Circuit Finds Law on Abortion Unconstitutional," read the front page of the next morning's *Post-Dispatch*. Sam clipped it, copied it, stapled it into his press packet, noted with satisfaction that his own remarks had made the fourth paragraph:

> Anti-abortion forces immediately urged the state to appeal the case to the U.S. Supreme Court
>
> Samuel H. Lee said, "Frankly we have been planning all along to get this to the Supreme Court, ever since we got the law passed in 1986."

Then he called the state attorney general's office to find out what came next. Sam was keyed up, full of questions. When would the lawyers begin the appeal? How was a petition to the Supreme Court constructed? How would the Court let them know if the case had been accepted? How were the lawyers going to word the part about using *Webster* v. *Reproductive Health Services*—or *Webster*, as the state lawyers were now referring familiarly to the case—as the vehicle to overturn *Roe* v. *Wade*?

At the state attorney general's office the lawyers replied to Sam, in effect: Take a number. Reporters were besieging the office, demanding interviews with William L. Webster himself. The Americans United for Life lawyers were calling from Chicago, offering to fly down to Jefferson City at once, proposing strategic lines of argument and themes for effective friend-of-the-court briefs. An envoy from the U.S. Attorney General's office in Washington telephoned Jerry Short, the Missouri assistant attorney general who had written most of the Eighth Circuit brief, to express the government's very keen interest in the *Webster* matter: were Mr. Short and his colleagues proceeding satisfactorily; could the Justice Department be of any use to them; had the Missouri attorneys decided—Jerry Short recalls this being phrased diplomatically at first, but the message was plain to everyone—what *approach* they intended to take with the Supreme Court?

After the telephone call came a letter from Washington, on Department of Justice stationery, and there were more calls from Washington, and more emphatic prodding from Americans United for Life, and the state right-to-life people pushing steadily with their letters and their expectant visits to Webster's office in Jefferson City. Sam could see what was happening; it was like watching advisers heap weapons and armor before a reluctant knight. The state of Missouri was not legally obliged to appeal *Webster* v. *Reproductive Health Services* at all. The state could accept the Eighth Circuit ruling and strike from its statute books all the court-overturned sections of H.B. 1596; or the state could approach the Supreme Court quietly, meekly, with a no-fanfare plea that said: we accept *Roe* v. *Wade*, we accept all the "progeny," we seek no modification to any of those rulings, but we believe the lower courts erred and that *under those rulings* this law is constitutional.

Or the state could force the larger question: since the lower courts found this law unconstitutional under *Roe*, the state could say, we ask you to examine not the lower courts' findings but rather *Roe* itself. That was what the Justice Department wanted—rather badly, as Sam understood it, U.S. deputy attorney general William Bradford Reynolds leaning hard on Webster and his assisting lawyers in Jefferson City—and certainly it was what Sam wanted, too, the test case before the new Supreme Court, with H.B. 1596 at the center of the case.

His bill. Sam kept telling himself not to swell up too much with pride of ownership; it was Lou DeFeo's bill, too, and Andy Puzder had written the life-begins-at-conception passage, and more to the point Sam could see that the content of the bill at issue didn't matter nearly as much as the timing. *Webster* v. *Reproductive Health Services* was an abortion case, that was all it had to be; it raised the words "abortion" and "state law" in the same lawsuit, and it happened to be first in line; and when Sam finally received a copy of the Jurisdictional Statement, the long, densely argued brief in which the state of Missouri formally requested appellate review from the United States Supreme Court, he turned anxiously to the opening pages to see how Webster and his attorneys had decided to make their pitch.

"QUESTIONS PRESENTED," the Statement read. Here were the summaries, the single-sentence distillations of the questions Missouri believed significant enough for the Supreme Court to answer in this case. Sam knew the phrasing of these questions was important. The first six made careful, specific references to H.B. 1596, with no mention of *Roe* v. *Wade*. Was the viability-testing requirement unconstitutional on its face? Was the passage declaring that the life of each human being begins at conception—the "preamble to a state abortion bill," the state's brief called it—unconstitutional on its face? Did physicians or other medical personnel have legal standing to contest the constitutionality of such a preamble?

Sam was disappointed to see that the Missouri state attorneys had not included the sixteen-week-hospitalization requirement in their questions to the Court; Webster and his assistants had decided to drop that issue from the appeal after losing on it in both the District Court and the Eighth Circuit. But there were three separate questions about the restrictions on public funds, public employees, and public facilities. ("6. Is a state civil statute facially unconstitutional that makes it 'unlawful for any public facility to be used for the purpose of performing or assisting an abortion not necessary to save the life of the mother'?")

Question number seven was worded somewhat differently. Sam read it slowly, to make sure he was getting it right:

> 7. Whether the *Roe* v. *Wade*, 410 U.S. 113 (1973), trimester approach for selecting the test by which state regulation of abortion services is reviewed should be reconsidered and discarded in favor of a rational basis test.

All those afternoons at the law library with the volumes of appellate opinions had paid off; Sam understood how to translate this. In layman's English, the seventh question posed by the state of Missouri in *Webster* v. *Reproductive Health Services* was this: Should the Supreme Court abandon *Roe* v. *Wade* and use instead a new standard—a "rational basis test" —to decide which abortion laws are acceptable and which are not?

Sam knew the vocabulary now, and how much it mattered even when it sounded dry as a fine-print entry in a textbook index. If the context was right and the legislators and judges responsive enough, the words "rational basis test" could shut down Reproductive Health Services more effectively than the grandest sit-in either Sam or Randall Terry could ever hope to lead. A state trying to defend its abortion laws under a rational basis test need only prove that its laws were, in the legal sense, "rational"—that they reasonably, rationally, served some legitimate state interest. No longer would states have to justify every new restriction with arguments so powerfully compelling as to warrant a cutback in women's fundamental right to abortion, for there would *be* no such fundamental right; abortion would no longer be afforded special protection under the Constitution, and a new and far more lenient standard would thus guide appellate judges as individual state abortion laws came up for review.

In his little office over the pawnshop Sam had thick file folders full of old legislation that even pro-abortion people would have to admit was "rational," whether or not they thought it was right. Surely spousal consent requirements served a legitimate state interest (father's rights, preservation of the family, protection of the unborn child). Hospitals-only requirements served a legitimate state interest (enhanced medical safety, probable drop in total numbers of abortions, protection of the unborn child). Mandatory waiting periods, mandatory two-parent consent forms for minors, mandatory pre-abortion descriptions of the baby and the rate of its heartbeat—every one of the regulations the Supreme Court had rejected over the last fifteen years might readily be upheld if all that was required of them was rationality; and more to the point, Sam had no trouble at all imagining a *ban* on abortion being upheld under a rational

basis test. Couldn't the protection of all unborn life reasonably be seen as
an interest of the state?

Sam called up his computerized list of Supreme Court reporters and
began adding names as fast as he could think of them: the *National Law
Journal, U.S. News & World Report, Newsday,* the Gannett papers, the
Copley News Service, the *Milwaukee Journal,* the *Houston Chronicle,* the
Baptist press. The cable network C-SPAN would want to know about
this, and also the Christian Broadcasting Network, and the conservative
newsletter *Family Protection Report.* "*National Catholic Register,*" Sam
typed into his computer. "*Washington Monthly.* Mutual Broadcasting System. *National Journal. The New Republic.*" He kept typing until he had
thought of sixty-six news organizations to which he could send the *Webster* v. *Reproductive Health Services* explanation he was about to write,
and then Sam composed his opening lines. *Missouri Attorney General
Asks High Court to Consider When Life Begins and Asks Court to Reconsider* Roe *v.* Wade. He knew this was too long and he knew he was only
writing a press release, but Sam made the words into a headline anyway,
capitalizing, imagining them stripped across a newspaper's front page.

THE UNITED STATES SUPREME COURT accepted *Webster* v. *Reproductive
Health Services* on January 9, 1989, with a two-line announcement that
was immediately dispatched to wire services and radio reports across the
country, and someone called Judy Widdicombe to tell her, and Judy
thought at once: Oh Jesus, I dropped the ball, I never went to Jefferson
City that year, I should have testified, I should have lobbied, I let that
goddamn bill go through.

She was living in Washington, D.C., when the Court took *Webster;*
six months earlier she had rented out her house in St. Louis and packed
all her clothing into suitcases and driven east to see what it would be like
to work full time in the capital. In Washington Judy found an adequate
place to live, a rental house near the National Zoo; she shared the house
with a friend who worked as a lobbyist, and every weekday morning Judy
got up, dressed in tailored clothing, combed and sprayed her hair into an
immaculate blond cap around her head, and took the bus to the office of
Voters for Choice, the political action committee that had hired her to
raise money for various campaigns around the country. Voters for Choice
occupied two plain rooms in a Dupont Circle building that housed a lot of
politically liberal advocacy groups, the kind of bare-bones enterprises
whose young volunteers tended to show up in bluejeans and Birkenstock
sandals, and Judy tried to act friendly in the building even though she
knew she must look elaborately composed and faintly out of place, like
someone's aunt dropping by from plusher corporate offices downtown.

She was miserable, really—Judy didn't let herself think about it much, she was always big-shouldered and efficient when she walked into the office, and people in the hallways could still hear her laughing or letting off a jovial profanity after hanging up on a frustrating telephone call. But Washington was not what she had expected. Judy had imagined a player's life, work weeks made dramatic by crucial policy decisions and high-dollar fund-raising; she had thought her resume would illuminate her name and that people would haul her in eagerly, attentive to a veteran's fine ideas.

But it wasn't working out that way at all. Nobody seemed to care that she had built the best clinic in the Midwestern states. Judy Widdicombe had co-*founded* the National Abortion Federation, the biggest clinics' and doctors' alliance; she had been president of NAF, valued consultant to Planned Parenthood, major board member at NARAL; some years back she had stepped in for a few months as interim director and run NARAL herself, and now when she called around Washington the young women on the telephone asked absently, "How do you spell your last name?" and "What organization did you say you were with?"

It was awful. Kate Michelman, the former regional Planned Parenthood director who had replaced Nanette Falkenberg as director of NARAL, was courteous around Judy but managed somehow to appear respectful and dismissive at the same time, as though Judy's ideas were certainly *interesting* and NARAL would certainly consider them along with many other interesting ideas. At multi-organization meetings Judy sat not at the center table but in the outside rows of folding chairs, where the legal assistants and junior policy analysts also were invited to sit. She felt useless and shrewish, and sometimes after hours she confided in her friend Frances Kissling, who had lived for almost a decade amid the Washington abortion advocacy groups. Frances directed a small but highly regarded enterprise called Catholics for a Free Choice, and Frances would remind Judy that working in Washington was a brutal business even among the high-minded social cause nonprofits. "History doesn't buy you a seat at the table," Frances would say: memories were short; turf was scarce and jealously guarded; nobody was deferred to without a big organizational name or a mention in last month's headlines; the front desks were staffed by earnest young people who were still sleeping with teddy bears while Judy was building Reproductive Health Services. It was not as though Judy herself was an easy package, Frances would observe gently; she was outspoken, she was opinionated, she still had the ability to intimidate just by walking into the room.

And perhaps the immediacy of Judy's own experience made people uncomfortable, there in the great city of policy statements and position papers; perhaps they didn't want somebody who had compared vacuum aspirator brochures and counted fetal parts in her bathtub. Perhaps they felt as though Judy were standing there in her lab coat shouting "Abortion!" at them while they were trying to steer the conversation toward the more delicate vocabulary of "choice" or "reproductive rights," Judy couldn't tell, but it made her feel worse and worse and after a while she tried to hunker down and do her midlevel job the best she could, raising money for Voters for Choice, cultivating foundations and donors who might be persuaded to give money to 1988 state political campaigns to restore abortion funding in Colorado and Michigan.

"I'm a bagman," Judy would say, joking only a little. And although the November election results were grim—voters rejected the Colorado and Michigan funding reinstatements; worse, the presidential election went to George Bush, whom the Antis had goaded into an even more rigid antiabortion position than Bush had held during his years as vice president —she tried to cheer herself up by observing that at least she was an accomplished bagman; she had raised nearly three hundred thousand dollars for the various failed campaigns.

That was how the year 1988 came to an end for Judy Widdicombe in Washington, D.C.. Nine days into 1989, over the span of one January morning, everyone she knew appeared suddenly to be repeating the name of the abortion clinic she had built.

Webster versus *Reproductive Health Services. Reproductive Health Services* versus *Webster.* The *Webster* case. It startled Judy to hear the Missouri names rattled off with such familiarity by people who had never been to St. Louis, never set foot over the threshold of 100 North Euclid Avenue. By midafternoon on January 9 the press releases and reaction statements were issuing from offices around Washington; somebody brought over Kate Michelman's, which was six pages long:

> The scales of justice have swung sharply to the right, and as I speak to you now, the fate of millions of American women quite literally hangs in the balance. In giving the anti-abortion movement a hearing in *Webster* v. *Reproductive Health Services,* the Supreme Court struck terror in the hearts of American women. This case will allow the Court to consider whether to restrict or even rescind women's most fundamental right: to retain control over our reproductive lives. Given the changed composition of the Court, this is without a doubt the most serious threat to reproductive choice in America in decades.

How long had Judy understood this without allowing herself to think about it—think seriously about it, not pushing it off to one side when it began to make her stomach hurt? It was *happening.* The new Court had its big case. The Supreme Court of the United States might genuinely be poised to overturn *Roe* v. *Wade.* And it was happening *now,* not in some alarmist fantasy brandished to make women write letters to Congress, and Judy had unwittingly helped it to happen, agreeing to offer Reproductive

Health Services as lead plaintiff in that lawsuit she had been so unenthusi-astic about to begin with. She knew that if the Court hadn't taken Frank's case it would have taken someone else's, that there were appeals from other state law challenges stacked up waiting in the lower courts. But it made Judy feel hollow, imagining the words "Reproductive Health Ser-vices" entering the history books in this way. The case that reversed *Roe:* was that how future generations would remember Judy's contribution to legal abortion in the Midwest?

As the news stories multiplied and the emergency meetings convened Judy felt stranger and stranger, as though she were in one of those dreams where she was shouting in a crowd and no one heard her or saw her, no sound came out of her mouth. People didn't walk up to her and make accusatory remarks, she supposed she should take some comfort in that; sometimes she imagined them hissing right at her *if we lose this we have you to thank* but nobody ever did, they went right on being polite and vague around Judy and not returning her telephone calls, and at the meet-ings she still sat in her outside tier and watched the women at the center table wrestle out large and complicated ideas about public response to the *Webster* case. Who would submit friend-of-the-court briefs? Who would run public advertising campaigns? Who would direct the big protest march to the Capitol and the Supreme Court? Which organizations would take credit, place names on letterheads, solicit support and new membership to prepare for what was coming?

It was a heady time, the way Washington can be when something dreadful is about to happen, and even in her gloom Judy could see that in the big rooms where these meetings were held there were ambitious pro-posals taking shape for the weeks before April 26, 1989, which was the date the Supreme Court had set for the *Webster* oral argument. NARAL had commissioned a set of polls and focus groups, hunting for the single-phrase summary that might be artfully compressed onto billboards and posters; the pollsters had snapped to attention at the offhand remark of one Florida focus group participant, and NARAL was now printing up a hundred thousand posters and bumper stickers with the words WHO DE-CIDES, YOU OR THEM? printed in large, feminine-looking cursive above the graphic upraised face of the Statue of Liberty. Political advertising experts were working late into the night to prepare television spots warning of the disastrous consequences of overturning *Roe*—a physician talking angrily into the camera about his own memories of treating women infected by illegal abortionists; a woman's footsteps clacking down a dark alley as she approached the rickety staircase that led to the abortionist's door. Women across the country were to be solicited for their participation in a lengthy friend-of-the-court brief that was to contain the abortion stories of many individual American women—"the voices of real people," as NARAL and NOW put it in the public Dear Friend letter asking women to join the brief.

The friend-of-the-court effort itself was turning into a massive organizational project. So many attorneys and organizations had asked to submit *Webster* friend-of-the-court briefs that Planned Parenthood and the American Civil Liberties Union had hired a new associate, the Philadelphia women's rights lawyer Kathryn Kolbert, solely to coordinate the preparation of the briefs: there was to be a historians' brief, a bioethicists' brief, a religious organizations' brief, a public health experts' brief, a social workers' brief, a public employees' brief, a members of Congress brief, an ethnic minorities' brief, briefs by law professors and state attorneys general and experts in international and comparative law. There were briefs about the right of privacy, briefs about women's equality, briefs about federalism and freedom of speech and uniformity of state law. Interstate travel was raised in the legal discussions, too; should someone prepare a brief exploring the legal status of travel agents who might be helping women across state lines in a post-*Roe* era of state-by-state criminalization? What about the legality of out-of-state clinic newspaper advertisements in states where abortion was prohibited?

Nothing remotely like this had happened before the oral argument in *Roe* v. *Wade*—Judy was sure of it. When Judy called Reproductive Health Services with her bulletins from Washington, B. J. Isaacson-Jones would tick off the list of reporters who had come through the clinic that week. ABC News had been there, and NBC News, and the BBC, *The Washington Post*, the *Boston Herald, USA Today*, Australian television, German television, Canadian television, South American television, some network people from Japan. The Belgian minister of health had come to visit, did Judy know that? B.J. had taken some professional coaching to help her appear more straightforward and natural on camera. The number of media visits was up over a hundred already; the receptionists kept permission cards in the waiting room now, so patients could indicate in writing whether they were willing to be interviewed or to have their procedures filmed, and sometimes the clinic accommodated two or three television crews in a single day. There were times when the hallways looked like obstacle courses, with the cameras and the crisscrossed electrical cords, B.J. told Judy; people had to watch where they were stepping, it was wild.

After these conversations Judy always hung up the phone carefully and stared out the window for a while, trying to sort out her feelings. B.J. was good at this part, maybe better than Judy would have been; the television people probably liked the delicacy B.J. exuded, that troubled, vulnerable expression that suggested not truculent resolve but rather the gravity of the moment, the imminence of threat. Some months earlier, B.J. had written quite a moving piece of poetry—Judy happened to have been visiting RHS the morning B.J. brought it in, B.J. saying she had been unable to sleep the night before and had risen from her bed in the middle of the night to write what was on her mind, and when Judy took the handwritten paper and read it she felt her throat tighten, the poem was so good. "Where are you?" B.J. had written, her opening line:

Where are you?
For over 16 years we have provided
you with choices
Painful choices
I remember—
I sometimes cried with you.
Choices, nonetheless when you
were desperate.

Remember how we protected your privacy and treated
*   you with dignity and respect when you*

were famous,
had been brought to us in shackles with an
armed guard, or were terrified that you
would run into one of your students?
I remember each of you.

B.J. had written about the firebombing, about paying for poor wom-
en's abortions, about pulling dollar bills from her own wallet to give
women taxi fare and dinner money, about opening the clinic after dark to
sneak in the daughters of right-to-life picketers. "Have you forgotten?"
she wrote.

I recall shielding your shaking
body, guiding you and
your husband through the
picket lines. They screamed
adoption, not abortion! You
wondered how you could explain your choice to your
four young children.

You broke our hearts.
You had just celebrated
your twelfth birthday when
you came to us. You clutched
your teddy bear, sucked
your thumb and cried out
for your mom who asked
you why you had gotten yourself
pregnant. You replied that you just wanted to be grown.
You're 20 today.
Where are you?

I pretend that I don't know you in the market,
at social gatherings and
on the street.
I told you I would.

After your procedure you told me that you would fight
for reproductive choices (parenthood, adoption, and
abortion) for your mother, daughters and
    grandchildren.
You will . . . won't you?

I have no regrets. I care about
each and every one of you and
treasure all that you've taught me.
But I'm angry.
I can't do this alone.
I'm not asking you to speak of your abortion, but
You need to speak out and you need to speak now.
Where are you?

B.J.'s poem had gone into the *Post-Dispatch,* under the title "An Open Letter to 21 Million Women," as a paid advertisement—but laid out as poetry, just the way B.J. wrote it. An outpouring of emotional letters had arrived at Reproductive Health Services over the next few weeks, many of them addressed personally to B.J. and folded around checks or twenty-dollar bills; it was right, what B.J. had written in her poem, the pride and grief and frustration all precisely right; they were in crisis, they did need the public support of those twenty-one million women, they could lose everything, it was true.

The television spots were right, too, and the friend-of-the-court briefs (Judy hadn't read them, but she was sure they were scholarly and urgent in their assessments of the stakes), and the frantic mailings from NARAL to its membership around the United States. It was all deeply serious, the constitutional right to abortion in genuine and immediate peril—and yet Judy found something disquieting in the *Webster* v. *Reproductive Health Services* uproar, which in Washington was growing to a volume unprecedented in modern Supreme Court litigation. Did either side genuinely want this battle to end? From the periphery to which she felt she had been shoved, watching at some remove now the furious planning and campaigning, Judy believed she could detect a kind of exhilaration in the combat, so many jobs made terribly important, so many people feeling purposeful and intense. Salaried employees, professional advertising campaigns, long-term office leases, full-time legal staffs, national fund-raising drives—they had rooted themselves into institutions, all of them, Antis and pro-choicers alike, and conflict, everyone's sincere intentions notwithstanding, was what kept these institutions alive. "They *love* all of this," Judy said more than once to Alison Gee, the younger woman who directed Voters for Choice and was becoming Judy's closest friend in Washington. "This is what they *do*. They don't know how to do anything else."

In her bleakest moments Judy wondered whether she would even go to the National Organization for Women march, the Washington protest planned for April 9, two weeks before the scheduled day of oral Supreme Court debate in *Webster*. For some weeks during late winter this march had been the source of heated, unfriendly argument, the NOW leaders championing it tirelessly, the more decorous and politically cautious Washington veterans dismissing it as too dated, too expensive, too nobody-cares-outside-Washington, too NOW—too In Your Face, in other words. That was still NOW's reputation even among women's organizations, NOW as the big raucous left-wing brute of an outfit, the only one with the sort of ground troops that might plausibly be summoned for protest purposes from all over the United States. NOW had been planning a 1989 Washington march since long before the Supreme Court took *Webster*, and in the unwieldy multi-organization meetings after the Court's announcement the NOW contingent insisted the march must proceed, advancing now upon the Supreme Court building as well as the Capitol. Molly Yard and Ellie Smeal, respectively NOW's president and former president, were famously thick-skinned veterans of political infighting, and they argued so vehemently against each alternative suggestion—break the protest into numerous regional marches, forget the march and save the money and volunteer effort for more sophisticated political work instead—that soon it was apparent to everyone that NOW intended to march whether other organizations joined in or not. Then there was a lot of contention about how the march was supposed to be handled, was it NOW's march or a coalition march, was it an abortion march or an abortion–welfare–Equal Rights Amendment march, but finally a manageable consensus had emerged and the date was set and even Judy could see that an extraordinary fervor was building as the word went out, that Molly Yard and Ellie Smeal had been right this time, that something massive was going to happen in Washington, like the old Vietnam War mobilizations or the great civil rights demonstrations of the 1960s.

Buses were being chartered in cities across the United States. Washington hotels were booking to overcapacity. Planned Parenthood of New York was reserving six cars on the 8:20 express train to Washington the morning of the march. The national NOW newsletter passed on triumphant reports of the fifty buses on reserve in New Jersey, the hundred buses on reserve in Pennsylvania, the thousands of airplane seats reserved out of California. St. Louis alone was sending seventeen buses to Washington, B.J. reported excitedly to Judy: there was March for Women's Equality/Women's Lives information everywhere you could imagine in Missouri, on college bulletin boards, on bookstore walls, on telephone poles, in church and neighborhood newsletters; people were standing up at community meetings to make march announcements; nearly the entire RHS staff was coming to Washington; she was closing the clinic so they could make the trip.

Closing the clinic, on what should have been a workday—not just skeleton-staffing it. That hit Judy hard. Of course Reproductive Health Services had scheduled no abortions for that day, April 8, the day the airplanes and charter buses were going to leave St. Louis for Washington. But April 8 was a Saturday, usually the busiest procedure day of the week, and over the telephone B.J. told Judy about the enormous hand-painted plywood CLOSED sign they intended to prop in the hallway against the locked clinic door: WE'VE NEVER CLOSED FOR ANTI-CHOICE ACTIVITIES, read the sign. WE HAVE CHOSEN TO TELL THE TRUTH ABOUT REPRODUCTIVE FREE-DOM BY JOINING THE NATION AT THE WOMEN'S MARCH IN WASHINGTON, D.C. B.J. was growing more and more adamant about the march: she had in-sisted that every clinic staffer who wanted to go should go; she had put out the word that anybody who needed financial help for the travel should come talk to her; they were all going to wear RHS T-shirts and they were going to march together, B.J. said, behind a canvas banner that would stretch the width of four women marching abreast: *Reproductive Health Services, St. Louis, Missouri*. Royal blue lettering, B.J. told Judy, visible a good distance away. They were painting the RHS logo, three silhouettes of women's faces in profile, right onto the canvas of the banner.

It was impossible to resist. There was march excitement in Judy's office, there were march messages on Judy's home answering machine (did she have any guest beds? Floor space? How far was her house from the rally site?), there were elaborate press and hotel arrangements underway for the planeloads of movie and television stars who were flying out from Los Angeles to join the march on behalf of the Hollywood Women's Political Committee. By the first days of April Judy was working the same frantic hours as everyone else she had come to know well in Washington; Voters for Choice and NARAL had agreed to organize the Hollywood contingent into a pre-march gala, a hotel-ballroom fund-raiser full of glamorous people, and the gala was a great success. Jane Fonda spoke, Marlo Thomas spoke, Alice Walker read a poem, Susan Sarandon read from *Roe*, flashbulbs popped, people jammed up against each other and stood on tiptoe to get a better look, the crowds were so thick that the hotel managers opened extra party rooms for the overflow. Judy arrived home long after midnight, exhausted and gratified. She tiptoed past her sleeping St. Louis houseguests (one on the couch, two side by side in the big guest bed), and in the morning, her ambivalence finally giving way before the relentless enthusiasm all around her, she went to the march.

She dressed in white, because the organizers had asked people to, in honor of the women's suffragists: clean white sweatpants, Judy decided, and a white sweatshirt. Someone had given her a yellow sash to put on and she draped the sash over one shoulder, the way she had seen it done in the old photographs of suffragists. She got in her car and picked up Alison Gee and drove out to a parking place some blocks from the Washington Monument, where the morning rally was supposed to begin. It was still quite early, the air fresh and cool; Judy and Alison had Voters for Choice T-shirts to sell, and as they climbed from the car with their boxes of T-shirts Judy saw that small tributaries of white had already begun flowing toward the monument, the sidewalks filling with women in white, women gathered at the stoplights, women spilling over into the traffic lanes, women striding up the side streets that led toward Independence Avenue and the monument grounds beyond.

There were so *many* of them, Judy thought; Alison was a politically demonstrative woman and had attended big Washington marches before, but Judy had never walked like this amid marchers collecting for a great event, and Alison could see Judy's astonishment growing as the women kept coming and coming, clambering out of taxicabs, piling off city buses, surging up from the train station, rising by escalator from the subway stops below. They carried NARAL signs and Planned Parenthood signs and coat-hanger pictures they had painted themselves; they crowded around Judy and Alison, demanding to see the T-shirts they were selling, pulling out ten-dollar bills to buy shirts on the spot; they hurried toward the monument, their pace quickening as they neared the grass of the grounds, and Judy hurried with them, looking for the RHS contingent, listening to the noise. The noise was swelling. Already women had begun to chant and sing, Judy heard applause from up ahead, there was microphone crackle from a platform stage she could not yet see, the signs bobbed closer and thicker overhead, she had to dodge and elbow to push through the people —older women and younger women together, Judy saw as she elbowed, were these mothers and daughters marching side by side? And men, really a remarkable number of men, and teenagers too young to remember life before *Roe* v. *Wade*, and exuberant lesbians in crew cuts and big shirts, and college students waving a WE WON'T GO BACK banner strung from a giant coat hanger, and grandmotherly ladies carrying a sign that read MENOPAUSAL WOMEN NOSTALGIC FOR CHOICE, and a blind woman with an ANOTHER DOG FOR CHOICE placard affixed to her guide dog, and women in wheelchairs and women in religious habits and women Judy realized must look very much like her, their startled, eager faces suggesting that until this day they had never thought of themselves as marchers.

How big *was* the crowd? From where Judy stood, mobbed on all sides now as she made her way along the grassy slope of the monument grounds, she could see no outside boundary to it. She lost Alison somewhere. She wanted to find B.J. She was anxious about the RHS banner, she wondered what people would say when they saw it passing by, she wondered how it would feel to see the banner without carrying it herself (for Judy had decided she would not; it was B.J.'s to carry, Judy would walk with them a little but would step away after that). When she finally saw the banner it was stretched out wide, B.J. and three other clinic women holding it high amid a thicket of Missouri signs, and Judy was surprised to see how unexceptional it looked, how straightforward, how easily and simply REPRODUCTIVE HEALTH SERVICES, ST. LOUIS, MISSOURI fit into the chanting, cheering throng as the march began its movement east toward the Supreme Court and the Capitol. What was it about this that made her feel so emotional? For a minute Judy was bewildered, watching B.J. and the clinic workers walking flushed and proud with the banner stretched between them; here before her was the splendid tableau that should have made Judy catch her breath: her clinic, her decade's work, the logo she commissioned and the name she created being lofted through the crowd to enthusiastic shouts of recognition and support (it had been like that all morning, B.J. said, women seeing the RHS name and screaming Way to go, Keep up the good work, What courage you people have!).

But in fact what moved her, what filled Judy now with unspeakable relief, was the very ordinariness of the banner and the women who carried it. On this day, in this place, they were not big or special at all. They were *tiny*. They were barely visible stitches in a vast, amazing quilt. Already the rumors of crowd size had begun to pass from the front end back as the march moved on up Constitution Avenue: it was two hundred thousand, no, it was three hundred thousand, no, somebody at the top said half a million, the march was packing Constitution for block after block, the chants rose and died away and rose again ("*pro choice*," clap-clap, "*pro choice*," clap-clap)—Judy worked her way over toward the base of a broad statue and climbed up for a look, steadying herself beside others who had scrambled up already, and as she straightened she could see for the first time the full breadth and length and she said aloud, "My *God*," knowing no one would hear her because the shouting was too loud. She put one hand to her mouth, and she began to cry. She had never seen so many people in her life.

ON THE AFTERNOON of April 25, eighteen hours before the scheduled oral argument in *William L. Webster* v. *Reproductive Health Services*, Sam Lee climbed out of his sister Emily's car at an intersection in Washington, D.C., and walked up the steps of the Supreme Court building with a sleeping bag, a pillow, a spiral notebook, and a small stack of books. Ahead of him, settled as comfortably as could be managed on the broad marble courtyard at the top of the steps, a lone young man with a sleeping bag of his own had already formed the beginning of a line.

Sam put his sleeping bag down and sat quietly on the marble for a while, hugging his knees to his chest. He said hello to his linemate, who told Sam he was a pro-life engineering student from Seattle, and then Sam got up and walked back and forth across the courtyard, looking up at the great white columns that flanked the central front doors of the Court. He nodded at the security guards, who nodded back—watchful but amiable, Sam thought; he looked like a respectable American citizen even if he was planning to sleep all night in the courtyard of a government building. More people arrived: someone with a backpack, someone with a beach chair, more sleeping bags, a group of undergraduates from Dartmouth. On the street below, a television truck pulled up and disgorged its crew. Cables snaked out across the sidewalk; satellite equipment was pushed into place around the base of the Supreme Court steps; the afternoon sun began to fade. More people arrived. Twenty people in the line, then thirty-five, then fifty—Sam kept counting; the evening stretched ahead of them, his sister had not even brought him his dinner yet, and already the sober esplanade of the United States Supreme Court was turning into an odd encampment of men and women sitting cross-legged on the marble, trying not to rumple their court clothes.

Sam had paid the equivalent of three weeks' household food budget for his airline ticket to Washington, and he began to worry that he was not going to get into the oral argument at all. He took his spiral notebook to one of the guards and asked if they would help him enforce a sign-up sheet, people writing their names in the order of their arrival at the line; the guards said they thought that sounded fair, and now Sam became the keeper of the sign-up, moving slowly down the Supreme Court line with his notebook and a pen. He felt a comforting surge of the familiar as he slipped back into character amid these grand surroundings: Sam was organizing, trying to coax a public gathering into some improvised form of order and decorum, and as he approached with his notebook out-stretched, people smiled at him and wrote in their names and the cities from which they had traveled to Washington. Rockwell, Texas; Anderson, South Carolina; Missoula, Montana; Dickinson, South Dakota—at first Sam thought they must be local law students or Washington tourists writing down their hometowns, but no, some of them really had flown or driven across many states just to stand in line for these arguments, not even knowing whether they would be allowed inside. The executive direc-tor of Planned Parenthood of Massachusetts had come down with her husband, and an Americans United for Life staff member had flown out from Chicago; there was a lawyer from Manhattan, and a lawyer from Terre Haute, Indiana, and as night came on and the line kept lengthening someone passed along the word about Sam, so that people began to ap-proach him (politely, it seemed to Sam, nobody with fists clenched or loathing in the eyes) to say, "Is it true you wrote the law?"

Sam would say yes, it was true, and there would be a conversation after that—warm and appreciative if the person was pro-life, cooler if the person was not. But in the moonlight Sam felt a kind of softness closing in around all of them and when he lay in his sleeping bag and looked up at stars he heard only the low voices of people murmuring to one another or praying the rosary. He slept, fitfully; a television truck generator sputtered up and Sam started awake, slept again. When the *Post-Dispatch* reporter Charlotte Grimes arrived at dawn, the Supreme Court plaza was so quiet that Grimes could hear birds in the trees nearby; she looked at the line, which stretched ragged but intact from the Supreme Court doors back out across the marble of the courtyard, and she watched men and women struggle sleepily to their feet, smoothing wrinkles from their clothes, pushing sleeping bags out of the way, passing donuts along the line—the waning hours of armistice, thought Grimes, who had reported from Central American war zones, and that was how she wrote her story: "Throughout the night they paced, walked, talked in peaceful coexistence," Grimes wrote. "And as the day itself dawned clear and warm, the fragile peace largely held, like an Easter truce among warring nations."

Sam's sister brought him a cup of coffee and a clean white shirt and took his sleeping bag away in her car. Sam combed his hair. A lot of people were arriving now, reporters, cameramen, line stragglers, preachers, protestors with signs, attorneys with special Court credentials allowing them to bypass the ordinary citizens' line—here was Andy Puzder, coming in through the lawyers' line, stopping first to say hello to Sam; and Frank Susman and Tom Blumenthal, Sam recognized them from Susman's Jefferson City visits and the District Court trial; and Roger Evans, the Planned Parenthood lawyer who had argued the case before the Eighth Circuit; and B. J. Isaacson-Jones, waiting beside a woman who had come with Frank Susman, that must be Susman's wife. Sam's line began to move. The Court door opened to him. Guards motioned him over the threshold, through metal detectors, down long marble hallways, along the length of a crowded wooden bench. Sam heard the bang of a gavel and a reedy male voice from the far end of the chamber crying *Oyez, Oyez, Oyez, All persons having business before the Honorable, the Supreme Court of the United States, are admonished to draw near and give their attention,* and Sam straightened in his seat and stared up at the interior marble columns, the massive American flags, the tall heavy curtains of deep red velvet, the high polished curve of mahogany desk that formed the most immediately visible barrier between himself, Samuel Henry Lee, and the nine justices of the United States Supreme Court.

Well, Sam thought. Well. I'm here.

And he thought: Justice Blackmun looks so small; you can barely see his head above the bench.

A momentarily disembodied voice—Sam realized after some confusion that it was the Chief, William Rehnquist, speaking so matter-of-factly that Sam had assumed the words must have been coming from a bailiff or a clerk—said, "We will hear argument now in number one eight six oh five William L. Webster versus Reproductive Health Services."

Attorney General William Webster, clearing his throat, got up and stood before the justices—tipping his head back just slightly to address his words respectfully upward, because the Supreme Court bench is positioned to make the justices loom above the lawyers.

"Mr. Chief Justice and may it please the Court," Webster said.

The attorney general is nervous, Sam thought. From where he sat, three quarters of the way back, pressed in from both sides by others also straining forward to hear every word, Sam could see only the back of Webster's head and what seemed to Sam the taut, anxious angle of his dark-suited shoulders. Sam listened to the attorney general defend piece by piece the now-contested sections of H.B. 1596 (Webster referred to each section by its statutory number, 188.205 and so on, but to Sam it would always be H.B. 1596), explaining particulars as the justices interrupted him with questions: yes, the state believed it was both constitutional and correct to prohibit the use of public funds, public facilities, and public employees in performing, assisting, encouraging, or counseling; no, the state had worked out no specific criminal sanctions for physicians or facilities found to have violated the law. "Arguably, it would be a misdemeanor," Webster said.

Sam shifted impatiently in his seat; the exchanges were so parched, so technical, as though Webster were explaining highway specifications or a revision of the state income tax. Sam had been told some weeks back that Webster intended to restrict his arguments to the specific merits of H.B. 1596, that he planned to leave the big debate—the *Roe* v. *Wade* debate—to the former U.S. Solicitor General Charles Fried, who would argue *Webster* on behalf of the United States. Still Sam had the feeling that he was watching a peculiar drama in which everyone present had agreed to speak in code, to talk about one thing when it was universally understood that they were all really talking about something else. Here was Webster, defending Andy Puzder's language about human life beginning at conception: "We contend this declaration doesn't affect anyone, that it was clearly improperly struck down at the lower court level and that legislative bodies around this country should be entitled at least in the nonabortion area to have a philosophical statement of when they contend life begins."

Let it stand, in other words; it doesn't mean anything anyway. For an instant Sam entertained a small urgent fantasy in which Andy Puzder rose from his seat to say *In fact it means quite a lot, it means the nation's laws are schizophrenic on this subject and we want to force the courts into confronting their own doubletalk,* but he understood Webster's strategy: avoid the broader implications, pretend this is simply about one state's effort at statutory revision, pretend people slept on the marble all night because of their intense interest in the fate of H.B. 1596. Webster and the justices trudged on, arguing about the viability-testing provision—did it or did it not require amniocentesis at twenty weeks, Webster sounded wobbly on this, he kept saying the state did not require physicians to perform "tests that would be unnecessary." Webster sat. Charles Fried rose. "Thank you Mr. Chief Justice and may it please the Court," Charles Fried began. And Sam couldn't believe it, the exchanges no longer sounded parched and technical; now they were *lofty*—distant, somehow, as though Mr. Fried were presenting an interesting academic problem. "Today the United States asks this Court to reconsider and overrule its decision in *Roe* v. *Wade*," Charles Fried said in his lofty, elegant voice, and Sam tried to feel transported by these words but found that he could not. Fried had cooled them. "At the outset," said Charles Fried,

> I would like to make quite clear how limited that submission is. First, we are not asking the Court to unravel the fabric of unenumerated and privacy rights which this Court has woven in cases like *Meyer* and *Pierce* and *Moore* and *Griswold*. Rather, we are asking the Court to pull this one thread. And the reason is well stated by this Court in *Harris* and *McCrae*: abortion is different.

Sam saw what Fried was doing—separating *Roe* v. *Wade* from the line of Supreme Court cases culminating in *Connecticut* v. *Griswold*, the 1965 ruling that married couples possess a constitutional right to privacy which includes their right to purchase and use contraceptives. We are not here fighting *Griswold*, Fried was saying; we believe abortion and contraception can and should be treated differently in constitutional law; we believe states should be allowed to prohibit one but not the other. It was a sensible argument to raise before the Court—Sam had read the other side's briefs and knew they intended to present this specter of an inexorable threat to the broader right of personal privacy—but Fried had led off with it straightaway, as though that was the most important thing he had to tell the justices, that he wanted only so much and no more. Where was the eloquence? Where was the passion? Where was the vigorous defense of unborn human life—or at least of every state's right to protect unborn human life? Maybe Sam didn't understand the point of Supreme Court oral argument as well as he thought he did; maybe the fervor was all in the written briefs and the lawyers were just supposed to get up there and anticipate the weak spots, answer the justices' questions as crisply and convincingly as they could. Even Fried's final remarks were so cautious, so crabbed, not: give us the right thing, but instead: give us *something*. "If the Court does not in this case in its prudence decide to reconsider *Roe*," said Fried, "I would ask at least that it say nothing here that would entrench this decision as a secure premise for reasoning in future cases."

Now *there* was a closer to inspire the troops. Sam glanced around to see if anyone else was wincing, but all eyes were fixed intently on the front of the courtroom; Frank Susman was rising to take his turn, and as he began to address the justices Susman lifted one shoulder, unaccountably, and appeared to be displaying the left arm of his jacket.

"Mr. Chief Justice and may it please the Court," Susman said. "I think the Solicitor General's submission is somewhat disingenuous when he suggests to this Court that he does not seek to unravel the whole cloth of procreational rights, but merely to pull a thread. It has always been my personal experience that when I pull a thread, my sleeve falls off."

An audible murmur spread quickly through the courtroom, people smiling and whispering to each other. Sam smiled himself and thought: Point, Susman. The line about Frank Susman's sleeve was going to show up in the next day's newspapers, Sam guessed; apparently the Supreme Court did not frighten Susman any longer, he was relaxed enough to be clever. He honed right in on Fried's opening argument. "It is not a thread he is after," Susman said:

> It is the full range of procreational rights and choices that constitute the fundamental right that has been recognized by this Court. For better or for worse, there no longer exists any bright line between the fundamental right that was established in *Griswold* and the fundamental right of abortion that was established in *Roe*. These two rights, because of advances in medicine and science, now overlap. They coalesce and they merge, and they are not distinct.

Justice Scalia, who in his relatively short tenure on the Court had already made himself famous for his own facility with words, immediately broke in for the first question. "Excuse me," Scalia said. "You find it hard to draw a line between those two, but easy to draw a line between first, second, and third trimester."

Susman said, "I do not find it difficult—"

Scalia pushed on. "I don't see why a court that can draw *that* line can't separate abortion from birth control quite readily."

Sam leaned forward in his seat, listening closely to every word. This was what he had imagined Supreme Court argument would sound like: sharp, compelling, delivered as though both parties had a deep and genuine feeling about the issues before them. In the right-to-life movement people liked to think Scalia was a real philosophical ally, a justice who believed not only that *Roe* was wrongly decided but that abortion itself was profoundly wrong, and as Susman sparred now with one justice after another—here were Kennedy and O'Connor, asking questions about late-trimester viability, and Rehnquist probing the significance of nineteenth- and early twentieth-century abortion laws—it was Scalia who seemed to Sam to force the argument into its grandest and most essential dimensions. When Scalia and Susman went at each other the rest of the room receded into the background; the spectators were utterly silent and still and for a moment Sam forgot where he was, so closely was he watching the white-haired lawyer and the black-haired justice. How could anyone talk about some fundamental right to abort, Scalia wanted to know, without first facing up to the question of whether the fetus is a human life? "It is very hard to say it is just a matter of basic principle that it must be a fundamental right," Scalia said, "unless you make the determination that the organism that is destroyed is not a human life. Can you, as a matter of logic or principle, make that determination otherwise?"

All right, thought Sam. We are coming to it now.

Susman said he believed that the basic question, whether life begins at conception, was verifiable not by fact but only by faith. "It is a question of labels," Susman said:

Neither side in this issue and debate would ever disagree on the physiological facts. Both sides would agree as to when a heartbeat can first be detected. Both sides would agree as to when brain waves can first be detected. But when you come to try to place the emotional labels on what you call that collection of physiological facts, that is where people part company.

Emotional labels, thought Sam. (How could they still be phrasing it this way? They *hear* the heartbeat, they can *measure* the brain waves; what astonishing levels of internal duplicity must be required to keep insisting that it is "emotional" to call this human life?) But Scalia said at once: "I agree with you entirely." Sam was startled. "But what conclusion does that lead you to?" Scalia asked, and Sam nodded now, seeing where the justice was going:

That therefore there must be some fundamental right on the part of the woman to destroy this thing that we don't know what it is? Or rather, that whether there is or isn't is a matter that you vote upon? Since we don't know the answer, people have to make up their minds the best they can.

Not an argument for full protection of life from conception on, perhaps, but still—leave it to the people, Scalia was saying, let the voters decide whether to protect the unborn. And Susman was nimble, taking the justice's lead, keeping it big. "The conclusion to which it leads me," he said, is that when you have an issue that is so divisive and so emotional and so personal and so intimate—that it *must* be left as a fundamental right to the individual to make that choice, under her then-attendant circumstances, her religious beliefs, her moral beliefs, and in consultation with her physician. The very debate that went on outside this morning, outside this building, and has gone on in various towns and communities across our nation, is the same debate that every woman who becomes pregnant and doesn't want to be pregnant has with herself.

He's good, Sam thought. He's wrong but he's good, they are worthy opponents, this is how the debate was supposed to be, here before the nation's highest court. Susman went right on:

Women do not make these decisions lightly. They agonize over them. And they take what we see out front and what we see in the media and they personalize it and they go through it themselves. And the very fact that it is so contested is one of those things that makes me believe that it must remain as a fundamental right with the individual, and that the state legislatures have no business invading this decision. Let me address particular sections, if I may for a moment. I would start with the public funding question. . . .

And then it was over, really, the moment was finished; they were into the particulars, the construction of the public funding language, the argument about whether the life-begins-at-conception passage would or would not have any practical impact on state law. Susman was arguing that it would, that it might be used to prevent *in vitro* fertilization or bring criminal causes of action against pregnant women who drank or otherwise treated their fetuses irresponsibly, but Sam had stopped listening quite so closely; he felt a little cheated, as though the important debate had been clipped as it was getting underway. If he had been scoring a college debate Sam would have noted that Frank Susman had gotten in the last word, but this was not a college debate and afterward Sam was angry at all the people who wanted to act as if it were, as if one side or the other could be determined to have racked up the most points. As the spectators and lawyers filed outside they were accosted at once by the reporters and demonstrators who now filled the Supreme Court plaza: How had it gone in there, who had the best arguments, who was questioned most roughly by the justices, who won?

Sam eased his way out of the crowd, the newspaper people with their notebooks and tape recorders, the television people thrusting their microphones forward, the photographers angling for the best composition of shouting protestors and warring picket signs. He saw Jack Willke, the National Right to Life Committee president, disappear amid a closing crush of reporters and network cameras; a few yards away, so that they all could have called to each other if the great noise around them had subsided long enough, a second crush surrounded Faye Wattleton of Planned Parenthood and Kate Michelman of NARAL. Reporters crowded around Janet Benshoof, of the American Civil Liberties Union; they crowded around Ellie Smeal, of the Fund for the Feminist Majority; they crowded around William Webster and they crowded around Frank Susman and for a while Sam stood at the edge of the plaza, watching, his hands in his pants pockets and his spiral notebook jammed up under one arm. Then he walked down the courthouse steps and headed for the subway by himself.

FOURTEEN

Zealots

1989

WILLIAM L. WEBSTER V. REPRODUCTIVE HEALTH SERVICES, the ruling, began arriving in Missouri at approximately 9:20 A.M. Central time on July 3, 1989, emerging one single-spaced page after another from fax machines in Kansas City and in Jefferson City and in St. Louis, where a roomful of television and newspaper reporters had been permitted onto the premises of Reproductive Health Services to record the reaction of the clinic director and her staff at the moment they received the news. The clinic director was observed to have stood for some moments in front of the television set that had been wheeled into her office, after which she began to weep and embrace the women around her. *The New York Times* put the photograph on the front page of the next morning's newspaper, *B. J. Isaacson-Jones, left, executive director of Reproductive Health Services, which challenged the Missouri law, and Judith Widdicombe, the clinic's founder, consoling each other after the decision;* and there was also a picture of B.J. slump-shouldered behind her desk, covering her face with both hands; and the *Post-Dispatch* had a picture of a man holding a little girl and a hand-drawn coat-hanger poster with the words I SWEAR THIS WILL NOT BE A PART OF MY DAUGHTER'S FUTURE scrawled across the top and bottom; and in Forest Park, the principal municipal park in St. Louis, five hundred demonstrators gathered at the World's Fair Pavilion to listen to speeches and shout and applaud. That was in the *Post-Dispatch,* too:

"We have to realize that this is the beginning of a war," said Judith Widdicombe, founder and chairwoman of Reproductive Health Services.

"We must shift our focus to the legislatures and demonstrate to them that we will not tolerate this. The goal of the anti-abortion movement is no abortion, with no exceptions. The next thing they will go for is birth control.

"This issue will become our Vietnam of the '90s. Our work is cut out for us. But we will prevail, because justice will prevail."

All week the newspapers repeated dramatic, vague headlines, with photographs of people looking angry or elated or anguished or sober. *Abortion Rights Curtailed. Abortion Is "Vietnam of 90s," Group Says. Decision Called "Near Mortal Blow" to Roe v. Wade Ruling. Abortion Foes See "Inevitable" Victory.* Here was how the *Post-Dispatch*, grappling with the text of the ruling itself, commenced two mornings later its explanation to the readership in St. Louis:

For the first time in 16 years, a majority of five justices on the Supreme Court either oppose or have reservations about core elements of the 1973 *Roe* vs. *Wade* decision establishing a fundamental right to abortion.

For the first time in 16 years, during which the Supreme Court has struck down almost every state law regulating abortion, the court has left standing all of the challenged provisions of a state abortion law—Missouri's.

And, for the first time in 16 years, the court implicitly has beckoned to states to pass new restrictive abortion laws even as the court takes up three new abortion cases next term that may whittle further the abortion right.

Those are three observations that legal experts on both sides of the abortion issue agree are significant about Monday's resonating Supreme Court decision upholding the Missouri law. . . .

The faint note of uncertainty in the carefully worded text was not accidental. What the ruling actually said—what practical, discernible change *Webster* v. *Reproductive Health Services* actually imposed upon abortion practice in Missouri and the other American states—was a matter of some conjecture even two days later, when "legal experts" (and lobbyists, and clinic directors, and former clinic directors) had been able to read and reread all eighty-two printed pages of ruling. The opinions themselves suggested why the Supreme Court had pushed the release of *Webster* back to the final day of the 1988 term: there were five of them, five different

Webster v. *Reproductive Health Services* opinions, and together they added up to such a muddle of quarrelsome, fragmented, contradictory argument that it required a tally sheet just to figure out which justice had agreed to what, and why.

The lead opinion, the first of the five to appear in the printed ruling, was written by Chief Justice Rehnquist. That in itself was a signal of defeat for one side and victory for the other, since Rehnquist was one of the two original No votes on *Roe,* but within the first few pages of Rehnquist's opinion the signal blurred and began to confuse. He led off with what the legal papers referred to as the H.B. 1596 "preamble"—the paragraphs asserting that human life begins at conception and that the unborn have "protectable interests in life, health, and well-being." The Eighth Circuit had erred in finding this passage unconstitutional, Rehnquist wrote; the language could remain in the Missouri statute books. But that did not mean the Supreme Court was formally deciding that it *was* constitutional; since no one had so far found this preamble to serve any purpose beyond expressing a philosophical viewpoint, the Court, for the time being, had decided not to examine its constitutionality at all. "It will be time enough for federal courts to address the meaning of the preamble should it be applied to restrict the activities of appellees in some concrete way," Rehnquist wrote.

Next: The Court was upholding as constitutional, Rehnquist wrote, both the Missouri law's section making it unlawful for any public employee "to perform or assist an abortion," and the section prohibiting the use of any public facility "for the purpose of performing or assisting an abortion not necessary to save the life of the mother." Previous Supreme Court rulings had already upheld a state's right to stop funding abortions, Rehnquist wrote; Missouri had merely extended the same principle to these provisions of its new statute.*

Next: The Court was upholding as constitutional the Missouri statutory provision requiring physicians to test for fetal viability before performing any abortion at twenty weeks or beyond. And it was true, Rehnquist wrote, that this provision appeared to violate the directives laid out in *Roe* v. *Wade;* since viability testing was intended to protect not the woman's

---

* Most of the H.B. 1596 prohibition on publicly supported "encouraging or counseling" for abortion was not directly addressed in the Supreme Court's ruling. When the bill was signed into law, the "encouraging or counseling" portion of the statute had contained three separate subsections, each of which was subsequently found unconstitutional by the District Court and the Court of Appeals. (One subsection declared that no public funds could be used for this "encouraging or counseling"; a second subsection said public employees could not engage in such speech while on duty; and a third prohibited "encouraging or counseling" in public facilities.) The state of Missouri chose to appeal to the Supreme Court only the overturning of the first of these subsections. But the justices' reading of both sides' *Webster* v. *RHS* appellate briefs, Rehnquist wrote, had convinced the majority that this issue was not moot and need not be taken up in this case.

health but rather the potential life of the fetus, and since some of this testing would obviously occur during the second trimester, then *Roe* and the Court's subsequent abortion rulings would seem to forbid it. That merely underscored the fatal weakness in the internal logic of *Roe*, Rehnquist wrote:

> In the first place, the rigid *Roe* framework is hardly consistent with the notion of a Constitution cast in general terms, as ours is, and usually speaking in general principles, as ours does. The key elements of the *Roe* framework—trimesters and viability—are not found in the text of the Constitution or in any place else one would expect to find a constitutional principle. . . . In the second place, we do not see why the State's interest in protecting potential human life should come into existence only at the point of viability, and that there should therefore be a rigid line allowing state regulation after viability but prohibiting it before viability.

Had the Rehnquist opinion proceeded in this vein, the harsh language of these paragraphs would appear to have been dismantling *Roe* v. *Wade*. But the opinion stopped short of that, pointing out that *Roe* v. *Wade* had overturned a Texas statute criminalizing nearly all abortions. The state of Missouri was merely asking the Court for increased powers of regulation. "This case therefore affords us no occasion to revisit the holding of *Roe* . . . and we leave it undisturbed," Rehnquist wrote. "To the extent indicated in our opinion, we would modify and narrow *Roe* and succeeding cases."

We cut back on *Roe* without reversing it entirely, in other words: we grant these increased powers of regulation; we reject the trimester divisions and the way they have been used to screen state abortion laws so strictly; but we do not revisit—not yet, not in this case—the central holding itself, that the Constitution protects women's right to abortion. That was the conclusion of Rehnquist's opinion, which was joined without comment by Justices White and Kennedy. But then Justice O'Connor presented her own separate opinion, agreeing with Rehnquist's *finding*, that the Missouri statute was constitutional, but disagreeing, significantly, with both Rehnquist's reasoning and his conclusions as to what the Missouri viability-testing passage actually required of abortion doctors and clinic personnel. O'Connor's reading of the viability-testing passage found it to present no conflict with *Roe*. Thus even though she had criticized the *Roe* trimester logic in the past, O'Connor wrote, there was no need in this case to reattack that logic or to reassess *Roe* v. *Wade* at all. "When the constitutional invalidity of a State's abortion statute actually turns on the constitutional validity of *Roe v. Wade*, there will be time enough to re-examine *Roe*," O'Connor wrote. "And to do so carefully."

Then to add one more layer of confusion to an already confusing decision, Justice Scalia weighed in with a barbed and angry opinion in which he agreed with the lead opinion's finding, that the Missouri statute was constitutional, but lashed out at his fellow justices for not going far *enough*. The *Webster* case had offered the Supreme Court its opportunity to reconsider the central holding in *Roe v. Wade*, Scalia wrote, and the Court owed it to the American people to take on that difficult and profoundly important task. But the justices had run for cover instead:

> We can now look forward to at least another Term with carts full of mail from the public, and streets full of demonstrators urging us—their unelected and life-tenured judges who have been awarded those extraordinary, undemocratic characteristics precisely in order that we might follow the law despite the popular will—to follow the popular will. Indeed, I expect we can look forward to even more of that than before, given our indecisive decision today.

Scalia directed much of his wrath specifically at Justice O'Connor, who had obviously infuriated him by declining to reconsider even the trimester framework that she had so quotably criticized in previous rulings. Just how defiant was a state going to have to be, Scalia wondered, to force such a timid Court into a real examination of *Roe v. Wade?*

> Given the Court's newly contracted abstemiousness, what will it take, one must wonder, to permit us to reach that fundamental question? The result of our vote today is that we will not reconsider [*Roe v. Wade*], even if most of the Justices think it is wrong, unless we have before us a statute that in fact contradicts it—and even then (under our newly discovered "no-broader-than-necessary" requirement) only minor problematical aspects of *Roe* will be reconsidered, unless one expects state legislatures to adopt provisions whose compliance with *Roe* cannot even be argued with a straight face. It thus appears that the mansion of constitutionalized abortion law, constructed overnight in *Roe v. Wade*, must be disassembled doorjamb by doorjamb, and never entirely brought down, no matter how wrong it may be.

Thus concluded the five-vote majority in *Webster* v. *Reproductive Health Services:* three votes for dismantling the trimester framework; one vote for leaving *Roe* undisturbed; and one vote for overruling *Roe* explicitly. The word "plurality" began showing up in the *Webster* analyses, leaving attorneys and lay readers alike to scramble for some clearer understanding. In an intently anticipated ruling by the U.S. Supreme Court, what weight did a "plurality" opinion carry? It was not a majority. It was not a minority. It was a majority of the majority, expressing the view of some justices without effecting any actual change in constitutional doctrine. The state of Missouri had won its appeal, that much was evident, but because no clear majority could agree on the reasoning behind the victory, all the rest was guesswork and speculation as to what the Court might approve the next time around. The Supreme Court appeared to be inviting the states to work up new abortion laws and come test out their constitutionality one by one, as Justice Harry Blackmun wrote in his long and extraordinarily passionate dissent:

> . . . to enact more and more restrictive abortion laws, and to assert their interest in potential life as of the moment of conception. All these laws will satisfy the plurality's non-scrutiny, until sometime, a new regime of old dissenters and new appointees will declare what the plurality intends: that *Roe* is no longer good law.

There was a second *Webster* dissent as well, written by Justice John Paul Stevens and arguing that Missouri's life-begins-at-conception "preamble" was indeed unconstitutional—both because it could be read to forbid post-fertilization contraceptive devices and because it violated the First Amendment by adopting religious beliefs as official state doctrine. But it was Harry Blackmun's dissent that resonated and was remarked upon afterward, for his prose was as mournful and outraged as an elegy for a murder victim. When the ruling was announced at the Court, Justice Blackmun had delivered much of his dissent aloud from the bench, reading the pages before him in a low, tremulous voice that sounded at certain moments as though he was near tears. "I fear for the future," Blackmun wrote:

> I fear for the liberty and equality of the millions of women who have lived and come of age in the sixteen years since *Roe* was decided. I fear for the integrity of, and public esteem for, this Court.

By dodging the question for the moment, while so plainly encouraging the states to keep trying, the Supreme Court—in particular, the three-justice plurality that proclaimed itself to be "narrowing" *Roe* while simultaneously leaving it "undisturbed"—had doomed the right to abortion without even having the decency to say so, Blackmun wrote:

Thus, "not with a bang, but a whimper," the plurality discards a landmark case of the last generation, and casts into darkness the hopes and visions of every woman in this country who had come to believe that the Constitution guaranteed her the right to exercise some control over her unique ability to bear children. The plurality does so either oblivious or insensitive to the fact that millions of women, and their families, have ordered their lives around the right to reproductive choice, and that this right has become vital to the full participation of women in the economic and political walks of American life.

Justice Blackmun's *Webster* v. *Reproductive Health Services* dissent was joined by Justices Thurgood Marshall and William Brennan, so that the last of the original *Roe* and *Doe* majorities—the remaining three justices of the seven who had decided sixteen years earlier that the abortion laws of Texas and Georgia violated the Constitution—now signed on together to a fervent plea that their work be left intact. "For today, at least, the law of abortion stands undisturbed," Blackmun wrote. "For today, the women of this Nation still retain the liberty to control their destinies. But the signs are evident and very ominous, and a chill wind blows."

SAM LEE read the opinions in *Webster* v. *Reproductive Health Services* many times, circling phrases and writing question marks into the margins of the printed pages that he had pulled with such eagerness from the fax machine in his office over the pawnshop. "The rigid *Roe* framework." "Disassembled doorjamb by doorjamb." "A chill wind blows." The separate opinions had appeared on first read to resonate with powerful, theatrical, end-of-an-era language, but when the elegant phrases were fit together into a single ruling, Sam could make no sense of it. He felt as though he were studying one of those obscure Scripture passages whose meaning was still being argued over in Bible class. The national right-to-life people had looked exultant that morning on the Supreme Court steps, triumphantly holding aloft their copies of *Webster* (Sam had watched on the evening news as James Bopp, the Indiana lawyer who served as general counsel to the National Right to Life Committee, was shown reading aloud so excitedly from the opening pages that Bopp's voice cracked midsentence), and the newspapers and magazines kept insisting that the ruling was a profoundly important signal from the Supreme Court to the states, that it was only a matter of time before the Court began permitting states to make major revisions in their abortion laws.

The post-*Webster* activity—that was the phrase Sam heard, "post-*Webster*," the way people had once said "post-*Roe*"—had started up within moments of the ruling, it seemed, and in simultaneous bursts of enthusiasm across the United States. The governor of Florida, a right-to-life Republican named Bob Martinez, called his state legislature into special session to draft new abortion legislation before the regular session got underway. Louisiana legislators began working up the legal mechanism for resurrecting the state's old criminal abortion law, which had been rendered useless by *Roe* v. *Wade* but had never formally been removed from the criminal code. A computer-based on-line political newsletter in Virginia began publishing a separate national *Abortion Report*, with daily bulletins from around the country: there were bills being drafted in Utah, Pennsylvania, Idaho, Michigan, South Carolina, and the United States territory of Guam; abortion arguments had grown suddenly sharper and more prominent in congressional races in Florida, Texas, and Maine; the only two 1989 gubernatorial races, in New Jersey and Virginia, were receiving new attention as contests between pro-life Republicans and pro-choice Democrats. In August, the political humorist Mark Russell sang about abortion on PBS:

> *California, here we come*
> *Where the babies started from*
> *The high court's decision, confusing to read.*
> *We're driving cross-country to find a law that fits our*
> *   need . . .*
>
> *Pregnancies to terminate,*
> *Varying from state to state.*
> *A fetus is safer in all of this strife*
> *Than a state legislator who votes against the right to life*
>
> *And pro-choice breaking down the doors*
> *Of your statehouse senators.*
> *Freedom of choice—you just lost yours!*
> *Duck for cover, here they come.*

But Sam was not euphoric, and the excitement made him uneasy. The Supreme Court had not reversed *Roe*, and had made no promise that it would; indeed, as Sam reread the individual opinions separately and then together, they left him with so many questions that he began to wonder whether the justices were dropping hints that they would *never* reverse *Roe*. Why had Sandra Day O'Connor held back? If she believed what the right-to-life movement still hoped she did, that *Roe* v. *Wade* was a destructive and badly argued ruling, then why had she sidestepped the chance to overturn it? And was it significant that Justice Kennedy had declined to put into writing his own reasons for voting with Rehnquist and White? And what had Rehnquist meant when he chided Justice Blackmun for Blackmun's insistence that abortion law must not be left to state legislative bodies? "The goal of constitutional adjudication is surely not to remove inexorably 'politically divisive' issues from the ambit of the legislative process," Rehnquist had written at the close of the plurality opinion:

> Justice Blackmun's suggestion that legislative bodies, in a Nation where more than half of our population is women, will treat our decision today as an invitation to enact abortion regulation reminiscent of the dark ages not only misreads our views but does scant justice to those who serve such bodies and the people who elect them.

"Reminiscent of the dark ages"? What kind of signal was *that*? Was the Chief Justice suggesting that he regarded the years before 1967, when state criminal codes protected both born and unborn human life, as the "dark ages"? Or was that line meant as encouragement to Justice O'Connor, who was plainly now the swing vote needed for a reversal of *Roe*? Did the Chief know something the public did not—perhaps that Sandra Day O'Connor would never join her vote to a Supreme Court opinion that might return abortion regulation to the "dark ages"?

Publicly, when reporters called him for help with their *Webster* follow-up articles, Sam tried to affect a demeanor that was buoyant but thoughtful; privately, he felt deflated, as though he had been deprived of both the jubilation of victory and the righteous anger of defeat. The weekend before the ruling Sam had driven up to Minneapolis for the annual National Right to Life Committee convention, and the workshops had been keyed up with anticipation: we'll do this after *Roe* is overturned, we'll do that after *Roe* is overturned, we must remind ourselves how much work we will still have before us after *Roe* is overturned. The right-to-life New Jersey congressman Chris Smith had given an inspirational speech about how he had run track during high school and had learned to push himself through mile races by thinking of them as two half miles. "We've gotten this far over sixteen years, but now we're just going into the next part of the race," Smith said. And then David O'Steen, the executive director of the National Right to Life Committee, had taken his turn to speak and said, no, the movement was not even at the half mile yet, maybe the quarter mile, but the finish line was plainly up ahead of them somewhere, the Court *would* reverse itself on *Roe*. And the Committee lawyer Jim Bopp had talked about different statutory approaches to consider once *Roe* was gone: criminal state laws were fine, Bopp suggested, but the movement ought also to propose civil statutes allowing individual citizens to seek monetary damages from abortionists. And between sessions Sam had relaxed in the hotel bar with his friends from other states, listening to excited arguments about the specifics of post-overturn abortion laws: was the civil statutes proposal a plausible one, what form of penalties should criminal laws contain, should they try for everything at once or proceed in stages? Look at us, Sam's friends had said, we've come to this point, *Roe* is going to be overturned, and we haven't even figured out what the new laws should say.

There was an interesting and moving moment in the midst of Chris Smith's speech, Sam remembered; it made him feel wistful now to think of it. Smith had urged the movement to commence this second half of the race with prayer, and the opening prayer, Smith said, should be a plea for wisdom. "Like Solomon," Smith said. "Collectively, our prayer should be: God, tell us what to do." The applause was so loud and sustained that Smith had to pause for a few moments before continuing with his speech, and Sam sometimes recalled this emotional ovation as the summer wore on and he worked his way through one post-*Webster* statutory proposal after another, looking for the model that would suit both grand long-range strategy and the legislators in Jefferson City. *God, tell us what to do.* Every national right-to-life organization with a staff lawyer or legislative analyst was drafting sample bills and mailing them around the country; by autumn Sam had received a Preborn Protection Act, a Preborn Victims Rights Act, a Death Pill Disinvestment Act, a Sample Constitutional Convention Call Resolution (evidently *Webster* had momentarily revived one branch of the Human Life Amendment stalwarts), and a complex eight-section bill entitled Sample Legislation to Bar Abortions Solely as a Means of Birth Control.

"GOALS FOR PROLIFE LEGISLATION 1990," Sam typed out, at the end of October, trying to keep things straight.

1. Stop as many abortions as possible, as soon as possible.
   —Public abortion is now stopped. Private abortion is the prime target.
   —Build positive inducements for childbirth. Remove inducements toward abortion.
   —Life of mother only assumed exception.

Sam's list proceeded briskly, full of resolve, an optimist's shorthanded instructions for the coming decade's work. "Capture the middle public." "Reaffirm and expand present prolife support in legislature." "Provide Supreme Court with opportunity for next major step in the ultimate rejection of *Roe* v. *Wade*." But something was going wrong; Sam could see it in the newspapers. The Florida special legislative session collapsed without producing a single workable right-to-life bill. The Illinois Pro-Life Coalition was making no headway during the fall legislative session in

Springfield. The right-to-life candidate for the governorship of Virginia, Marshall Coleman, appeared to be slipping in his campaign against the Democrat Douglas Wilder. The New Jersey Republican gubernatorial candidate, U.S. representative James Courter, was sounding less and less enthusiastic about the pro-life pledges that were supposed to be a cornerstone of his own campaign; Courter's opponent, the Democrat James Florio, was using attack phrases like "extremist idealogue," and accusing Courter of endorsing "intrusion by the government into the most personal aspect of people's lives."

"CW," the *Abortion Report* called it—Conventional Wisdom. Suddenly, the CW was placing all political momentum with the other side. Except for the gamely upbeat roundups in the *National Right to Life News,* nearly every news-from-many-fronts article Sam read was now sounding the same theme, that the "choice" activists had captured public attention, that *Webster* v. *Reproductive Health Services* had shaken from their complacency all the ordinary Americans who believed abortion ought to remain legal, that politicians were holding their fingers to the wind and finding that it blew toward NARAL. THE ABORTION RIGHTS MOVEMENT HAS ITS DAY, *The New York Times* headlined in mid-October, with a photograph of an exuberant looking legal-abortion rally and some pithy quotes from a formerly pro-life Florida state legislator who had just switched allegiances ("Until the Court ruled, there was a silent majority out there . . . I'd be a fool not to listen to my constituents"). Even the right-wing Catholic newspaper *The Wanderer* repeated the CW in its gloomy weekly reports ("As of Friday, October 13, the news coming in from nearly all sources was bad for pro-lifers . . ."), and Sam dug around in his files one afternoon for a memo National had sent out to the affiliates back in early summer, when everyone was waiting so eagerly for *Webster.* The lawyers Jim Bopp and Burke Balch had prepared this memo together, explaining at length six possible scenarios in the Court's impending ruling, from "Complete Reversal" to "Majority Reaffirms *Roe* and Strikes Down Statute." In retrospect it was Scenario IV that now seemed to Sam to fit most precisely, "Majority Avoids *Roe* But Upholds Statute," and he read again the warnings Bopp and Balch had included as they advised groups like Missouri Citizens for Life on the proper reaction to Scenario IV:

> The most important thing to get across in this scenario is that it represents a significant prolife victory. Because of all the hoopla about the possibility the Court would use *Webster* to reverse *Roe,* the press may be inclined to represent this as a partial defeat for the prolife movement. But it merely means that the Court did not yet feel itself called upon to address *Roe,* and it did uphold a prolife statute that had been struck down by two lower federal courts. *Ironically, we may be aided in countering this perspective by our opponents, whose extremism and de-*

sire to avoid complacency in their ranks may well lead them to empha-
size the danger to women's abortion rights inherent in the provisions
upheld. (emphasis added)

Had either Bopp or Balch imagined how prescient they would prove to have been? Sam and Loretto Wagner drove to Jefferson City the morning the new statewide coalition called Missouri Alliance for Choice held its November 12 march through the capital; from inside the capitol building, where they were preparing a press conference to announce *their* new coalition, the Missouri Pro-Life Network, Sam and Loretto watched as chanting crowds advanced toward the statehouse behind a "March for Women's Lives" banner so wide that it was carried by ten men and women walking abreast. Loretto snorted when she heard the march leaders declare that they had counted fourteen thousand people at their rally that day— the police estimated only half that number, Loretto said, and she knew her march sizes well by now, she could tell that it was not so big as all that. (Besides, everybody inflated march numbers; the March for Life leaders probably tilted toward the optimistic in their annual Washington, D.C., January crowd counts, and the big April pro-abortion march organizers were still insisting that their march had drawn six hundred thousand, though the U.S. Park Police had issued a firm estimate of three hundred thousand.) But even Loretto had to admit that she had never seen so many people at a legal abortion rally in Missouri and that the marchers were certainly boisterous, with their purple shirts and sashes and their speeches about the unleashing of the great sleeping giant and so on. They were making some Jefferson City politicians nervous. They were making *Sam* nervous. An old Capitol hand standing in the statehouse with Sam that day looked out at the crowd and said companionably, nodding at Sam, one veteran to another: Well, look at this, now we have two teams on the football field instead of one.

From outside Missouri, as winter approached, bad news continued to accumulate. By early December the movement seemed to Sam to have reached a kind of slow-motion freefall, the Virginia and New Jersey governors' races both won by pro-abortionists, most of the good statehouse bills defeated or imprisoned in hostile committees, the U.S. House of Representatives faltering in its loyalty. (The once reliably pro-life House had voted to restore funding for Medicaid rape and incest abortions; only President Bush's veto kept the funding ban intact.) The Supreme Court had accepted three new abortion cases for the 1989–90 term, but one of the three settled shortly before its scheduled December argument, taking with it the brightest hopes for a powerful follow-up to *Webster;* the state of Illinois had been defending a set of licensing laws that imposed specific facilities and equipment requirements on all doctors performing abortions. But Illinois attorney general Neil Hartigan was campaigning for higher office and evidently determined not to drag an abortion dispute around with him, as Sam's *National Right to Life News* grimly observed (*Pro-Abort AG Running for Governor*), and Hartigan's out-of-court settlement of the licensing case left the Court only with two mandatory-parental-involvement cases, challenges to Ohio and Minnesota laws that required doctors to provide advance notice to the parents of underage patients.

And although it was theoretically possible for the Court to overturn *Roe* in a parental involvement case—theoretically, if a majority of the justices was ready to do it, the Supreme Court could overturn *Roe* in *any* case involving state abortion laws—the chances were remote. The Court had already drawn a distinction between minors' and adults' abortion rights; since the early 1980s, the Supreme Court had been ruling that states could require parental involvement in minors' abortions as long as the laws contained escape clauses allowing local judges to issue waivers on a case-by-case basis. The justices might refine this principle during the 1989–90 session, but Sam knew they were unlikely to alter it substantially or to make some doctrinal leap from parental notice to the validity of *Roe* v. *Wade*. The next important ruling was at least a year away, Sam could see; the justices were waiting, watching to see what *Webster* stirred up, looking for new legislation that placed the big questions more squarely before them. Justice O'Connor could scarcely have made the invitation more direct, Sam thought: *When the constitutional invalidity of a State's abortion statute actually turns on the constitutional validity of* Roe v. Wade, *there will be time enough to reexamine* Roe. *And to do so carefully.*

Sam began composing again, in his office over the pawnshop. It was as though O'Connor were commissioning something, but Sam wasn't sure what: *actually turns on the constitutional validity of* Roe v. Wade. The Pennsylvania state legislature had lived up to its combative pro-life reputation by adopting the country's first post-*Webster* legislation, an omnibus regulatory bill called the Abortion Control Act; in mid-November Pennsylvania governor Robert Casey signed the Abortion Control Act into law, meaning Supreme Court–bound litigation would surely follow at once, but when Sam read the Act he felt sure Missouri could do better. Pennsylvania's new law was simply a more advanced version of H.B. 1596, Sam thought: another kitchen-sink bill, a lot of regulatory language strung together without anything that looked like an open challenge to *Roe*. Informed consent was in Pennsylvania's law (doctor required to offer each patient materials that "describe the unborn child"), and spousal notification (married women required either to notify husbands or sign sworn statements as to why this was inadvisable), and a mandatory waiting period (women required to wait twenty-four hours between informed consent and abortion). There were some potentially interesting restrictions on certain types of abortion—here was the sex-selection prohibition again, and a provision that third-trimester abortions be allowed only to prevent death of the mother or "substantial and irreversible impairment of a major body function"—but Sam was tired of the scattershot approach. He wanted a bill that actually placed into writing the words Sam believed ought to appear in the Missouri criminal code. He wanted the bill that he liked to imagine Justice O'Connor had been asking for. He wanted a *ban*.

DRAFT. Sec. 188.017.
Performing an abortion on another is a Class B felony.
Attempt to perform an abortion on another is a Class B felony.

Sam worked carefully on the opening sentences of his 1990 abortion bill, trying to give them the simplicity and clarity he had learned to recognize in good legislation. He numbered his subsections, One through Four. Definition of Abortion. Classification of the Crime of Abortion. Definition of Attempted Abortion. Classification of the Crime of Attempted Abortion. Then Sam wrote Subsection Five: Defenses.

5. It shall be a defense to the crime of performing an abortion on another, or to the crime of attempt to perform an abortion on another, that:

Sam's draft abortion bill contained four defenses, written definitions of exceptional circumstances under which a physician could legally perform an abortion. The first was life of the mother: *would have been endangered if the unborn child were carried to term*. Sam felt comfortable writing this phrase; it belonged in the criminal code, he believed. All but a very few holdouts in the right-to-life movement had come to agree that life-of-the-mother exceptions made sense under the movement's own moral logic: surely defending the right to life, the right to *physical* life, meant that no person should be forced to die in order to give life to another.

Then Sam wrote in his second defense: *severe and long lasting physical health damage to the mother*. This was somewhat more controversial, Sam knew; there were movement purists who argued that *life* must always hold a higher value than *health*, that a woman with pregnancy-induced health damage is at least *alive*, whereas an infant who dies in an abortion is not. And the word "health" had proved so treacherous when it began appearing in pre-*Roe* abortion statutes and litigation—with "physical health" giving way to "mental health," and "mental health" giving way to "emotional health," and "emotional health" turning out to mean anything the pregnant woman and her physician wanted it to mean. Sam himself felt some uneasiness about the potential for slippage here, but he hoped that writing the words "physical" and "severe and long lasting" would help ensure that the exception apply to those rare cases involving genuine medical peril. And he could see the moral soundness in this form of exception to an abortion ban; he felt less wholehearted about this than he did about a life-of-the-mother exception, but Sam told himself that it was appropriate to make the additional declaration in law that no woman should have to suffer lifelong physical impairment in order to give life to another.

Then Sam wrote in his third and fourth defenses, neither of which he believed in at all.

> (3) ... the pregnancy resulted from forcible rape, which was reported to a law enforcement officer within 48 hours after the victim was physically capable of reporting the rape. ...
>
> (4) the pregnancy resulted from incest which, prior to the abortion, was reported to a law enforcement officer. ...

The rape-and-incest exception: Sam was writing it into his own abortion ban. He was giving up the rape-and-incest babies. In practical terms there were not very many rape-and-incest babies to begin with—Missouri kept no statistics specifically on rape-and-incest abortions, but no one believed them to make up more than a minute fraction of the total numbers—but that was quite beside the point, quite irrelevant to the gravity of what Sam understood he was doing. With the Supreme Court apparently inviting states to offer up their most aggressive abortion bills, bills no longer constrained by the limitations of *Roe*, Sam was about to propose that Missouri Citizens for Life approach the legislature with the very language that for twenty years most right-to-life activists had understood to destroy the moral crux of their argument.

And Sam agreed with them: it *did* destroy the moral crux of the argument. The old reasoning had always seemed wholly consistent and honorable to Sam: If these are truly living babies we are defending, they are babies regardless of the circumstances of their conception. If we are truly in this battle to defend these babies (and not, as our opponents like to imagine, to punish women for their sexual activity), then we must either defend them all or admit aloud that we are willing to permit a few of them to be legally killed. And if we permit the legal killing of certain babies because their birth and upbringing would cause their mothers too much distress, then what moral barrier truly distinguishes us from abortionists?

But state legislators were not philosophy students. The precision of this logic was going to appear as little more than an irritant to most of the legislators Sam had worked so assiduously for the last three years. Sam knew his pro-life votes in the Jefferson City statehouse; he knew which legislators understood the principled opposition to Hard Cases abortions and which did not, and which legislators understood it but told Sam it didn't matter that they understood it, their constituents would never support a ban without a rape-and-incest exception. Sometimes it had made Sam feel slightly ill, nodding attentively in back offices or Jefferson City restaurants while certain legislators leaned in as though sharing some special confidence and said, Come on, Sam, what about the white woman raped by a black guy? I'm supposed to vote for something that says *she* can't have an abortion? And Sam would think: Great, I'm compromising my principles and catering to racists at the same time. It was at moments like these that Sam realized the piercing and particular way in which he did miss the sit-ins, and the rides in the police wagon, and jail.

God, tell us what to do. In December Sam presented to the board of Missouri Citizens for Life his working draft of a 1990 bill to outlaw abortion in Missouri. Angrily, and by an overwhelming margin, the MCL board rejected it. The bill was unthinkable, Loretto Wagner and the other board members told Sam; the bill suggested by its wording that MCL supported abortions for rape and incest. The bill was a sellout. The bill was a surrender before the battle had even commenced, Loretto told Sam: *handing* rape-and-incest to the legislature, putting it on a silver *platter* for them. What had happened to Sam? Did he need a look at the war pictures to refresh his memory? Was this Sam's big post-*Webster* offering, his big legislative hope for a world without *Roe?*

And Sam answered Loretto, more heatedly than he remembered ever having answered her before: They won't *vote* for it without rape-and-incest. I know these people, I've done the counts, Sam said: a quarter of them will support us no matter what we do, a few of them are dead set against us, and the rest are in the middle, waiting for cues. You'll lose the middle if you come in without rape-and-incest, Sam argued to Loretto and the other MCL board members: you'll look cruel, you'll look intractable, you'll look anti-woman. You'll give political ground to the other side.

So maybe we'll settle for it *later*, Loretto argued back, when it's *forced* upon us, when it's not *our* argument. We're not going to go in there *waving* it. Loretto was powerfully disappointed in Sam, and she told him so; around MCL the word had spread rapidly, that Sam Lee was pushing rape-and-incest exceptions, and in the ranks indignation was growing. Was Sam serious, expecting them to back this bill? Did he understand how very deeply it offended them? Could Sam be trusted as MCL's lobbyist during such a crucial legislative year?

Long before Loretto and the other board members finally said it to him directly, Sam understood what was coming: if he lobbied his bill, if he argued its merits to the legislators in Jefferson City, Missouri Citizens for Life would fire him. Sam told himself that he did not mind that part so very much; Missouri Citizens for Life had never paid him anything but expenses anyway. What difference did it make, being fired by an organization that expected a person to solicit donations in order to buy groceries for his family?

But the suggestion that he was insufficiently committed, that he had forgotten the central meaning of the words "right to life," that he had lost his way amid the unsavory demands of political bartering—this was nearly unbearable for Sam, and he tried to teach himself how to stop thinking about it as the legislators arrived in Jefferson City and Sam set about lobbying his bill.

JUDY WIDDICOMBE flew back to Washington two days after *Webster* v. *Reproductive Health Services* was announced, and then turned around three weeks later and flew back to St. Louis. She was vigorous with resolve again. She had a plan.

> From: Judy Widdicombe and Laura Cohen
> Date: July 21, 1989
>
> The *Webster* decision turns the spotlight on the Missouri legislature as the battle ground for the next two to four years. We must have the ability to bring irresistible pressure on key legislators and to organize electorally in targeted, swing districts. To make this happen, we need to organize a statewide campaign Strategy Committee that brings together the best skills and abilities that our movement has to offer.

The beginning of a war, Judy had cried into the microphone three hours after the *Webster* announcement, the crowd at Forest Park clustered angrily and eagerly before her: the Vietnam of the Nineties. Both phrases had occurred to Judy as she was driving her car from Reproductive Health Services to the park, and she thought they made for satisfactory sound bites, a term she had heard frequently during her months in Washington. "We will prevail because justice will prevail." The crowd had cheered her on, just as they cheered Frank Susman and the other speakers who climbed to the podium with their warnings and exhortations, and Judy felt at once, listening to the cheers, the pleasure of an urgent sense of purpose, of homecoming. The applause made her more eloquent. Her sound bites quickened and soared. "I want you in the *streets*," Judy cried. "We will all go to the ballot box, one person, one vote, and we will make a *difference*," Judy cried. "The battle is going to be *long*," Judy cried, "and it's going to be *arduous*, and it starts *right now*."

What was it, after all, that the Supreme Court justices had done? By the time she drove to Forest Park that morning Judy had still not read the *Webster* ruling itself; she had sat in B.J.'s office at Reproductive Health Services, Judy and B.J. at one side of the room and at the other all those newspeople, their cameras poised, their pens lifted in anticipation, watching Judy and B.J. as Judy and B.J. watched the television broadcast from Washington. And there on the television was the NBC man, Carl Stern, flipping rapidly through the printed pages and saying yes, the Court had upheld *Roe* v. *Wade*—"it looks like the basic idea, that the state may not absolutely prohibit abortion, that that idea from *Roe* versus *Wade* remains intact, that issue, issue number one," still turning the pages, trying to see what came next; and Rita Braver on CBS skimming the synopsis at the top and saying yes, the Court was still going along with the *Roe* v. *Wade* decision of 1973—"as far as we can see, the answer to that question is yes, it may very well be unanimous," talking into her microphone about how complicated it was, how extremely complicated. And Judy had risen from her chair and given B.J. a hug, evidently that was now the thing to do, the camera strobes flashing, the voices from the television set explaining from Washington the specific ways in which abortion rights advocates had lost this highly significant case, and B.J. was weeping, and Judy wanted to say, They didn't overturn *Roe!* They left *Roe* alone! Why are you crying? The cameras were pointed at both of them, still, and then B.J. delivered that astonishing line, which the reporters wrote down at once—"I assume it's still our constitutional right to cry," B.J. had said, after which she excused herself, looking dignified, to retreat into temporary seclusion with the RHS staff. And Judy had thought, Whoa, Ms. Isaacson-Jones, you have come into your own. But she was still not sure what *Webster* v. *Reproductive Health Services* said.

Thus Judy took her early cues from the television networks, which for some time pressed on with their live broadcasts, as though reporting from the scene of a political assassination or a natural disaster. On television the New Jersey representative Chris Smith, whom Judy recognized as a devoted Anti, was using phrases like "erosion of *Roe* v. *Wade*" and "substantial victory for the unborn." On television Faye Wattleton was announcing that women's access to abortion was about to become "hostage to geography, as states enact a patchwork of laws." On television Randall Terry was promising to introduce an avalanche of legislation, that was his word, "avalanche," and Kate Michelman was declaring that America's political landscape would never be the same again, and the screen filled momentarily with colored maps of the United States, the color-coded boundary lines indicating States Expected to Restrict Abortion, and States Expected to Keep Abortion Legal, and Battleground States. At the Missouri NARAL office, just up the street from Reproductive Health Services, a bulletin arrived by fax from Washington, the national NARAL field director dispatching instant advisories to the affiliates:

RE: *Today's Webster Decision*—Talking Points

—*Roe* has been gutted by today's decision. This decision is as bad as it could have been, short of *Roe* being reversed.

—The Court has taken from women the right to choose abortion and has given that right to the state legislatures.

—5 to 4 decision (O'Connor and Kennedy with majority).

—*Roe* v. *Wade was neither overturned nor affirmed.* . . . Chief Justice Rehnquist stated that this was not the appropriate case to reconsider *Roe.*

Some days afterward Judy finally read the *Webster* opinions herself —read all five of them, working her way slowly through the thick stapled printout in the Washington Voters for Choice office, and then, mystified, going back to the beginning to read them again. Judy was not skilled at assessing Supreme Court opinions; she never had managed to read *Roe* v. *Wade* and *Doe* v. *Bolton* all the way through, and since 1973 she had learned her Court doctrine mostly by listening to Frank Susman's explanations and reading newsletter analyses from NARAL or the ACLU. But these *Webster* pages were not law at all, Judy thought; they read like the transcript of a family feud, Scalia laying into O'Connor, Rehnquist laying into the trimester framework, Blackmun laying into Scalia and Rehnquist. What was that remarkable language Blackmun had used to describe the Rehnquist assertion that *Roe* was being left undisturbed? "This disclaimer is totally meaningless," Blackmun had written. "The plurality opinion is filled with winks, and nods, and knowing glances to those who would do away with *Roe* explicitly, but turns a stone face to anyone in search of what the plurality conceives as the scope of a woman's right under the Due Process Clause to terminate a pregnancy free from the coercive and brooding influence of the State."

For sheer fervor of sentiment this was certainly memorable prose, Judy thought, but she was a practical person and she had imagined, amid all the commotion and speechmaking the *Webster* decision had inspired, that she would find in the decision some practical instructions for a provider of abortions in Missouri—some measurable cutback, some new set of rules that would change the way RHS must now conduct its business. But there was nothing. The two provisions of H.B. 1596 that had once worried Judy most, the sixteen-week-hospitalization requirement and the twenty-week viability-testing requirement, had vanished as *Webster* ascended to the Supreme Court; once the Missouri state attorneys dropped the hospitalization requirement from their appeal, it had become apparent that the RHS doctors could go on performing their second-trimester abortions in the clinic procedure rooms, just as they had before. And even though the viability-testing requirement had been upheld, the appellate court arguments had suggested that the state was going to be satisfied with the sort of second-trimester ultrasound screening that was routine at RHS anyway.

What else? The ban on abortions in public hospitals had survived; that was a disservice to the small number of women whose doctors might have treated them at the publicly funded hospitals in Kansas City or Columbia, but those doctors ought to be able to relocate the procedures at private hospitals or send the patients to St. Louis to be accommodated at RHS. The "life of each human being begins at conception" language had survived, but the Supreme Court and the Missouri state attorneys appeared to have agreed that this passage was in the nature of throat-clearing —a "value judgment," as Rehnquist's *Webster* opinion said, a philosophical statement with no immediate application to abortion law. Really Justice Blackmun had articulated it best, Judy thought; the core of the *Webster* ruling was winks, nods, and knowing glances, all of them directed at the Sam Lees of the fifty state legislatures: Try it again. Bring us your best effort. Give us more substance next time.

Now Judy understood—for the first time in many, many months, she realized; how long had it been since she had felt this way?—exactly what she was supposed to do. She was supposed to go home. The Supreme Court had offered up an extraordinary gift, Judy thought, a warning that might rouse the complacent without directly affecting most abortion patients at all. The work required to make use of this warning—the organizing, the choreographing, the summoning of volunteers—must take place not in Washington but out in the states from which the next H.B. 1596 would emerge, and here was work so deeply familiar that at fifty-one years of age Judy realized she had been doing it nearly half her life. Sheepdogging a single state, raising money from men and women who knew her name well, bullying, prodding, cajoling, demanding, urging people along at a pace considerably brisker than they seemed inclined to assume on their own—Judy knew how to do this in Missouri, and as she prepared to fly back to St. Louis at the end of July it occurred to her that her field of vision was narrowing, uncharacteristically, and that for the time being, at least, a single state was enough.

She couldn't fix everything. Judy recognized those multicolored maps she had seen on the television networks the morning *Webster* was handed down; there were versions of the same maps in the Washington offices of NARAL and Voters for Choice, one color for the safe states, in which the legislatures or the governor or both could be relied upon to keep abortion fully legal even in the face of the Court's apparent invitation to challenge *Roe* again, and another color for the dangerous states. The tallies on the total number of dangerous states depended on relative degrees of guesswork and pessimism (CBS had predicted nine, the morning of the *Webster* ruling, and NBC nineteen), but every map outlined Missouri in the color signifying danger: Missouri, it was widely understood, would try at once for the next run at *Roe*.

There was nothing Judy could do about Pennsylvania, which was also highlighted as a dangerous state by everyone who kept such summaries; Louisiana, South Dakota, Wisconsin, Alabama, Utah, South Carolina, Kentucky, all dangerous states, all plausible sources of new anti-*Roe* legislation, and all out of Judy's reach, she saw now, all other people's domain. But Missouri she could help. Missouri could be excised from the dangerous list. That man in the *Post-Dispatch* picture, with his coat-hanger sign that said, I SWEAR THIS WILL NOT BE A PART OF MY DAUGHTER'S FUTURE, and his arm wrapped sweetly and protectively around his little girl's back—that man was ready to be put to work, Judy thought; she had no idea who he was, perhaps he had brought a sister or a girlfriend to RHS one day, perhaps he had hurried right past all the postcards and bulletin board flyers saying Call Your Legislator, Your Right Is Endangered. Perhaps he had never believed it before. But he believed it now. At the beginning of August Missouri NARAL published its first post-*Webster* newsletter, with a remarkable black-and-white photograph of women grappling with a waist-high tangle of unfurling rolls of paper. Eighteen thousand women had signed petitions of support for legal abortion during the months before *Webster*, the accompanying article explained, and the women in the photograph were inside a room at the statehouse, preparing to drape from the rotunda balconies a massive length of the petitions taped end to end. "Supporters in the Rotunda took up chants that sent waves of sound thundering through the halls," the article read:

> SAFE AND LEGAL and WE WON'T GO BACK echoed through the capitol. ... Small groups armed with lobbying packets supplied by NARAL headed off to visit their state senator or representative.... Secretaries were taken by surprise as twos, threes—or in some cases dozens—filed in to see their legislators. When asked if this happened often, State Senator Wayne Goode's secretary slowly shook her head and said, "Never. Never in my life."

Well, there it was: *petitions*. Once or twice during the years before *Webster* Judy had talked about petitions—was it with Frank Susman? Or Sylvia Hampton? Judy no longer remembered; she was running a clinic then, traveling around the country, she had other business to attend to. But the idea had occurred to someone, and Judy recalled remarking upon it back in the mid-1980s, saying if only we had time, if only we could get people's attention, if only we had a prayer of winning. In Missouri signed petitions could be used for more than simple expressions of conviction to public officials; the state of Missouri, like approximately half the American states, permitted ordinary citizens to compose initiatives, written proposals for changing state law, and then to mount petition drives to collect enough signatures to place these initiatives directly on the ballot. Voter-approved ballot initiatives could enact new statutes, repeal existing statutes, or amend the state constitution—all the while bypassing entirely the machinery of statehouse politics.

In theory, in other words, a petition drive and ballot initiative were a direct expression of the people's will, unsullied by lobbyists or backroom trade-offs or nervous legislators trying to protect their own reelection campaigns. And *now* was the time, Judy decided for herself during the final weeks of July 1989, to express the people's will in the state of Missouri— to remove Missouri from the roster of dangerous states, now and forever, by placing on the November 1990 ballot a state constitutional amendment guaranteeing women the right to legal abortion in Missouri. A telephone conversation with Cathryn Simmons, a Kansas City political consultant who had worked on occasion for RHS while Judy was director, convinced Judy that a ballot initiative was the best idea Judy could possibly bring back to St. Louis. "Throw this sucker out on a petition drive," said Simmons, who tended toward indelicacy when she was impatient (she and Judy got along extremely well). "Put this goddamn thing to a vote, *once and for all.*"

Cathryn Simmons had made money from petition drives before; she had coordinated drives on matters as diverse as trucking and pari-mutuel gambling, and she charged for her services. But when she expounded on the merits of a pro-choice petition drive she was eloquent, and everything she said made sense to Judy. Simmons was a veteran Missouri political operator, with strong ties to the Speaker of the House, Bob Griffin; when Judy had retained Simmons back during the mid-1980s on behalf of RHS, Simmons' assignment—which she had delivered quite competently, Judy thought, until the 1986 session when H.B. 1596 steamed past them all— had been the arrangement of some low-key Jefferson City lobbying services, some monitoring of legislation, some persuasive contact with the Speaker. Simmons had told Judy many times that Bob Griffin did not want abortion on the Missouri legislative agenda at all. Abortion votes were nothing but trouble for most of the state legislators; the only legislators who genuinely cared about abortion were the partisans on either end of the spectrum, the True Believer Antis and the Kamikaze Squad. The rest would be gratified to watch the entire subject vanish from their personal workloads and reappear directly on the general ballot, Simmons declared, so that every legislator who had spent a statehouse career being intimidated by the Antis might someday be able to turn away the Missouri Citizens for Life interrogators with a regretful shrug: I can't vote with you anymore, legal abortion is now guaranteed in the state constitution, it's not up to me.

The more Judy thought about it, the more she came to believe Cathryn Simmons was right. A ballot initiative writing freedom of choice into the Missouri constitution: it was a novel idea, a big brave gamble, and Judy could imagine how the very novelty of it would carry it forward, how eagerly women and men would volunteer their signatures and their time and their money. Regional headquarters would open around the state, fund-raising parties would generate donations and enthusiasm, petitions would be proffered at supermarkets and churches and community meetings—Judy could envision this, all of it, and she was ready to go to work.

THE PROPOSED AMENDMENT

BE IT RESOLVED BY THE PEOPLE OF THE STATE OF MISSOURI THAT THE CONSTITUTION BE AMENDED BY ADDING A NEW SECTION 32 TO ARTICLE I AS FOLLOWS:

That no government or law shall restrict a female's use of contraception or decision to continue or terminate a pregnancy.

Judy moved into an upstairs bedroom at the St. Louis house she still owned, this time as a temporary tenant, sharing the house with the woman who had been renting it while Judy was in Washington. There was a great deal to accomplish now, and a brief window of opportunity in which to accomplish it—both Judy and Cathryn Simmons had begun repeating this phrase often now, "window of opportunity," describing, as though it was about to slam to a close, the strategically crucial period after the *Webster* ruling and before whatever viciousness the Antis next had in mind for the Missouri legislature, which would convene in January in Jefferson City. With Cathryn, who laid out the instructions for drawing up legally acceptable petitions, Judy worked up a plan for beginning a petition drive by the opening weeks of 1990. She commissioned a state survey and a series of focus groups, to be conducted by a Washington, D.C., polling firm; she began calling St. Louis friends and political allies to muster support and potential funding sources; she readied herself for a set of strategic meetings with other pro-choice activists, during which Judy intended to explain why a petition drive was the surest and boldest strategy for the post-*Webster* era.

But at the meetings things went badly from the start. Judy had imagined presenting her case to a group of five or six, Laura Cohen of NARAL and Mary Bryant of Planned Parenthood and a manageable assortment of other Missouri women who had taken up the pro-choice cause during the years of Judy's tenure at Reproductive Health Services. But there were dozens of people who now wanted to listen and argue and plan—forty or fifty, Judy had never encountered a lineup of these proportions in Missouri before, and each of these individual people representing organizations with policies and pressing agendas of their own. The St. Louis Freedom of Choice Council, the League of Women Voters, the American Association of University Women, the National Council of Jewish Women, the American Civil Liberties Union East, the American Civil Liberties Union West, the Women's International League for Peace and Freedom, Project Choice of Springfield, the Cole County Citizens for Choice, the Choice Coalition of Kansas City, Planned Parenthood of the Central Ozarks—where had all these people *come* from? And they didn't like the petition drive. At first they were skeptical, not quite seeing it, willing to consider a ballot initiative but only as part of a larger and more conventional plan, maybe a petition drive managed alongside the lobbying of legislators and the targeting of key electoral districts. Then they formed Steering Committees and Strategy Sessions and decided that a ballot initiative might not be a good idea at all, that it was expensive and complicated and dangerously easy to lose at the polls.

And Judy said: Right, it's risky. But we can win it. *I know how to do this.* And she would not back off.

And the Missouri Alliance for Choice—that was what the large coalition of organizations had named itself—met again, the younger members in particular somewhat baffled by the information passing swiftly from one telephone conversation and private caucus to another. Was Judith Widdicombe under the impression that she ran the Missouri pro-choice community—that she could fly back from Washington, like some military commander returning from overseas, and set about issuing orders to the rest of them? Had she failed to notice the very size of their coalition, the impressive dimensions of their new presence in Missouri, the importance of consensus? Perhaps Judy was not getting the message clearly: many members of the Alliance were opposed to a ballot initiative, and these were women who must be listened to, their reservations considered seriously and with respect.

Judy went to the Sunnen Foundation, her oldest and most reliable source of funding, and was promised two hundred thousand dollars for a ballot initiative campaign. She drove to Jefferson City; she met with state legislators; she wrote out a budget that included a salary for Cathryn Simmons. Thirty-eight separate organizations the Missouri Alliance for Choice was now claiming as members; Judy was not going to sit on her hands awaiting consensus from *that*. Consensus was cumbersome. Consensus exasperated Judy; inefficiency exasperated Judy; any delay exasperated Judy, when she set her mind to something, and they knew that about her, those Alliance women who were old enough to remember two decades back. Judy kept working, collaborating with Cathryn, drafting memos for the coming campaign. "The window of pro-choice political opportunity open now could soon close," Judy wrote. "Of all the avenues available to secure the broadest range of protections for Missouri women to obtain an abortion, a Constitutional Amendment is the *safest*."

Now the Missouri Alliance for Choice began imploring Judy directly to let the ballot initiative idea drop. Harriett Woods, the former Missouri lieutenant governor and Democratic state senator, was recruited to assume diplomatic duties on behalf of the Alliance. Woods had weighty pro-choice credentials, having established herself as a leading Kamikaze during her state Senate years, and in her approaches to Judy, Harriett worked at pressing the Alliance's position. If a ballot initiative lost it would set them back irreparably, Harriett Woods argued—the people's will coalescing *against* legal abortion, think how the Antis could make use of that. And Cathryn Simmons was a problem; there were those in the Alliance who mistrusted her, who said she was motivated only by money, or that she spent too much time courting legislators who voted regularly with the Antis. And pro-choice initiatives in other states had set a lousy track record in recent years; the Alliance members were passing around copies of a scholarly magazine article in which the author assessed the 1988 initiatives in Colorado and in Michigan, both of which had been defeated at the polls.

Did Judy actually shout at Harriett Woods during any of these meetings? Did she actually stalk out and slam any doors behind her? Afterward she didn't think so, but she knew there were others who said she had— and she did leave the room a few times to compose herself, almost certainly she did that, but she was so frustrated by what was happening, so taken aback. Judy knew all about the Colorado and the Michigan initiatives; at Voters for Choice she had helped raise money for both campaigns, and there was a crucial difference between those failed 1988 ballot measures and the one Judy was now proposing for Missouri. The Michigan and Colorado initiatives had been *funding* measures—ballot referenda asking voters to restore state public funding for poor women's abortions. Thus the question presented (and in one state, presented rather badly at that; the Colorado initiative was worded so elliptically that it never mentioned the word "abortion") was not: Shall women in this state be guaranteed the right to seek an abortion?, but rather: Shall the taxpayers be obliged to pay for abortions for the poor?

Anybody who had ever read the fine print in a 1980s abortion poll knew the difference between the answers to those two questions. For some years now the complicated signals of survey results had made for useful and unsettling information around the headquarters of organizations like NARAL and Voters for Choice; it was widely understood that general audience abortion polls tended to produce results that were supportive in principle but that slipped on some of the specifics, and one of the specifics was public funding. Most people—particularly when the question was presented in the impersonal, artless phrasing of a poll question or a ballot measure—did not want the government providing abortions for free. Most people wanted minors to be required to obtain their parents' consent, too;

indeed, the hostility most people showed when presented with various *hypothetical* unwanted-pregnancy situations—the poor family unable to afford another child, the single woman trying to finish college, the married woman with already-grown children—had given rise to a movement wise-crack that Judy had been hearing and appreciating for years. I support legal and readily accessible abortion only for the Four Hard Cases, the adage went: health-of-the-mother, rape, fetal deformity—and me.

There was no point any longer in trying to label this phenomenon, Judy thought ("ambivalence" was a benevolent word for it, although in acerbic moments she inclined more toward "hypocrisy"); you built what you could with the material available to you, and as a political objective, public abortion funding was just exceedingly difficult to sell. Not impossi-ble—in a few states, well-managed pro-choice campaigns had helped keep funding intact by defeating right-to-life-sponsored ballot initiatives to end it—but difficult. And what Judy had in mind was aimed precisely at those no-on-funding-but-yes-on-abortion voters, the legal-with-limits voters, the ambivalent majority. Pushing ahead now, convinced the Missouri Alli-ance for Choice members would come around once they understood her proposal's careful accommodation to the realities of public sentiment, Judy produced the full wording that was to appear on the initiative petitions:

THE PROPOSED AMENDMENT

BE IT RESOLVED BY THE PEOPLE OF THE STATE OF MISSOURI THAT THE CONSTITUTION BE AMENDED BY ADDING A NEW SECTION 32 TO ARTICLE I AS FOLLOWS:

That no government or law shall restrict a female's use of contraception or decision to continue or terminate a pregnancy.

House Bill 1596 (1986), which Bill was referred to in *Webster v. Reproductive Health Services,* and laws that are inconsistent with this section shall have no force or effect *except for other state statutes in force on the effective date of H.B. 1596.* Such statutes shall remain effective only if they are not hereafter amended or repealed.

Translation: This amendment freezes Missouri abortion law—sets it into the state constitution, out of reach of legislating politicians—at pre-cisely where it was before H.B. 1596 took effect last summer, which is to say: abortion is legal in Missouri; the state does not pay for abortions for the poor; post-viability abortions are subject to certain extra precautions;

minors must first obtain the consent of one parent or a judge. Surely Harriett Woods and the Missouri Alliance for Choice women must now see what Judy was doing—how deftly and effectively she was going to deflect the fundamental problem that had beset the Colorado and Michigan initiatives. Judy's constitutional amendment prevented any further incursion on the right to legal abortion in Missouri. It prevented the Missouri legislature from producing the next *Roe* challenge to the Supreme Court. It removed abortion from the annual political posturing that was now such a reliable source of frustration and defeat; it put an end to the Litmus Test. It sidestepped the more hazardous issues of public funding and parental consent; its final sentence said, in effect: We will let those restrictions be for the time being, although in the future the legislature may amend those restrictions in keeping with our broad constitutional guarantee of the right to terminate a pregnancy.

It was a compromise. It spoke to the middle. There was no other way to do it. Couldn't they see it now—how very sensible Judy's ballot initiative was?

But most of the Missouri Alliance for Choice leadership hated it. They said the wording was confusing, that people would not understand what they were voting on and thus would vote No because that is what people tend to do when confronted in a polling booth with mysterious legalese. They said their own Missouri survey had produced evidence that a ballot initiative was likely to lose, that the electorate needed more education and competent organizing before it could be relied upon to cast a statewide Yes vote for the principle of legal abortion. And they hated the proposed amendment itself, with its implicit acceptance of funding cutoffs and parental consent requirements. How could a coalition calling itself Alliance for Choice support an amendment that wrote abortion restrictions *into* the state constitution?

But the amendment says those restrictions can be relaxed later, Judy said—look at the last sentence.

But the amendment writes the restrictions into the constitution *now*, the Alliance leaders said. How could they urge their organizations' members to promote it? Why was Judy so eager to accommodate this kind of public waffling on abortion? Was she wavering in her own convictions about reproductive freedom and choice?

But what you are proposing instead is useless, Judy said—"lobbying," "mobilizing pro-choice voters," "targeting key races." For sixteen years it was precisely these tactics that had failed Missouri pro-choice women at every state House and Senate abortion vote. Did they *like* the annual Litmus Test? Were they *attached* to it in some way Judy had failed to understand? Judy was sniping now, drawing into a defensive crouch, marshaling supporters on her own; the split was deepening, Judy on one side, the Alliance leadership on the other, and she knew they were closing ranks against her. It infuriated her and it broke her heart. How dare some of these women pass along the suggestion—no one would say such a thing publicly, but Judy heard the remarks deflected back from mutual acquaintances—that Judy was deserting them deliberately, that she was no longer genuinely dedicated to the pro-choice cause? How many of *them* had stood at the backs of airplanes and instructed flight attendants in emergency uterine massage? How many years had Judy spent playing the role of the big Missouri bulldozer, organizing the Clergy Service, coaxing out the Sunnen money, bankrolling the lawsuits, mopping women's foreheads while they aborted into her toilet, building the damn clinic very nearly by herself? Judy knew she could be arrogant, she knew she tended to knock people sideways and leave them spluttering in her wake; but it ought to be evident to anybody that this arrogance grew from a passion about women and the public health, that Judy had never tried to turn Reproductive Health Services into a profit-making clinic, never driven an elegant car, never lived in the neighborhoods of the affluent funders she had spent so many years learning to cultivate. All she had done was *push,* that was how it seemed to Judy now, she had pushed and pushed and been admonished in various ways by various people that what she was trying to do was too difficult, too risky, too unsuited to the limited range of a registered nurse. This exchange was so familiar to her by now that she had cultivated a way of thinking about it, of phrasing it to herself: *Tell me I can't do something and I'll show you that I can.*

In December Judy made the two-hour drive down to Columbia for a statewide Missouri Alliance for Choice meeting; she carried with her a sheaf of papers filled with polling data and prospective timelines on the petition drive. When she had finished her presentation, the Alliance members voted formally and unanimously against Judy's ballot initiative. "Non-support" was what they called their agreed-upon posture, in the written memoranda afterward: "a position of non-support."

In January 1990, Judy filed the required paperwork for a ballot initiative political action committee, identifying herself as president. She gave her political action committee a name, People Working to STOP! Government Interference, and a logo, a red octagonal stop sign with the word STOP! emblazoned in white capitals across the middle. She bought a fax machine and plugged it in beside her dining-room table. She called her Washington friend Alison Gee, who had made several visits to Missouri during the months since the *Webster* oral argument, and asked Alison to move to St. Louis for a while to help.

The printed petitions arrived by delivery truck from Kansas City, where Cathryn Simmons had overseen their final wording and duplicating: *We, the undersigned, registered voters of the state of Missouri and _____ County (or "city of St. Louis") respectfully order that the following proposed amendment to the Constitution shall be submitted to the voters of the state of Missouri, for approval or rejection.* Twenty signatures per petition. Separate signature and stamp line for the notary.

One hundred twelve thousand verified Missouri voter signatures were required in order to qualify the initiative for the ballot in November.

From New York, at the national offices of the American Civil Liberties Union, the ACLU attorney Kathryn Kolbert wrote a personal letter to Judy and Alison, urging them to abandon the ballot initiative in its current form and observing that its wording was confusing, legally problematic, and unsatisfactory to those pro-choice voters opposed to public funding restrictions and parental consent laws. "I think that the language you are proposing as an initiative for next fall presents a number of problems," Kolbert wrote. "I would strongly recommend a different approach."

Judy carried plastic milk crates into the dining room and set them side by side along the floor to hold file folders and petitions and newspaper clippings. She made lists of likely donors, copying names from her Voters for Choice records, from old RHS files, from typewritten Pregnancy Consultation Service memos dating back to 1972. She worked the telephones, tracking down women who might be willing to volunteer as regional petition coordinators, or to invite people into their homes for meetings and fund-raisers. In February, the *Post-Dispatch* ran a story about her ballot initiative on the front page, under the headline ABORTION BATTLE AT KEY POINT; the article was written by Martha Shirk, a veteran *Post-Dispatch* reporter Judy trusted and liked, and as Judy read the opening paragraphs she felt her spirits lift. Here was vindication in print, and from the most unlikely source:

LINES ARE DRAWN OVER AMENDMENT

The ink is barely dry on a proposal for a constitutional amendment to outlaw further restrictions on abortion in Missouri. But the proposal already is being called potentially the most important development on the local abortion front since the U.S. Supreme Court legalized abortion in 1973.

"If it passed, it would be all over for the pro-life movement in this state," said Samuel Lee, lobbyist for Missouri Citizens for Life, the state's most active anti-abortion group.

"There would be no way to stop abortions in this state," he said. "It's the most serious challenge we've faced since Roe vs. Wade."

That's the intent, says Judith Widdicombe, architect of the proposal. "It would preserve the legality of abortion for generations to come, whatever happens at the Supreme Court," Widdicombe said last week.

Judy read on quickly; here was the Missouri Alliance for Choice critique, "vigorous opposition to the strategy by all the abortion-rights groups in the state," Laura Cohen saying the more prudent approach was the targeting of key political candidates. Then Sam Lee saying he was "thrilled" by the prospect of Missouri Alliance for Choice refusing to support Judy's signature-gathering effort, that discord among abortion supporters could only bolster his side. And then another wrinkle entirely, startling Judy as she read it; apparently the statehouse gossip she had heard was true, Sam was in some difficulty of his own, battling his colleagues, perhaps about to be exiled within the capitol:

> But the anti-abortion movement itself is increasingly factionalized and has not yet agreed on a strategy to fight the petition drive.
>   Lee and two activists with Lawyers for Life formed a political action committee this month—Campaign Life Missouri—to organize against the amendment proposal and to lobby for an anti-abortion bill introduced by Lee in the Missouri Legislature. Missouri Citizens for Life is not a party to that political action committee and is considering forming a separate one.

The world was a rich and perverse place, Judy thought: there they were, she and Sam, eleven years after he had first walked into her clinic, the two of them bookended now beneath ABORTION BATTLE AT KEY POINT. Absently Judy wondered what the problem was, there in the inner circles of Missouri Citizens for Life; had Sam's intransigence proved too much for them? All that solemnity, that blinding moral certainty, it must grow tiresome after a while. Zealots, Judy thought to herself, with a small lurch of recognition: crusaders, martyrs, pigheaded people, we make useful leaders but exhausting colleagues, we *know* we can see what is right. She cut out the *Post-Dispatch* article and added it to her stack of press clippings.

At least Judy had said nothing overtly hostile, there on the front page of the newspaper, about the Missouri Alliance for Choice; indeed, she and Laura Cohen appeared to have selected their words judiciously, avoiding personal insult or direct political attack. But for two months now Judy had been trying to work the territory she thought she knew so intimately, calling her old supporters for funding and volunteer help, and she was beginning to understand how badly she had underestimated the Alliance's unfriendly reach. Every morning she put on her Judy the Clinic Director clothes and drove out into suburban Missouri, ringing doorbells at the end of long curving driveways, and when she drove home in the evening Alison would watch from the front steps and know by the slope of Judy's shoulders that the affluent suburban ladies had told her no again: No, I've been contacted by this Missouri Alliance for Choice organization, I believe I'll be directing my donation to their political action committee; or No, I understand there's some conflict here, I don't want to get caught in the middle, I'm sure you understand. Sometimes they said Yes, I'll arrange a petition-distributing meeting for you, call me next week; and Judy would call the next week and listen to the most sincere and regretful apologies: I tried to set this up, I really wanted to help, but everybody begged off, I couldn't get anyone to come.

Judy began waking earlier and earlier in the morning, climbing from bed before dawn, so that by the time her housemates awoke she was already dressed and flipping restlessly through the newspaper downstairs, waiting for the hour when she could begin the day's telephone calls. Prospective volunteers began telephoning Judy and Cathryn Simmons on their own, offering to address envelopes or collect signatures on the street; here and there a fund-raising party was offered, a good donation collected, an encouraging set of telephone calls logged, and Judy would take heart again, her animation revived. She began pulling money from her own savings account to help pay the signature-gathering campaign's expenses. She contacted political columnists and editorial boards; she drove to Springfield, five hours to the southwest, and St. Joseph, six hours to the northwest. She looked in downtown St. Louis for a rental office into which she could relocate the operations center that was now overflowing from her dining room into the kitchen and hallway.

She wrote text for the newsletters Cathryn was laying out, which were four glossy pages each, with the red STOP! logo and the words *Action Report* stripped in urgent black italics across the top.

STOP! PAC requires both Money and People-Power. Your donation can help move our grassroots effort forward!

If each of us just do what we can . . . join the committee, make a contribution, or obtain petition signatures, our goal of taking the issue of abortion directly to the voters is achievable.

If you have any signatures, please mail them to us. When we receive your petitions we will send you a new supply of blank petitions *immediately*.

One morning in early spring Judy drove down to Jefferson City to spend the day meeting with legislators at the statehouse. A few state legislators had publicly endorsed the petition drive; Judy knew there were others in the capitol who were on her side but had remained discreet about saying so publicly, since it did not sound especially principled or courageous to declare aloud what Judy was sure many of them were thinking, which was: Take this issue *away* from me. Cathryn Simmons had been right about that part; only the Kamikazes and the True Believers could summon any enthusiasm at all about the prospect of another round of General Assembly abortion votes. The rest of them were fed up with it, Judy could see, and desperate for any excuse to avoid casting a vote one way or the other. And who could blame them? Maybe they were afraid of the Antis. Maybe their mothers telephoned them every week to remind them what the pope had said about innocent human life. Maybe the entire topic made their heads hurt and they had just cast all those votes the way they thought a lot of Missouri people probably wanted them to: keep it out of the criminal code but regulate it in a special way, it is *not* like setting a broken ankle, it is different from other medical procedures, it is different from anything else we do.

Judy had a collection of printed materials that she liked to pass around when she was out promoting her ballot initiative, and on this particular morning at the statehouse she had brought with her a satchel filled with these handouts: the sample petitions, the stop! PAC newsletters, the press packets with their laudatory editorials from newspapers around the state. She had no steady company in the capitol as she made her rounds; she felt faintly like a traveling salesman, working her way by herself from one legislative office to another, and she was about to push open the next glass and mahogany office door when she heard someone calling her name. "*Judy.*"

Judy looked over her shoulder and saw Sam Lee at the far end of the hall, carrying a briefcase. He was wearing a good herringbone jacket, his dark pants pressed, his hair trimmed and combed, his tie knotted neatly over his white shirt. He had a way of moving noiselessly, as though he were walking the marble floors in socks, and Judy remembered this about him now, how Sam had always seemed to make small pockets of quiet around himself, how oddly soothing it had been to stand near Sam amid the rancorous clamor of the waiting-room sit-ins. He came toward her in the statehouse hallway in that same noiseless way, smiling a little, his brown eyes amused behind the wire-rimmed glasses.

"I guess we're in the same kind of trouble, aren't we?" said Sam.

"Hello, Sam," Judy said. She saw that he had reached his arm forward to shake hands with her, but she pushed past Sam's hand and folded one arm abruptly and affectionately around his shoulder, and for an instant, the briefest of moments, the legislators and secretaries and lobbyists hurrying by without so much as a turn of the head, Sam and Judy bent toward each other in the capitol hallway and embraced.

Epilogue
Spring 1990 to Summer 1999

BOTH JUDY WIDDICOMBE's abortion ballot initiative and Sam Lee's abortion bill failed during the spring and early summer of 1990. Judy's initiative campaign was unable to collect enough verified signatures to secure a place on the November ballot. Sam's bill, introduced by a state representative from suburban St. Louis, was never released from legislative committee; by advance agreement of the Missouri House and Senate leadership, all abortion legislation introduced during the 1990 General Assembly was blocked in committee so that the legislature could bypass abortion entirely for the duration of the session.

IN JANUARY 1992, the Supreme Court agreed to hear *Planned Parenthood of Southeastern Pennsylvania* v. *Casey*, the challenge to the abortion legislation adopted by the state of Pennsylvania during the autumn of 1989. By 1992, only two of the original *Roe* majority opinion justices remained on the Supreme Court; Justices William Brennan and Thurgood Marshall had resigned, and although their replacements, federal appeals court judges David Souter and Clarence Thomas, declined during their confirmation hearings to discuss their views of abortion or the merits of *Roe* v. *Wade*, it was widely assumed—particularly after Thomas's conservative views were examined during his confirmation hearings—that the Court now had its five votes for reversing *Roe*. But the ruling in *Planned Parenthood* v. *Casey*, handed down in June 1992, confounded attorneys and activists from both sides by explicitly reaffirming what the lead opinion described as the "essential holding" of *Roe* v. *Wade:* that "the woman has a right to choose to terminate or continue her pregnancy before viability."

Justices O'Connor, Kennedy, and Souter were identified in *Planned Parenthood* v. *Casey* as joint authors of the lead opinion, an unusual display of judicial solidarity; in most Supreme Court opinions only one justice is listed as the author, and those justices who agree with the opinion "join" by signing on. But on the morning the *Casey* ruling was announced at the Court, all three justices took turns reading aloud from their lengthy opinion, which declared that the *Roe* trimester framework ("which we do not consider to be part of the essential holding of *Roe*," the justices had written) would no longer serve as the guideline for reviewing state abortion laws. Instead, the justices wrote, an "undue burden" standard would guide them in future cases: the Supreme Court would strike down as unconstitutional any state regulation that placed "a substantial obstacle in the path of a woman seeking an abortion before the fetus attains viability."

Under this new standard, the justices wrote, most of the challenged portions of the Pennsylvania Abortion Control Act—for example, the requirement that doctors inform patients of materials describing the fetus and that patients wait a minimum of twenty-four hours between receiving this information and obtaining the abortion—were constitutional. But the husband notification requirement was not, the justices ruled; the Pennsylvania law had required every married abortion patient either to notify her husband in advance or to sign a statement certifying that she was exempt for one of certain specific reasons, such as a fear of bodily injury if her husband were to learn of her abortion plans. This requirement, the justices wrote, *was* unduly burdensome and was therefore unconstitutional: "The husband's interest in the life of the child his wife is carrying does not permit the State to empower him with this troubling degree of authority over his wife."

Like *Webster* v. *Reproductive Health Services*, the *Casey* ruling was complicated by split opinions and vehement dissents. Justices Stevens and Blackmun joined certain sections of the lead opinion but dissented from others; neither Stevens nor Blackmun agreed that the trimester framework, which in the past had led to strict Supreme Court review of state abortion laws, should be discarded in favor of this less demanding "undue burden" test for guiding the Court in future cases. And in separate dissents Justices Rehnquist and Scalia, with the support of Justices Thomas and White, repeated their argument that the Court should reverse *Roe* and abandon entirely the concept of abortion as a constitutional right. (Over the course of Scalia's scathing dissent, the justice at intervals described the arguments in the O'Connor-Souter-Kennedy opinion as "outrageous," "rootless," "contrived," "Orwellian," "Nietzschean," and "frightening.")

The *Casey* opinions left unanswered many questions about the particular form of state abortion regulation that might in the future be characterized as "unduly burdensome." But the O'Connor-Souter-Kennedy opinion made it plain that no state would be permitted to prohibit abortion, and the justices borrowed from the vocabulary of the abortion rights movement when they wrote of the profound social consequences of reversing *Roe* v. *Wade:*

> . . . [F]or two decades of economic and social developments, people have organized intimate relationships and made choices that define their views of themselves and their places in society, in reliance on the availability of abortion in the event that contraception should fail. The ability of women to participate equally in the economic and social life of the Nation has been facilitated by their ability to control their reproductive lives. The Constitution serves human values, and while the effect of reliance on *Roe* cannot be exactly measured, neither can the certain cost of overruling *Roe* for people who have ordered their thinking and living around that case be dismissed.

IN NOVEMBER 1992, following a campaign during which he promoted himself as an advocate of legal abortion, Bill Clinton was elected president of the United States. Seven months later, after the resignation of Supreme Court Justice Byron White, President Clinton named as White's replacement the Washington, D.C., federal appeals court judge Ruth Bader Ginsburg, who declared unambiguously during her confirmation hearings that she believed the Constitution protected the right to abortion; Ginsburg was confirmed by the Senate in August 1993, giving the Supreme Court an apparently firm six votes in support of the abortion right. Justice Harry Blackmun announced his retirement the following spring; his replacement, Boston federal appeals court judge Stephen Breyer, was not questioned extensively about *Roe* during his confirmation hearings, but did in those hearings refer to the right to abortion as "settled law."

Justice Blackmun died on March 4, 1999.

IN MARCH 1993, an abortion doctor, David Gunn, was shot to death outside his clinic in Pensacola, Florida. A clinic protestor, Michael F. Griffin, who was heard to shout "Stop killing babies!" as he fired at Dr. Gunn's back, was charged in the shooting; the following spring Griffin was found guilty of first-degree murder and sentenced to life in prison. Gunn was the first American fatality in a clinic attack by abortion opponents, but the following year, four months after an intensely emotional Chicago meeting at which some right-to-life activists debated the morality of killing abortionists, Pensacola doctor John B. Britton and his bodyguard were shot to death outside a clinic by a former minister, Paul Hill, who was reported to have attended the Chicago meeting with a petition justifying violence. Then three weeks after Hill was convicted of both murders and sentenced to die in the electric chair, two receptionists were shot to death and five other persons wounded in separate attacks within ten minutes of each other at two abortion clinics in Brookline, Massachusetts. A New Hampshire man, John C. Salvi III, was convicted in the shootings, was sentenced to two consecutive life terms, and subsequently committed suicide by hanging himself in his prison cell.

By mid-1999, the death toll had risen to seven. In January 1998, an off-duty police officer working as a guard was killed in the bombing of an abortion clinic in Birmingham, Alabama, and nine months later an upstate New York doctor, Barnett Slepian, who performed abortions at Buffalo's only clinic, was killed by a shot that came through the kitchen window of his home as he stood talking to his wife and sons. Slepian's name, photograph, and home address, along with those of many other abortion doctors, had appeared on a Web site that was widely regarded as a form of nationally broadcast hit list; within hours of Slepian's death, his name had been crossed out on the list. Although that particular Web site shut down after the publicity that followed Slepian's death, other comparable lists remain on the Internet, and the National Abortion Federation has recently added to its staff a full-time personal security expert who visits clinic staff members' homes to advise them on the safety precautions they ought to follow even when they are no longer at work.

IN MAY 1994, Congress gave final approval to a new statute called the Freedom of Access to Clinic Entrances Act (FACE), which makes it a federal crime to "injure, intimidate or interfere with" any person providing reproductive health services. The statute, which was signed into law by President Clinton, specifies criminal penalties for FACE violations—fines of up to twenty-five thousand dollars, for example, and life imprisonment for an assault resulting in death—and in addition makes available civil remedies such as injunctions and the assessing of monetary damage awards against clinic protestors. Paul Hill, the convicted murderer of Dr. John Britton and his bodyguard, James Barrett, was the first person convicted under FACE for violations resulting in injury and death; even before he received his death sentence for the murders in Florida state court, Hill had been sentenced to two concurrent life terms in prison for violating FACE.

As of the summer of 1999, Justice Department attorneys around the country had brought four dozen civil actions and criminal prosecutions under FACE, some of these cases involving multiple defendants.

IN OCTOBER 1997, the U.S. House of Representatives voted final congressional approval for the first national legislation prohibiting a particular method of surgical abortion. The method, which is used for mid- to late-term abortions and came to be known in the popular press as "partial birth abortion," had been described by the physicians who developed it as "dilation and extraction," or D and X. The mechanics of this procedure, which involves aspirating brain tissue to collapse the fetal skull and bring an otherwise intact fetus through the birth canal, had set off intense public argument ever since a physician's published description of it was reprinted in right-to-life journals in 1993. Two days after the final House vote, President Clinton vetoed the congressional ban, declaring as he did so that although the D and X procedure "appears inhumane," he was basing his veto on the fact that the ban contained no exception for D and X abortions performed to protect women's health.

By the winter of 1999, state versions of the congressional ban had been adopted by twenty-eight states, but two thirds of those bans had been blocked or limited by legal challenges or judicial rulings.

BOB GRIFFIN, the former Missouri Speaker of the House, and Cathryn Simmons, the political consultant who helped organize the abortion ballot initiative petition drive, were indicted in October 1996 on federal bribery and racketeering charges that alleged a pattern of kickbacks and illegal influence between 1992 and 1994. (None of the allegations detailed in the indictment involved abortion legislation or the ballot initiative.) In June 1997, after a trial in Kansas City, a federal jury convicted Simmons of eighteen counts of racketeering, bribery, and mail fraud. Griffin was acquitted on three counts of bribery; the jury was unable to reach a verdict on the other six counts with which he was charged. Two months later, as Griffin was scheduled to stand trial again on the unresolved charges, he pled guilty to one count each of mail fraud and bribery, and the remaining charges against him were dismissed. In December 1997, Griffin and Simmons were each fined and sentenced to four years in prison.

FORMER MISSOURI attorney general William L. Webster, his political ambitions felled by a two-year investigation into allegations that he misused public funds and committed campaign fraud during his run for the Missouri governorship, pled guilty in 1993 to conspiracy to commit public corruption and embezzlement of public resources. He entered a federal prison camp in early 1994 and served nineteen months of a two-year sentence.

In September 1996, after research trials in five American cities, the Food and Drug Administration declared the French abortifacient RU-486, or mifepristone, to be a safe and effective method for terminating pregnancies before the eighth week. Commercial manufacture and distribution of mifepristone for the American market has been delayed by a series of business complications, but as of the summer of 1999 a New York–based nonprofit run by NARAL co-founder Lawrence Lader was producing and dispensing a limited number of pills for continued research purposes, and wider production of American-market mifepristone was expected to get under way by the end of the year.

A second pharmaceutical abortion method was made public in 1994, when a New York gynecologist reported success in inducing early abortions by giving patients both a chemotherapy drug called methotrexate and an additional medicine commonly used for ulcer treatment. Since both medications have already received FDA approval, no law bars their use for other medical purposes, and as of the summer of 1999, National Abortion Federation data indicated that approximately eighty clinics around the United States were offering methotrexate abortions to patients early enough in pregnancy to qualify.

In 1996, Reproductive Health Services formally went out of business and reopened at the same site, with many of the same staff members and physicians, as Reproductive Health Services of Planned Parenthood of the St. Louis Region. In April 1997, a twenty-two-year-old second-trimester patient died at the clinic when she suffered an amniotic fluid embolism during her abortion. The fatality was the clinic's second; the first occurred in 1981, when a severely retarded woman died following an allergic reaction to anesthetic. By the summer of 1999, as the clinic was entering its twenty-seventh year as the largest abortion facility in Missouri, physicians at Reproductive Health Services had performed approximately 180,000 abortions.

Judy Widdicombe lives in Jefferson City and works for the Missouri Department of Social Services. Her fourth grandchild was born in 1997.

SAM LEE works as a lobbyist during the Missouri state legislature's annual five-month session, writing and promoting legislation on behalf of Campaign Life Missouri, the one-man organization he incorporated after being fired by Missouri Citizens for Life. From July to January, when the legislature is out of session, Sam fund-raises for Campaign Life, takes temporary jobs for extra income, and cares for the Lees' four children while Gloria works as director for the pregnant women's residential facility Our Lady's Inn.

FRANK SUSMAN practices law in St. Louis. TOM BLUMENTHAL has left Susman's firm and established his own practice in St. Louis.

LORETTO WAGNER resigned in 1992 as an officer of Missouri Citizens for Life. For several years after her resignation she continued to organize the St. Louis bus contingent to the annual March for Life in Washington, D.C.; she also volunteers as a fund-raiser and board member for Our Lady's Inn, which she co-founded.

SYLVIA HAMPTON moved to Southern California and volunteers for the League of Women Voters and the Coalition for Quality Health Care.

LOUIS DEFEO is counsel and chief lobbyist for the Missouri Catholic Conference in Jefferson City.

MARY BRYANT left Planned Parenthood in 1991. She and her husband live near Jefferson City and own a business that provides state-certified educational services for persons convicted of driving while intoxicated.

KATHY EDWARDS, whose name after her second marriage was Kathy Wise, died of cancer in 1996.

MICHAEL FREIMAN is emeritus professor of obstetrics and gynecology at the Washington University School of Medicine. He no longer performs abortions.

MATT BACKER divides his time between St. Louis, where he is professor and chairman emeritus of obstetrics and gynecology at the Washington University School of Medicine, and San Diego, where he is a staff consultant at the Naval Medical Center. He has not lectured on abortion since 1974, but keeps current his membership in the St. Louis–based organization Doctors for Life.

B. J. ISAACSON (she divorced and stopped using the surname Isaacson-Jones) resigned from Reproductive Health Services in 1994, after a period of financial and administrative difficulty at the clinic, and is now executive director of the Indiana Primary Health Care Association. ANDREW PUZDER moved in 1991 to California, where he now works as a business executive in Santa Barbara. From late 1989 through much of the next decade, Isaacson and Puzder participated in a series of meetings that began after Puzder wrote a commentary piece for the *St. Louis Post-Dispatch;* in his commentary, which he entitled "Common Ground on Abortion," Puzder cited statistics about the high number of Missouri children living in poverty and suggested that abortion opponents and advocates alike might "redirect to other social problems the enormous energy we expend on this single problem."

Because B. J. Isaacson shared these sentiments, she invited Puzder to Reproductive Health Services after reading his essay in the newspaper. Puzder accepted her invitation, although he asked that the meeting take place after business hours so that he could avoid entering the office while abortions were being performed, and for some months afterward Puzder and Isaacson continued on a regular basis what they described as "common ground" conversations, accompanied in their meetings by Loretto Wagner and RHS public affairs director Jean Cavender. In June 1991, the *Post-Dispatch* published a commentary jointly written by Puzder, Isaacson, Wagner, and Cavender. Under the headline "Common Ground," the four implored opponents in the abortion conflict to remember their common humanity. "Demonizing the opposition in the abortion debate serves no useful political purpose and resolves no differences," they wrote:

The disparaging labels used by both sides might make good rallying cries or media copy, but in the long run, they accomplish nothing more than to erect walls of bitterness, walls that obscure the areas where each side's objectives and motives overlap or, at least, run parallel to each other.

For example, neither side wants to see poor women in situations where they feel economically compelled to have abortions. From the pro-life side, the desire to save lives should obviously direct society's energies to assisting such women. From the pro-choice side, one of the choices made must be the choice to give birth. . . . We believe it is possible to maintain respect for each other's right to hold one's position and to work within our political and judicial system without rancor or bitterness. We also believe that talking to one another and listening to one another is the only path to a lasting resolution of the issue.

The authors cited two local examples of efforts that they argued deserved joint support from both sides of the abortion conflict: one was Our Lady's Inn, and the other was Adoption Associates, a nonprofit agency set up by Reproductive Health Services to help women who had decided to place their babies for adoption rather than having abortions. Adoption Associates was eventually closed, but the public disclosure of the common ground talks attracted local and national press attention, first in St. Louis and later in Buffalo, Boston, Madison, and San Francisco, where activists from both sides had begun similar efforts to talk to each other in an atmosphere of mutual respect.

In 1993, Isaacson, Wagner, Puzder, and Cavender were invited to Buffalo to help establish a national Common Ground Network for Life and Choice, which now offers mediation help and coordinates projects among approximately a dozen regional groups meeting at regular intervals around the United States. The meetings, which in many communities have pressed ahead despite skepticism from partisans on both sides, have so far produced heartfelt conversation and very modest tangible results. Isaacson and Puzder together wrote a paper on adoption, one of a planned series of Network pamphlets on subjects about which both sides might agree, and in 1996 the first national Reach for Common Ground conference convened over three days at the University of Wisconsin in Madison. The principal visual symbol inside the conference brochure, superimposed over the text of the program schedule, was a shaded line drawing of a single outstretched hand.

Sources

The primary sources of information for this narrative were approximately five hundred personal interviews, most of them conducted between 1991 and 1996, and the documentation and personal records many of those interview subjects had accumulated over their years of involvement in the abortion conflict. In some cases these records had been archived in libraries, but more often the old papers—meeting minutes, memoranda, pamphlets, news clippings, personal letters, and the other souvenirs of volunteer careers in pro-life or pro-choice activism—had been piled into home file cabinets or cardboard boxes by people who had guessed, correctly, that somebody might one day take interest in the story these papers helped to tell. For access to particularly useful collections stored both in formal and informal fashion, I extend my thanks to the following libraries and individuals:

LIBRARIES AND ARCHIVES:

Mercantile Library, St. Louis (archives of the now-defunct *St. Louis Globe-Democrat*)
Missouri Catholic Conference archive
New York Public Library (Lawrence Lader Papers)
Reproductive Health Services archive
St. Louis Post-Dispatch Library
St. Louis Review Library
Arthur and Elizabeth Schlesinger Library on the History of Women in America, Radcliffe College, Cambridge, Massachusetts (National Abortion Rights Action

League Papers; Mary Steichen Calderone Papers; Records of the Society for Humane Abortion and the Association to Repeal Abortion Laws; Ruth Proskauer Smith Papers; Family Planning Oral History Project Records)

University of Missouri, St. Louis, Western History Collection (Reproductive Health Services papers; Committee for Legal Abortion in Missouri papers)

Washington University Medical Center Manuscripts Archive (John Vavra Papers)

PERSONAL COLLECTIONS:

Burke Balch, Edward Epstein, Samuel Lee, Lana Phelan, Dr. Keith Russell, Ruth Proskauer Smith, Loretto Wagner, Kathy Wise

The list of abortion books published over the last twenty-five years is extensive, but certain volumes proved especially helpful. *Abortion and the Politics of Motherhood*, by Kristin Luker (Berkeley: University of California Press, 1984), uses California as a base for a sociological study that is both highly informative and scrupulously respectful to both sides' views; in a similar vein, *Contested Lives*, by Faye Ginsburg (Berkeley: University of California Press, 1989), examines a North Dakota community during the early 1980s. *Liberty and Sexuality: The Right to Privacy and the Making of Roe v. Wade*, by David J. Garrow (New York: Macmillan, 1994), contains the most detailed and comprehensive national account of the campaigns toward abortion legalization in state legislatures and the courts; and *When Abortion Was a Crime*, by Leslie J. Reagan (Berkeley: University of California Press, 1997), uses Illinois historical records to provide unprecedented detail on abortion practice and prosecutions during the century before *Roe v. Wade*.

OTHER BOOKS, ARTICLES, UNPUBLISHED PAPERS, AND ORAL HISTORIES (SELECTED LIST):

Alan Guttmacher Institute, "Safe and Legal: 10 Years' Experience with Legal Abortion in New York State" (New York: Alan Guttmacher Institute, 1980).

Alinsky, Saul, *Rules for Radicals* (New York: Random House, 1971).

American Law Institute, *Model Penal Code*, Proposed Official Draft, Sec. 230:3;2 (Philadelphia: American Law Institute, 1962).

Andrews, Joan, with John Cavanaugh-O'Keefe, *I Will Never Forget You* (San Francisco: Ignatius, 1989).

Anonymous, "Does Anybody Care?" *American Journal of Nursing*, September 1973. Student nurse's first-person account of saline abortion.

Baars, Conrad W., M.D., "The Psychiatrist—Friend or Foe of the Pregnant Woman?" *Linacre Quarterly*, November 1971, p. 261.

Baird, Robert M., and Stuart E. Rosenbaum, *The Ethics of Abortion* (Buffalo: Prometheus, 1989).

Barno, Alex, M.D., "Criminal Abortion Deaths, Illegitimate Pregnancy Deaths, and Suicides in Pregnancy: Minnesota 1960–1965," *American Journal of Obstetrics and Gynecology* 96:3 (1966).

Bart, Pauline, "Seizing the Means of Reproduction: An Illegal Feminist Abortion

Collective—How and Why It Worked," *Qualitative Sociology* 10:109 (Winter 1987).

Bartlett, Robert H., M.D., and Clement Yahia, M.D., "Management of Septic Chemical Abortion with Renal Failure," *New England Journal of Medicine* 281:14 (1969).

Bates, Jerome E., and Edward S. Zawadzki, M.D., *Criminal Abortion: A Study in Medical Sociology* (Springfield, Ill.: Charles P. Thomas, 1964).

Baulieu, Etienne-Emile, with Mort Rosenblum, *The "Abortion Pill"* (New York: Simon & Schuster, 1990).

Berger, Gary S., M.D., et al., "Termination of Pregnancy by 'Super Coils': Morbidity Associated with a New Method of Second-Trimester Abortion," *American Journal of Obstetrics and Gynecology* 116:3 (1973).

Blake, Judith, Ph.D., "Negativism, Equivocation, and Wobbly Assent: Public 'Support' for the Prochoice Platform on Abortion," *Demography* 18: 309–20 (August 1981).

Bopp, James, ed., *Restoring the Right to Life: The Human Life Amendment* (Provo, Utah: Brigham Young University Press, 1984).

Bosgra, Tj., *Abortion, the Bible and the Church* (Toronto: Life Cycle Books, 1987).

Boston Women's Health Course Collective, *Our Bodies, Our Selves: A Course by and for Women* (Boston: Boston Women's Health Course Collective and New England Free Press, 1971).

Bourke, Vernon J., *Ethics: A Textbook in Moral Philosophy* (New York: Macmillan, 1951).

Breuer, William B., *An American Saga: The Fabulous Tale of Sunnen Products Company* (St. Louis: Pierre Laclede, 1979).

Brown, Judie, *It Is I Who Have Chosen You* (Stafford, Va.: American Life League Publishers, 1992).

Brownmiller, Susan, "Abortion Counseling: Service Beyond Sermons," *New York* magazine, Aug. 4, 1969, p. 31.

Byrn, Robert M., "An American Tragedy: The Supreme Court on Abortion," 41 *Fordham Law Review* 807 (1973).

Byrnes, Timothy A., *Catholic Bishops in American Politics* (Princeton: Princeton University Press, 1991).

———, and Mary C. Segers, eds., *The Catholic Church and the Politics of Abortion: A View from the States* (Boulder, Colo.: Westview, 1992).

Calderone, Mary Steichen, M.D., "Illegal Abortion as a Public Health Problem," *American Journal of Public Health* 50:948 (1960).

———, ed., *Abortion in the United States* (New York: Hoeber-Harper, 1958).

Callahan, Daniel, *Abortion: Law, Choice and Morality* (New York: Macmillan, 1970).

Carmen, Arlene, and Howard Moody, *Abortion Counseling and Social Change: From Illegal Act to Medical Practice* (Valley Forge, Pa.: Judson, 1973).

Carney, Bruce H., M.D., "Vaginal Burns from Potassium Permanganate," *American Journal of Obstetrics and Gynecology* 65:1 (January 1952).

Cates, Willard, Jr., M.D., "Legal Abortion: The Public Health Record," *Science* 215:26 (1982).

———, et al., "Legalized Abortion: Effect on National Trends of Maternal and Abortion-Related Mortality (1940 Through 1976)," *American Journal of Obstetrics and Gynecology*, Sept. 15, 1978.

Cavanagh, Denis, M.D., and Allan G. McLeod, M.D., "Septic Shock in Obstetrics and Gynecology," *American Journal of Obstetrics and Gynecology* 96:7 (1966).

Cheek, Jeannette B., oral history interview with Patricia Maginnis, Nov. 16, 17, 18, 1975, from Schlesinger-Rockefeller Oral History Project.

——, oral history interview with Estelle Griswold, March 17, 1976 (at Schlesinger Library).

Chernyak, A. A., "Artificial Termination of Pregnancy with a Vacuum Aspirator," *Zdravookhranenie Belorussii* (Minsk) 9:28–30, May 1983, in Russian; English translation provided by Donn Casey, Bibliography of Reproduction, September 1964.

Chesler, Ellen, 1976 oral history interview with Arlene Carmen, Schlesinger Library.

——, 1976 oral history interview with Constance Cook, Schlesinger Library.

——, 1976 oral history interview with Dr. Lonny Myers, Schlesinger Library.

——, *Woman of Valor: Margaret Sanger and the Birth Control Movement in America* (New York: Simon & Schuster, 1992).

Clark, Tom C., "Religion, Morality, and Abortion: A Constitutional Appraisal," *Loyola University Law Review* 2:1 (1969).

Connery, John, S.J., *Abortion: The Development of the Roman Catholic Perspective* (Chicago: Loyola University Press, 1977).

Constitutional Amendments Relating to Abortion, transcript, U.S. Senate Judiciary Committee Constitutional Subcommittee hearings, 1981.

Cooke, Robert E., M.D., et al., eds., *The Terrible Choice: The Abortion Dilemma* (New York: Bantam, 1968). Based on the proceedings of the International Conference on Abortion sponsored by the Harvard Divinity School and the Joseph P. Kennedy Jr. Foundation.

Crowley, Ralph M., M.D., and Robert W. Laidlaw, M.D., "Psychiatric Opinion Regarding Abortion: Preliminary Report of a Survey," *American Journal of Psychiatry* 124:4 (1967).

Cuneo, Michael W., *Catholics Against the Church: Anti-Abortion Protest in Toronto, 1969–1985* (Toronto: University of Toronto Press, 1989).

Dailey, the Reverend Thomas G., "The Catholic Position on Abortion," *Linacre Quarterly,* August 1967.

Davis, Flora, *Moving the Mountain: The Women's Movement Since 1960* (New York: Simon & Schuster, 1991).

Donovan, Patricia, "The 1988 Abortion Referenda: Lessons for the Future," *Family Planning Perspectives,* September/October 1989.

Drinan, Robert F., S.J., "The Inviolability of the Right to Be Born," 17 *Western Reserve Law Review* 465 (1965).

Droegemueller, William, M.D., et al., "The First Year of Experience in Colorado with the New Abortion Law," *American Journal of Obstetrics and Gynecology* 103:5 (1969).

D'Souza, Dinesh, *Falwell: Before the Millennium* (Chicago: Regnery Gateway, 1984).

Edgington, Steven, "Securing Liberal Legislation During the Reagan Administration," 1982 interview with Anthony Beilenson, Oral History Program, UCLA Governmental History Documentation Project.

Eisenberg, Howard, "The Mad Scramble for Abortion Money," *Medical Economics,* Jan. 4, 1971.

Ely, John Hart, "The Wages of Crying Wolf: A Comment on *Roe v. Wade*," 82 *Yale Law Journal* 920 (1973).

Falwell, Jerry, *Strength for the Journey* (New York: Simon & Schuster, 1987).

Faux, Marian, *Roe v. Wade* (New York: Macmillan, 1988).

FitzGerald, Frances, "A Disciplined, Charging Army," *The New Yorker*, May 18, 1981.

Flannery, Austin, O.P., ed., *Vatican Council II: The Conciliar and Post Conciliar Documents* [Vol. 1] (Grand Rapids, Mich.: Eerdmans, 1975).

———, *Vatican Council II: More Post-Conciliar Documents* [Vol. 2] (Northport, N.Y.: Costello, 1982).

Fox, Leon Parrish, M.D., "Abortion Deaths in California," *American Journal of Obstetrics and Gynecology* 98:645 (1967).

Frankfort, Ellen, *Vaginal Politics* (New York: Quadrangle, 1972).

Fried, Marlene Gerber, ed., *From Abortion to Reproductive Freedom* (Boston: South End, 1990).

Friedan, Betty, *The Feminine Mystique* (New York: Norton, 1963).

———, *It Changed My Life* (New York: Norton, 1991).

Friedman, Leon, ed., *The Supreme Court Confronts Abortion: The Briefs, Argument, and Decision in Planned Parenthood v. Casey* (New York: Farrar, Straus, and Giroux, 1993).

Gage, Suzann, *When Birth Control Fails: How to Abort Ourselves Safely* (San Diego: Speculum, 1979).

Galebach, Stephen, "A Human Life Statute," *The Human Life Review*, Winter 1981 (reprinted as booklet from the Human Life Foundation, New York, 1981).

Geijerstam, Gunnar K. af, M.D., *An Annotated Bibliography of Induced Abortion* (Ann Arbor, Mich.: Center for Population Planning, 1969).

Glendon, Mary Ann, *Abortion and Divorce in Western Law* (Cambridge, Mass.: Harvard University Press, 1987).

Goldsmith, Sadja, M.D., and Alan J. Margolis, M.D., "Aspiration Abortion Without Cervical Dilation," *American Journal of Obstetrics and Gynecology* 110:4 (1971).

Goldstein, Michael S., "Abortion as a Medical Career Choice: Entrepreneurs, Community Physicians, and Others," *Journal of Health and Social Behavior* 25:211–29 (1984).

———, "Creating and Controlling a Medical Market: Abortion in Los Angeles After Liberalization," *Social Problems* 31:5 (1984).

Goodno, John A., Jr., M.D., et al., "Management of Infected Abortion: An Analysis of 342 Cases," *American Journal of Obstetrics and Gynecology* 85:1 (1963).

Gornel, Daniel L., M.D., and Ralph Goldman, M.D., "Acute Renal Failure Following Hexol-Induced Abortion," *Journal of the American Medical Association* 201:2 (1968).

Greely, Henry T., "A Footnote to 'Penumbra' in *Griswold v. Connecticut*," 6 *Stanford Law Review* 251 (1989).

Greenhouse, Linda, "After July 1, an Abortion Should Be as Simple to Have as a Tonsillectomy, but—," *The New York Times Magazine*, June 28, 1970.

Grimes, David A., M.D., "Second-Trimester Abortions in the United States," *Family Planning Perspectives* 16:6 (December 1984).

Grisez, Germain, *Abortion: The Myths, the Realities, and the Arguments* (New York: Corpus, 1970).

Grobstein, Clifford, *Science and the Unborn* (New York: Basic Books, 1988).

Guttmacher, Alan, M.D., ed., *The Case for Legalized Abortion Now* (Berkeley: Diablo, 1967).

Hadden, Jeffrey K., and Anson Shupe, *Televangelism: Power and Politics on God's Frontier* (New York: Holt, 1988).

Hall, Robert, M.D., "Abortion in American Hospitals," *American Journal of Public Health* 57:11 (1967).

Harting, Donald, M.D., and Helen J. Hunter, M.P.H., "Abortion Techniques and Services: A Review and Critique," *American Journal of Public Health* 61:10 (1971).

Hausknecht, Richard U., M.D., "Free Standing Abortion Clinics: A New Phenomenon," *Bulletin of the New York Academy of Medicine* 49:11 (1973).

Hern, Warren M., M.D., "Midtrimester Abortion," *Obstetrics and Gynecology Annual*, 1981.

——, "First and Second Trimester Abortion Techniques," *Current Problems in Obstetrics and Gynecology* VI:11 (July 1983).

——, *Abortion Practice* (Philadelphia: Lippincott, 1990).

——, and Bonnie Andrikopoulos, eds., *Abortion in the Seventies: Proceedings of the Western Regional Conference on Abortion, Denver, Colorado, 1976* (New York: National Abortion Federation, 1977).

——, and Billie Corrigan, R.N., M.S., "What About Us? Staff Reactions to D&E," *Advances in Planned Parenthood* XV:1 (1980).

Heymann, Philip B., and Douglas E. Barzelay, "The Forest and the Trees: *Roe v. Wade* and Its Critics," 53 *Boston University Law Review* 765 (1973).

Higgins, George C., "The Prolife Movement and the New Right," *America*, Sept. 13, 1980.

Hilgers, Thomas, M.D., Dennis J. Horan, and David Mall, *New Perspectives on Human Abortion* (Frederick, Md.: Aletheia, 1981).

Hilgers, Thomas W., and Dennis J. Horan, eds., *Abortion and Social Justice* (Thaxton, Va.: Sun Life, 1972).

Hill, Edward C., M.D., and John M. Thomas, M.D., "Potassium Permanganate Ulcers of the Vagina," *Obstetrics and Gynecology* 18:6 (1961).

Hodgson, Jane E., *Abortion and Sterilization: Medical and Social Aspects* (London: Academic Press, 1981).

Horan, Dennis J., Edward R. Grant, and Paige C. Cunningham, eds., *Abortion and the Constitution: Reversing Roe v. Wade Through the Courts* (Washington, D.C.: Georgetown University Press, 1987).

Huffman, John W., and Gerald B. Holzman, eds., "Implementation of Legal Abortion: A National Problem," *Clinical Obstetrics and Gynecology* 14:4 (1971). Special issue.

Hurst, Jane, *Abortion: The Development of the Roman Catholic Perspective* (Washington, D.C.: Catholics for a Free Choice, 1989).

Irons, Peter, *The Courage of Their Convictions* (New York: Free Press, 1988).

Jain, Sagar C., Ph.D., and Steven Hughes, "*California Abortion Act 1967: A Study in Legislative Process*" (Chapel Hill, N.C.: Carolina Population Center, University of North Carolina at Chapel Hill, 1968).

Jain, Sagar C., Ph.D., and Steven W. Sinding, *North Carolina Abortion Law 1967: A Study in Legislative Process* (Chapel Hill, N.C.: Carolina Population Center, University of North Carolina at Chapel Hill, 1968).

Jeffries, Liz, and Rick Edmonds, "Abortion: The Dreaded Complication," *The Inquirer Magazine*, Aug. 2, 1981.

Jessee, R. W., M.D., and Frederick J. Spencer, M.D., "Abortion—The Hidden Epidemic," *Virginia Medical Monthly*, August 1968.

Joffee, Carol, *Doctors of Conscience: The Struggle to Provide Abortion Before and After Roe v. Wade* (Boston: Beacon, 1995).

Julienne, Mark, M.D., "Suddenly I'm a Legal Abortionist," *Medical Economics*, Nov. 23, 1970.

Jung, Patricia Beattie, and Thomas A. Shannon, *Abortion & Catholicism: The American Debate* (New York: Crossroad, 1988).

Kaltreider, Nancy B., et al., "The Impact of Midtrimester Abortion Techniques on Patients and Staff," *American Journal of Obstetrics and Gynecology* 135:2 (1979).

Kelly, James R., "A Dispatch from the Abortion Wars: Reflections on 'Common Ground,' " *America*, Sept. 17, 1994.

Kerslake, Dorothea, and Donn Casey, "Abortion Induced by Means of the Uterine Aspirator," *Obstetrics and Gynecology* 30:1 (1967).

Kleinknecht, Colonel G. H., and Major Gerald O. Mizell, "Abortion: A Police Response," *FBI Law Enforcement Bulletin*, March 1982.

Knapp, Robert Charles, M.D., et al., "Septic Abortion: Five-Year Analysis at the New York Hospital," *Obstetrics and Gynecology* 15:3 (1960).

Koop, C. Everett, M.D., *The Right to Live; the Right to Die* (Wheaton, Ill.: Tyndale, 1976).

―――――, *Koop* (New York: Random House, 1991).

Lader, Lawrence, *Abortion* (Boston: Beacon, 1966).

―――――, "The Scandal of Abortion Laws," *The New York Times Magazine*, April 25, 1965.

―――――, *Abortion II: Making the Revolution* (Boston: Beacon, 1973).

Lamm, Richard D., et al., "The Legislative Process in Changing Therapeutic Abortion Laws: The Colorado Experience," *American Journal of Orthopsychiatry*, July 1969.

Landwehr, Carl J., *Organizing for Community Pro-Life Action* (Jefferson City, Mo.: Carl J. Landwehr & Associates, 1979).

Lovitt, W. V., M.D., "Pulmonary Embolism by Powdered Household-Mustard Incident to Self-Induced Abortion," *American Journal of Clinical Pathology* 23:914 (September 1953).

Lucas, Roy, "Federal Constitutional Limitations on the Enforcement and Administration of State Abortion Statutes," 46 *North Carolina Law Review* 730 (1968).

Lynch, John J., S.J., "Ectopic Pregnancy: A Theological Review," *Linacre Quarterly*, February 1961.

Maginnis, Patricia, and Lana Clark Phelan, *The Abortion Handbook* (North Hollywood, Calif.: Contact Books, 1969).

Maguire, Daniel C., "Abortion: A Question of Catholic Honesty," *The Christian Century*, Sept. 14–21, 1983.

Manabe, Yukio, M.d., "Danger of Hypertonic-Saline-Induced Abortion," *Journal of the American Medical Association* 210:11 (1969).

McBride, D. E., D.O., "Septic Abortion Followed by Septic Shock," *Journal of the American Osteopathic Association*, March 1961.

McKeegan, Michele, *Abortion Politics: Mutiny in the Ranks of the Right* (New York: Free Press, 1992).

Means, Cyril, Jr., "The Law of New York Concerning Abortion and the Status of the Foetus, 1664–1968: A Case of Cessation of Constitutionality," XIV *New York Law Forum* 411 (1965).

———, "The Phoenix of Abortional Freedom," 17 *New York Law Forum* 335 (1971).

Minkler, Donald H., M.D., "Abortion—The Role of Private Foundations," *Clinical Obstetrics and Gynecology* 14:4 (1971).

Model Penal Code, Tentative Draft No. 9, May 8, 1959, The American Law Institute, Philadelphia.

Mohr, James C., *Abortion in America: The Origins and Evolution of National Policy* (New York: Oxford University Press, 1978).

Monroe, Keith, "How California's Abortion Law Isn't Working," *The New York Times Magazine*, Dec. 29, 1968.

Moody, Howard, "Man's Vengeance on Woman: Some Reflections on Abortion Laws as Religious Retribution and Legal Punishment of the Feminine Species," *Renewal Magazine*, February 1967.

Morgan, Robin, ed., *Sisterhood Is Powerful: An Anthology of the Women's Liberation Movement* (New York: Vintage, 1970).

Nathanson, Bernard N., M.D., *Aborting America: A Doctor's Personal Report on the Agonizing Issue of Abortion* (Garden City, N.Y.: Doubleday, 1979).

———, *The Abortion Papers: Inside the Abortion Mentality* (New York: Fell, 1983).

National Conference of Catholic Bishops, *Documentation on the Right to Life and Abortion* (Washington, D.C.: U.S. Catholic Conference Publications Office, 1974).

Nilsson, Lennart (photographer), "Life Before Birth," *Life*, April 30, 1965 (no writer identified).

Noonan, John T., Jr., *A Private Choice: Abortion in America in the Seventies* (New York: Free Press, 1979).

———, ed., *The Morality of Abortion: Legal and Historical Perspectives* (Cambridge, Mass.: Harvard University Press, 1970).

Ohio State Medical Association Committee on Maternal Health, "Maternal Deaths with Septic Shock After Criminal Abortion," *The Ohio State Medical Journal*, June 1969. Report compiled and published annually.

Olasky, Marvin, *Abortion Rites: A Social History of Abortion in America* (Wheaton, Ill.: Crossway, 1992).

Osofsky, Howard J., and Joy D. Osofsky, eds., *The Abortion Experience* (New York: Harper & Row, 1973).

Packer, Herbert L., and Ralph J. Gampell, "Therapeutic Abortion: A Problem in Law and Medicine," *Stanford Law Review* 11:417 (1959).

Pakter, Jean, M.D., et al., "A Review of Two Years' Experience in New York City with the Liberalized Abortion Law," in Osofsky, pp. 48–50.

Pertschuk, Michael, and Wendy Schaetzel, *The People Rising: The Campaign Against the Bork Nomination* (New York: Thunder's Mouth, 1989).

Peterson, William F., M.D., et al., "Second-Trimester Abortion by Dilation and Evacuation: An Analysis of 11,747 Cases," *Obstetrics and Gynecology* 62:2 (August 1983).

Pilpel, Harriet, "The Right of Abortion," ASA reprint from *The Atlantic Monthly*, June 1969.

Polgar, Steven, and Ellen Fried, "The Bad Old Days: Clandestine Abortions Among the Poor in New York City Before Liberalization of the Abortion Law," *Family Planning Perspectives* 8:3 (1976).

Potts, Malcolm, Peter Diggory, and John Peel, *Abortion* (New York: Cambridge University Press, 1977).

Rabinowitz, Paul, M.D., et al., "Management of Postabortal Infections Complicated by Acute Renal Failure," *American Journal of Obstetrics and Gynecology* 84:6 (1962).

Raymond, Janice G., et al., "RU 486: Misconceptions, Myths and Morals" (Cambridge, Mass.: Institute on Women and Technology, 1991).

Rock, John, M.D., *The Time Has Come* (New York: Knopf, 1963).

Roraback, Catherine G., "*Griswold v. Connecticut: A Brief Case History,*" 16 *Ohio Northern University Law Review* 395 (1989).

Rosen, Harold, M.D., "Psychiatric Implications of Abortion: A Case Study in Social Hypocrisy," 17 *Western Reserve Law Review* 435 (1965).

————, ed., *Therapeutic Abortion* (New York: Julian, 1954).

Rosenblum, Victor G., *Law as a Political Instrument* (New York: Random House, 1955).

Ruzek, Sheryl Burt, *The Women's Health Movement: Feminist Alternatives to Medical Control* (New York: Praeger, 1978).

Savage, David G., *Turning Right: The Making of the Rehnquist Supreme Court* (New York: Wiley, 1992).

Schaeffer, Frank V., and C. Everett Koop, M.D., *Plan for Action: An Action Alternative Handbook for Whatever Happened to the Human Race?* (Old Tappan, N.J.: Revell, 1980).

Scheidler, Joseph M., *Closed: 99 Ways to Stop Abortion* (Chicago: Regnery, 1985).

Schoeneck, F. J., M.D., "Fatalities Associated with Abortions," *New York State Journal of Medicine*, May 15, 1964.

Sharp, Gene, *The Politics of Nonviolent Action* (Boston: Porter Sargent, 1973), p. 315.

Silver, Marvin D., M.D., and Tommy N. Evans, "Air Embolism: A Discussion of Maternal Mortality with a Report of One Survivor," *Obstetrics and Gynecology* 31:3 (1968).

Smith, Ronald, M.D., "Soap-induced Abortion," *Obstetrics and Gynecology* 20:2 (1962).

Smith, Roy G., M.D., et al., "Physicians' Attitudes on the Abortion Law," *Hawaii Medical Journal* 29:3 (1970).

Solinger, Rickie, *The Abortionist: A Woman Against the Law* (Berkeley: University of California Press, 1994).

Speroff, Leon, M.D. (panel moderator), "Is There a Best Way to Do Midtrimester Abortions?" *Contemporary Obstetrics/Gynecology*, May 1979.

Steinhoff, Patricia G., and Milton Diamond, *Abortion Politics: The Hawaii Experience* (Honolulu: University Press of Hawaii, 1977).

Stevenson, Lee B., M.D., "Maternal Death and Abortion: Michigan 1955–1964," *Michigan Medicine*, March 1967.

Studdiford, William E., M.D., and Gordon Watkins Douglas, M.D., "Placental Bacteremia: A Significant Finding in Septic Abortion Accompanied by Vascular Collapse," *American Journal of Obstetrics and Gynecology*, 1956.

Taussig, Frederick J., *Abortion, Spontaneous and Induced* (St. Louis: Mosby, 1936).

Teeter, Richard R., M.D., and Theodore Watson, M.D., "Psychiatric Indications for Abortions," *Minnesota Medicine* 50:49 (January 1967). Dual editorials, pro and con.

Thompson, M.D., et al., "California Maternal Mortality Survey," pamphlet, 1964.

Thompson, Judith Jarvis, "A Defense of Abortion," *Philosophy and Public Affairs*, Fall 1971, pp. 44–66.

Tietze, Christopher, "The Public Health Effects of Legal Abortion in the United States," *Family Planning Perspectives* 16:1 (1984).

Tribe, Laurence H., *Abortion: The Clash of Absolutes* (New York: Norton, 1990).

United States Catholic Conference, *Documentation on Abortion and the Right to Life II* (Washington, D.C.: USCC Publications Office, 1976).

Walker, Marlan C., and Andrew F. Puzder, "State Protection of the Unborn After *Roe v. Wade:* A Legislative Proposal," *Stetson Law Review* XIII:2 (1984).

Wall, Leonard A., M.D., "Abortions: Ten Years Experience at Kansas University Medical Center," *American Journal of Obstetrics and Gynecology* 79:3 (1960).

Wang, Jack, M.S., et al., "Body Composition Studies in the Human Fetus After Intra-amniotic Injection of Hypertonic Saline," *American Journal of Obstetrics and Gynecology* 117:1 (1973).

Wattleton, Faye, *Life on the Line* (New York: Ballantine, 1996).

Weddington, Sarah, *A Question of Choice* (New York: Grosset/Putnam, 1992).

White, Robert, "Induced Abortions: A Survey of Their Psychiatric Implications, Complications, and Indications," *Texas Reports on Biology & Medicine* 24:4 (1966).

Williams, Glanville, *The Sanctity of Life and the Criminal Law* (New York: Knopf, 1957).

Willke, J. C., M.D., *Abortion and Slavery: History Repeats* (Cincinnati: Hayes, 1984).

Willke, Dr. and Mrs. J. C., *The Wonder of Sex: How to Teach Children* (Cincinnati: Hiltz, 1964).

———, *Handbook on Abortion* (Cincinnati: Hiltz, 1971, subsequently reprinted and updated as *Abortion: Questions and Answers*, Cincinnati: Hayes, 1985).

———, *How to Teach the Pro-Life Story* (Cincinnati: Hiltz & Hayes, 1973).

Wolff, Jennifer, "In Search of Common Ground," *Glamour*, November 1992.

Woodward, Bob, and Scott Armstrong, *The Brethren* (New York: Simon & Schuster, 1979).

Wulff, George, M.D., and Michael Freiman, M.D., "Elective Abortion: Complications Seen in a Free-Standing Clinic," *Obstetrics and Gynecology* 49:3 (1977).

Zakin, David, M.D., et al., "Foreign Bodies Lost in the Pelvis During Attempted Abortion with Special Reference to Urethral Catheters," *American Journal of Obstetrics and Gynecology* 70:2 (1955).

Principal Interviews:

Ladd Alexander, Willard Allen, Gail Anderson, David Andrews, Matt Backer, Burke Balch, Anne Bannon, Jim Barnes, Thea Rossi Baron, Marie Bass, Carl Bauman, Anthony Beilenson, Janet Benshoof, Jean Berg, Elizabeth Bilgere, Beatrice Blair, Tom Blumenthal, James Bopp, Richard Bott, Diane Boulware, John Boyce, Curtis Boyd, Elizabeth Boyer, Mary Bryant, Ceil Callahan, William Cameron, Elizabeth Canfield, Arlene Carmen, Denis Cavanagh, John Cavanaugh-O'Keefe, Emily Champagne, Michael Chastain, Sherri Finkbine Chessen, Lucinda Cisler, Richard Cizik, Dick Clark, Laura Cohen, Manuel Comas, Guy Condon, Colleen Connell, Constance Cook, William Cox, Robert Crist, Mary Kay Culp, Paige Comstock Cunningham, David and Nancy Danis, Louis DeFeo, Charles Donovan, Carol Downer, Jean Doyle, William Drake, Carl Dreyer, Edward Epstein, Roger Evans, Vivian Eveloff, James Ewing, Nanette Falkenberg, Lisa Fierstein, Tim Finnegan, Dan Foley, Peter Forbes, Clark Forsythe, Erika Fox, Maureen McCarthy Franz, Michael and Sarijane Freiman, Sybille Fritzsche, Jean Garton, Penn Garvin, Georgie Gatch, Jules Gerard, Carolyn Gerster, Ira Gerstley, Wayne Goode, the Reverend Ken Gottman, Edward Grant, Michael Greenwald, Sadja Greenwood, Bob Griffin, Charlotte Grimes, David Grimes, Rowena Gurner, Kim Haddow, Crystal Williams Hagevik, Margie Pitts Hames, Sylvia Hampton, Mildred Hanson, Gwyndolyn Harvey, Gloria Heffernan, Warren Hern, Paul Hetrich, Tony Hiesberger, the Reverend Allen Hinand, Herbert Hodes, Jane Hodgson, William Hogan, Joe Holt, Delores Horan, Mary Frances Horgan, Steven Hurst, Robert Iles, B. J. Isaacson, Mildred Jefferson, Mary Ann Johanek, Laura Kaplan, Sandra Keifer, Thomas Kerenyi, Frances Kissling, the Reverend George Knight, John Knox, Kathryn Kolbert, Stephen Kovac, Kermit Krantz, Lawrence Lader, Richard Lamm, Richard Land, Carl Landwehr, Uta Landy, Eugene Lang, Edward Lazarus, Gloria Lee, Mark Lee, Richard Lee, Robert G. Lee, Samuel Lee, Sophie Lee, Harry Levin, Lois Lipton, Patricia Maginnis, Mario Mandina, Alan Margolis, Roberta Lee Marsh, Connaught Marshner, Thomas Marzen, Ed McAteer, Karen McCarthy, James McClure, Amelia McCracken, J. P. McFadden, Mary Meehan, Gerald Meisel, Vivian Diener Meyer, Kate Michelman, Lucia Miller, Samuel Montello, the Reverend Howard Moody, the Reverend Christian Moore, Alfred Moran, Stewart Mott, Ben Munson, Jerry Murphy, William Murr, Bernard Nathanson, Ann O'Brien, the Most Reverend Edward J. O'Donnell, Joe Ortwerth, Marilyn Backer Parker, the Reverend E. Spencer Parsons, Alberta Passanante, Eve Paul, Emily Lee Peckman, the Reverend Vince Petersen, William Peterson, Lana Phelan, Louis Prange, Mildred Prange, Raymond Probst, Andrew Puzder, the Reverend Tom Raber, Jane Rabon, Curtis Reitz, Samuel Rodgers, Ruth Roemer, Catherine Roraback, Victor Rosenblum, Lorraine Rothman, Dorothy Roudebush, Keith Russell, Sheryl Burt Ruzek, Howard Schwartz, Louis Schwartz, Melvin Schwartz, Jeff Shaeperkotter, Sue Shear, Paul Shively, Jerry Short, Cathryn Simmons, Judy Slatopolsky, Elizabeth Smith, Rose-Lynn Sokol, Abby Soven, Joseph Stanton, Nancy Stearns, the Reverend Charles Straut, Frank Susman, the Reverend Bob Terry, Charlotte Thayer, Carolyn Thompson, Carlean Turner, the Reverend Foy Valentine, Albert Van Amberg, Kenneth Van Derhoef, Norma Vavra, John and Helen Ann Wagner, Loretto Wagner, James Wagoner, William Walden, Herbert Wechsler, Paul Weyrich, Miles Whitener, Art Widdicombe, Judy Widdicombe, Marcy Wilder, Buck Williams, John and Barbara Willke, Kathy Edwards Wise, Christine Wolf, Harriett Woods, George Wulff, the Reverend Evelyn Yeo, the Reverend Curtis Young.

Notes

CHAPTER ONE. CODES.

22 Medical detail and list of abortion methods principally from medical journal articles: Barno, Bartlett, Carney, Fox, Gornel, Hill, Lovitt, McBride, Ohio State Medical Association Committee on Maternal Health, Rosen, Schoeneck, Silver, Ronald Smith, Stevenson, Studdiford, Teeter, Wall, White, Zakin.

22 Description of markers left by specific methods: air embolism: Silver, p. 405; potassium permanganate craters like butter: Hill, pp. 747–51, Gerstley interview; odor of violets: Gornel, p. 168. Kansas University Medical Center abortions: Wall. Ohio 1955–59 deaths: OSMA Committee, pp. 257–58. California 1957–65 deaths: Fox.

23 Taussig numerical calculations: Taussig, pp. 24–28.

23 Calderone et al. studies: Calderone, *Abortion in the United States.* Other death and statistical estimates appear in Geijerstam, Bates and Zawadzki, and Lader, *Abortion.* Debate over the correct number of annual abortion deaths continued long after Calderone cited these National Center for Health Statistics figures. Although there appeared to be widespread consensus among legal-abortion advocates that the actual death rate was higher than three hundred per year, the lack of reliable national reporting obliged researchers to rely on statistical extrapolation or educated guesses, most of which estimated the national death rate before the mid-1960s at one thousand to eight thousand per year.

24 "Abortion is no longer a dangerous procedure": Calderone, "Illegal Abortion as a Public Health Problem."

24 "preventable disease": personal correspondence, 1954, from Calderone to Dr. Christopher Tietze, archived at the Schlesinger Library.

28 "DECISION . . . IN THE WRONG HANDS": St. Louis County Medical Society *Bulletin*, unsigned editorial, Oct. 21, 1966, p. 10.

33 "Twenty-one Protestant ministers . . .": Edward B. Fiske, *New York Times*, May 22, 1967. (Moody recalls that the clergy service actually began with twenty-six pastors.)

CHAPTER TWO. THE WEDGE.

general newspaper references 1965–69: *St. Louis Post-Dispatch, St. Louis Globe-Democrat, San Francisco Chronicle, Los Angeles Times, New York Times, Denver Post, Charlotte News and Observer, National Catholic Reporter, Our Sunday Visitor, St. Louis Review, The Wanderer*.

40 "Catholic obstetrician and gynecologist, father of 12": Connie Rosenbaum, "Abortion: Modern Dilemma," *St. Louis Post-Dispatch*, March 21, 1971.

44 "Medically speaking . . .": Calderone, "Illegal Abortion as a Public Health Problem," p. 948.

44 1959 California hospital survey: Packer and Gampell.

44 Therapeutic abortion committee criticisms: ibid., Joffee, Reagan.

45 "Inevitably there will be strong feelings . . .": Model Penal Code, p. 156.

46 "greater than the alleged evil . . .": Williams, p. 233.

48 Debate over ALI proposal: 36th Annual Meeting, American Law Institute, Philadelphia, *Proceedings, 1959*, pp. 252–83; "ALI Audiovisual History No. 1," Herbert Wechsler (interview with Paul A. Wolkin, Aug. 30, 1990). © 1990 by the American Law Institute; Model Penal Code and Commentaries, the American Law Institute, 1980, pp. 368–444; Herbert Wechsler, "The Model Penal Code and the Codification of American Criminal Law," in Roger Hood, ed., *Crime, Criminology and Public Policy: Essays in Honor of Sir Leon Radzinowicz* (New York: Free Press, 1974), pp. 419–68.

49 Knox/Beilenson bill background: Jain and Hughes; general coverage in *San Francisco Chronicle* and *Los Angeles Times;* Beilenson interview.

49 "under-the-rug kind of subject": Edgington oral history interview with Beilenson.

49 Sherri Finkbine material: interview; general coverage *New York Times, St. Louis Post-Dispatch, Life* magazine; Sherri Finkbine memoir, "The Lesser of Two Evils," in Guttmacher.

51 Gallup poll results: Luker, p. 82.

53 "A group of San Francisco's most distinguished physicians . . .": David Perlman, "Doctors Reply in Abortion Dispute," *San Francisco Chronicle*, May 18, 1966.

55 DECISION . . . IN THE WRONG HANDS: *St. Louis County Medical Society Bulletin*, unsigned editorial, Oct. 21, 1966, p. 10.

56 "THE DECISION TO KILL" . . . "WHO WANTS IT?": Matt H. Backer (published letter), *Missouri Medicine* 64:479–80 (June 1967).

57 North Carolina background on Jones: Jain and Sinding, p. 15.

57 "I was writing papers": Kelly Gruson, "Abortion Law Challenged by Author of Bill," *The News and Observer*, May 18, 1970.

57 New York legislative hearing detail: Sydney H. Schanberg, "Abortion Hearing Is Marked by Bitter Clashes," *New York Times*, Feb. 9, 1967.

57 one of "the greatest speaking marathons": Jain and Hughes, p. 63.

57 "I cannot justify": Jerry Gilliam, "Reagan Says He Would Sign Bill on Abortion—'If,' " *Los Angeles Times*, May 24, 1967, pt. 1, p. 3.

58 "reform" defined: *Funk & Wagnalls Standard College Dictionary* (New York: Funk & Wagnalls, 1963).

59 "Abortion: Once a Whispered Problem, Now a Public Debate": Jane E. Brody, *New York Times*, Jan. 8, 1968.

59 The American Civil Liberties Union "civil right" resolution, December 3 and 4, 1967, reads: "The ACLU asserts as a civil right the right of a woman to have an abortion, that is: to have a termination of pregnancy during the period prior to the viability of the fetus, which the medical professional currently accepts as extending during the first twenty weeks of pregnancy, and the right of a licensed physician to perform an abortion, without the threat of criminal sanctions." A further motion asserted the civil right of a woman to terminate a pregnancy at *any* point for cases of endangerment to health, serious fetal deformity, rape or incest, or unmarried pregnant girls under fifteen.

59 Lamm complaints about Colorado bill: *New York Times* and *Denver Post* coverage; interview.

60 *Social Justice Review* citations: "hot irons": William Drake, M.D., "The Land of Lincoln's Free Men Confronts the Soviets," April 1967; "weeds of modern paganism": Harvey Johnson, *Warder's Review*, September 1967, p. 155.

CHAPTER THREE. HEAVEN AND EARTH.

68 Evans sermon: "The Problem of Unwanted Pregnancies," April 12, 1970. Evans' private papers.

70 "God has so constituted life": the Reverend Ken Gottman at "Freedom of Choice: The Ethics of Abortion" conference in Grand Rapids, Michigan, Sept. 17, 1989.

70 "Such a futuristic explorer": Moody, p. 6.

70 "Liberalization doesn't mean nuthin' ": Brownmiller, p. 31; reiterated in interview with Moody.

71 North Carolina 1968 hospital abortion data: Elizabeth Tornquist, "Abortion Business in N.C.," *North Carolina Anvil*, March 7, 1970; Pat Borden, "Our 'Liberal' Law Has Little Effect on N.C. Abortion," *Charlotte Observer*, Dec. 28, 1969.

71 "For every therapeutic": California Clergy Counseling Service for Problem Pregnancies internal memo, June 15, 1968.

73 "I am sure you have heard": personal letter, Lana Phelan to Dr. W. J. Bryan Henrie, June 7, 1979.

74 " . . . noses up our skirts": Bill Wagner, "National Association to Repeal Abortion Laws Forms Local Chapter After Californian's Rapid City Speech," *Rapid City Journal*, Sept. 26, 1969.

74 "ARE YOU PREGNANT?" advice on sterile technique: SHA flyers in personal Lana Phelan collection and Schlesinger Society for Humane Abortion archive.

75 "The third time I was in the San Diego area": Cheek, Maginnis oral history interview.

76 "He said, 'I told you not to get pregnant' ": Phelan interview. (Phelan repeated this account for an early 1990s documentary, and the manner in which she described the abortion on camera, repeating every detail nearly verbatim, made it apparent that she had told this story the same way many times.)

78 "my great Catholic uterus": Cheek, Maginnis oral history interview; repeated in interview.

79 "If it doesn't work, I'll get an abortion anyway": Alan Cunningham, "Woman Plans Abortion Test on Self," *Redwood City Tribune*, May 20, 1967.

80 "bunch of men deciding for a woman": Chesler, Myers oral history interview.

80 1967 psychiatrists' survey: Crowley and Laidlaw. See also Roy G. Smith et al. 1970 *Hawaii Medical Journal* article reporting, in a state with comparatively liberal attitudes toward abortion, that approximately one third of the surveyed physicians favored repeal of the state law, while more than half favored either a Model Penal Code system or a Model Penal Code system with *extra* controls built in.

81 questions from Washington conference proceedings: Cooke et al.

81 " . . . 'repeal' position was virtually ignored": Lader, *Abortion II*, p. 33.

81 Drinan argument and quote: Cooke, p. 57; amplified in Drinan, "The Inviolability of the Right to Be Born."

82 "Women suffer needless grief and pain": Lader, "The Scandal of Abortion Laws."

83 "Please call me collect in Houston" and other letter excerpts: Society for Humane Abortion archive files, 1966 and 1967, Schlesinger Library.

83 ASA objectives: Association for the Study of Abortion, Inc., founding flyers and newsletters, April 1965 (printed at the organization's formal name change from the Association for Humane Abortion).

84 "to protect itself against the invading horde": HBA press release, June 19, 1964, from the files of Ruth Proskauer Smith at Schlesinger Library.

84 " . . . pointless and bound to fail": personal memo, Lader to Lonny Myers and two other conference organizers, Sept. 3, 1968 (from Lader archive at New York Public Library).

85 "What Al kept doing was compromising": Chesler, Cook oral history interview.

85 "He kept adding things . . .": Chesler, *Woman of Valor*. Further citations principally from oral history, interview with Cook, Lader's *Abortion II*, and *New York Times* coverage of bill passage.

86 "tailored, matronly manner": Francis X. Clines, "Abortion Bill Advocate," *New York Times*, April 9, 1970.

86 "For many years I had been": Chesler, *Woman of Valor*.

87 "They asked about labor law changes": ibid.

88 "Abortion reform is something dreamed up by men": Friedan, *It Changed My Life*, p. 157.

88 "Without the full capacity": Lucinda Cisler, "Unfinished Business: Birth Control and Women's Liberation," in Robin Morgan, ed., *Sisterhood Is Powerful*, p. 274.

89 "At a certain stage": "Abortion Law Repeal (Sort of): A Warning to

Women," handout from New Yorkers for Abortion Law Repeal, reprinted from *Notes (from the Second Year): Radical Feminism.* Also chronicles, newsletters of the Committee for the Cook-Leichter Bill: newsletters from New Yorkers for Abortion Law Repeal.

90 "two-day crisis": Lader, *Abortion II*, p. 136.

90 "I said, 'Look, it's this or nothing' ": interview.

90 " 'I hope you win' ": interview.

91 "I continue to believe very strongly": *New York Times,* Jan. 8, 1970.

92 " 'Murderer,' she called out": Bill Kovach, "Abortion Reform Is Voted by the Assembly, 76 to 73," *New York Times,* April 10, 1970.

92 "My own son": *New York Times;* also Lader, *Abortion II.*

93 "I was deeply grateful": interview.

CHAPTER FOUR. THE WRITING IN THE HEART.

96 "The Mad Scramble for Abortion Money": Howard Eisenberg, *Medical Economics,* Jan. 4, 1971, p. 35.

96 Women's Services numbers: Nathanson, *Aborting America.*

97 New York City resident/nonresident 1970 abortion figures: Pakter et al., pp. 48–50.

97 Hawaii residency requirement detail: Steinhoff and Diamond, p. 180. Steinhoff and Diamond, both University of Hawaii professors at the time of the passage of the 1970 law, reported that hospitals initially required that each patient's sworn affidavit be signed off by a notary public. Nevertheless, they reported, between 8 and 13 percent of abortions in Hawaii were performed on nonresidents until the New York law went into effect in July 1970, after which the Hawaiian nonresident abortion rate dropped to one percent.

98 *Linacre Quarterly* articles: John B. Gent, "Proposed Abortion Laws: 'Slaughter of the Innocents,' " February 1969, p. 47.

98 Eugene F. Diamond, M.D., "The Physician and the Rights of the Unborn," May 1967, p. 174.

98 Diamond, "Who Speaks for the Fetus?" February 1969, p. 58.

98 N. M. Camardese, M.D., "Man Plays God," February 1968, p. 42.

98 "prenatalism": Grisez, p. 470.

98 *every* distressed pregnant woman: Baars, p. 261.

100 "some mothers don't want a new baby": Willke, *The Wonder of Sex,* pp. 26–27.

104 Willke slide show guidelines: Willke, *How to Teach the Pro-Life Story,* pp. 9–10.

107 *Humanae Vitae* line ("life must be guarded"...): Flannery, Vol. 1, p. 995.

108 *"Expose the tax-deductible lobbying efforts":* minutes of NARAL annual meeting, September 1970, Boulder, Colorado.

110 adult Baltimore Catechism reference: Confraternity of Christian Doctrine, *This We Believe/By This We Live, Revised Edition of the Baltimore Catechism* (Patterson, N.J.: St. Anthony Guild Press, 1957), p. 206.

111 "One of the most basic of human rights": Bourke, p. 351.

113 changing Catholic abortion doctrine: Published studies include many

scholarly examinations of evolving Church teaching: Callahan, Connery, Dailey, Grisez, Hurst, Maguire, Noonan ("An Almost Absolute Value in History," in his *The Morality of Abortion: Legal and Historical Perspectives*).

114 ectopic pregnancy notes: Lynch, pp. 9–10.

116 The "written in the heart" reference appears to have been lifted from the words of Paul in Romans 2:14–15: "For when the Gentiles who do not have the law by nature observe the prescriptions of the law, they are a law for themselves even though they do not have the law. They show that the demands of the law are written in their hearts, while their conscience also bears witness . . ." *(New American Bible for Catholics* [Iowa Falls: World Bible Publishers, 1970]). Backer felt satisfied by his friend's explanation and did not press him as to precisely where it originated.

116 " . . . natural means of fertility control": Rock, p. 167.

117 "responsible parenthood": *Humanae Vitae* (Encyclical Letter on the Regulation of Births), translation by Vatican Press Office, July 25, 1968. Reprinted (with slightly different translation than the one cited here, which appeared in July 1968 in the archdiocesan newspaper the *St. Louis Review*) in Flannery, Vol. 2, pp. 397–416.

CHAPTER FIVE. EXHIBIT A.

125 *"Arizona's* bill": *NARAL Newsletter,* July 1970.

126 Plaintiff's brief from *Rodgers* v. *Danforth,* Petition for Declaratory Judgment and Injunctive Relief, Circuit Court of St. Louis County, Oct. 2, 1970. (Gwyndolyn Harvey's name was added to the Circuit Court case; in the initial U.S. District Court case, filed five months earlier, Planned Parenthood of Kansas City president Jean Green was the third "married woman of child-bearing age." Green's name did not appear on the Circuit Court suit.)

133 "drastically interferes with a woman's right . . .": *The People of the State of California* v. *Leon Phillip Belous,* California Supreme Court, *Amicus Curiae* Brief on Behalf of Medical School Deans and Others in Support of Appellant, Zad Leavy et al., attorneys for *amici curiae,* p. 21.

133 "common intelligence must necessarily guess": *Connally* v. *General Construction Co.,* 269 U.S. 385 (1926).

134 "Moreover, a definition requiring certainty of death": *People* v. *Belous,* 80 Cal. Rptr. 354 (1969).

134 "This is the first State Supreme Court decision": Oct. 28, 1969, letter from Dr. Keith Russell to signers of the *Belous* physicians' brief (from personal files of Russell).

135 "slow, expensive, and possibly disastrous": Clark, p. 9.

136 "very idea is repulsive" and other excerpts: *Griswold* v. *Connecticut,* 381 U.S. 479 (1965).

137 "You could really publicize that": Cheek oral history interview with Griswold.

137 "their right to enjoy the privacy of their marital relations . . .": *Poe* v. *Ullman,* 367 U.S. 497 (1961).

138 "partially shaded region": Greely, p. 251.

139 "The values implicit in the Bill of Rights . . .": Lucas, pp. 761 and 759.

142 Details on Hodgson case: interview; Hodgson oral history in Irons, pp. 255–79.

143 Details on preparation of *Roe* and conflicts with Roy Lucas: Weddington, Garrow, Faux, Hames interview.

151 Wisconsin ruling quote: *Babbitz v. McCann,* 310 F. Supp. 293 (1970); Louisiana ruling quote: *Rosen v. State Board of Medical Examiners,* 318 F. Supp. 1217 (1970); Vuitch: *United States v. Vuitch,* 402 U.S. 62 (1971).

152 "a procedure so cumbersome": *Doe v. Bolton,* appellants' brief, Aug. 6, 1971.

156 "We continue to maintain . . .": Pilpel, p. 2.

157 Illinois opinion: *Doe v. Scott,* 321 F. Supp. 1385 (1971). Florida opinion: *Florida v. Barquet,* 262 So. 2d 431 (1972). Vermont opinion: *Jacqueline R. v. Leahy,* 130 Vt. 164 (1972). Kansas opinion: *Poe v. Menghini,* 339 F. Supp. 986 (1972). (The ruling voided that portion of Kansas' Model Penal Code law requiring that abortions be restricted to hospitals accredited by the Joint Commission on Accreditation of Hospitals, which at the time of the ruling accredited only 80 of 169 hospitals in Kansas. "So substantial a limitation of facilities," wrote the federal court, "constitutes a significant encroachment upon the exercise of a fundamental right.")

158 "I could sit here and not recall": Susman speech, March 1972, from papers of Dr. John Vavra.

159 "without merit," "the issues in this case," etc.: *Rodgers v. Danforth,* 486 S.W. 2d 258 (1972).

160 "The reasoning of this decision": letter, Susman to Samuel Rodgers and other plaintiffs, Oct. 13, 1972.

CHAPTER SIX. THE RULING.

161 "The Supreme Court barred the states": *St. Louis Post-Dispatch,* Jan. 22, 1973 (afternoon edition story from first wire service reports).

163 "1. A state criminal abortion statute" and all subsequent *Roe* excerpts: *Roe v. Wade,* 410 U.S. 113 (1973).

164 "The medical judgment": *Doe v. Bolton,* 410 U.S. 179 (1973).

167 Background on *Roe* court deliberations: The Supreme Court justices' deliberations and evolving positions in *Roe* and *Doe* are described in well-documented detail in several recent books about the Court. Woodward and Armstrong's *The Brethren* provided the earliest insider account of the deliberations; *The Brethren* included no published source notes, but more fully documented accounts, drawing on material in judicial papers that had not yet been made available to researchers in 1979, can be found in Garrow, in James Simon's *The Center Holds: The Power Struggle Inside the Rehnquist Court* (New York: Simon & Schuster, 1995), and in John J. Jeffries, Jr.'s *Justice Lewis F. Powell, Jr.: A Biography* (New York: Scribner, 1994).

167 "I have seen them operate": quoted by Garrow, p. 550.

172 "together with our observation": In a footnote, Blackmun observed that the Texas state attorneys faced a logical dilemma in arguing that Texas could extend Fourteenth Amendment protection to fetuses. The very law the attorneys were defending suggested otherwise, Blackmun wrote. For example, the law, like

those of other states, contained built-in exceptions: Texas made abortion legal to save the life of the mother—without offering the fetus due process of law, which would seem to be required if the fetus really were a protected person.

175 "Accustomed as they were": Luker, pp. 140–41.

176 "The Supreme Court's decision today": New York Times, Jan. 23, 1973.

176 "I think it's fantastic": Les Pearson, "Abortion Ruling Draws Cheers and Dismay Here," St. Louis Globe-Democrat, Jan. 24, 1973.

176 "A few unresolved issues": John P. MacKenzie, "Supreme Court Allows Early-Stage Abortions," Washington Post, Jan. 23, 1973.

176 "In a historic resolution": Warren Weaver, Jr., "High Court Rules Abortions Legal the First 3 Months," New York Times, Jan. 23, 1973.

177 St. Louis Review stories: "ABORTION—The Battle Lines Are Drawn," Feb. 9, 1973.

177 "If I had been alive in Nazi Germany": "Rep. Hogan Begins Fight on Abortion," Washington Post, Jan. 31, 1973.

178 Hogan wire service story: "Antiabortion Move Proposed," UPI story in New York Times, Feb. 1, 1973.

184 Sanger birth control battles: Chesler, Woman of Valor, pp. 313–35.

184 "both Catholic and American": Byrnes, p. 38.

185 "In wonder at their own discoveries": Flannery, Vol. 2, p. 904.

186 "sum total of social conditions": ibid., p. 927.

188 "The U.S. Supreme Court": W. Barry Barrett, "High Court Holds Abortion to Be 'a Right of Privacy,' " news release from the Baptist Press, Jan. 31, 1973.

188 "I refute it thus": James Boswell, The Life of Samuel Johnson (New York: Penguin, 1979), p. 122.

CHAPTER SEVEN. THE TARGET.

196 "Next January the Legislature": Greenhouse, p. 27.

197 Washington, D.C., doctor story: Dr. Peter Forbes interview.

197 early published vacuum aspirator reports: Hodgson, pp. 225–77; Kerslake and Casey; Chernyak.

197 Chinese abortion report footnote: "Suction 'Curettage' for Artificial Abortion: Preliminary Report of 300 Cases," China Journal of Obstetrics, 6:26 (1958). (Translated summary from af Geijerstam, 1969.) Other vacuum studies appeared during the early and middle 1960s in medical journals from Czechoslovakia, Denmark, Germany, Hungary, Italy, Poland, the U.S.S.R., and Yugoslavia.

200 "To cope with even 50,000 abortions": Greenhouse, p. 27.

201 Nathanson memorandum: "The Nathanson Proposals for NARAL," unpublished NARAL memo, 1970, from Lader papers.

201 "There are 22,000 OBG men in this country": Julienne, p. 93.

203 "Raped by Medical Bureaucracy" sign and other details of clinic vs. hospital conflict: Lader, Abortion II, pp. 149–57; also Greenhouse, Pakter, New York Times daily coverage.

203 "Of the some 26-odd clinics": Hausknecht, p. 986.

206 "the rest of us milled through the audience": Ain't I a Woman, July 10, 1970.

207 Our Bodies, Ourselves intro: p. 13, 1979 Simon & Schuster edition, preface (lifted from the 1973 edition).

209 "Now you are in the psychiatrist's office": Maginnis and Phelan, p. 112.

209 Karman and "supercoil" method: Berger et al.

210 "What man would have to spend": Ruzek, p. 58. (Ruzek notes that a colleague, Coleen Wilson, was arrested with Downer, but pleaded guilty to fitting a diaphragm and was fined $250 and placed on two years' probation.)

211 Rothman patent application: Patent #3,828,781, "Method for Withdrawing Menstrual Fluid," granted Aug. 13, 1974.

212 forms of vacuum aspiration, including milking machines: Gage, pp. 14–18.

213 "I would still hesitate": Frankfort, p. 226.

214 "We believed that information was power": Fried, p. 99. Other "Jane" info: Bart; Diane Elze, "Underground Abortion Remembered," reprinted from *Our Paper*, Maine, in *Sojourner*, Cambridge, Massachusetts, April and May 1988; Laura Kaplan interview.

215 "The clinic was paying physicians": Nathanson, *Aborting America*, pp. 112–13. According to Nathanson, the clinic's formal name was Center for Reproductive and Sexual Health. But it was widely known as Women's Services.

221 "miserable and frightening consequences": Breuer, p. 69. Other Sunnen background from Minkler; Boulware and Canfield interviews; and memos, NARAL minutes, and letters from Lader papers, including April 13, 1970, letter from Lader to the Hopkins Fund requesting funding and describing the Sunnen Foundation as so far the largest single contributor to the effort to open a NARAL New York abortion clinic. Emko newsletter quote from Dec. 1, 1964, issue; in Lader's papers.

PART TWO. A HOUSE AFIRE.

CHAPTER EIGHT. THE EDUCATION OF THE PHARISEE.

238 "No amount of statistical calculation": testimony of United States Catholic Conference on constitutional amendments protecting human life before the Subcommittee on Civil and Constitutional Rights of the House Committee on the Judiciary, March 24, 1976, p. 15.

239 "Is it morally incumbent": Thompson, p. 45.

243 *"18–24 days Has a heartbeat":* Many of these details of timing, including the very early onset of the heartbeat, were reported in "Life Before Birth," the April 30, 1965, *Life* article featuring the Lennart Nilsson pictures. The vocabulary of this right-to-life flyer was less nuanced than the article's; for example, the three-and-a-half-week embryo photographed in *Life* was described in the magazine's text as possessing "the *beginnings* of eyes, spinal cord, nervous system," etc. Other assertions from the flyer were probably culled from "The Early Biography of Everyman," an extensively footnoted essay by the Chicago physician and right-to-life leader Bart T. Heffernan, included in the 1972 anthology *Abortion and Social Justice.*

246 "no rights which the white man": *Dred Scott v. Sanford*, 19 How. (60 U.S.) 393 (1857).

247 "Was it surprising": Willke, *Abortion and Slavery*, pp. 12–13, 21.

248 details on 1975 Rockville sit-in: private communication, Cavanaugh-O'Keefe; and Joe Green, "Six Md. Abortion Protesters Convicted of Trespassing," *Washington Post*, Sept. 9, 1975.

248 "Protestors, Trial in Cleveland": *National Right to Life News,* February 1977.

250 "revolution by the Have-Nots": Alinsky, p. 9.

250 "Self-exposure to the elements" and other excerpts: Sharp, p. 315 and table of contents.

255 Ann O'Donnell stories: Boyce and Bannon interviews.

• 261 *Post-Dispatch* coverage of convention: James E. Adams, "Dilemma of the Pro-Life Activists," *St. Louis Post-Dispatch,* July 3, 1978.

261 Cleveland clinic firebombings: regional and wire newspaper coverage generally from February 1978, e.g.: "No Fire Link Found to Abortion Foes," *Cleveland Press,* Feb. 21, 1978.

CHAPTER NINE. A LESSER OF EVILS.

270 "This summer I approached": Sam Lee, *Circle of Life,* newsletter of St. Louis University Pro-Life, Dec. 11, 1978.

274 S.O. flyer: *Involving Significant Others: A Health Rights Advocacy Program Guide,* published by RHS in cooperation with the National Abortion Federation, 1979.

278 "Abortion is a real feminist issue": Elaine Viets, "The Pro-Abortion Conscience," *St. Louis Post-Dispatch,* April 15, 1974.

279 "Minors, as well as adults": *Planned Parenthood of Missouri v. Danforth,* 428 U.S. 52 (1976).

281 "Stripped of all rhetoric": *Poelker v. Doe,* 515 F2d 541 (1975).

282 "We certainly are not unsympathetic": *Mather v. Roe,* 432 U.S. 464 (1977).

282 "only where an abortion is medically indicated": Missouri Code of State Regulation, 13 C.S.R. 40-81.100.

283 Akron ordinance detail and letters-to-editor page expanded: Reginald Stuart, "Akron Divided by Heated Abortion Debate," *New York Times,* Feb. 1, 1978.

284 "If this legislation": letter, Eileen Roberts to Judy Widdicombe, Sept. 18, 1978.

285 "Please return this information": confidential Missouri Catholic Conference memorandum on H.B. 902, 1974.

288 *Chicago Sun-Times* "Abortion Profiteers" series: Nov. 12–28, 1978.

289 "Table 2 summarizes": Goldstein, "Abortion as a Medical Career Choice," p. 216.

295 saline detail: Kerenyi interview; Kerenyi, "Midtrimester Abortion," chapter 18 in Hodgson; Hern, "First and Second Trimester Abortion Techniques"; Speroff; Wang.

296 New York saline data: Pakter et al., in Osofsky, p. 53.

296 *live baby:* Jeffries and Edmonds, p. 13.

297 "The State . . . would prohibit": *Planned Parenthood of Missouri v. Danforth,* 428 U.S. 52 (1976).

300 D and E detail: Hern, "First and Second Trimester Abortion Techniques," "Midtrimester Abortion"; Kaltreider et al.; Peterson et al.; Speroff; P. G. Stubblefield, "Midtrimester Abortion by Curettage Procedures: An Overview," chapter 11 in Hodgson.

301 "visual impact": Hern, "Midtrimester Abortion," p. 416; "physician who experiences much of the anguish": Stubblefield, "Midtrimester Abortion," p. 293; "Physicians reported": Kaltreider, p. 235.

304 "detrimental effect on the health interests of women": *Planned Parenthood Association of Kansas City, Missouri v. Ashcroft,* 483 F. Supp. 679 (1980).

CHAPTER TEN. A HOUSE AFIRE.

311 "No person, without lawful authority": s. 795.010, Revised Code of the City of St. Louis (1961).

314 "The veneer of scholarship": Byrn, p. 859.

314 "*Roe* v. *Wade* seems like a durable decision"; "I do wish 'Wolf!' hadn't been cried"; "The problem with *Roe*"; "Dogs are not 'persons' ": Ely, pp. 947, 932, 943, 926.

315 *Lochner* v. *New York:* 198 U.S. 45 (1905).

317 "The Court's opinion in *Roe*"; "Striking the necessary balance": Heymann and Barzelay, pp. 765, 775–76.

317 "Nobody *has* to kill": Tribe, pp. 114–15.

319 "We hope the police": "Anti-Abortion Protestors Are Arrested at Clinic," *St. Louis Globe-Democrat,* Nov. 19, 1979.

324 "To facilitate the large number": Kleinknecht and Mizell, p. 20.

326 "The Pastoral Plan seeks to activate": *Pastoral Plan for Pro-Life Activities,* promulgated by National Conference of Catholic Bishops, Dec. 5, 1975.

329 "The fate of this nation": Carolyn Gerster, "From the President's Desk," *National Right to Life News,* January 1980.

330 "If we can do it in Iowa"; "We go to *all* churches": audiotaped proceedings, 1979 National Right to Life Committee convention.

333 "In that respect you are not a 'one-issue' voter": *Missouri Citizens for Life News,* September 1980.

336 "disappointed," "duped", etc.: *National Right to Life News,* July 21, 1980.

337 "Never before has the cause you espouse": "Dear Friends" letter from Ronald Reagan to Dr. J. C. Willke and The National Right to Life Convention 1980, June 26, 1980, reprinted in the *National Right to Life News,* July 7, 1980, and the *Missouri Citizens for Life News,* September 1980.

340 "obstetricians in the same hospital . . .": Koop, *The Right to Live; the Right to Die,* p. 17.

340 "Late that evening . . .": ibid., p. 267.

341 "The only way we know . . .": Schaeffer and Koop, pp. 189–90.

342 "the only source . . .": Bosgra, p. vii.

343 "I feel that the dignity of life . . .": "An Interview with the Lone Ranger of American Fundamentalism," *Christianity Today,* Sept. 4, 1981. (These remarks, according to Falwell's autobiography and other accounts, accurately reflect Falwell's views two years earlier.)

343 "Attention to so-called social issues . . .": "The New Right: A Special Report," *Conservative Digest,* June 1979.

345 "to fight together . . .": Falwell *Christianity Today* interview.

346 "All across the land": Falwell, p. 362.

347 "able simply to wave his hand": Hadden and Shupe, p. 174.

348 "If this situation is allowed to continue" and subsequent excerpts: Higgins, pp. 107–10.

350 "PROLIFE GAIN": *National Right to Life News,* Nov. 10, 1980.

353 "The basic idea": J. P. McFadden, *Lifeletter,* Jan. 30, 1981.

354 "informed by Congress's judgment": Galebach, p. 24.

358 "The Catholic *apparat*": personal correspondence from J. P. McFadden to colleague, Oct. 2, 1981.

358 "PRO-CHOICE AMENDMENT": Kansas Right to Life "Bulletin—Alert," November 1981; *"Who among us is willing . . . ?"*: *PHL-SW News,* Pennsylvanians for Human Life, Southwest Region, January 1982; "destructive activism": "Dear Pro-Life Leader" open letter, Aug. 4, 1981, Curtis J. Young, Christian Action Council; lemmings on the march: Mary Meehan, "Lemmings to the Sea," *National Catholic Register,* Jan. 18, 1982

359 *"Isn't it a twisted logic"*: Willke, *Handbook on Abortion,* p. 36.

360 1979 Gallup poll: Feb. 23, 1979–Feb. 26, 1979, poll of 1,534 people conducted for the Roper Center for Public Opinion Research, University of Connecticut. *New York Times* poll: E. J. Dionne, Jr., "Abortion Poll: Not Clear-Cut," *New York Times,* Aug. 18, 1980. Similar data on polling results from numerous Gallup and other polls; also Blake, "Elective Abortion and Our Reluctant Citizenry: Research on Public Opinion in the United States," in Osofsky; Blake, "Negativism, Equivocation, and Wobbly Assent"; Raymond J. Adamek, study conducted for National Right to Life Educational Trust Fund.

362 "The average politician": Bill Peterson, "Worries for New Right," *Washington Post,* Feb. 16, 1982.

363 "We are in the situation": John T. Noonan, "Protect Life Amendment: Questions, Answers, and Axioms," open memorandum, September 1981.

364 "a compromise of the very principle": open letter, National board members to Willke, Jan. 29, 1982.

364 "What are we doing to ourselves?": Feb. 12, 1982, Oklahomans for Life memo, Anthony J. Lauinger to directors and state officers of National Right to Life Committee.

369 "In *Roe* v. *Wade*": *City of St. Louis* v. *Klocker,* No. 44077, June 1, 1982.

369 "I thought that breaking the injunction": Andrews, p. 86.

PART THREE. THE SEVENTH QUESTION.

CHAPTER ELEVEN. THE LITMUS TEST.

395 "On October 27" and other bombing/arson data: National Abortion Federation member alerts 1984–86.

397 "the child is extremely agitated": full text of film in *The Silent Scream* (Anaheim, CA.: American Portrait Films Books, 1985).

398 "mouth of the fetus": *THE FACTS SPEAK LOUDER: Planned Parenthood's Critique of "The Silent Scream,"* Planned Parenthood Federation of America brochure, 1985.

399 "Do they think the fetus . . . ?": Ellen Goodman, "Forgotten in Abortion Debate," *Boston Globe,* Jan. 31, 1985; remark repeated in Falkenberg interview.

400 "meeting shock": January 1973 Lader memos to NARAL board members, responding to political use and December 1972 television broadcast of "Catholic 'horror' fetus pictures" (from NARAL files).

400 "Whores." "Tramps," etc.: excerpts from letters in NARAL file.

402 "Change the *message*": memo, Falkenberg to NARAL affiliates, Feb. 27, 1985.

403 "It was an evening": Rusty Brown, published column, March 25, 1985, newspaper unknown (clipping from papers of Judy Widdicombe).

405 RU-486 material: Raymond Baulieu et al., *The Case for Antiprogestins*, report of the Reproductive Health Technologies Project, Washington, D.C., 1991. (The report, which critiques RU-486 as an understudied and possibly dangerous medical intervention, urges a more cautious approach to pharmaceuticals and advocates instead expanding the pool of those permitted to perform vacuum abortions—for example, training and licensing paramedics.)

411 "I looked up from whatever I was doing": Tom Blumenthal, "Bad Fiction: A True Account of Webster v. R.H.S." (unpublished manuscript, 1991), p. 10.

CHAPTER TWELVE. THE COUNT.

416 "We're planning on appealing . . .": "Abortion Foes Begin Jail Terms for Contempt," *St. Louis Globe-Democrat*, June 9, 1983.

418 "*If you remain indifferent*": *The New American Bible for Catholics* (Iowa Falls: World Bible Publishers, 1970).

419 "As a matter of strategy": Walker and Puzder, p. 239.

423 "It is truly surprising": *Planned Parenthood* v. *Danforth*, 428 U.S. 52 (1976).

423 O'Connor background and voting record testimony: Savage, pp. 113–15; *New York Times* general coverage of O'Connor nomination, especially Linda Greenhouse, "O'Connor Hearings Open on a Note of Friendship," Sept. 10, 1981.

424 "pro-abortionist": John Cavanaugh-O'Keefe, "O'Connor Memo Is Deceptive," *National Right to Life News*, July 12, 1981.

424 "fallacy inherent in the *Roe* framework": *City of Akron* v. *Akron Center for Reproductive Health*, 462 U.S. 416 (1983).

426 "numbers that really count": Douglas Badger, quoted in Thomas D. Brandt, "Abortion Foes Prepare for Next Round in Battle," *Washington Times*, June 28, 1983.

431 "If American crematoriums . . .": Richard Cowden Guido, "Morality and the Pensacola Four," *The Wanderer*, Jan. 17, 1985. (Other articles of similar tone and argument appeared in *The Wanderer*, particularly after the Pensacola bombings and arrests, e.g.: "The Bomber as Victim," Jan. 24, 1985; "Marylander Found Guilty in Abortuary Bombings," May 30, 1985; "Abortuary Bombing and the Justification Defense," June 27, 1985.)

432 "citizen's arrest on a mass murderer": Charles Bosworth, Jr., Zevallos kidnap coverage, *St. Louis Post-Dispatch*, July 31, 1983.

432 "While we might respect the zeal": Scheidler, p. 298.

433 Ryan's Appleton remarks recorded on film in the documentary *Holy Terror*, Victoria Schultz and Nancy Kanter, dirs., Hudson River Productions (Helsinki Films, 1986).

435 "Urological Complications," etc.: Hilgers et al., *New Perspectives on Human Abortion.*

436 "That historic decision . . .": Rosenblum, p. 38.

437 "They accepted the current state . . .": Richard S. Myers, "Prolife Litigation and Civil Liberties," in Horan et al. p. 25. This paper was first prepared in a 1984 version co-authored by Kenneth F. Ripple, then a law professor at the University of Notre Dame.

438 "The collapse and reversal": Victor Rosenblum and Thomas Marzen in Horan et al., p. 195.

440 "In the years since this Court's decision": *Thornburgh* v. *American College of Obstetricians and Gynecologists* 476 U.S. 747 (1986).

441 "not just wrong turns": Brief for the United States as *Amicus Curiae* in Support of Appellants, p. 2. Fried's brief was submitted in two Supreme Court cases simultaneously: *Thornburgh* v. *A.C.O.G.* and *Diamond* v. *Charles,* a challenge to certain provisions of the Illinois abortion law.

444 "Personhood is our absolute" and "We testified, we prayed . . .": Brown, pp. 72 and 68.

447 *Wall Street Journal* item on Gephardt switch: quoted in full in Jo Mannies, "Gephardt Still Anti-Abortion, Staffer Says," *St. Louis Post-Dispatch,* May 4, 1986.

449 "a significant barrier to a woman's right": *Reproductive Health Services, et al.,* v. *William L. Webster, et al.,* No. 86-4478-CV-C-5, U.S. District Court, W.D. Missouri, C.D., March 17, 1987 (as amended April 30, 1987).

CHAPTER THIRTEEN. THE SEVENTH QUESTION.

450 DEPARTURE IS HAILED, MOURNED: *St. Louis Post-Dispatch,* June 27, 1987.

455 Bork observations generally from *New York Times, St. Louis Post-Dispatch, Kansas City Star,* and *National Right to Life News* coverage of nomination and subsequent three months; additional detail from Pertschuk and Schaetzel. Bork's *Roe* criticisms were reprinted in various forms in many publications: 1981 "unconstitutional decision" remark came during Bork's testimony on the Human Life Bill and appeared, among other places, in Stuart Taylor, Jr., "A Committed Conservative: Robert Heron Bork," *New York Times,* July 2, 1987. "The deformation of the Constitution": statement of Professor Robert Bork, Yale Law School, before the Commitee on the Judiciary, United States Senate, during hearings on S. 158 (Washington, D.C.: U.S. Government Printing Office, 1982).

459 Nathanson invention of statistics: Nathanson has repeated this assertion many times (the declaration that the figures were inflated is borne out by national health statistics), and writes about it in *Aborting America,* p. 193: "I confess that I knew the figures were totally false, and I suppose the others did too if they stopped to think of it."

462 "We have sinned as Christians" and other Terry remarks: Randall A. Terry, *Higher Laws,* 1987–88 pamphlet reprinted from article in *The Rutherford Institute Magazine,* March–June 1987; also Operation Rescue training videos, 1987–89.

464 "Anti-abortion forces": Michael D. Sorkin, "8th Circuit Finds Law on Abortion Unconstitutional," *St. Louis Post-Dispatch,* July 14, 1988.

479 "Throughout the night they paced": Charlotte Grimes, "Before the Hearing, a Truce," *St. Louis Post-Dispatch*, April 27, 1989.

CHAPTER FOURTEEN. ZEALOTS.

487 "We have to realize": Martha Shirk and Tom Uhlenbrock, "Abortion Is 'Vietnam of '90s,' Group Says," *St. Louis Post-Dispatch*, July 4, 1989.

487 "For the first time": William H. Freivogel, "Ruling Ends Court's 16-Year Practice," *St. Louis Post-Dispatch*, July 5, 1989.

488 "It will be time enough" and subsequent ruling excerpts: *Webster* v. *Reproductive Health Services*, 492 U.S. 490 (1988).

494 "We've gotten this far" and other remarks from convention: National Right to Life Committee 1989 convention recordings.

496 "Until the Court ruled": Richard L. Berke, "The Abortion Rights Movement Has Its Day," *New York Times*, Oct. 15, 1989.

496 "As of Friday . . .": Robert L. Mauro, "GOP Adviser: 'Anti-Abortionists Stunningly Inept,'" *The Wanderer*, Oct. 26, 1989.

496 "The most important thing . . .": internal National Right to Life Committee memo, June 1989.

498 *Pro-Abort AG . . .* : David Andrusko, *National Right to Life News*, Nov. 30, 1989.

514 "ABORTION BATTLE AT KEY POINT": Martha Shirk, *St. Louis Post-Dispatch*, Feb. 11, 1990.

EPILOGUE

519 "essential holding" and other excerpts: *Planned Parenthood of Southeastern Pennsylvania* v. *Casey*, 112 Sup. Ct. 2791 (1992).

525 "The disparaging labels": B.J. Isaacson-Jones, et al., "Common Ground," *St. Louis Post-Dispatch*, June 9, 1991.

Index

abolitionism, 247, 361, 367
abortion:
  as civil rights issue, 242–43
  facilities for. *See* abortion clinics
  fetal defects and. *See* fetal deformity
    abortions
  as human rights issue, 163, 191, 242–
    243
  illegal. *See* illegal abortions
  lawsuits and. *See* abortion litigation
  legislative definition of, 165
  opinion surveys on, 51, 345, 359–61,
    510–11
  opposition to. *See* right-to-life
    movement
  photographs of. *See* abortion pictures
  to preserve health of woman, 43–45,
    47–48, 56, 58, 60, 151, 164, 172n,
    361, 499–500. *See also* mental
    health abortions; therapeutic
    abortions
  procedures for. *See* dilation and
    curettage; dilation and evacuation;
    dilation and extraction;
    hysterotomy; saline abortions;

    suction abortions; vacuum
    aspiration
  protests against. *See* protests,
    antiabortion; sit-ins, at abortion
    clinics
  public funding of. *See* public funding
    of abortion
  safety of, 170
  statutes on. *See* abortion statutes
  therapeutic. *See* therapeutic
    abortions
  ultrasound videotape of, 396–99
  woman's fundamental right to, 129–
    134, 139–41, 143, 160–76, 282,
    316–17, 356, 357, 466, 520
Abortion and Slavery (Willke), 247
abortion clinics, 451
  "abortion mill" imagery and, 200–201,
    202
  bombing and arson attacks on, 261,
    274, 374n, 394, 395–96, 431–32,
    452–53, 521
  economic concerns and, 96–97, 215–
    216, 289–90, 291
  FACE and, 521–22

abortion clinics, *continued*
  hospitals-only requirements and, 199, 200, 202–3
  kidnap of doctor from, 373–75, 376, 432
  male domination in, 214–15
  organization for. *See* National Abortion Federation
  protests at, 225–26, 240, 252, 394–95. *See also* sit-ins, at abortion clinics
  requirements for, 195–96
  RU-486 and, 405–6
  shootings at, 520–21
  substandard, 288–89, 291, 292
  vacuum aspiration technology and, 197–200
  see also Reproductive Health Services
abortionists:
  in era of illegal abortions, 27–28
  kidnap of, 373–75, 376, 432
  murders of, 520–21
  as plaintiffs in lawsuits, 130–34, 141–144
  stigma attached to, 202, 216, 218
  underground referral systems and, 29–37, 62–70, 122–26
abortion litigation, 120–21, 126–78, 388
  Akron, 410, 411, 421, 424–26, 438, 440, 456, 457
  Ashcroft, 285–88, 292–93, 299, 301–4, 407, 410, 421–22, 439
  AUL strategy for, 435–42
  Beal v. *Doe,* 282, 421, 439
  Bellotti v. *Baird,* 421
  Belous, 130–34, 135, 141, 153
  Casey, 518–20
  and changes in medical practice and technology, 424–25
  Colautti v. *Franklin,* 439
  common law and, 168–69, 314
  Connecticut v. *Menillo,* 421
  courts chosen for, 143–44
  Danforth, 278–79, 284, 297, 421, 423, 426, 439, 440
  declaratory judgment suits, 142–43
  Doe. See Doe v. *Bolton*
  financing of, 221, 222, 393, 407, 409
  gender of attorneys in, 154, 155
  guardians ad litem in, 120–21, 145
  Harris v. *McRae,* 421, 481
  Maher v. *Roe,* 282, 421
  national coordination of, 155–57
  plaintiffs chosen for, 141–43, 144
  Poelker v. *Doe,* 281–82, 421, 439

  privacy rights and. *See* privacy rights
  rational basis test and, 466–67
  Rodgers v. *Danforth. See Rodgers* v. *Danforth*
  Roe. See Roe v. *Wade*
  Simopolous v. *Virginia,* 421–22
  Supreme Court decisions and, 161–78, 188, 190, 191, 193–95, 238, 244–46, 248, 252, 278, 279, 282, 283, 297, 333, 335, 339–40, 346, 353, 369, 407–11, 413, 419–26, 437–41, 451, 455–56, 518–20
  Thornburgh, 439–41, 456, 457
  trimester approach and, 163–64, 173–174, 314, 408–9, 438, 424–25, 466, 482, 519
  "undue burden" test and, 425, 426, 519, 520
  vagueness doctrine and, 133–34, 135
  Vuitch, 141–42, 151, 153, 154, 164
  Webster. See Reproductive Health Services v. *Webster; Webster* v. *Reproductive Health Services, et al.*
  see also specific issues and states
"abortion mill," use of term, 200–201, 202
abortion pictures, 99–106, 162, 163, 172, 177, 180, 241, 242, 243, 254, 257, 261, 277
  dead-woman photos, 400–401, 402
  evangelicals and, 342
  NARAL's response to, 399–400
  Nilsson's *Life* photos, 101, 179, 188–189, 332
  The Silent Scream and, 396–99
Abortion Report, 493, 496
abortion statutes, 38–61, 70–94
  abortion on demand and, 41, 96, 97, 119–20
  ballot initiatives and, 180, 242, 469, 507–17, 518
  Chicago conference on (1969), 84
  Clergy Service ministers' views on, 70–72
  constitutional test as point of, 391–92, 419–20
  counseling restrictions and, 381, 407–408, 413, 444, 488n
  feminists' opposition to, 87–90, 208
  history of, 42–43, 75–76, 82, 168–70
  hospitals-only requirements and. *See* hospitals-only requirements
  informed consent requirements and, 283–84, 286–88, 304, 421, 425, 440, 498

invited by *Webster* ruling, 489, 505–6
justifiable abortion notion and, 47–48
lawsuits and. *See* abortion litigation;
   specific cases
life-begins-at-conception notion and.
   See beginning of life
as litmus test, 386–87, 388, 512
Model Penal Code and, 45–49, 51, 56–
   58, 62, 70–74, 80, 84, 85, 91, 107,
   119, 131, 188, 336, 339
noncooperation sanctions and, 383,
   426–27, 443, 458
opposition to liberalization of, 95–121.
   See also right-to-life movement
parental consent requirements and,
   165, 278–79, 284, 288, 293, 304,
   383, 421, 440, 466, 498, 510, 512,
   514
pathology-report requirements and,
   422
physicians-only requirements and, 89,
   90, 421
post-*Roe* bills, 165–66, 189–90, 278–
   279, 283–88
post-viability abortions and, 163, 165,
   174, 422, 439n, 440
post-*Webster* bills, 492–93, 495–501
and pre-abortion descriptions of baby,
   466
public facilities restrictions and, 380-
   381, 388, 391, 407, 465–66, 480,
   505
public funding and. *See* public funding
   of abortion
repeal vs. reform of, 72–94
residency requirements and, 91, 97,
   152, 162, 200, 201
restrictive procedural requirements
   and, 152, 162–63
and rights of unborn children, 382–83,
   388, 419–20, 488
right-to-life activists' strategy and,
   241–44
Roe reversal and, 458–59
saline abortion prohibitions and, 190,
   278–79, 421
sex-selection prohibitions and, 438,
   443, 498
spousal notification or consent
   requirements and, 278–79, 421, 422,
   498, 519
state's interest and, 170–71, 173
struck down by *Roe* and *Doe* decisions,
   161–78

Thalidomide and rubella scares and,
   49–54, 59
therapeutic abortions and, 43–45, 47–
   48, 56–58
third-trimester abortions and, 89, 90,
   96, 498–99
viability-testing requirements and,
   408–9, 465, 480–81, 488–89, 504–5
waiting periods and, 284, 288, 293,
   304, 421, 422, 425, 498, 519
Washington, D.C., symposium on
   (1967), 81
and woman's right to control her own
   body, 74, 83, 87–90
written consent requirements and,
   279, 421
see also specific states
Abzug, Bella, 178
Ad Hoc Committee in Defense of Life,
   352–53
adoption, 381, 391, 449, 526
Adoption Associates, 526
Aeromedic, 97
Akron, Ohio, restrictive abortion
   ordinance in, 283–84, 286–88, 410,
   411
Akron v. *Akron Center for Reproductive
   Health*, 410, 411, 421, 424–26, 438,
   440, 456, 457
Alabama, 506
Alaska, abortion statutes in, 91, 94, 97,
   107–8
Alinsky, Saul, 250
America, 348–49
American Baptist Convention, 72
American Civil Liberties Union (ACLU),
   59, 129–30, 141, 176, 281, 284, 292,
   389, 410, 435, 438, 455, 471, 504,
   514
   Susman's relationship with, 410–11
American Journal of Public Health, 24,
   44
American Law Institute (ALI), 45, 46, 48,
   84, 187
   see also Model Penal Code
American Life League, 444–45
American Life Lobby, 347
American Medical Association (AMA),
   80, 206
Americans United for Life (AUL), 107,
   435–42
   Webster and, 446–47, 457, 464
amniocentesis, 408, 409
Anderson, Don Benny, 374n, 432

Anderson, Gail, 25
Andrews, Joan, 324, 366, 369–70, 414,
    415, 417, 431
"Antis:"
    use of term, 273
    see also right-to-life movement
antiwar movement, 34, 248, 250, 311,
    312, 321
Appeals Court. *See* Eighth Circuit Court
    of Appeals
Arizona, abortion statutes in, 107, 108,
    125, 423*n*
Arkansas, abortion statutes in, 41
Army of God, 374–75
Ashcroft, John, 285–86, 392–93, 435
Ashcroft. See Planned Parenthood v.
    Ashcroft
Assemblies of God, 385
Association for the Study of Abortion
    (ASA), 83–84, 140, 156, 221

Backer, Laverne, 54, 115–16
Backer, Matthias H., Jr., 39–42, 48, 53–
    56, 58, 95–99, 185, 186, 525
    background and education of, 109–12,
        113
    birth control as viewed by, 115–18
    Catholic question and, 106–18
    ethical quandaries of, 112–13, 114–15
    family life of, 40, 54, 115–16
    mail in support of, 95–96, 118–19
    as public speaker, 95–96, 98–99, 101,
        102, 106, 113, 118, 119
    Rodgers v. *Danforth* and, 120–21, 145,
        165
Baker, Msgr. Joseph W., 192, 429
Balch, Burke, 496–97
ballot initiatives, 180, 242, 469
    in Missouri, 507–17, 518
Bangladesh, family planning in, 273
Baptists, 72, 178, 188
    see also Southern Baptist Convention
Baptists for Life, 332
Barger, John, 261–62
Barnes, Jim, 381, 391–92
Barrett, James H., 521
Barzelay, Douglas E., 317–18
Bayh, Birch, 350
Beal v. *Doe*, 282, 421, 439
Beasley, Dorothy, 154
beginning of life, 505
    Catholic doctrine on, 113–14
    as civil rights question, 242–43

conception as, 113–14, 170–74, 177,
    238, 284, 321, 353–54, 382–83, 388,
    406, 413, 419–20, 443, 465, 480,
    483–84, 488, 491, 505
    congressional legislation on, 353–54
    passages in Missouri statutes on, 284,
        382–83, 388, 406, 413, 419–20, 443,
        449, 465, 480, 483–84, 488, 491
    Roe findings and, 170–74, 177, 312–
        313, 318, 353
Beilenson, Anthony, 49, 51, 57–58, 73
Bellotti v. *Baird*, 421
Belous, Leon Phillip, 130–34, 141
Benshoof, Janet, 292, 410, 411, 485
Bensing, Sandra, 143
Berkeley Bio-Engineering (formerly
    Berkeley Tonometer Co.), 199, 204,
    263–64
Bible, 67, 68, 69, 338–39, 341, 342, 367,
    418
Biden, Joseph, 455
birth control, 81–84, 100, 151, 156, 160,
    179, 184, 213, 487, 508
    Catholic posture on, 115–18, 340
    Emko contraceptive foam and, 221–22
    legal distinction between abortion and,
        481, 482
    post-fertilization, 173, 405–6, 491, 523
    privacy rights and, 136–39, 317, 456
    right-to-life movement's goals and,
        179
birth control pill, 116–17, 179, 213
birth defects. *See* fetal deformity
    abortions
Birthright, 309, 325, 376
Black, Hugo, 151, 159
Blackmun, Harry A., 282, 439*n*, 442,
    451, 479, 520
    Casey and, 519
    Danforth and, 279, 297
    Roe and, 161–62, 166–74
    Thornburgh and, 440
    Webster and, 491–92, 494, 504·
Blumenthal, Albert, 85
Blumenthal, Tom, 411–12, 413, 479, 524
Bond, Christopher, 278, 335, 350
Bopp, James, 492, 494, 496–97, 455–57,
    460
"born-again" Christians, 345
Boston University Law Review, 317–18
Boston Women's Health Collective, 207–
    208
Boswell, James, 188–89
Bott, Richard, 338, 339, 340

Bourke, Vernon J., 112
Braver, Rita, 503
Brennan, William J., 282, 442, 451, 492, 518
Breyer, Stephen, 520
Britton, John B., 521
Brody, Jane E., 59
Brown, John, 367
Brown, Judie, 347, 444–45
Brown, Paul, 347
Brown v. *Board of Education*, 436–37
Bryant, Mary, 379–89, 392, 393, 413, 508, 524
Buckley, James, 177
Burger, Warren, 423, 426, 440, 441
Bush, George, 336, 497
Buxton, Lee, 137, 138
Byrn, Robert M., 314
Byrnes, Timothy A., 184

Calderone, Mary Steichen, 23, 24, 25, 44
California, 22, 162, 350
  abortion facilities in, 196, 198–99, 289–90
  abortion litigation in, 130–34, 135
  abortion statutes in, 41, 49, 57–58, 71, 73, 125, 126, 131–34, 196, 199, 241, 336
  clergymen's counseling service in, 31–32, 35, 71
  rubella abortions in, 52–53
  screening for therapeutic abortions in, 44, 59, 62, 97, 336
  women's health movement in, 208–13, 214
California Committee for Therapeutic Abortion, 131, 134
California Supreme Court, 131–34
Callahan, Ceil, 428, 433–34, 445, 447
Calloway, DeVerne, 119, 120
Camel, Marvin, 28
Campaign Life Missouri, 515, 524
capital punishment, 160, 179, 250, 367
Carberry, Cardinal John, 119, 176, 182–183, 192
Carmen, Arlene, 32–34, 397
Carter, Jimmy, 335, 336, 337, 345, 350
Casey. See Planned Parenthood of Southeastern Pennsylvania v. *Casey*
Casey, Robert, 498

Catholic Church, 75, 106–18, 140, 163, 176–93, 285, 346
  and canon law on abortion, 113–14
  charismatic strain in, 233–34
  contraception opposed by, 115–18, 340
  and directives for hospital physicians, 114–15
  ethical principles and, 108–9, 111–12
  evangelicals and, 340, 342, 348–49
  Kennedy presidency and, 184
  mobilization capabilities of, 180–83, 187, 191–92, 334
  political arena entered by, 183–84, 185–87, 191
  Protestants' lack of cooperation with, 187–89
  Roe and *Doe* decisions denounced by, 176–78
  Second Vatican Council and, 176, 184–187, 188, 235
Catholics for a Free Choice, 468
Cavanagh, Denis, 40–42, 48, 59, 96, 107
Cavanaugh-O'Keefe, John, 260–62, 263, 309, 310
Cavender, Jean, 525–26
Center for Constitutional Rights, 176, 281
Chastain, Mike, 366, 367–68
Chicago Sun-Times, 288–89, 291
Chicago Women's Liberation Union, 213–14
China, abortions in, 197*n*, 273
Christian Action Council, 426
Christianity Today, 340, 341
Christian Life Commission, 339–40
Christians:
  "born-again," 345
  See Catholic Church; evangelicals; Protestants; *specific denominations*
Church, Frank, 350
Circle of Life, 270–72
Cisler, Lucinda, 88–89
Citizens' Committee for Humane Abortion Laws, 74
civil disobedience, 232, 237, 248–49, 250–51, 260
  underground referral services and, 32–35
  see also nonviolence; sit-ins, at abortion clinics
civil rights, abortion as issue of, 242–43
civil rights movement, 33, 34, 191, 247, 260, 436–37
Clark, Dick, 330–33, 334, 349

Clark, Tom C., 135
Clergy Consultation Service of Missouri, 62–68, 69–70, 91, 122–26, 128, 129, 146, 150, 158, 217, 219, 223–24
  Sunnen's funding of, 221, 222–23
  see also Pregnancy Consultation Service
Clergymen's Consultation Service on Abortion, 32–35, 70–71, 72, 83, 397
  abortion clinic set up by, 96–97
Clinton, Bill, 520, 521, 522
coat-hanger imagery, 81, 402, 476, 486, 506
Coffee, Linda, 130, 142–43, 154
Cohen, Laura, 403, 404, 502, 508, 514–515
Colautti v. Franklin, 439
Coleman, Marshall, 496
Colorado, 62
  abortion statutes in, 41, 56, 59, 71, 73, 80, 107
  ballot initiative in, 469, 510, 512
Columbia Planned Parenthood, 279
Committee for the Survival of a Free Congress, 344
"Common Ground," 525–26
Common Ground Network for Life and Choice, 526
common law, 168–69, 314
conception. See beginning of life
Concerned for Life, 243
Congress, U.S., 521, 522
  election to. See electoral politics
  Hatch Amendment and, 355–58, 361–365, 375, 422, 445, 460
  Human Life Amendment and, 351–52
  Human Life Bill and, 352–58, 362–65, 375, 457
  as target of right-to-life movement, 177, 190, 279–80, 329, 353–54
Congressional Record, 398
Connecticut:
  abortion litigation in, 282
  abortion statutes in, 42
  birth control law in, 136–39
Connecticut v. Menillo, 421
consciousness-raising, 89, 291
consent. See father's consent requirements; informed consent requirements; parental consent requirements; spousal notification or consent requirements; written consent requirements
Conservative Digest, 343–44

Constitution, U.S., 137, 169, 314, 383–384, 388
  Bork's approach to, 455–56, 457
  fetal rights and, 172, 177, 313–14, 316–17, 353–54, 355, 419–20
  Hatch Amendment and, 355–58, 361–365, 375, 422, 445, 460
  privacy rights and. See privacy rights
  right-to-life amendment to. See Human Life Amendment
  and women's right to abortion, 129–134, 139–41, 143, 160–76, 282, 316–17, 356, 357, 466, 520
  see also specific amendments
contraception. See birth control
Cook, Constance, 85–86, 87, 89, 90, 92, 93
Cook-Leichter bill, 85–86, 89–94
the Corpsman, 63–64, 71, 123, 149–50, 196
counseling restrictions, 381, 407–8, 413, 444, 488n
Courter, James, 496
Cox, Bill, 161–63, 164–65, 176, 186, 187
Cranston, Alan, 350
Crist, Robert, 302–3, 304–5
Csapo, Arpad, 294–95
Culp, Mary Kay, 245
Culver, John, 338, 350

D'Amato, Alfonse, 338
Danforth, John C., 120–21, 279, 338
Danforth. See Planned Parenthood of Central Missouri v. Danforth; Rodgers v. Danforth
Danis, David, 319, 321, 325
Danis, Nancy, 323–24
Davis, Father James, 324, 325, 326
dead-woman pictures, 400–401, 402
death penalty, 160, 179, 250, 367
"Defense of Abortion, A," 239–40
DeFeo, Louis, 164–67, 168, 174, 175, 176, 182, 183, 186, 188–90, 384, 388, 391, 392, 443–44, 446, 465, 524
Del-Em, 211–13
Democratic Party, 336, 344, 348
dilation and curettage (D&C), 17, 20, 197, 214, 218, 294, 297, 406, 424
  pictures of, 103, 105, 106, 163
  post-Roe statutes on, 189, 190
  right-to-life movement's literature on, 243

dilation and evacuation (D and E), 293, 299–303, 304–7
dilation and extraction (D and X), 522
Direct Action, 431–33
Direct Action News, 461
District Court, U.S., 143–44, 151, 279, 281, 299, 303–4, 439
    Webster and, 413, 446–47, 449, 464
Dobson, James, 342
Doe v. *Bolton,* 151, 157, 159, 160, 224, 279, 284, 492, 504
    in Georgia courts, 129, 130, 142–43, 144, 154
    statute tested in, 152
    Supreme Court decision in, 162, 164, 165, 166, 167, 175, 177, 195, 238, 244, 413, 423
Dolan, John (Terry), 344
Donovan, Charles (Chuck), 351, 354, 362
double effect, 113
Douglas, William O., 136, 137, 138–39, 423
Downer, Carol, 208–13
Drake, William, 102, 107, 243
Dred Scott v. *Sanford,* 246–47
Drinan, Rev. Robert F., 81*n*
Ducey, Mary, 240
Duemler, Robert, 15–17, 28, 37, 65
due process of law, 315–16
    see also Fourteenth Amendment
Duryea, Perry, 92

Eagleton, Thomas, 335, 338, 348
ectopic pregnancy, 114
Edelin, Kenneth, 297*n*
Edwards, Kathy, 255, 352, 443, 524
    electoral politics and, 329, 333, 335, 336
    Hatch Amendment and, 362–63
    New Right and, 348, 349
    and sit-ins at abortion clinics, 261, 329
    statehouse politics and, 258, 433–34
Egan, Helen Ann, 253
Eighth Circuit Court of Appeals, 281, 446, 457–58, 461, 463, 464, 465, 488
electoral politics, 329–38
    lit drops in, 331–33, 334
    New Right and, 343–50
    in 1976, 337
    in 1978, 330–33, 334, 349
    in 1980, 335–38, 349–51
    in 1982, 375–76
    in 1988, 469

    in 1989, 493
    PACs in, 331, 334–35, 336
    prerequisite voters and, 333–34, 350
    right-to-life movement's entry into, 329–33
Ely, John Hart, 314–17
embryo implantation, 173
Emko, 221–23
Episcopalians, 188
Equal Rights Amendment, 344
Escobedo, Bolivar, 252, 263, 323, 324, 326, 328, 368, 394
eugenics, 179
euphemisms, 79*n*–80*n*, 200
euthanasia, 179, 340
evangelicals, 338–49, 384, 385
    anti-Catholicism and, 340
    defined, 338–39
    Moral Majority and, 345–47
    New Right politics and, 343–50
    right-to-life position adopted by, 340–343
    Roe decision as viewed by, 339–40, 346
Evans, Rev. J. Claude, 31, 68–69
Evans, Roger, 412–13, 479
Ewing, Rev. James W., 29–31, 36

Fahey, Gloria. *See* Lee, Gloria Fahey
Falkenberg, Nanette, 399–402, 468
Falwell, Rev. Jerry, 343, 345–47, 348
father's consent requirements, 156, 166, 176, 190
    see also spousal notification or consent requirements
Federal Bureau of Investigation (FBI), 373, 374
Feminine Mystique, The (Friedan), 86–87, 205
fetal deformity abortions, 38, 41, 47, 49–54, 56, 57–58, 72, 81, 298, 404
    debate in right-to-life movement over, 358–61
    rubella and, 51–54, 142, 198, 221
    Thalidomide and, 49–51
fetus, 398–99
    constitutional rights of, 172, 177, 313–314, 316–17, 353–54, 355, 419–20
    quickening of, 169, 305–6
    use of word, 101, 189, 191, 247
    viability of. *See* viability; viability-testing requirements
    see also beginning of life

Fifth Amendment, 413
Fight for Life, 60, 187, 191
Finkbine, Sherri, 49–51, 66
Finnegan, Tim, 322, 328
First Amendment, 388, 413, 491
Florida, 338, 493
    abortion litigation in, 157
    abortion statutes in, 493
Florio, James, 496
Floyd, Jesse J., 297n
"Focus on the Family," 342
Foley, Dan, 234, 235
Food and Drug Administration (FDA),
    49, 523
Ford, Betty, 337
Ford, Gerald, 337
Fourteenth Amendment, 171, 172, 313–
    314, 315–16, 353–54, 413
Franciscans, 230–37, 240, 248, 249–
    252
Francis of Assisi, Saint, 236
Frankfort, Ellen, 212–13
Freedom of Access to Clinic Entrances
    Act (FACE), 521–22
Freiman, Michael, 279, 299–301, 302,
    525
Freind, Stephen, 439
Fried, Charles, 441, 480, 481, 482
Friedan, Betty, 86–88, 205
Frieman, Michael, 217–20, 223, 224,
    226
Frieman, Sarijane, 217, 218–19, 225
Fritzsche, Sybille, 155
Furman v. *Georgia*, 160

Galbraith, John, 125
Galebach, Stephen, 353–54
Gallup polls, 51, 345, 360–61
Gandhi, Mohandas K., 236, 237, 321,
    366, 367, 415, 426
Gatch, Georgie, 124
Gee, Alison, 476, 513, 514, 515
Georgia, 350
    abortion litigation in. *See Doe* v.
        Bolton
    abortion statutes in, 57, 152
Gephardt, Richard, 447–48
Gerard, Jules, 312, 313–14, 318
German measles. *See* rubella
Gerster, Carolyn, 107, 108, 178, 179, 241,
    245, 261, 329–30, 355
Gerstley, Ira, 26
Ginsburg, Douglas, 460

Ginsburg, Ruth Bader, 520
Goedeker, Lillian, 323
Goldberg, Arthur, 139
Golden, Edward, 119
Goldsby, Matthew, 431
Goldstein, Michael S., 289–90
Goodman, Ellen, 399
Gottman, Rev. Ken, 18, 19, 63, 67–68,
    69–70
Grassley, Charles, 338
Greece, ancient, 168
Griffin, Bob, 389–91, 434, 507, 522
Griffin, Michael F., 520–21
Grimes, Charlotte, 479
Grisez, Germain, 98
Griswold, Estelle, 137, 138
Griswold v. *Connecticut*, 135–39, 141,
    153, 155, 171, 455–56, 481
Grossman, Susan, 155
Guam, abortion statutes in, 493
Gunn, David, 520–21
Gurner, Rowena, 74–75, 79–80
Guttmacher, Alan, 156
gynecologists, abortions in offices of,
    201–2, 203

Hall, David, 279
Hall, Robert, 196, 200
Hallford, James Hubert, 143n
Hames, Margie Pitts, 130, 141, 142–43,
    154
Hampton, Sylvia, 265–72, 275, 282, 285,
    306–7, 524
    Lee's relationship with, 265, 268–72,
        307, 308, 310
    Significant Others handled by, 266–68,
        307
    sit-ins and, 310
Handbook on Abortion (J. Willke and
    B. Willke), 103, 359, 417, 435
Hard Cases Exceptions:
    debate in right-to-life movement over,
        358–61
    Lee's legislation and, 499–501
    see also fetal deformity abortions;
        incest, abortions for victims of;
        mental health abortions; rape,
        abortions for victims of
Harlan, John Marshall, 137, 138, 159
Harper, Fowler, 137
Harris v. *McRae*, 421, 481
Hartigan, Neil, 497–98
Harvard Divinity School, 80–81

Harvey, Gwyndolyn, 19, 20–21, 36–37, 63, 67, 122, 123, 126, 128
Harvey, Hale, 215
Hatch, Linda, 269–70
Hatch, Orrin, 355, 356, 362, 375, 422
Hatch Amendment, 355–58, 361–65, 375, 422, 445, 460
    Two Step Plan and, 357–58
Hawaii, abortion statutes in, 91, 97
Helms, Jesse, 352, 353, 356, 357, 358, 362, 365, 375
Henrie, W. J. Bryan, 27, 73
Hern, Warren, 303
Herzog, George, 52
Heymann, Philip B., 317–18
Hiesberger, Anthony, 164–65, 185–88
Higgins, Msgr. George C., 348–49
Hill, Paul, 521
Hiltz Publishing, 242
Hippocratic Oath, 167, 168, 170
Hitler, Adolf, 166, 367
Hodgson, Jane, 142
Hogan, Lawrence J., 177, 178
Hogan, William, 107
Holocaust, 48, 56, 431–32
homosexuality, 343, 344, 346, 348
Hope Clinic for Women, 373
Hopkins Charitable Fund, 221
Horan, Dennis, 436
Horgan, Mary Frances, 192–93
hospital abortions:
    abortion mills vs., 200–201, 202
    vacuum aspiration as alternative to, 197–200, 210
hospitals-only requirements, 190, 199, 200, 202–3, 466
    for second-trimester abortions, 284, 293–94, 299, 302, 304, 407, 421, 422, 425, 440, 446–47, 449, 465, 504
    sixteen-week cutoff and, 383, 407, 413, 446–47, 449, 465, 504
    twelve-week cutoff and, 407, 422
How to Choose an Abortion Facility, 291–92
Human Betterment Association for Voluntary Sterilization, 83–84
Human Life Amendment, 248, 258, 275, 279–80, 329, 330, 335, 340, 348, 349, 351–58, 360n, 422, 423n, 495
    evangelicals' support of, 342
    Gephardt defection and, 447–48
    Hatch Amendment and, 355–58, 361–364, 365

infighting over, 356–58, 361–65
as issue in 1980 elections, 336, 338, 350
ordinary legislation vs., 352–57, 362, 365
post-*Roe* call for, 177, 178, 190
Two Step Plan and, 357–58
Human Life Bill (HLB), 352–58, 362–65, 375, 457
    reasoning behind, 353–54
    states' rights and, 354–55
Human Life Journal, 353–54
Human Life Review, 353, 354
human rights, 163, 191, 242–43
Humphrey, Gordon, 398
Hunter, Elmo, 304
husband's consent requirements. *See* father's consent requirements; spousal notification or consent requirements
Hyde, Henry, 280, 292, 353
Hyde Amendment, 280–82, 361, 421
hysterotomy, 101, 165, 190, 294, 295, 298
    live-birth complications in, 297
    pictures of, 102, 103, 105–6
    right-to-life movement's literature on, 243

Idaho, 350, 493
illegal abortions, 20–37, 46, 47, 59, 82, 149–50, 153, 155, 208, 212, 219, 270
    classes for women on, 75–79
    clergy counseling and, 30–31, 32–35, 67–70
    coat-hanger imagery and, 81, 402, 476, 486, 506
    consciousness-raising on, 89
    dead-woman pictures and, 400–401, 402
    deaths attributable to, 22, 23, 24, 59, 459–60
    doctors' experiences with, 16–17, 25–27, 28–29, 40–41, 217–18, 302–3
    euphemisms for, 79n–80n
    hospital wards for complications from, 25–26
    by Jane collective, 214
    lawsuits and, 129–39
    methods used in, 21–22
    national conference on (1955), 23–25

illegal abortions, *continued*
  possibility of return to, 451, 470
  statistics on, 22–24
  underground referral systems and, 29–
    37, 62–70, 122–26. *See also* Clergy
    Consultation Service of Missouri;
    Clergymen's Consultation Service
    on Abortion; Pregnancy
    Consultation Service
Illinois:
  abortion litigation in, 143, 151, 155,
    293, 497–98
  abortion statutes in, 436, 495–96
Illinois Citizens for Medical Control of
  Abortion, 80
Illinois Pro-Life Coalition, 495–96
Illinois Right to Life, 107, 244
incest, abortions for victims of, 29, 38,
    49, 56, 58, 72, 339, 500
  debate in right-to-life movement over,
    358–61
  public funding of, 280, 497
Indiana, 338, 350
  abortion statutes in, 57
infanticide, 48, 107, 176, 340, 346
informed consent requirements, 283–
    284, 286–88, 304, 421, 425, 440,
    498
intrauterine devices (IUDs), 179, 213
Iowa, electoral politics in, 330–33, 334,
    338, 349, 350
Iowans for Life, 330–33, 334
Isaacson-Jones, B. J., 454–55, 471–77,
    479, 486, 503, 525–26

Jane collective, 214
Japan, abortions in, 98, 156, 198, 294–
    295
Javits, Jacob, 338
Jefferson, Mildred, 247, 354
Jepsen, Roger, 330–33, 334
Johanek, Mary Ann, 244
John XXIII, Pope, 184
Johnson, Harvey, 60
Johnson, Lyndon B., 162
Johnson, Samuel, 188–89
Jones, Arthur, 57
Judaism, 178, 332
Judson Memorial (New York City), 32–
    35, 89
Justice Department, U.S., 440–41, 464,
    465, 522
justifiable abortions, 47–48

Kansas, 162
  abortion litigation in, 157
  abortion statutes in, 62, 97–98, 125
Kansas University Medical Center, 22,
    23, 97–98, 302
Karman, Harvey, 209
KCCV (Kansas City Christian Voice),
    338, 339, 342
Kennedy, Anthony, 460–61
  Casey and, 519, 520
  Webster and, 482, 489, 493–94, 504
Kennedy, John F., 184
Kennedy, Ted, 448
Joseph P. Kennedy Jr. Foundation, 80–81
Kentucky:
  abortion litigation in, 143, 151
  abortion statutes in, 506
Kerenyi, Thomas, 294–95, 296, 298
King, Rev. Martin Luther, Jr., 33, 236,
    247, 250, 321, 366, 367
Kissling, Frances, 290, 291, 468
Klein, Jack, 299–300
Kleinknecht, G. H., 324
Knox, John, 49
Kolbert, Kathryn, 471, 514
Koop, C. Everett, 340–42
Kovac, Stephen, 325
Krol, Cardinal John J., 176

Lader, Lawrence, 33, 81–83, 84, 88, 90,
    142, 400, 523
Ladies' Center (St. Louis County), 368
Lalor Foundation, 221
Lamm, Richard, 56, 59
Landsman, Stephan, 410
Landwehr, Carl, 328–29, 330, 333, 335
Landy, Uta, 140, 291–92
Lasky, Herbert, 144–45, 146–48,
    150–51
Lauinger, Anthony J., 364
Lavin, Lucille, 323
Law Enforcement Bulletin, 324
Lawyers for Life, 249, 515
Lee, Gloria Fahey, 370–73, 375–76, 414–
    416, 418, 427–29, 433, 445, 456,
    460, 524
Lee, Miriam, 428, 429
Lee, Nicholas, 446
Lee, Samuel Henry, 229–40, 261–64,
    322, 365–76, 395, 414–35, 442–47,
    453, 456–67, 492–502, 517
  abortion issue studied by, 237–40
  ballot initiative and, 514, 515

in dialogue with pro-choicers, 269–72, 307
essay written by, 270–72
Fahey's relationship with, 370–73, 375–76
family background of, 232–33
fired by MCL, 501–2
full-time ministry in pro-life begun by, 428–29
legal studies of, 416, 420–27, 429
legislation written by, 426–27, 434–435, 443, 498–501, 518
and post-*Webster* statutory proposals, 492–93, 495–96, 498–501
pragmatism of, 459–60
priesthood considered by, 230–31, 233, 370, 371
Puzder's legislative proposal and, 418–420, 426
religious experiences of, 233–34
Reproductive Health Services infiltrated by, 264, 265, 268–69
rescue movement and, 433, 461–62, 463
Roe reversal and, 458–59
sentence served by, 388–89, 414–27
in sit-ins at abortion clinics, 231–32, 236–37, 251–53, 254, 260, 307, 388–89, 430–31, 433, 308–11, 319–323, 327–29, 365–73, 375, 451, 463
statehouse politics and, 388–89, 392, 433–34, 443–46, 449, 460, 461, 500–502, 515, 524
Supreme Court reconfiguration and, 450, 451, 456–57, 460–61
violent tactics as viewed by, 431–33
Webster and, 442–43, 446–47, 449, 457–58, 461, 463, 464–67, 477–85
wedding of, 415
Zevallos kidnap and, 374–75, 376
Leichter, Franz, 85, 89
Levin, Harry, 194, 195, 205
life. *See* beginning of life
Life, 51, 101, 104, 243
Life Amendment Political Action Committee, 347
Lifeletter, 353
"Life or Death," 242, 254
Linacre Quarterly, 98, 114
lit drops, 331–33, 334
litigation. *See* abortion litigation; *specific cases and courts*
Lochner v. *New York*, 315, 316
London, Ephraim, 33

Los Angeles County General Hospital, 25
Los Angeles Feminist Health Center, 211
Louisiana:
    abortion litigation in, 151
    abortion statutes in, 284, 493, 506
Lucas, Roy, 139–41, 142, 152, 154–55
Luker, Kristin, 174–75
Lutherans, 178
Lutherans for Life, 332

McAteer, Ed, 345*n*
McAuliffe, Bishop Michael, 176
McCarthy, Karen, 386
McCarthy, Maureen, 269–70, 272, 292
McCormack, Ellen, 280, 331
McCorvey, Norma, 143
McFadden, James P., 352–53, 354, 358
McGovern, George, 350
McHugh, Msgr. James, 107
McNary, Gene, 120, 335
McNicholas, Bishop Joseph, 193
James Madison Constitutional Law Institute, 141
Maginnis, Patricia, 35, 74–75, 78–80, 81, 208–9
Maher v. *Roe*, 282, 421
Maine, 493
March for Life, 246, 248, 256, 351, 448, 497
March for Women's Equality/Women's Lives (1989), 474–77
Margolis, Alan, 52, 198–99, 202
Marshall, Thurgood, 282, 442, 451, 492, 518
Marshner, Connie, 362
Martinez, Bob, 493
Marzen, Thomas, 438
Massachusetts, abortion statutes in, 107
May, Archbishop John L., 326–28, 365
Mayo Clinic, 167
Mazzoli, Romano, 353
Means, Cyril, 154, 168, 170
Mecklenberg, Marjorie, 178, 179
Medicaid, 275, 280–82, 497
Medical Economics, 96, 201–2, 203
menstrual extraction, 211–13
mental health abortions, 44, 47, 58, 59, 60, 62, 72, 97, 98, 151, 336, 499
    advice on qualifying for, 208–9
    debate in right-to-life movement over, 358–61
Methodists, 188

methotrexate, 523
Mexico, abortion services in, 35, 63, 71, 73, 74–75, 78, 123–24, 125, 131, 153
Meyers, Richard S., 437, 438
Michaels, George, 92–93, 94
Michelman, Kate, 468, 469, 485, 503
Michigan:
    abortion statutes in, 493
    ballot initiatives in, 180, 469, 510, 512
mifepristone (RU-486), 405–6, 523
Miller, Keith, 91n, 94
Miller, Lucia, 403, 404
Minnesota:
    abortion litigation in, 154
    abortion statutes in, 57, 142
Minnesota Citizens Concerned for Life, 439
minors, parental consent requirements for, 165, 278–79, 284, 288, 293, 304, 383, 421, 440, 466, 498, 510, 512, 514
Missouri:
    abortion litigation in, 278–79, 281–82, 284. *See also Reproductive Health Services* v. *Webster; Rodgers* v. *Danforth*
    abortion statutes in, 20, 28–29, 30, 38–43, 55, 58–61, 67, 95–96, 102, 119–20, 120–21, 123, 125, 126, 128, 135, 151, 157–58, 165–66, 176, 181–83, 187, 189–90, 195, 224, 278–79, 283–88, 292–93, 301–4, 313, 380–93, 406–9, 439, 443–44, 446–47, 498–501, 505–6
    ballot initiative in, 507–17, 518
    elections of 1980 in, 335, 338, 350
    ingredients for Supreme Court challenge in, 439
    Medicaid abortions in, 282
    statehouse politics in, 258–60, 278, 283–85, 443–44, 379–93, 507–8, 512, 516–17
Missouri Alliance for Choice, 497, 509–515
Missourians Opposed to Liberalization of Abortion Laws, 58–59, 60, 102, 107
Missouri Catholic Conference, 161–66, 175, 176, 182, 183, 185–90, 328, 406
    first post-*Roe* bills and, 165–66, 189–190
    origins of, 185–86
    Protestants and, 187–89

statehouse politics and, 259, 278, 283–285, 381, 385, 388, 391
Missouri Citizens for Life (MCL), 244, 250, 253–60, 355, 374, 406, 428, 434, 450, 500, 515
    courthouse rally of (1973), 246, 253
    electoral politics entered by, 329–38
    Hatch Amendment and, 363–65
    Human Life Amendment and, 447–48
    leadership of, 253–56
    Lee fired by, 501–2
    and sit-ins at abortion clinics, 253–54, 260, 328–29
    statehouse politics and, 258–60, 278, 283–85, 381, 385, 386, 388–89, 443, 446, 449, 461, 507
Missouri Citizens Opposed to the Liberalization of Abortion Laws, 182
Missouri Court of Appeals, 369, 415
Missouri House Bill 902, 283–88, 292–293
Missouri House Bill 1596, 380–93, 406–413, 422, 511
    litigation over. *See Reproductive Health Services* v. *Webster; Webster* v. *Reproductive Health Services, et al.*
    statehouse politics and, 443–44
    writing of, 418–20, 426–27, 434–35, 443
Missouri Medical Association, 56
Missouri Medicine, 56, 60, 61
Missouri Pro-Life Network, 497
Missouri Supreme Court, 151, 157–58, 159–60
Mizell, Gerald, 324, 325
Model Penal Code, 45–49, 51, 70–74, 80, 84, 107, 119, 188, 339
    state legislation based on, 56–58, 62, 85, 90–91, 131, 336
Mohr, James, 42, 43n, 79n–80n
Mondale, Walter, 336
Montello, Samuel, 22
Moody, Rev. Howard, 32–35, 63, 69, 70–72
Moore, Father Christian, 233–34, 235
Moral Majority, 345–47
Mormons, 178
morning-after pill, 173
motherhood, 257
Mott, Stewart, 221
Mulhauser, Karen, 280
Murr, William, 198–99
Myers, Lonny, 80, 84

Nathanson, Bernard, 201, 202, 215–16, 396–99, 401, 402, 459
National Abortion Federation (NAF), 286, 290–92, 299, 302, 303, 307, 393, 394, 395, 405, 468, 521, 523
National Abortion Rights Action League (NARAL), 274, 276–77, 279, 280, 283, 292, 331, 385, 393, 441, 396, 455, 456, 458, 475, 476, 496, 510, 523
  hate mail sent to, 277, 400–401
  Missouri office of, 403–4
  post-*Webster* activities of, 506
  Silent No More campaign of, 402–4
  visuals dilemma and, 399–401, 402
  Webster and, 470–77, 503–4, 505
National Association for Repeal of Abortion Laws (NARAL), 84, 87–88, 90, 93, 108, 125, 180, 201, 203, 216, 221, 459
  pro-lifers as viewed by, 273–74
  renamed National Abortion Rights Action League, 274
National Association for the Advancement of Colored People (NAACP), 437, 438, 456
National Association of Evangelicals, 338, 339
National Catholic Register, 358
National Catholic Reporter, 57
National Catholic Welfare Conference (NCWC), 184
National Center for Health Statistics, 24
National Committee on Maternal Health, 24–25
National Conference of Catholic Bishops (NCCB), 327–28
National Conservative Political Action Committee, 344, 347, 348, 349–50
National Council of Catholic Bishops, 358, 362
National Federation of Catholic Physicians' Guilds, 177
National NOW Times, 396
National Organization for Women (NOW), 86, 87, 89, 279, 441, 456
  march organized by (1989), 474–77
National Pro-Life Political Action Committee, 347
National Right to Life Committee, 107, 177, 178–80, 241, 245, 247–48, 256, 284, 335, 435, 441, 444, 492, 496
  debate over purpose of, 179
  electoral politics and, 329–33, 336–37, 338

Hatch Amendment and, 362, 363–64
  Human Life Amendment and, 351–55
  New Right and, 347
  1978 convention of, 260–63
  as nonsectarian organization, 178
  PAC of, 336
  post-*Webster* convention of, 494–95
  rescue movement and, 462–63
  sit-ins at abortion clinics and, 260–63
  Supreme Court reconfiguration and, 457, 461
National Right to Life News, 248, 329, 336, 350–51, 424, 457, 460, 498
National Youth Pro-Life Coalition, 177
Nazi Germany, 177, 300, 321–22
necessity defense, 319–21, 325–26, 368, 369, 422
New England Journal of Medicine, 397
New Jersey:
  abortion litigation in, 143, 154
  electoral politics in, 493, 496, 497
New Jersey Right to Life, 178
New Mexico, abortion statutes in, 41, 44, 57
New Right, 343–50, 362
  debate over right-to-life alliance with, 347–50
  Moral Majority and, 345–47
New York, 162
  abortion facilities in, 96–97, 196, 199–203, 214–16, 220, 223–24
  abortion litigation in, 143, 155
  abortion statutes in, 57, 85–86, 88, 89–94, 96, 97, 99, 125, 126, 129, 155n, 169, 180, 196, 200, 201, 241
  electoral politics in, 338, 350
  out-of-state women's abortions in, 97, 98, 123, 124–26, 158, 159
New Yorkers for Abortion Law Repeal, 89
New York Right to Life Committee, 99, 107, 119, 178, 180
New York Right to Life Party, 331, 338
New York Times, 33, 34, 59, 176, 196, 200, 401, 403–4, 438, 486, 496
New York Times Magazine, 71, 81–82, 83
Nilsson, Lennart, 101, 179, 188–89, 332
Ninth Amendment, 171, 413
Nixon, Richard M., 159, 167
noncooperation sanctions, 383, 426–27, 443, 458
nonviolence, 250–51, 252, 261–63, 415, 426
  see also civil disobedience

Noonan, John T., 363
North Carolina, 350
  abortion litigation in, 151
  abortion statutes in, 41, 56–57, 62, 71,
    73, 154
North Carolina Law Review, 139–41
North Dakota, ballot initiative in, 180

O'Brien, Ann, 366, 414–17
O'Connor, Sandra Day, 438, 440, 442,
    448, 452
  Casey and, 519, 520
  Roe and, 423–24, 455
  Webster and, 482, 489, 490, 493, 494,
    498, 499, 504
O'Donnell, Ann, 255–56, 259, 272, 328,
    334, 335, 352, 365, 374, 428
O'Donnell, Edward, 190–93, 256
Ohio, 22
  abortion litigation in, 143, 151, 154, 283
  abortion statutes in, 100, 102, 125, 283
  see also Akron, Ohio
Ohio Right to Life, 103, 105
Oklahoma, 350
  abortion statutes in, 57
O'Malley, Mary, 249
Operation Rescue, 461–63
opinion surveys, 51, 345, 510–11
  on Hard Cases Exceptions, 359–61
Oregon, abortion statutes in, 41
Orthodox Church, 178
Orthodox Rabbinical Council of America,
    178
O'Steen, David, 494
Our Bodies, Ourselves, 207–8
Our Lady's Inn, 370, 376, 415, 429, 460,
    524, 526
Overstreet, Edmund, 198

"packing," 63, 64–65, 294
parental consent requirements, 165, 278–
    279, 284, 288, 293, 304, 383, 421,
    440, 466, 498, 510
  Missouri ballot initiative and, 512, 514
Parsons, Rev. E. Spencer, 31, 63–64, 72,
    80
partial birth abortion, 522
Passanante, Bertie, 276, 307
passive resistance, legislative protection
    for, 383, 426–27, 443, 458
Pastoral Constitution on the Church in
    the Modern World, 185–86, 191

Pastoral Plan for Pro-Life Activities,
    327–28
pathology-report requirements, 422
Paul VI, Pope, 117
Paul, Eve, 157
Pennsylvania:
  abortion litigation in, 282, 439–40,
    518–20
  abortion statutes in, 125, 439, 493,
    498, 506
  ingredients for Supreme Court
    challenge in, 439–40
People Expressing Concern for Everyone
    (PEACE), 249
People v. *Belous,* 130–34, 135, 141, 153
People Working to STOP! Government
    Interference, 513
Peters, Donald, 132, 133
Petersen, Vince, 231, 236–37, 248, 249–
    252, 260, 263, 310–11, 366, 433
Peterson, William, 299–300, 301
petition drives, pro-choice, 506–7
  Missouri ballot initiative and, 507–17
Phelan, Lana Clarke, 73–80, 81, 208–9
Philadelphia General Hospital, 25, 26
Phillips, Howard, 344
Philosophy and Public Affairs, 239–40
physicians-only requirements, 89, 90,
    421
pictures. *See* abortion pictures
Pilpel, Harriet, 156–57
Pius IX, Pope, 114
Planned Parenthood Affiliates of
    Missouri, 379–89
Planned Parenthood Federation of
    America, 83–84, 88, 156, 157, 176,
    218, 247, 286, 290, 396, 398, 404,
    435, 438, 444, 468, 471, 476
  litigation unit of, 412–13
  1955 abortion conference of, 23–25
Planned Parenthood League of
    Connecticut, 136–37
Planned Parenthood of Central Missouri
    v. *Danforth,* 278–79, 284, 297, 421,
    423, 426, 439, 440
Planned Parenthood of Kansas City, 412
Planned Parenthood of New York, 474
*Planned Parenthood of Southeastern
    Pennsylvania* v. *Casey,* 518–20
Planned Parenthood v. *Ashcroft,* 285–88,
    292–93, 299, 301–4, 407, 410, 421–
    422, 439
Plessy v. *Ferguson,* 436, 437
Poelker, John, 281

Poelker v. *Doe*, 281–82, 421, 439
Poe v. *Ullman*, 138
political action committees (PACs), 331,
    334–35, 336, 349
population control, 55–56, 84, 156, 221–
    222
Powell, Lewis F., 159, 282, 442, 449,
    451
  resignation of, 450–52, 455
Prange, Robert, 58, 60, 187
Pratter, Leigh, 306
Pregnancy Consultation Service
    (formerly Clergy Consultation
    Service of Missouri), 158–59, 194–
    196, 222–23, 514
  Widdicombe's clinic and, 204, 220
prerequisite voters, 333–34, 350
Preterm, 97, 205
privacy rights, 134, 135–39, 141, 157,
    317–18, 369, 481
  Bork's refutation of, 456
  Griswold and, 135–39, 141, 153, 155,
    171, 455–56, 481
  Roe v. *Wade* and, 161–62, 171, 284
"pro-choice," use of term, 401–2
Pro-Life Direct Action League, 429–
    430
Pro-Life Legal Defense Fund, 311, 318,
    319–22, 368, 429
pro-life movement. *See* right-to-life
    movement
prostaglandin injection, 297, 298
Protestants:
  right-to-life movement and, 187–89,
    332
  see also evangelicals; *specific
    denominations*
protests, antiabortion:
  annual March for Life, 246, 248, 256,
    351, 448, 497
  civil disobedience and, 232, 237, 248–
    249, 250–51, 260
  nonviolence and, 250–51, 252, 261–63,
    415, 426
  at Reproductive Health Services, 225–
    226, 240, 252, 394–95
  at vacuum aspirator factory, 263–64
  violence in, 261–62, 274
  see also sit-ins, at abortion clinics
protests, pro-choice, 470, 474–77
public facilities restrictions, 380–81, 388,
    391, 407, 465–66, 480, 505
public funding of abortions, 248, 274,
    275, 444, 497, 511

ballot initiatives on, 469, 510, 512, 514
Hyde Amendment and, 280–82, 361,
    421
Missouri legislation on, 380–81, 388,
    391
opinion surveys on, 510
Supreme Court decisions on, 248, 282,
    421, 422
Puzder, Andrew, 311–13, 322, 325, 326,
    365, 371, 446, 449, 525–26
  legislation written by, 418–20, 422,
    426–27, 443, 465, 480
  sit-in participants defended by, 311,
    318–22, 422
  Webster and, 463, 479
Puzder, Lisa, 322, 365, 371

quickening, 169, 305–6

Raber, Rev. Tom, 63, 69
racial segregation, 33, 436–37
radio stations, Christian, 335, 342, 345,
    346
rape, 240
  statutory, 48, 58, 152
rape, abortions for victims of, 27, 28, 29,
    38–39, 41, 47, 48, 49, 56, 58, 59, 71,
    72, 81, 152, 339, 500
  debate in right-to-life movement over,
    358–61
  public funding of, 280, 497
rational basis test, 466–67
Reagan, Leslie, 80*n*
Reagan, Ronald, 57, 58, 73, 346, 350,
    374, 396, 398, 420, 441, 448
  presidential campaign of (1980), 335–
    337
  Supreme Court appointments of, 423,
    426, 451, 452, 455, 460–61
Regency Park Gynecological Center,
    252–53, 263, 323, 324, 326, 328,
    368, 394
Rehnquist, William, 159, 424, 441, 442,
    452, 519
  Roe and *Doe* and, 175, 423
  Webster and, 479, 482, 488–89, 494,
    504, 505
Reitz, Curtis, 45, 46
Religious Coalition for Abortion Rights,
    389
Reproductive Freedom Project, 410–11,
    412

Reproductive Health Services of Planned
  Parenthood of the St. Louis Region,
  523
Reproductive Health Services (RHS),
  265–78, 286, 291, 393–95, 407, 404,
  412, 444, 450–51, 452, 473, 523,
  525, 526
  atmosphere of, 205
  bomb threats at, 395
  counseling of clients at, 205, 266, 276,
    287–88, 393
  doctors recruited for, 217–20, 302–3,
    304
  and elimination of public funding, 281,
    282
  financing for, 220–23
  firebombing of, 452–53
  founding of, 194–96, 203–6, 214, 216–
    226
  hate mail received by, 277–78
  Lee's contact with, 264, 265, 268–72,
    307
  media visits to, 471
  new director of, 454–55
  nursing staff for, 223–24
  office space for, 223, 225
  opening of, 224–26
  as plaintiff in *Webster*, 393, 406–9,
    412, 456, 469–70, 471, 486–87
  pro-choice literature at, 274–75
  pro-life patients at, 275–76
  protests at, 225–26, 240, 252, 394–95
  second-trimester abortions at, 293,
    297, 298–307
  Significant Others programs at, 266–
    268, 307, 393
  sit-ins at, 307, 308–11, 366, 368, 369,
    430
  suburban branch of, 393–94, 452–53
  Washington march and, 475, 476–77
  Webster ruling and, 504–5
  Widdicombe's management style at,
    265–66, 272–73, 299, 393
  Widdicombe's resignation from, 453,
    454–55
Reproductive Health Services v. *Webster*,
  388, 391–92, 393, 406–13, 435,
  442–43, 446–47, 449
  Widdicombe's support in, 406–9
  see also *Webster* v. *Reproductive
    Health Services, et al.*
Republican Party, 344, 347–48
  abortion stance of, 336, 337, 350
rescue movement, 433, 461–63

residency requirements, 91, 97, 152, 162,
  200, 201
Reynolds, William Bradford, 465
rhythm method, 116, 117
right-to-life movement, 95–121, 161–93,
  229–64, 273–88, 308–76, 414–49
  abortions performed on members of,
    275–76
  Catholic question and, 106–18, 163,
    178–83, 340, 342
  constitutional amendment sought by.
    See Human Life Amendment
  debate over Hard Cases Exceptions in,
    358–61, 501–3
  dual suppositions of, 241
  electoral politics and, 329–38, 349–51
  evangelicals and, 338–49
  handouts of, 242–44, 254
  hate mail sent by, 277–78, 400–401
  Human Race series and, 340–42
  infighting in, 351–65
  legal think tank of, 435–42
  lit drops of, 331–33, 334
  Missouri statehouse politics and, 258–
    260, 278, 283–85, 379–93, 443–44,
    446, 449, 461, 507
  nationalization of, 177, 178–80. *See
    also* National Right to Life
    Committee
  New Right politics and, 343–50
  PACs of, 331, 334–35, 336
  post-*Roe* legislation and, 165–66, 189–
    190, 278–79, 283–88, 292–93
  post-*Roe* mobilization of, 178–93
  pre-*Roe* strategy of, 241–44
  presidential candidate of, 280
  pro-choicers' views on, 271–74
  prohibition of abortion as goal of, 190
  Protestants' reluctance to work with,
    187–89
  protests of. *See* protests, antiabortion;
    sit-ins, at abortion clinics
  race against time perceived by, 245–
    246
  rescue movement and, 433, 461–63
  Roe and *Doe* rulings as viewed in,
    161–78, 244, 246, 255, 258
  The Silent Scream and, 396–99
  slavery analogies and, 246–47, 336,
    355, 361, 367
  violent acts of, 261, 274, 373–75, 376,
    394, 395–96, 431–33, 452–53, 520–
    521
  war pictures of. *See* abortion pictures

see also specific organizations and individuals; anti-abortion protests
Right to Life of Southern California, 99, 107, 243
Ritter, Archbishop Joseph E., 191
Robertson, Pat, 347
Robison, James, 347
Rock, John, 116
Rockefeller, Nelson, 91, 93, 94, 96, 180
Rodgers, Samuel U., 144
Rodgers v. Danforth, 120–21, 126–48, 195, 410
  Backer as guardian ad litem in, 120–121, 145, 165
  legal precedents and, 130–39
  in Missouri Supreme Court, 151, 157–158, 159–60
  plaintiffs chosen for, 143, 144
  in U.S. Circuit Court, 120–21, 126, 144–48, 150–51
  in U.S. District Court, 143–44
Roe v. Wade, 151–55, 157, 159, 160, 224, 251, 279, 283, 319, 392, 448, 449, 461, 471, 482, 493, 500
  Bork's criticism of, 455–56, 457
  Casey and, 518–19
  denunciations of decision in, 176–78
  dissenting opinions in, 175–76
  Dred Scott analogy and, 246–47
  evangelicals' reaction to, 339–40, 346
  legal experts' critiques of opinions in, 312–18
  likely outcome of reversal of, 458–59, 470
  O'Connor's views on, 423–24, 455
  overturning of, 419–20, 422, 426, 437–42, 481, 487–96, 498, 519–20
  right-to-life movement's response to, 177, 178–93, 244, 246, 255, 258
  statute tested in, 151–52
  Supreme Court decision in, 161–76, 188, 191, 195, 238, 244, 319, 330, 353, 369, 408–9, 413, 423, 451, 455–56
  in Texas courts, 129, 130, 142–43, 144
  Webster and, 465, 466, 470, 480, 481, 487–92, 493, 496, 503–4, 505
Roman Catholic Church. See Catholic Church
Roraback, Catherine, 137, 154, 155
Rosenberg, Vivian, 223–24, 225, 226
Rosenblum, Victor, 107, 108, 436–37, 438
Rothman, Kenny, 259

Rothman, Lorraine, 211–12
Roussel Uclaf, 405
RU-486 (mifepristone), 405–6, 523
rubella (German measles), 51–54, 59, 142, 198, 221
Russell, Keith P., 134
Russell, Mark, 493
Ruzek, Sheryl Burt, 213
Ryan, John, 366–67, 369, 414–18, 427, 429–33, 444, 459, 461
Ryan, Msgr. John A., 184

St. Louis Archdiocese, 58
  organizing meeting of (1973), 192–93
  Pro-Life Committee of, 327
St. Louis County Circuit Court, Rodgers v. Danforth and, 120–21, 126, 144–148, 150–51
St. Louis County Medical Society Bulletin, 54–55
St. Louis Globe-Democrat, 176, 224, 225, 415–16
St. Louis Post-Dispatch, 93–94, 158–59, 224, 261, 276, 278, 319, 326, 396, 401, 450, 479, 486–87, 506, 514–15, 525
St. Louis Review, 58, 176–78, 182–83, 190, 191, 192, 328
St. Louis Speak Out (1985), 403–4
St. Louis Suicide Prevention, 17–18, 19–21
  abortion callers and, 20–21, 36–37
St. Louis University Pro-Life, 251
saline abortions, 101, 165, 294–98, 301, 303
  live-birth complications in, 296–97
  pictures of, 102–3, 105, 242
  prohibitions on, 190, 278–79, 421
Salvi, John C., III, 521
Sanger, Margaret, 82–83, 156, 184
Save Our Unwanted Life (SOUL), 250
Scalia, Antonin, 441, 442, 448, 452, 455, 519–20
  Webster and, 482–84, 490, 504
Schaeffer, Francis A., 340–42
Scheidler, Joe, 431, 432
Schlafly, Phyllis, 344
Schneider, John, 382, 383, 386, 392, 445
Schwartz, Dr., 268–69
Schwartz, Louis B., 45–48, 58
Schwartz, Melvin, 26–27, 28–29, 37
second-trimester abortions, 209, 279, 286
  hospitals-only requirements for, 284, 293–94, 299, 302, 304, 407, 413, 421, 422, 425, 440, 446–47, 449, 465, 504

second-trimester abortions, *continued*
procedures for, 293–301, 303
at RHS, 293, 297, 298–307
Roe and, 163–64, 173–74, 424–25
timing of, 297–98
viability-testing requirements and,
408–9, 465, 480–81, 488–89, 504–5
Second Vatican Council, 176, 184–87,
188, 235
Seiler, Robert, 160
self-defense concept, 262
self-examination groups, 210–11, 212
Senate, U.S., elections of 1980 and, 335,
338
Senate Judiciary Committee, 456
sex-selection prohibitions, 438, 443,
498
Sharp, Gene, 250–51, 426
Shenker, Morris, 318–19, 320
Shields, Karen, 441
Shirk, Martha, 514
Shively, Paul, 52, 142
Short, Jerry, 464
Silent No More campaign, 402–4
Silent Scream, The, 396–99
Simmons, Cathryn, 507–10, 513, 516,
522
Simmons, James, 431
Simopolous v. *Virginia,* 421–22
sit-ins, at abortion clinics, 231–32, 236–
237, 247–54, 260–63, 264, 308–11,
319–29, 365–70, 388–89, 430–33,
451, 466
archbishop's criticisms of, 326–28, 365
arrests and legal proceedings in, 248–
249, 251, 253, 311, 323–26, 328,
365, 366
civil disobedience and, 232, 237, 248–
249, 250–51, 260
conversations with clients in, 309–10
defense attorneys for, 311–13, 318–22,
325, 365, 368
emotionality in, 430–31
entering clinic in, 309
escalation of tactics in, 366–70, 373–
376, 430, 431
first, 248
first in St. Louis, 252–53
Harvard pamphlet on, 248, 261, 309,
310
Lee's leadership role in, 310–11, 365–
366, 367, 368
Lee's prison sentence and, 388–89,
414–27

minimum number of people needed
for, 418, 433
necessity defense and, 319–21, 325–
326, 368, 369, 422
opposition within right-to-life
movement to, 328–29
philosophical groundwork for, 247–48,
250–51, 252
praying before, 308–9
rescue movement and, 433, 461–63
restraining orders and, 368–73, 375, 376
St. Louis as center of, 324
sit-ins, of civil rights movement, 260
sixteen-week-hospitalization
requirement, 383, 407, 413, 446–47,
449, 465, 504
slavery analogies, 336, 355, 361, 367
Dred Scott decision and, 246–47
Slepian, Barnett, 521
Smeal, Eleanor, 474, 485
Smith, Chris, 494, 495, 503
Smith, Judy, 153
Society for Humane Abortion, 73–80,
81, 83
Souter, David, 518, 519–20
South Carolina, abortion statutes in, 493,
506
South Dakota, 338, 350
abortion litigation in, 129
abortion statutes in, 506
Southern Baptist Convention, 188, 338,
339–40, 342, 385
Soven, Abby, 132–34, 143
Speak Outs, 402–4
Special Needs Adoption Tax Credit, 449
speculum, 210–11
Spencer, Robert, 27
spousal notification or consent
requirements, 278–79, 421, 422,
498, 519
states' rights, 354–55
Hatch Amendment and, 355–58, 361–
364, 365
statutory rape, 48, 58, 152
Stearns, Nancy, 155
sterilization, 108, 115, 138
Stern, Rabbi Shira, 404
Stevens, John Paul, 423, 442, 451–52
Casey and, 519
Webster and, 491
Stewart, Potter, 423
suction abortions, 189, 190, 214, 220
hand-operated devices for, 451
Karman's device for, 209–10, 211

pictures of, 103, 105, 242
Rothman's device for, 211–13
ultrasound videotape of, 396–99
vacuum aspiration and, 197–203,
  210
"suicidal" women, 44, 45, 59, 158
summary judgments, 145–46
Sunnen, Joe, 220–23
Sunnen, Robert, 223
Sunnen Foundation, 221, 509
supercoil abortions, 209*n*
Supreme Court, U.S., 121, 133, 134, 140,
  160, 191, 356, 383–84, 388, 446
  abortion litigation and, 144, 151–57,
    159, 160, 161–78, 188, 190, 191,
    193, 194, 195, 196, 238, 244, 245,
    246, 248, 252, 274, 275, 278, 279,
    282, 283, 297, 312–18, 319, 330,
    333, 335, 339–40, 346, 353, 369,
    407, 408–9, 410, 411, 413, 419–26,
    437–42, 451, 455–56, 464–67, 469–
    497, 497–98, 518–20. *See also*
    abortion litigation; *specific cases*
  abortion statutes targeted at, 391–92,
    419–20
  Bork's nomination to, 455–57, 460
  civil rights cases and, 436–37
  Dred Scott and, 246–47
  gradual reconfiguration of, 338, 423–
    426, 433, 448, 449, 450–52, 455–57,
    460–61, 518–19
  Human Life Bill and, 352, 354
  Lee's studies of, 416, 421–22
  Lochner and, 315, 316
  privacy rights and, 134, 135–39
Susman, Frank, 127–30, 132, 154, 176,
  204, 283, 285–88, 388, 389, 420,
  455, 502, 504, 506, 524
  ACLU's relationship with, 410–11
  Ashcroft and, 285–86, 292–93, 304
  background of, 128
  Poelker v. *Doe* and, 281, 282
  post-*Roe* legislation and, 278–79
  Rodgers v. *Danforth* and, 127, 130,
    135–36, 141, 143–47, 150–51, 157–
    158, 159–60, 195, 410
  Webster and, 393, 406–13, 435, 442,
    443, 458, 470, 479, 482–84, 485

Taney, Roger B., 246
Taussig, Frederick J., 23, 24
Teasdale, Joseph P., 283, 335, 350
television ministries, 345–46

Terry, Randall, 461–63, 466, 503
Texas, 493
  abortion litigation in. *See Roe* v. *Wade*
  abortion statutes in, 151–52
Thalidomide, 49–51, 59
Thayer, Charlotte, 128, 129–30, 141,
  143–48, 150–51, 159–60
therapeutic abortions, 43–45, 59, 98, 151,
  339
  broadening of, 47–48, 56–61
  after enactment of reform legislation,
    71–72
  hospital procedures and, 196–97
  hospital review committees and, 44,
    71, 90, 208–9
  Thalidomide and rubella scares and,
    49–54, 59
third-trimester abortions, 408–9
  legislative restrictions on, 89, 90, 96,
    498–99
  Supreme Court decisions and, 162–64,
    173, 174
Thomas, Clarence, 518, 519
Thompson, Carolyn, 330–33, 334
Thompson, Judith Jarvis, 239–40
Thornburgh, Richard, 439
Thornburgh v. *American College of
  Obstetricians and Gynecologists,*
  439–41, 456, 457
Tietze, Christopher, 156
Timanus, G. Lotrell, 27
Tribe, Laurence, 317*n*
trimester approach, 314, 408–9, 438, 466,
  482, 519
  O'Connor's criticism of, 424–25
  Roe decision and, 163–64, 173–74
  see also second-trimester abortions;
    third-trimester abortions
Turner, Carlean, 305
twelve-week hospitalization
  requirements, 407, 422
Two Step Plan, 357–58

ultrasound, 408, 505
  videotape of abortion, 396–99
"unborn children":
  legal definitions of, 284
  rights of, 382–83, 388, 419–20, 488
  see also fetus
"undue burden" test, 425, 426, 519,
  520
Unitarians, 332
United Methodist Church, 188

United States v. *Vuitch,* 141–42, 151, 153, 154, 164
University of California at San Francisco (UCSF), 52, 198–99
U.S. Catholic Conference, 238
Utah, abortion statutes in, 493, 506

vacuum aspiration, 197–203, 210, 251, 297
   development of technology for, 197–199
   and nature of abortion businesses, 199–203
   RU-486 vs., 405–6
vagueness doctrine, 133–34, 135
Valentine, Foy, 339–40
Value of Life Committee, 107
Vavra, John, 30–31, 36, 37
Vermont, abortion litigation in, 157
viability, 316, 438, 519
   abortions performed after, 163, 165, 174, 422, 439n, 440
viability-testing requirements, 408–9, 465, 480–81, 488–89, 504–5
Vietnam War, 34, 35, 248, 250, 311, 312, 321
Viguerie, Richard, 344, 347
Village Voice, 212–13
Vinyard, Mrs. (Mrs. Gertrude and Mrs. Martha), 27–28, 30, 149, 218, 294
Virginia, 493, 496, 497
Voice for the Unborn, 107
Volk, Valerie, 323
Voters for Choice, 467, 469, 474, 475, 476, 504, 505, 510, 514
Vuitch, Milan, 141–42

Waddill, William, Jr., 297n
Wade, Henry, 120
Wagner, Loretto, 256–58, 370, 453, 497, 524, 525–26
   electoral politics and, 334–38, 350
   Gephardt defection and, 447–48
   Hatch Amendment and, 355–57, 362, 365
   Human Life Amendment and, 447–48
   New Right and, 348
   and sit-ins at abortion clinics, 253–55, 260, 322–28, 335, 365, 366
   statehouse politics and, 259
waiting periods, 284, 288, 293, 304, 421, 422, 425, 498, 519

Walden, Bill, 215, 216
Wall Street Journal, 447
Wamble, Hugh, 188, 189
Wanderer, 431–32, 496
Washington, D.C.:
   abortion statutes in, 97, 135, 151, 152
   ballot litigation in, 135, 141–42, 151, 154, 164
Washington Post, 176, 362
Washington State:
   abortion initiative in, 180, 242
   abortion statutes in, 97, 107, 142, 151, 241–42
Washington University, abortion referral service at, 29–31, 35–36
Wattleton, Faye, 412, 485, 503
Webster, William L., 464; 465, 479–81, 522–23
Webster v. *Reproductive Health Services, et al.:*
   and abortion movement mobilization, 470–77, 496–97
   cases after, 497–98
   choice activists' ascendancy after, 496–497
   in Eighth Circuit Court, 457–58, 461, 463, 464, 465
   oral arguments in, 477–85
   pro-choice march and, 474–77
   ruling in, 486–92, 493–94, 503–5
   statutory proposals after, 492–93, 495–96, 497, 498–501
   in Supreme Court, 464–67, 469–97
   see also Reproductive Health Services v. *Webster*
Wechsler, Herbert, 45
Weddington, Ron, 153
Weddington, Sarah, 130, 142–43, 152–155, 157
Weyrich, Paul, 343–48, 350
Whatever Happened to the Human Race?, 340–42
White, Byron, 175–76, 423, 424, 442, 520
   Webster and, 489, 494
Widdicombe, Arthur, 17, 18, 19, 63, 64–65, 67, 122–23, 148–50, 205, 218, 225, 273, 394, 454
Widdicombe, Judith McWhorter, 15–21, 214, 393–96, 404–5, 450–56, 467–477, 502–17, 523
   abortion clinic run by, 194–96, 203–5, 214, 216–26, 265–66, 269–78, 282, 286, 393–94, 407. *See also* Reproductive Health Services

abortion referrals and, 20–21, 36–37,
    62–63, 64–67, 94, 122–26, 128, 149,
    150, 158–59, 194–95, 217, 220,
    222–23
background of, 18–19
ballot initiative of, 507–17, 518
bomb threats and, 395–96
clinic firebombing and, 452–53
demonstrations and, 225–26, 394–95
divorce of, 273, 454
family life of, 17–18, 19
feminism as viewed by, 205–7
hate mail and, 277–78
Lee and, 269–72, 308, 395, 451, 517
management style of, 265–66, 272–73,
    299
personal crisis of, 453–54
post-Roe legislation and, 278–79, 282,
    283–84, 286–88, 292–93
pregnancy and abortion of, 126, 148–
    150
and quality of abortion providers, 288–
    292
relocated to Washington, D.C., 467–69
resignation of, from RHS, 453, 454–55
Roe and Doe decisions and, 194, 195
RU-486 and, 405–6
and second-trimester abortions at
    RHS, 293, 297, 298–307
statehouse politics and, 507–8, 516–
    517
suicide prevention and, 17–18, 19–21
Supreme Court reconfiguration and,
    450–52, 455–56
surveillance of, 122–23
Washington march and, 474–77
Webster and, 393, 406–9, 442, 456,
    467, 469–71, 473–77, 406–9, 486–
    487, 503–5
Wilder, Douglas, 496
Williams, Glanville, 46, 72
Willke, Barbara, 100–106, 107, 177, 180,
    238, 242, 277, 359
Willke, John (Jack), 107, 456, 457
    abortion pictures and, 100–106, 172,
        177, 180, 238, 242, 261, 277
    books and pamphlets written by, 103,
        246, 247, 359, 417

electoral politics and, 350
Human Life Bill and, 354
infighting and, 364
as president of National Right to Life
    Committee, 336, 364
Webster and, 485
Wisconsin, 338
    abortion litigation in, 129, 151
    abortion statutes in, 506
women:
    abortions to preserve mental health of.
        See mental health abortions
    abortions to preserve physical health
        of, 43–45, 47–48, 56, 58, 60, 151,
        164, 172n, 361, 499–500. See also
        therapeutic abortions
    fundamental right of, to abortion,
        129–34, 139–41, 143, 160–76, 282,
        316–17, 356, 357, 466, 520
Women's Equity Action League, 89
women's health activists, 207–14
    alternative abortion methods and,
        209–10, 211–13
    self-examination and, 210–11, 212
women's movement, 86–90, 205–14,
    256–57, 270–71, 385
    abortion providers and, 290, 291
    evangelicals' hostility toward, 343,
        348, 349
Women's Services (New York City), 96–
    97, 205, 215–16, 396
Woods, Harriett, 510, 512
Wright, Scott O., 411, 446, 449
written consent requirements, 279, 421
    see also informed consent
        requirements; parental consent
        requirements; spousal notification
        and consent requirements
Wulff, George, 37, 219

Yale Law Journal, 314–17
Yard, Molly, 474
Yeo, Eleanor, 34–35
Y.W.C.A. v. Kugler, 143

Zevallos, Hector, 373–75, 376, 395, 432